Magnetic Resonance and Related Phenomena

Magnetic Resonance and Related Phenomena

Volume 2

PROCEEDINGS of the 18th AMPERE CONGRESS,
Nottingham, 9–14 September, 1974

EDITED BY
P.S. ALLEN, E.R. ANDREW and C.A. BATES,

DEPARTMENT OF PHYSICS, THE UNIVERSITY,
NOTTINGHAM NG7 2RD, ENGLAND

NORTH-HOLLAND PUBLISHING COMPANY - AMSTERDAM · OXFORD
AMERICAN ELSEVIER PUBLISHING COMPANY, INC. - NEW YORK

North-Holland ISBN for this Series: 0 7204 1200 5
North-Holland ISBN for this Volume: 0 7204 1210 2
American Elsevier ISBN: 0 444 10859 9

Published by:
NORTH-HOLLAND PUBLISHING COMPANY — AMSTERDAM
NORTH-HOLLAND PUBLISHING COMPANY, LTD. — OXFORD

Distributors for the U.S.A. and Canada:
American Elsevier Publishing Company, Inc.
52 Vanderbilt Avenue
New York, N.Y. 10017

PRINTED IN ENGLAND

PREFACE

The 18th Ampere Congress on Magnetic Resonance and Related Phenomena was held at the University of Nottingham, England, 9-14 September, 1974. There were 463 scientific participants and some 70 non-scientific participants from 33 countries. The conference sessions were held in the Department of Physics of the University and most of the participants were accommodated in Cripps Hall and Hugh Stewart Hall on the University Campus.

The scientific programme, which covered the full range of magnetic resonance and radiospectroscopy and its applications in physics, chemical physics and biophysics, consisted of seventeen invited papers and 259 contributed papers. The 18th Congress coincided most closely with the bicentenary of the birth of Ampère. This was marked by a special lecture at the Congress on the life and work of Ampère by Professor A. Kastler, Honorary President of the Groupement Ampere. These invited and contributed papers form the body of these two volumes which constitute the Proceedings of the 18th Ampere Congress.

At the Opening Ceremony the Chairman of the Organizing Committee, Professor E. R. Andrew, welcomed the participants. The Congress was then declared open by the Vice-Chancellor of the University of Nottingham, Dr. J. H. Butterfield, after which the President of the Groupment Ampere, Professor A. Lösche, gave an address. During the Congress a reception was held in the Nottingham Council House by the Nottingham City Council where guests were welcomed by the Sheriff of Nottingham. The Congress Dinner took place in the Portland Building of the University of Nottingham; the After-Dinner Speaker was Professor A. Abragam. At the Closing Ceremony the Secretary-General, Professor G. J. Béné, outlined the future programme of the Groupement Ampere and both he and the President, Professor Lösche, thanked the Local Organizing Committee of the Congress at Nottingham for all their work.

Twelve exhibitors took part in the Instrument Exhibition held during the Congress on the ground floor of the Physics Building. The Exhibition was a joint venture of the Ampere Congress Local Organizing Committee, Dr. W. S. Moore acting as Exhibition Secretary, and The Institute of Physics, whose Exhibition Officer was Mr. N. A. Walter assisted by Miss S. J. Elvey.

The research laboratories of the Department of Physics of the University of Nottingham were open to visitors throughout the Congress, and arrangements for this were co-ordinated by Dr. P. Mansfield. Exhibits relating to the life and work of Ampère and of Green, who lived and worked in Nottingham (1793-1841), were mounted in the Entrance Hall by Mr. N. Green of the University Library.

In addition to the reception in the Council House and the Congress Dinner, social activities in the evenings included a University Reception, a piano concert by Alan Schiller in the Great Hall of the University, a Lecture (with demonstrations) on Explosives by Dr. B. D. Shaw, and a Reception in the Nottingham Castle Museum. A day-time social programme primarily intended for non-scientific participants included visits to Stratford-on-Avon and the Shakespeare Memorial Theatre, Oxford and Blenheim Palace, Newstead Abbey and Southwell Minster, Chatsworth House and the Derbyshire Dales, Lincoln, and local sights and industries. The Social Programme was organized by Dr. M. Heath.

The 18th Ampere Congress received sponsorship from The International Union of Pure and Applied Physics, The European Physical Society, The Royal Society and The Institute of Physics. Support was also given by ICI Ltd., Newport Instruments Ltd., Oxford Instrument Company Ltd. and Unilever Ltd.

All arrangements for the lecture theatres, laboratories and projection facilities were undertaken by the technical staff of the Department of Physics under the responsibility of the Laboratory Superintendent, Mr. D. R. Ringer. All art work for programmes, publications and signs was done by Mr. D. Cameron and Mr. A. Bezear of the Department of Physics photographic unit. The correspondence and typing were handled by the Department of Physics secretarial staff and especially Miss Maya Jackson, who also dealt with travel arrangements. The printing of the Programme, Book of Abstracts and Conference Proceedings was carried out by the University of Nottingham Printing Unit under Mr. J. Houldgate.

The work relating to registration, accommodation, finance, scientific programme, manuscripts and general organization was undertaken by the General Secretaries, Dr. P. S. Allen and Dr. C. A. Bates.

On behalf of the Organizing Committee I wish to express to everyone who helped in the successful organization of the 18th Ampere Congress at Nottingham our most cordial thanks for their hard work, enthusiasm and co-operation. I should like to add a further personal word of thanks to my colleagues on the Organizing Committee for their tremendous efforts and their friendly collaboration these past two years.

E. R. ANDREW
Chairman - Organizing Committee
18th Ampere Congress, Nottingham.

25 September, 1974.

OFFICERS OF THE AMPERE GROUP

PRESIDENT: Professor A. Lösche
Karl Marx University, Leipzig, DGR

SECRETARY-GENERAL: Professor G. J. Béné
University of Geneva, Switzerland

* * *

INTERNATIONAL ADVISORY COMMITTEE

S. A. Altshuler	USSR	A. Lösche	DGR
E. R. Andrew	UK	K. A. Müller	Switzerland
P. Averbuch	France	N. J. Poulis	Netherlands
B. Bleaney	UK	J. G. Powles	UK
R. Blinc	Yugoslavia	T. A. Scott	USA
M. Bloom	Canada	E. F. W. Seymour	UK
M. F. Deigen	USSR	K. J. Standley	UK
D. Fiat	Israel	K. W. H. Stevens	UK
A. Gozzini	Italy	K. Tompa	Hungary
K. H. Hausser	FGR	I. Ursu	Rumania
J. W. Hennel	Poland	L. Van Gerven	Belgium
V. Hovi	Finland	J. S. Waugh	USA
G. R. Khutsishvili	USSR		

* * *

LOCAL ORGANIZING COMMITTEE

CHAIRMAN: Professor E. R. Andrew

GENERAL SECRETARIES: Dr. P. S. Allen
Dr. C. A. Bates

EXHIBITION SECRETARY: Dr. W. S. Moore

SOCIAL SECRETARY: Dr. M. Heath

* * *

THE INSTRUMENT EXHIBITORS

Brookdeal Electronics Ltd.
Bruker Spectrospin Ltd.
Bryans Southern Instruments Ltd.
Cambridge Scientific Instruments Ltd.
Dale Electronics Ltd.
Dillon's Nottingham University Bookshop Ltd.
Mossbauer Group (Harwell)
Newport Instruments Ltd.
Oxford Instrument Co. Ltd.
Spin-Lock Electronics Ltd.
Thor Cryogenics Ltd.
Varian Associates Ltd.

* * *

GENERAL CONTENTS

VOLUME I AND VOLUME II

* * * * *

DETAILED CONTENTS: VOLUME II

CONTRIBUTED PAPERS

SECTION H: FERROELECTRICS AND PHASE CHANGES

SESSION CHAIRMAN: Z. Pajak (Poland)

SECTION I: METALS, ALLOYS AND SEMICONDUCTORS

SESSION CHAIRMEN: B. I. Kochelayev (USSR)
K. Tompa (Hungary)
D. G. Hughes (Canada)
L. E. Drain (UK)

SECTION J: MOTION IN MOLECULAR SOLIDS

SESSION CHAIRMEN: A. Lösche (DGR)
T. A. Scott (USA)

SESSION K: POLYMERS

SESSION CHAIRMAN: J. G. Powles (UK)

SESSION L: ELECTRON SPIN-LATTICE RELAXATION

SESSION CHAIRMAN: K. W. H. Stevens (UK)

SESSION M: NMR IMAGE FORMATION

SESSION CHAIRMAN: S. Fujiwara (Japan)

SESSION N: PAIR INTERACTIONS

SESSION CHAIRMEN: Y. Merle d'Aubigné (France)
G. J. Troup (Australia)

SESSION O: THEORY OF NMR RELAXATION AND LINESHAPES

SESSION CHAIRMEN: E. L. Hahn (USA)
G. R. Khutsishvili (USSR)

SESSION R: THE LIQUID STATE

CHAIRMEN: A. Tzalmona (Israel)
G. Bonera (Italy)

SECTION S: HIGH RESOLUTION NMR IN SOLIDS

SESSION CHAIRMEN: J. Jeener (Belgium)
H. Pfeifer (DGR)
B. Schnabel (DGR)

SECTION H

Ferroelectrics and Phase Changes

NUCLEAR MAGNETIC DOUBLE RESONANCE VIA RF INDUCED COUPLING BETWEEN SPIN SYSTEMS

R.Blinc, J.Seliger, M.Mali, R.Osredkar and A.Prelesnik
University of Ljubljana, Institute J.Stefan, 61001 Ljubljana, Yugoslavia

Abstract. A new nuclear double resonance technique based on strong RF magnetic field induced coupling between spin systems is described. The technique overcomes "spin quenching" and thus allows the measurement of the pure NQR spectra of integer spin nuclei in zero static magnetic field via double resonance with protons in the laboratory frame.

It is well known that dipolar coupling between integer nuclear spins in an asymmetric electric field gradient (B spin system) and any other non-resonant spin system (A) is highly reduced in zero static magnetic field. It is this "spin quenching" effect which prevents the detection of the pure nuclear quadrupole resonance (NQR) spectra of integer spin nuclei (like deuterium of N 14) via proton-deuteron or proton-N 14 double resonance in the laboratory frame. The "level crossing" technique where the A and B spin systems couple in a non-zero static magnetic field is therefore often used to measure the pure NQR spectra of integer spin nuclei in powdered samples if the NQR frequencies are too low for the classical NQR techniques to be applicable.

It is the purpose of this communication to describe a new double resonance technique based on the "solid effect" the sensitivity of which is higher than the sensitivity of the level crossing technique. The proposed technique is particularly valuable when the level crossing signals cannot be obtained. This is the case when

(i) The spin-lattice relaxation times of the B-system are much shorter than the time the sample stays in zero field.
(ii) The level crossing times are too short for sufficient energy exchange between the two spin systems to take place.
(iii) The quadrupole resonance frequencies of the B nuclei are too high for the level crossing to occur.

The technique described in this letter is based on the fact that a strong radiofrequency magnetic field $H_1 \cos\omega_0 t$ induces in the frequency spectrum of the A spin dipolar fields new peaks at $\omega_A \pm \omega_0$ and in the dipolar spectrum of the B spins new peaks at $\omega_B \pm \omega_0$. The strong RF field thus couples the two spin systems A and B even though spin quenching occurs if

$$\omega_0 = \omega_A \pm \omega_B \qquad (1)$$

where ω_A is one of the transition frequencies of the A and ω_B one of the transition frequencies of the B spin system. If the above condition is satisfied the A spins driven by the RF magnetic field induce transitions between the energy levels of the B spins and vice versa. Under the influence of the RF magnetic field the two spin systems relax to a common equilibrium state in a frame of reference in which the energy levels of the A and the B spins are equally spaced. The A-B cross relaxation times are however longer than the ones for rotating frame double resonance by a factor $(\omega_0/\gamma H_1)^2$.

The above technique can be applied to Zeeman-Zeeman, quadrupole-quadrupole or Zeeman-quadrupole double resonance. In the following we shall concentrate on the case of proton-deuteron or proton N 14 double resonance in zero external static magnetic field.

The proton (A) spin system is first polarized in a high magnetic field, and then adiabatically demagnetized by moving the sample out of the magnet. The sample is now irradiated with a strong RF magnetic field ($H_1 \approx$ 10-40 Gauss) at a frequency ω_0 for a time τ. After the irradiation the sample is moved back into the magnet and the A spin magnetization is determined from the free induction decay following a 90° pulse. The whole sequence is then repeated without the strong RF magnetic field irradiation at ω_0. The difference between the A spin free induction decay signals with and without the RF irradiation is recorded as a function of ω_0. The double resonance signal is non-zero whenever eq.(1) is satisfied.

The sensitivity and the resolution of this technique is increased when one uses instead of a single coupling pulse a train of coupling pulses at a frequency $\omega_A \pm \omega_B$ with weak RF field pulses at the frequency ω_B in between. The purpose of the weak RF pulses is to saturate the particular quadrupole transition in the B system. The weak RF field pulses at ω_B can be obtained by simultaneous attenuation and modulation of the carrying signal at the frequency $\omega_A \pm \omega_B$ with the known frequency ω_A, while sweeping the frequency of the carrying signal. Here ω_A should lie within the A spin zero field linewidth. The double resonance spectra consist in this case of broad single coupling lines-the width of which equals the A spin zero field linewidth - with strong narrow lines added at frequencies $\omega_A \pm \omega_B$. The width of these additional lines equals the B spin linewidths.

Using this technique the N^{14} NQR spectra of solid p-azoxyanisole (PAA), diheptyl-oxyazoxybenzene (DOAB), p-anisalazine and polyglycine have been measured. The room temperature N^{14} quadrupole coupling constant of p-anisalazine is $e^2qQ/h = 620$ kHz and the asymmetry parameter is $\eta = 0.23$. In PAA ($T = 246°K$) and DOAB ($T = 250°K$) $e^2qQ/h = 570$ kHz, $\eta = 0.21$ and $e^2qQ/h = 600$ kHz, $\eta = 0$ for one nitrogen, whereas for the other, to which the oxygen atom is bound, $e^2qQ/h = 1200$ kHz, $\eta = 0.13$ for PAA and $e^2qQ/h = 1170$ kHz, $\eta = 0.17$ for DOAB. For the peptide nitrogens in polyglycine at $T = 77°K$ $e^2qQ/h = 3.097$ MHz and $\eta = 0.76$.

References

Blinc, R., Mali, M., Osredkar, R., Prelesnik, A., Seliger, J., Zupančič, I., and Ehrenberg, L., J.Chem.Phys. 57, 5087.

Edmonds, D.T., Hunt, M.J., and Mackay, A.L., 1973, J.Mag.Res. 9, 66.

Goldman, M., 1971, Spin Temperature and Nuclear Magnetic Resonance in Solids, Oxford University Press, London.

Koo, J., Ph.D.Thesis, University of California, Berkeley (unpublished).

Leppelmeier, G.W., and Hahn, E.L., 1966, Phys.Rev. 141, 724

NEW PHASE TRANSITIONS IN $RbCdCl_3$ DETECTED BY ^{35}Cl-NQR

R. Kind and J. Roos
Laboratory of Solid State Physics, Swiss Federal Institute of Technology, CH 8049 Zürich, Switzerland

Abstract. The temperature dependence of the ^{35}Cl-NQR frequencies in water-free $RbCdCl_3$ crystals confirmed the occurence of the phase transition cubic to tetragonal at 114.5°C. In addition two unknown phase transitions were found at 90°C and at 67.5°C. Preliminary results of the ^{87}Rb-NMR-NQR allow a resonable explanation of the transition cubic-tetragonal by one soft mode.

1. Introduction. The search for ferroelectrics in ABX_3-compounds revealed dielectric anomalies in $RbCdCl_3$ (Natarajan et al.,1971) evidently due to a complicated polymorphic behaviour. Many contradictory observations were found, until the great influence of water on the structure was proved (Bohac et al.,1973). Water-free crystals show a perovskite-type structure and a reversible phase transition from cubic to tetragonal at 114.5°C. Due to the great affinity to water the colourless crystals become milky within 10min once exposed to free air. Within few hours the whole bulk is affected by a reconstructive phase transition. For the cubic and the tetragonal phases superstructure with z=8 and a transition temperature of 190°C was reported (Natarajan et al.,1971), but X-ray data were taken without any precautions concerning humidity and thus are rather doubtful. Since the presence of cadmium excludes neutron diffraction and as the problem of crystal preparation for measurements of the dielectric constant and the optical properties has not been solved up until now, NMR-NQR was the only tool for further investigations.

2. Specimen. A stoichiometric mixture of RbCl and $CdCl_2$ was dissolved in 5% hydrochloric acid and $RbCdCl_3$ powder was obtained by evaporization of the solvent. The powder was dehydrated at 500°C in a HCl gas flow and sealed afterwards under vacuum into a quartz ampoule. The dimensions of the tube and the amount of powder were chosen to obtain an optimal filling factor for the probe head of the pulse spectrometer. The single crystal was grown by zone melting. It was colourless and untwinned. Attempts to grow the crystal by a Bridgman technique failed, because the very pure melt exhibited extraordinary large supercooling.

3. ^{35}Cl-NMR-NQR. The temperature dependence of the pure ^{35}Cl-NQR (Fig.1) revealed a first order transition at 114.5°C. On lowering the temperature, the single NQR-line (11.23MHz) splits suddenly into two lines corresponding to the two chemically inequivalent chlorine-sites of the tetragonal phase. The behaviour in the tetragonal phase indicates, that the transition is not far from being of second order. At 90°C a very small splitting of the upper line indicates a second order transition to an orthorhombic phase with three different Cl-sites. This splitting allows the assignement of the higher resonance frequencies to the chlorine atoms Cl_2 in the Cd-(001)-plane, whereas the lower frequency belongs to Cl_1 in the Rb-(001)-plane. Another first order transition to an unknown phase with again two chemically different chlorine-sites occurs at 67.5°C where the small splitting collapses. The three phase transitions show no temperature hysteresis within the experimental error of .1° and neither in line width nor in T_1 critical behaviour has been observed. Despite the knowledge of the NQR-results all attempts failed in detecting the phase transitions at 90°C and at 67.5°C by X-ray diffraction or by differential thermoanalysis. The behaviour of the Cl-NQR is very similar in $CsPbCl_3$ for which the following progression of space groups is postulated: cubic Pm3m - tetragonal P4/mmm - orthorhombic Pmmm - monoclinic P2/m (Cohen et al.,1971).
In order to obtain the orientation and the asymmetry parameter η of the efg-tensors, rotation patterns in a low magnetic field of 60G were taken for the cubic, tetragonal and orthorhombic phases. In all these phases the z-axes of the efg are parallel to the cubic <100>-directions. The tetragonal axis was almost perpendicular to the rotation axis (i.e. growth direction) of the crystal. For none of the phases and for none of the sites η was found to differ from zero within the experimental error, i.e. η must be smaller than .05 for the Cl_2-sites in the tetragonal phase. This indicates, that the Cl-octahedra remain practically undistorted. Since the chlorine resonances gave no further indications about the nature of the phases, the search for the ^{87}Rb signals was started.

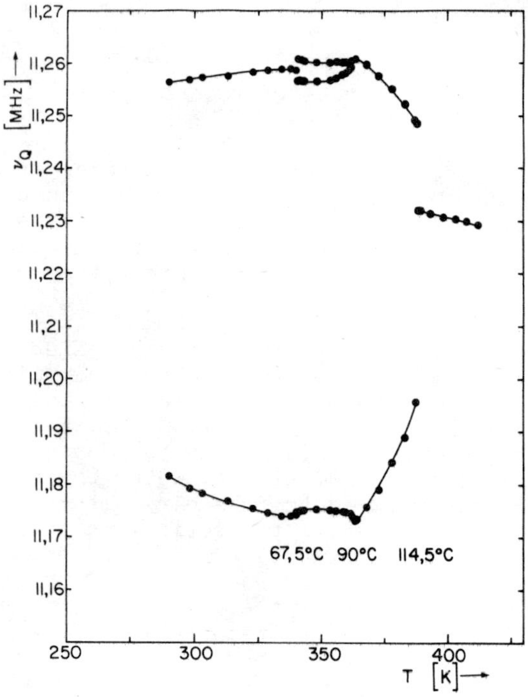

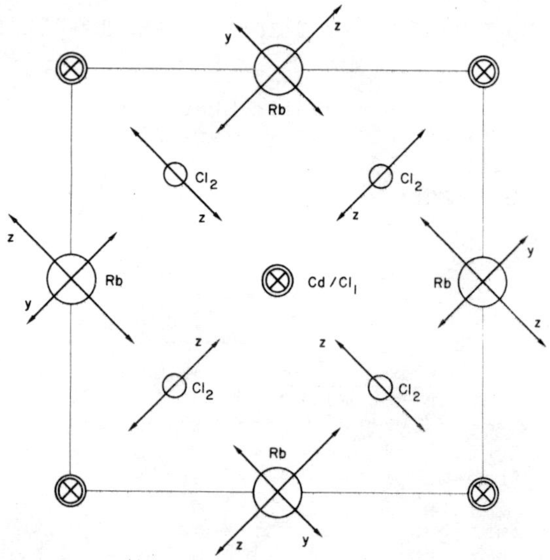

Fig.2 Orientation of the efg-tensors at the Cl- and Rb-sites in the tetragonal Phase of $RbCdCl_3$ (projection parallel to the tetragonal axis).

←Fig.1 Temperature dependence of ^{35}Cl-NQR frequencies in $RbCdCl_3$

4. ^{87}Rb-NMR-NQR. In the cubic phase at 124°C e^2qQ/h was found to be zero for the Rb-sites. The signal at 10MHz was a pure magnetic resonance of 2kHz line width and neither a splitting nor an orientation dependence of the resonance frequency was observed. In the tetragonal phase at 96°C rotation patterns were taken. The measuring frequency was 10MHz and the magnetic field was swept between 5kG and 9kG. The measurements revealed two sets of resonance lines belonging to chemically equivalent Rb-sites,but with different orientation of the efg-tensors. The analysis of the curves was carried out by comparing the measurements with calculated rotation spectra obtained by numerical diagonalization of the hamiltonian. For the two sets the following results were found: e^2qQ/h= 4.4±.1MHz, η= .40±.01 and $V_{xx} \parallel [001]$. $V_{zz_1} \parallel V_{yy_2} \parallel [100]$, $V_{zz_2} \parallel V_{yy_1} \parallel [010]$,i.e. all axes are parallel to cubic <100>-directions.

5. Conclusions. The results for the cubic phase of $RbCdCl_3$ are in all details in agreement with the space group Pm3m with one formula unit in the unit cell. A superstructure would be announced by a non-vanishing efg-tensor at the Rb-site and probably by a line splitting in the rotation pattern of ^{35}Cl.
The results for the tetragonal phase indicate a superstructure with z=2 or z=4. The most probable unit cell with the orientation of the efg-tensors is depicted in Fig. 2. Because of the orientation of the efg-tensors at the Rb-site the space group P4/mmm ($CsPbCl_3$) is not allowed for $RbCdCl_3$. Assuming a Landau transition from Pm3m to the space group with the highest symmetry which is in agreement with the NQR-results, the space group must be P4/mbm (D^5_{4h}) with z=2. In this case the soft mode transforms according to the M_3-representatuon with $\vec{k}= \frac{1}{2}(\vec{g}_1+\vec{g}_2)$ and consists of a pure rotation of the chlorine octahedra around the tetragonal axis. Contrarily to the case of $SrTiO_3$ the rotations of adjacent octahedra are in phase along the c-direction.

The autors are very indebted to Dr. H. Arend and W. Huber for the crystal growth and to Dr. J. Petzelt and P. Bohac for helpful discussions. This work was partially financed by the Swiss National Science Foundation.

6. References.

Bohac,P.,Gäumann,A. and Arend,H., Mat.Res.Bull. 8,1299 (1973).
Cohen,M.I.,Young,K.F.,Chang,Te-Tse and Brower,W.S., J.Appl.Phys. 42,5267 (1971).
Natarajan,M. and Prakash,B., phys.stat.sol.(a) 4,K167 (1971).

LASER EXPOSITION EFFECT ON THE NMR OF Nb^{93} AND EPR Fe^{3+} IN $LiNbO_3$

M.A.Bogonoszev, V.A.Golenishchev-Kutuzov,A.A.Monachov and B.M.Khabibullin
Kazan Physico-Technical Institute of the Academy of Sciences, USSR.

Abstract. The laser exposition effect in the single domain $LiNbO_3$ crystals with different concentration of iron has been studied. The shift and broadening the Nb^{93} NMR and Fe^{3+} EPR lines under the action of the He-Ne laser illumination was observed. The experiments showed that this effect is strongly correlated with laser damage.

Nowadays the mechanism of laser induced optical damage in $LiNbO_3$, that is optically induced changes of refraction indeces, is not sufficiently undestand. Some models describing the process of optical damage have been proposed. Chen (1969) showed that optical damage can be explained by the drifting of the photoexcited electrons out of the illuminated region along the c axis of the crystal which followed their retrapping at the laser beam. The space--charge field between the retrapped electrons and the positive ionized centers gives the change of the refraction indeces through the electro-optic effect.

Johnston (1970) proposed that the photoexcitation of electrons within the illuminated region leads to local variations of the macroscopic spontaneous polarisation density and to the creation of an electric field in the direction of opposite to the spontaneous polarisation.

The Japan autors (Ohmori at al,1974) showed, that the internal drift field originally presents in the poled crystals which have large spontaneous polarisation (Ps); its value for $LiNbO_3$ is about 45kv/cm and its direction is antiparallel to the Ps.

In this paper we give the results of the experimental investigations of the spectra NMR Nb^{93} and EPR Fe^{3+} in lithium niobate before and after the laser illumination. It is known, that the nuclei which possess the quadrupole moments are strongly coupled with the gradients of crystal electric field. The shift and broadening of linewidths is created by such interactions. The laser exposition which gives a strong electric field can contribute to the additional shift and broadening of lines.

For lithium niobate the Nb^{93} nuclear spin is 9/2, the quadrupole moment (Q) is 0,2 barn and quadrupole coupling constant (e^2qQ/h) is about 22MHz. The energy of the interaction between nuclei and electric field of axial symmetry in the main axes of field tensors may be written (Abragam,1960).

$$E_{Iz} = E_{-Iz} = \frac{e^2qQ}{4I(2I-1)}\left[3I_z^2 - I(I+1)\right], \qquad (1)$$

where $q = V_{zz}/e$.

For an NMR magnetic field of the order of 1,4 10^4 gauss ($\frac{1}{2} \leftrightarrow -\frac{1}{2}$) the magnetic field effect is more then the quadrupole effect; thus $\nu_H > \nu_Q$ where $\nu_H = \gamma H_0/2\pi$ and $\nu_Q = e^2qQ/2\pi h$.

The quadrupole shift of line may be expressed by the formula

$$\nu_{\pm 1/2} = \frac{\nu_Q^2}{\nu_H}\cdot\frac{9}{256}\cdot\frac{2I+1}{I^2(2I-1)}(1-\cos^2\theta)(1-9\cos^2\theta), \qquad (2)$$

where θ - is the angle between the magnetic field H_0 and local symmetry c' axis of axial electric field.

The lattice defects will modify both the direction (θ) and magnitude of the electric field gradient (q)

$$\theta = \theta_0 + \theta_1 \;;\quad q = q_0 + q_1, \qquad (3)$$

here θ_0 and q_0 are constants while θ_1 and q_1 are random variables.

For $c \parallel H_0$ θ_0 and q_0 are equal zero and a set of random values of local electric field gradient q_1 and angle θ_1 which is due to lattice defects producesthe broadening $(\Delta\nu)^2$ and the shift ($\Delta\nu$) of NMR lines.

The random values $\pm q_1$ and $\pm\theta_1$ have equal probability in crystals without laser exposition, therefore broadening and shift are proportional

$$\langle\theta_1^2\rangle = \int_{-\pi}^{\pi}\rho(\theta_1)\,\theta_1^2\,d\theta_1 \;;\qquad \langle q_1^2\rangle = \int_{-\infty}^{\infty} f(q_1)\,q_1^2\,dq_1, \qquad (4)$$

where $\rho(\theta_1)$ and $f(q_1)$ are the functions, which signify the expected values θ_1 and q_1. For the crystals without laser illumination $\langle q_1\rangle = \langle\theta_1\rangle = 0$. In a crystals when underwent the laser action the drifting electric field is created and thus $\langle q_1\rangle \neq 0$; $\langle \theta_1\rangle \neq 0$.

At these conditions the shift and broadening of NMR line are given by:

$$\langle \Delta \nu\rangle = \frac{11}{288}\cdot\frac{\nu_Q^2}{\nu_H}\langle\theta_1^2\rangle\langle\Delta\nu_Q\rangle \ , \quad where \ \langle\Delta\nu_Q\rangle = \frac{e^2Q}{2\pi h}\langle q_1\rangle,$$

$$\sqrt{\langle\Delta\nu^2\rangle} = \left[\left(\frac{\partial\nu}{\partial\theta}\right)_{\theta_0=0}\langle\theta_1^2\rangle + \left(\frac{\partial\nu}{\partial q}\right)_{q_0=0}\langle q_1^2\rangle\right]^{1/2}. \qquad (5)$$

The next experimental setup to study the laser exposition effect on the NMR was applied. The light from 25mw He-Ne laser was irradiated into the sample along the x-axis.

The shift and broadening of the Nb^{93} NMR ($1/2 \leftrightarrow -1/2$) line under the action of the laser illumination were observed.

		before illumination	after illumination
line width	$\sqrt{\langle\Delta\nu^2\rangle}$	4.25 kHz	5.7 kHz
shift	$\langle\Delta\nu\rangle$	0	1.5 kHz

The shift of the Fe^{3+} EPR ($1/2 \leftrightarrow 3/2$) line under the laser action was also observed: $\langle\Delta\nu\rangle \sim 3\,kHz$.

The shift and broadening completely vanished after heating for 10 hours at 200°C.

The susceptibility of crystal laser illumination decreases with decreasing Fe^{3+} concentration and is independent from laser power. The positive shift of Nb^{93} and Fe^{3+} lines show that the change in the density of the electric field $\langle q_1\rangle$ must be opposite to the spontaneous polarisation. The value of internal electric field according to our calculations is about $5\cdot10^4\ V/cm$.

Thus our experiments showed that the shift and broadening of the NMR and EPR lines in lithium niobate were strongly correlated with the laser-induced refractive changes (laser damage). We proposed that these effects were due to the displacement of the photoelectrons beyond the illuminated region and the creation of the gradients of the electric field in the antiparallel direction to the sp ntaneous polarisation.

The dependence of the shift on the Fe^{3+} concentration can be explained if we assume that the Fe ions are the source of the photoelectrons and traps.

References

Abragam, A. 1960, Principles of Nuclear Magnetism, Oxford Nniversity Press, London.
Chen, F., 1969, Journ.Appl.Phys., 40, 3389.
Johnston, W., 1970, Journ.Appl.Phys., 41, 3279.
Ohmori, Y., 1974, Technology Reports of the Osaka Univ., 24, N 1166.

STUDY OF THE CRITICAL SLOWING-DOWN OF FLUCTUATIONS AT STRUCTURAL PHASE TRANSITIONS BY QUADRUPOLAR RELAXATION IN THE LABORATORY AND IN THE ROTATING FRAME

G. Bonera, M. Mali* and A. Rigamonti
Istituto di Fisica dell'Università di Pavia, 27100 Pavia, Italy
*On leave of absence from "J. Stefan" Institut, Ljubljana, Yugoslavia

Abstract. A study of the critical slowing-down of the fluctuations at structural phase transitions in $NaNO_2$ and $NaNbO_3$ by quadrupolar spin-lattice relaxation of Na^{23} and Nb^{93} in the laboratory and in the rotating frame is reported.

1. Introduction. The nuclear quadrupole moment is an excellent probe to study the microscopic critical dynamics near structural phase transitions. In fact the enhancement and slowing down of the collective fluctuations in the atomic positions often drive the spin-lattice relaxation process through nuclear quadrupole interaction with the crystalline electric field gradients (efg). In particular, since by T_1 and $T_{1\rho}$ measurements one can probe spectral densities typically from $2\pi.2$kHz to $2\pi.50$MHz, it appears possible to investigate the low-frequency part of the dynamical structure factor $S(\underline{q},\omega)$, having in mind the problem of the shape and the width of the "central peak" in the soft modes. In this paper two typical crystals are considered: $NaNO_2$ (as order-disorder Ising-type system) and $NaNbO_3$ (as displacive-type system with rotational fluctuations of the NbO_6 octahedra).

2. Theoretical expression. For the cubic phase of $NaNbO_3$ the Na^{23} and Nb^{93} relaxation rates due to a fluctuating efg can be written:

$$T_1^{-1}=A\{J_1(\omega_o)+4J_2(2\omega_o)\} \qquad T_{1\rho}^{-1}=(A/4)\{9J_o(2\omega_1)+10J_1(\omega_o)+4J_2(2\omega_o)\} \qquad 1)$$

where $A=(2I+3)e^2Q^2/40I^2(2I-1)\hbar^2$ and $J_\mu(\omega)=\int\exp(-i\omega t)\langle V_\mu(0)V_{-\mu}(t)\rangle dt$, where V_μ are the well-known efg functions. For a $NaNO_2$ single crystal with the ferroelectric $\underline{c}$ axis $\perp \underline{H}_o$ only the central line was irradiated; the recovery law is given by two exponentials with time constants (for I=3/2)

$$(T_1^{-1})_a=5AJ_1(\omega_o);\ (T_1^{-1})_b=5AJ_2(2\omega_o);\ (T_{1\rho}^{-1})_a=\tfrac{5}{2}A\{J_1(\omega_o)+2J_2(2\omega_o)\};\ (T_{1\rho}^{-1})_b=\tfrac{5}{16}A\{9J_o(2\omega_1)+8J_1(\omega_o)+2J_2(2\omega_o)\}$$

By relating the efg components to the local critical variable $\phi_\ell(t)$ (angle of rotation of NbO_6 octahedra) or $s_\ell(t)$ (orientation of NO_2^- dipole) and by introducing the collective variable $\phi_{\underline{q}}(t)$ or $s_{\underline{q}}(t)$ one obtains (Avogadro et al., 1974):

$$J_\mu(\omega)=\Sigma_{\underline{q}} A_{\underline{q}}^\mu S(\underline{q},\omega) \qquad \text{where} \qquad S(\underline{q},\omega)=\int\exp(-i\omega t)\langle\phi_{\underline{q}}(0)\phi_{-\underline{q}}(t)\rangle dt \qquad 3)$$

The $\underline{q}$-dependent $A_{\underline{q}}^\mu$ factor reflects the fact that the relaxation process is sensitive to the auto and pair correlation and it gives direct information about the dimensionality of the fluctuations. The $A_{\underline{q}}^\mu$ have been evaluated and one obtains the following results, useful for the discussion:

in powdered $NaNbO_3$, for Nb^{93} $\qquad A^0=1/4A^1=4/11A^2=32/15E\{1-\cos(q_x-q_y)a\}$ $\qquad$ 4)

for Na^{23}, for $\underline{q}$ at the zone boundary ($\underline{q}_B=\pi/a,\pi/a,q_z$) $\qquad A^0=2/3A^1=2/3A^2=4/15(1-u)$ $\qquad$ 5)

where $u=\cos q_z a$, $E=144e^2(1-\gamma_\infty)^2/a^6$ (a lattice constant, γ_∞ antishielding factor);
in $NaNO_2$ single crystal, for fluctuations with $\underline{q}$ along the $\underline{a}$-axis ($\underline{q}=0,0,q_z$) and 3D correlated:

for $\underline{H}_o /\!/ \underline{a}$ $\quad A^0=A^2=0 \quad A^1=3.6D(1-u)$ $\qquad$ for $\underline{H}_o /\!/ \underline{b}$ $\quad A^0=A^1=0 \quad A^2=3.6D(1-u)$ $\qquad$ 6)

while assuming complete uncorrelation between the fluctuation waves propagating along the different $\underline{a}$-chains of dipoles:

for $\underline{H}_o /\!/ \underline{a}$ $\quad A^0=D(1-.39u)^2 \quad A^1=D(1.49-.54u-.95u^2) \quad A^2=D(2.72+.43u+.027u^2)$ $\qquad$ 7)

for $\underline{H}_o /\!/ \underline{b}$ $\quad A^0=D(1+.17u) \quad A^1=D(.47+.47u) \quad A^2=D(3.47-.65u-.9u^2)$

where $D=9\mu^2(1-\gamma_\infty)^2/(c/2)^8$ and μ effective dipole moment of NO_2^-.

By extending the theory of phase transitions and critical phenomena some predictions about $S(\underline{q},\omega)$ can be made. For the dynamics of the collective fluctuations in an Ising-type system, as $NaNO_2$, on the basis of the Glauber model one can write, in the M.F.A.:

$$\langle s_{\underline{q}}(0)s_{-\underline{q}}(t)\rangle=\langle|s_{\underline{q}}|^2\rangle\exp(-t/\tau_{\underline{q}}) \qquad \langle|s_{\underline{q}}|^2\rangle=\tau_{\underline{q}}/\tau_o=kT\chi(0) \qquad 8)$$

where τ_o is the fluctuation time in absence of interactions. Regarding the q-dependence we can use a O.Z. anisotropic expression

$$\tau_{\underline{q}}=\tau_{q_c}\cdot\kappa^2/(q^2+\kappa^2+\delta\cos^2\theta) \qquad \tau_{q_c}=\tau_o(T/T_o)\,\varepsilon^{-\gamma} \quad \text{where} \quad \varepsilon=(T-T_o)/T_o \qquad \theta=\widehat{\underline{q}\,\underline{c}} \qquad 9)$$

and for the inverse of the correlation length $\kappa=\kappa_o\varepsilon^\nu$. As regards the NbO_6 rotational fluctuations, one can assume (Schwabl, 1972) for the correlation function at long times:

$$\langle\phi_{\underline{q}}(0)\,\phi_{-\underline{q}}(t)\rangle=\langle|\phi_{\underline{q}}|^2\rangle\exp(-\Gamma_{\underline{q}}t) \qquad \{\langle|\phi_{\underline{q}}|^2\rangle\}^{-1}\propto\Gamma_{\underline{q}}=\Gamma_{q_c}\kappa^{-2}\{q^2+\kappa^2-(1-\Delta)q^2\cos\theta\} \qquad 10)$$

$$\text{and} \qquad \langle|\phi_{q_c}|^2\rangle=\langle\phi_o^2\rangle(T/T_c)\varepsilon^{-\gamma} \qquad \Gamma_{q_c}=(T_c/T)\,\Gamma_o\,\varepsilon^{\gamma} \qquad 11)$$

3. Experimental results and discussion. $NaNO_2$. The Na^{23} recovery plot was practically exponential both in the laboratory and in the rotating frame: one can infer that the two relaxation constants in Eq. 2) do not differ by more than a factor of 2. According to earlier measurements (Bonera et al., 1970), $(T_1^{-1})_{exp}=\frac{1}{2}\{(T_1^{-1})_a+(T_1^{-1})_b\}$ is characterized, for T around T_c, by a temperature behavior of the type $\propto -\ln\varepsilon$. In the light of section 2, taking into account that the critical wave-vector is $q\simeq0$ (q =0 for the ferroelectric transition and $q\simeq\pi/10a$ for the transition to the sinusoidal phase) the existence of a critical contribution to T_1 and $T_{1\rho}$ allows one to conclude that the critical fluctuations of the NO_2^- dipoles in adjacent a-chains must be uncorrelated. In fact, no critical

raising in T_1^{-1} and $T_{1\rho}^{-1}$ can be detected (see Eq. 3) if $A_{q=q_c}=0$, since only for $q \sim q_c$ one has a divergence in $S(q,\omega_o)$. In addition it could be observed that from Eq. 7) for $q \sim 0$ and complete uncorrelation, one has $T_{1\rho}(H_o /\!/ a)/T_{1\rho}(H_o /\!/ b)=1.1$, that agrees with the experimental results.
As regards the temperature dependence one has only for $\tau_{q_c} \ll (2\omega_o)^{-1}$:

$$T_{1\rho}^{-1}=T_1^{-1}=31.4AD\tau_o(T/T_o)^2\kappa_o^4\delta^{-\frac{1}{2}}\ln \varepsilon^{-\nu} \qquad 12)$$

Since $T_{1\rho} \sim T_1$ also for $T \sim T_c$ one can conclude that the slowing-down of the fluctuations do not reach the rf range or that the width of the central peak is greater than $4\pi\omega_o$=250MHz also for $T \sim T_c$.

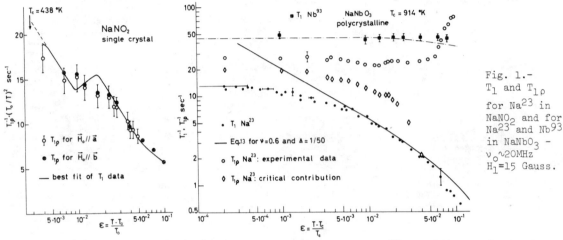

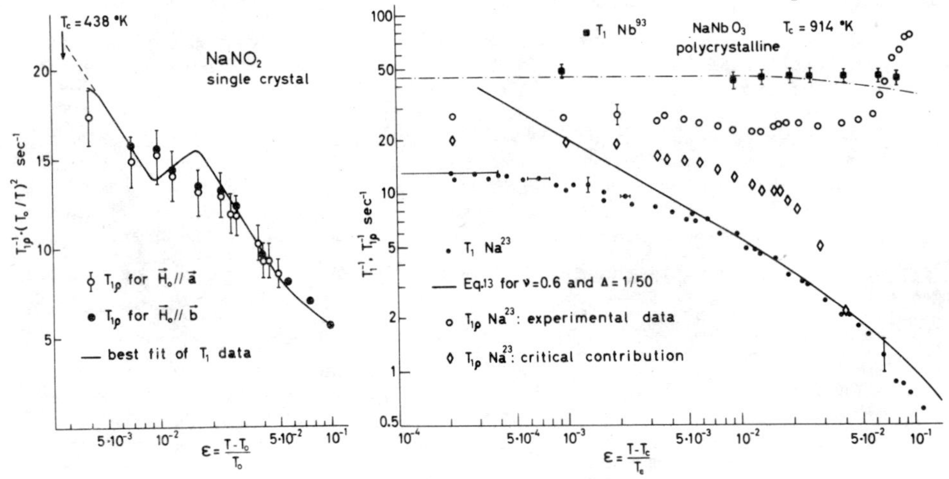

Fig. 1.- T_1 and $T_{1\rho}$ for Na^{23} in $NaNO_2$ and for Na^{23} and Nb^{93} in $NaNbO_3$ - $\nu_o \sim 20$MHz H_1=15 Gauss.

NaNbO$_3$ - A critical contribution to T_1 is present at T_c only for Na^{23}. From 4) and 5) one concludes that the transition must involve the softening of M_3 mode ($aq_c=\pi,\pi,0$) or of a large part of the M-R branch and not of the R_{25} mode ($aq_c=\pi,\pi,\pi$) as in $SrTiO_3$. A pure M_3 mode corresponds to 3D correlations in the rotational fluctuations of the oxygen octahedra and Δ=1 in Eq. 10); the softening of the whole M-R branch corresponds to rotations of the octahedra only 2D correlated in a (0,0,1) plane, without correlation in adjacent planes and Δ=0.
As regards the temperature dependence of T_1 for Na^{23}, from section 2 the complete expression can be obtained (Rigamonti, 1974): for $\Gamma_{q_c} \gg 2\omega_o$ one has

$$T_1^{-1}=4AE\langle\phi_o^2\rangle\Gamma_o^{-1}(T/T_c)^2\kappa_o^3 \,\mathrm{arctg}\{\pi\Delta^{\frac{1}{2}}/a\kappa_o\varepsilon^{\nu}\}/4\pi^2\Delta^{\frac{1}{2}}\varepsilon^{\nu} \qquad 13)$$

while if Γ_{q_c} slows-down below ω_o a flattening in the temperature behavior of T_1^{-1} occurs. Therefore the experimental values for Na^{23} indicate that for $\varepsilon=(4\pm2).10^{-3}$ the width Γ_{q_c} of the central peak should become approximately 280MHz. [A similar result has been obtained from the broadening of the E.P.R. line in $SrTiO_3$, (Müller, 1974)]. Moreover a fitting of the T_1 values of Na^{23} for $\varepsilon > 5.10^3$ on the basis of Eq. 13) suggests for the critical parameters the indicative values ν=0.6 and Δ=1/50. These results are confirmed also by T_1 measurements for Nb^{93}. In fact the theoretical expression for Nb^{93} (Rigamonti, 1974) for Δ=1 exhibits a cusped-shape behavior for T close to T_c while for $\Delta \ll 1$ is practically temperature independent (see fig. 1., dotted line, for ν=0.6 and Δ=1/50). While the numerical values of ν and Δ cannot be taken too litterally, it is however possible to deduce the 2D nature of the rotational fluctuation, as also indicated by the strongly anisotropic X-ray diffuse scattering. In presence of slowing down around ω_o, the $T_{1\rho}$ measurements become, in principle, particularly significant. In fact, for Na^{23} in a polycrystalline sample, one has

$$\text{for } \Gamma_{q_c} > 2\omega_o \quad T_{1\rho}=T_1 \quad \text{and for } \Gamma_{q_c} < \omega_o \quad T_{1\rho}^{-1} = 14/20(T_1)^{-1} + AE\langle\phi_o^2\rangle\Gamma_o^{-1}\kappa_o^3/5\Delta^{\frac{1}{2}}\pi\varepsilon^{\nu} \qquad 14)$$

i.e. close to T_c, while T_1 reaches a practically constant value, $T_{1\rho}$ exhibits a divergence of the type $T_{1\rho} \propto \varepsilon^{-\nu}$ that should allow a direct unambiguous evaluation of ν. Unfortunately a further mechanism of relaxation for $T_{1\rho}$ is present (see fig. 1). A possible explanation of the $T_{1\rho}$ data stands on a relaxation process based on the diffusion of oxygen vacancies, since D.T.A. measurements reveal a loss of oxygen for T>700°C. The $T_{1\rho}$ values far from T_c can be justified by oxygen diffusion with activation energy of 1.8eV/atom and with oxygen vacancies less than 1:10^5. By extrapolation to lower temperatures one could extract the critical contribution to $T_{1\rho}$ around T_c (see fig. 1). If the results of this very crude procedure could be taken with confidence, serious modifications in the commonly accepted idea about a central peak of lorentzian shape have to be considered. In order to clarify the mechanism of relaxation in the rotating frame in $NaNbO_3$ $T_{1\rho}$ measurements vs H_1 and for Nb^{93} are under way.

References

Avogadro A., G. Bonera, F. Borsa and A. Rigamonti, Phys. Rev. B9, 3905 (1974).
Bonera G., F. Borsa and A. Rigamonti, Phys. Rev. B2, 2784 (1970).
Müller K.A., W. Berlinger, C.H. West and P. Heller, Phys. Rev. Lett. 32, 160 (1974).
Rigamonti A., in "Local Properties at Phase Transitions", Proc. LIX Course Int. School Phys. "E. Fermi" (Varenna 1973) - Edited by K.A. Müller and A. Rigamonti - S.I.F. (1974).
Schwabl F., Phys. Rev. Lett. 28, 500 (1972).

EPR STUDY OF PHASE TRANSITION IN $[Ni/NH_3/_6]/NO_3/_2$ UNDER HIGH PRESSURE

J.Stankowski and L.Laryś
Div. of Solid State Radiospectroscopy, Polish Academy of Sciences, Poznań, Poland

Abstract. In nikel-hexammino-nitrate the transition pressure between two phases, characterized by different linewidths, was determined by EPR. The phase transition at room temperature occurs at p_c = 2.85 kbar

1. Introduction. Compounds of the composition $Me^{2+}/NH_3/_6 Y_2$ characterized by a temperature T_c where anomalies of specific heat C_p, EPR linewidth, second moment of NMR line quadrupole splitting of Mössbauer line and other electric field gradient dependent properties occur, have been studied by numerous authors. For $[Ni/NH_3/_6]/NO_3/_2$, Long and Toettcher [1] from their specific heat studies, report a phase transition point T_c lying closest to room temperature.

This compound was accordingly chosen for synthesis, carried out by the procedure of King [2] .
As expected, the point of steep change in EPR linewidth was found to lie nearest the room temperature. Such a high transition temperature favours stimulation of a low symmetry state at room temperatureby applying high hydrostatic pressure.

2. Experimental. EPR investigation was performed with a JEOL-3BX microwave spectrometer in the X band. Temperature was varied with nitrogen vapour and measured by means of a Cu-constantan thermocouple.
High pressure was applied to the system investigated in a special device, constructed at our Institute and described in detail by Stankowski et al [3] .
The high-pressure window consisted of a sapphire cylinder, acting as microwave resonator with TE_{111} mode and thus providing for direct coupling of the resonator microwave line from the adjusted waveguide. The resonator had a Q-factor of 2500, ensuring correct operation of the spectrometer. Compared of Bloembergen's resonator [4] , ours had moreover the advantage of a very weak pressure-dependence of the resonance frequency, as expressed by a coefficient of 6 MHz/Kbar.
At room temperature, 293°K, $[Ni/NH_3/_6]/NO_3/_2$ exhibits an EPR linewidth amounting to $2\Delta B_{1s}$ = 864 Gs and a spectroscopic splitting factor g = 2.20.
In hexammino-niclous nitrate, likewise to other compounds with the general structure $Me^{2+}/NH_3/_6/ Y^{-1}/_2$, a jump - wise change in EPR was observed at 233°K. The linewidth amounted to 1000 Gs and 2400 Gs above and, respectively, below 233°K. With decreasing temperature, a constant increase in linewidth was observed below the phase transition at rate of $\alpha = -2.61 \cdot 10^{-3}$ deg^{-1}. At constant temperature /293°K/and variable pressure, a pressure-induced low-temperature phase, characterized by a broad EPR line, was obtained. The curve presented on the Fig.1

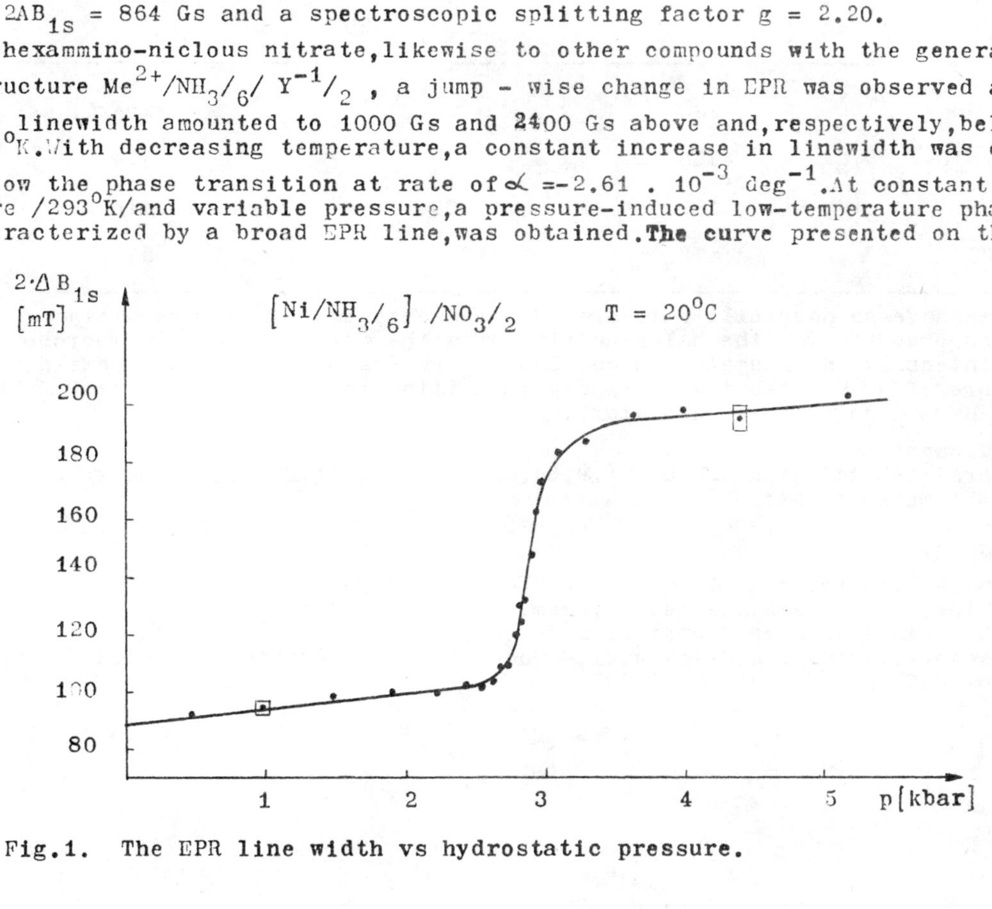

Fig.1. The EPR line width vs hydrostatic pressure.

shows the isotherm obtained for $Ni/NH_3/_6\ /NO_3/_2$ at room temperature.The phase transition was obtained at a pressure of $p_c = 2850$ bar.Also,pressure-dependent measurements revealed linearity of EPR linewidth vs.pressure,at rate of $\alpha = 6.29 \cdot 10^{-5}\ bar^{-1}$ .The jump in linewidth in the pressure-induced phase transition amounts to 860 Gs.

3. Discussion of the results. Pressure investigation permits to follow the low-temperature phase and its dynamics already at room temperature.Moreover, with respect to phase transition occuring at the very lowest temperatures as is the case for NiA_6Cl_2 , higher pressures permit to performs investigations of the low-temperature phase i.e. of the phase lying in the near 0° Kelvin temperature region,at low pressures.Measurements of the influence of high pressures on T_c make it possible to calculate the pressure coefficient of the phase transition temperature from the simple relation:

$$T_c^p = T_c^0 \left(1 + \varkappa p_1\right) \qquad 1$$

where T_c^p is the transition point under the pressure p and T_c^0 that in normal conditions.
Since in $NiA_6/NO_3/_2$ the transition takes plase at $T_c^p = 293^oK$ under a pressure of $p = 2850$ bar and $T_c = 233^oK$ in normal conditions,the coefficient amounts to $\varkappa = 9.03 \cdot 10^{-5}\ bar^{-1}$.On the assumption that the same coefficient is applicable to other hexammino-compounds,we obtain transition the pressure values for the substances with the highest transition temperatures T_c studied by us.

Table I

Substance	T_c^0	P_c bar at $T = 293^oK$	Remarks
$NiA_6/NO_3/_2$	233	2850	experm.
$NiA_6/ClO_4/_2$	168	8240	calc.
$NiA_6/BF_4/_2$	135	13000	calc.
NiA_6Cl_2	76	31600	calc.

Whereas we have no possibility to proceed to a checking of the transition under high pressure for the halogenodrivatives,the perchlorate and fluoroborate are now intensely investogated in our Laboratory.The present results explain the failure of dr M.Maćkowiak to find a transition in NiA_6Cl_2 up to 10.5 Kbar in his EPR studies at this Laboratory.

Acknowledgement.
The authors wish to thank M.Krupski,M.Sc,and H.Gierszal,M.Sc,for their kind help in the measurements.

4.References.

1.E.A.Long,F.C.Toettcher,J.Am.Chem.Soc. 64,629 /1942/
2.H.J.S.King,A.W.Cruse,F.A.Angell,J.Chem.Soc.2928 /1932/
3.J.Stankowski et al / in preparation for press/
4.J.Stankowski,S.Waplak,A.Gałęzewski,M.Maćkowiak,Acta Physica Polonica, A 43,367 /1973/

NUCLEAR SPIN-LATTICE RELAXATION NEAR THE CRITICAL POINT OF COLEMANITE

H.Theveneau and P.Papon
Laboratoire de Résonance Magnétique - Ecole Supérieure de Physique et Chimie Industrielles et Université de Paris VI, 10, Rue Vauquelin - 75231 PARIS Cedex 05 -

Abstract . We report results of nuclear spin-lattice time measurements on B^{11} nuclei on single crystals of colemanite Ca B_3 O_4 $(OH)_3$ H_2O near the ferroelectric phase transition.

Introduction :

NMR has long been used to study critical phenomena in ferroelectrics and particularly nuclear spin-lattice relaxation measurements have been performed on many nuclei in the vicinity of the Curie point. In ferroelectrics, without paramagnetic impurities, the main source of relaxation for nuclei with I > 1/2 is the interaction of the nuclear quadrupole moment with the electrical field gradient (e f g) of the crystal which is modulated by the lattice motions. Therefore the study of the nuclear spin-lattice relaxation time T_1 in function of temperature gives us informations about the dynamical behavior of the ferroelectric.

We report results of the investigation of the critical behavior of a ferroelectric near T_c with $2.10^{-5} \leqslant \varepsilon \leqslant 10^{-1}$, $\varepsilon = \frac{\Delta T}{T_c}$ and $\Delta T = |T - T_c|$. We have performed nuclear spin-lattice relaxation time measurements on B^{11} nuclei (I = 3/2) on single crystals of colemanite. Colemanite Ca B_3 O_4 $(OH)_3$ H_2O presents a second order phase transition at $T_c = -1.70°C$, which is particularly convenient. The crystal is monoclinic centrosymmetric in the high temperature phase and loses its center of symmetry below T_c. Furthermore although its crystalline structure is rather complex it has two kinds of Boron sites, some being located at the center of a BO_3 triangle the others at the center of $BO_3(OH)$ or of $BO_2(OH)_2$ tetrahedra.

Techniques :

To change and control temperature of the sample very near T_c , we have built a system including

- a metallic cryostat with four successive enclosures. The outer evacuated enclosure contains in its upper part, a bath of cryogenic liquid (a solution of alcohol and solid CO_2) and in its lower part three others which can be evacuated or filled with helium gas to permit thermal contact with the bath. The last enclosure is a copper block containing in two symmetrical positions the sample in its resonant coil and the thermometer.
- the thermometer is a four leads miniature platinum resistance standard developed by MINCO and calibrated by the N.B.S. . Its resistance is determined by comparison with the one of a TINSLEY reference resistor in a ROSEMOUNT comparison bridge.

To obtain the required temperature we use the non-equilibrium voltage of the bridge to monitor the power input of a heater resistor sealed on the copper block. The whole system enables varying and controling temperature by paths of less than 10^{-2} Kelvin.

Single crystals of colemanite were obtained from mineral samples of Death Valley (USA). They were oriented by X-Rays techniques and cut perpendicularly to the $\vec{b}$ and $\vec{c}$ axes.

The spin-lattice relaxation rate was measured with a BRUKER pulse spectrometer operating at a frequency of 15 MHz. B^{11} magnetic resonance was observed in a field of 11 KGauss.

An extensive study of the B^{11} spectrum in colemanite has been made by Holuj and Petch (Holuj et al. 1960). They determined the quadrupolar splittings of the lines for all the orientations of the sample in the magnetic field.

So we have chosen orientations where the lines are quasi-superimposed to be able to saturate and detect all the spin populations with a strong R.F. magnetic field. The relaxation rate measurements were performed by the 90° - t - 90° pulse sequence method. (We did not use a comb of 90° pulses to saturate the lines to avoid a too large increase of the temperature of the sample). Knowing in the spectrum the position of the lines corresponding to the two kinds of Boron sites we could identify their relative contributions in the free precession decay signal and measure the corresponding relaxation rates.

Results :

The magnetization recovery law was exponential for the two kinds of Boron sites, but we have obtained two different relaxation times T_1 for the Boron in triangular and tetrahedric configurations ($T_1^{-1} = \frac{2}{5}\left(W_1 + 4 W_2\right)$, where W_1 and W_2 are the quadrupole spin-lattice relaxation transition probabilities).

Our results for T_1^{-1} are given on figure 1.

We have plotted these results on a semilogarithmic scale also and they clearly show a logarithmic behavior of T_1^{-1} in the whole temperature range for all B sites. This logarithmic behavior have been observed by various authors on H in colemanite (Brosowski et al, 1974) and of course in other materials but not so close to T_c .

One can account for the logarithmic divergence of (T_1^{-1}) with a simple model (Bonera et al. 1970). We admitt that near T_c the nuclear relaxation is caused by the fluctuations of the components $V_\mu(t)$ of the e f g tensor induced by the fluctuations of the electric dipoles, the process being a direct relaxation mechanism. $V_\mu(t)$ being linear in $p(t,\vec{r})$,where p is the local polarization, and using a Landau type expansion for the Free Energy in terms of p one derives the form of the susceptibility $\chi(\vec{q},\omega)$ associated with $p(\vec{q},\omega)$ which is of the diffusive form :

$$\chi(\vec{q},\omega) = \frac{\chi_q}{1 + \frac{i\omega}{\omega_q}} \quad \text{with} \quad \omega_q = \tau\chi_q^{-1} .$$

T_1^{-1} is related to the spectral density function of $V_\mu(t)$ and thus to $\chi''(q,\mu\omega_L)$ through the fluctuation-dissipation theorem (μ = 1 or 2) where ω_L is the Larmor frequency of B^{11}.

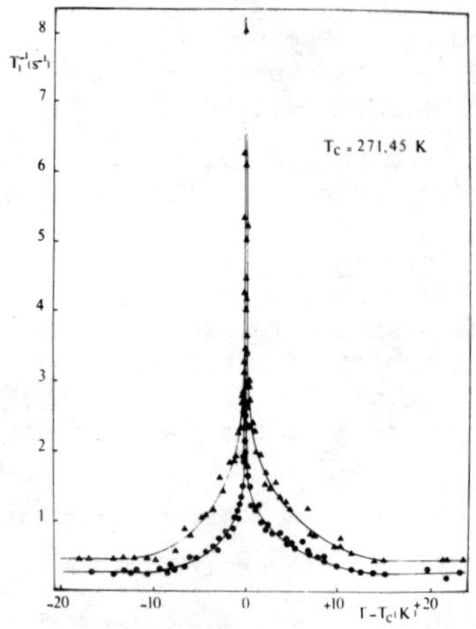

Temperature dependence of the B^{11} reciprocal spin-lattice relaxation time T_1^{-1} for Boron in triangular (▲) and tetrahedric (•) sites

Assuming $\omega_L < \omega_q$ and integrating on $\vec{q}$ one finds in the case of an anisotropic diffusive mode ω_q that the divergent part of T_1^{-1} near T_c behaves as Log $| T - T_c |$ which is in agreement with our experimental results.

We have thus been able to extend previous NMR measurements in ferroelectrics up to $\pm$ 0.01 Kelvin of T_c.

Our results do not reveal a deviation of the critical behavior from the one given by a mean field type theory.

References

Bonera G., Borsa F., Rigamonti A., 1970, Physical Review B 2 2784

Brosowski G., Buchheit W., Müller D., Petersson J., 1974, Physica Status Solidi (b) 62-1-93

Holuj F., Petch H.E., 1960, Canadian Journal of Physics, 38 515

SPIN-LATTICE RELAXATION OF ^{35}Cℓ in NH_4Cℓ DUE TO CRITICAL EFFECTS

M.J.R. Hoch*, S.P. McAlister+ and M.I. Gordon‡

* Department of Physics, University of the Witwatersrand, Johannesburg, South Africa.
\+ Department of Physics, Simon Fraser University, B.C., Canada.
‡ Department of Physics, University of South Africa, Pretoria, South Africa.

Abstract. A molecular field theory of nuclear relaxation due to critical effects in order-disorder type systems is outlined. The theory is used to analyze ^{35}Cℓ spin-lattice relaxation time measurements in NH_4Cℓ and ND_4Cℓ. An anomalously low order parameter critical exponent, β = 0.15, is found to be consistent with the results for both systems.

I. Introduction. The λ-transition in NH_4Cℓ has been studied by a wide variety of methods and it is well established that the NH_4^+ ions are ordered below the λ-point and disordered above it. Theoretically NH_4Cℓ may be described as an Ising system on a compressible lattice (e.g. Slichter et al. 1971). Yelon (1974) has recently discussed tricritical behaviour in this material at elevated pressures.

Itoh and Yamagata (1962) have shown that ^{35}Cℓ line-width and relaxation time measurements provide information on the order-disorder flips of the NH_4^+ ions. We have made further relaxation time measurements in NH_4Cℓ and ND_4Cℓ and in Section 3 analyze these results using a theory which allows for critical fluctuations of the order parameter.

2. Theory. For ferro-electrics it appears that a molecular field approach is adequate for $\varepsilon = \left|\frac{T - T_c}{T_c}\right| > 10^{-4}$. NH_4Cℓ is similar to certain ferro-electric systems and it is highly likely that this approach will be satisfactory for data discussed in this paper. Since the λ-transition is of first order Ising model dynamics modified using the approach of Slichter et al (1971) will be adopted.

Two molecular field approaches to Ising model dynamics have been given, one represented by the work of Burgess (1958) and the other by the work of Suzuki and Kubo (1968). The essential conclusions are that the relaxation time for order parameter (η) fluctuations increases near T_c and also that relaxation becomes non-linear very near T_c. As an approximation we ignore non-linear effects and adopt the solutions to the linearized equations of motion. The characteristic relaxation time τ is given by (Burgess 1958).

$$\tau = \frac{\tau_o}{2\left[\frac{1}{1-\eta^2} - \frac{T_c}{T}\right]\sqrt{1-\eta^2}} \qquad (1)$$

where τ_o may be expected to follow an Arrhenius type behaviour $\tau_o = \tau_\infty e^{E/kT}$. For $T > T_c$ $\eta = 0$ and $\tau = \tau_o/2\tilde{\varepsilon}$ where $\tilde{\varepsilon} = |T - T_c|/T$. The approach of Slichter et al (1971), as applied to the present situation, replaces T_c by $T_o(1 - n\Delta V/V_o)$ where T_o is fitted transition temperature, n is a fitting parameter and $\Delta V/V_o$ gives the fractional change in volume of the crystal referred to the volume at T_o.

The ^{35}Cℓ nucleus has I = 3/2 and spin-lattice relaxation proceeds via a quadrupolar mechanism in which order-disorder flips of the NH_4^+ ions are of dominant importance (Itoh and Yamagata 1962). In order to obtain an expression for the spin lattice relaxation time, T_1, we proceed along conventional lines but use modified probability expressions in evaluating the correlation functions. Since the $Cℓ^-$ ion site has, on average, cubic symmetry the spin temperature approach may be used and we can write (e.g. Abragam 1960)

$$1/T_1 = 2/5\,(eQ/\hbar)^2\,\{j_1(\omega_o) + 4j_2(2\omega_o)\} \qquad (2)$$

$$\text{where } j_1(\omega_o) = \sum_\alpha 9/4 \sin^2\theta_\alpha \cos^2\theta_\alpha \int_{-\infty}^{\infty} \overline{V_{zz}^\alpha(0)\,V_{zz}^\alpha(t)}\, e^{-i\omega_o t}\,dt \qquad (3)$$

A similar expression gives $j_2(2\omega_o)$. eQ is the nuclear quadrupole moment and the sum over α includes the eight NH_4^+ ions surrounding a Cℓ ion (cross-correlation terms are not important). z is the principal axis of charge distribution α which is assumed to have axial symmetry and θ_α is the angle between z and H_o. V_{zz}^α is the electric field gradient (e.f.g.) due to ion α. Introducing an ordered orientation a and a disordered orientation b, for each ion, and putting $V_{zz}^\alpha = eq^\alpha$ we have

$$\overline{V_{zz}^\alpha(0)\,V_{zz}^\alpha(t)} = e^2 \sum_{u,v=a,b} p(u,o)q_u^\alpha\, p(v,u,t)q_v^\alpha \qquad (4)$$

where q_u^α is the q^α appropriate to the u-orientation and p(u,o) is the probability of finding ion (spin) α in the u-orientation at t=0. p(v,u,t) is the probability of finding spin α in the v orientation at t = t given that it was in the u orientation at t = 0. It is easily seen that $p_a = (1 + \eta)/2$, $p_b = (1 - \eta)/2$ while $p(a,a,t) = (1 - \eta)/2\,\{\frac{1+\eta}{1-\eta} + e^{-|t|/\tau}\}$ with similar expressions for the other probabilities. τ is given by Eq (1) with the modified form for T_c.

Carrying through the calculation yields

$$I/T_I = I6/5\left(\frac{e^2Q\Delta q}{\hbar^2}\right)^2\left(\frac{I-\eta^2}{4}\right)\left\{(\lambda_I^4+\lambda_2^4+\lambda_3^4)\frac{\tau}{I+\omega_o^2\tau^2} + \left\{3-(\lambda_I^4+\lambda_2^4+\lambda_3^4)\right\}\frac{\tau}{I+4\omega_o^2\tau^2}\right\} \quad (5)$$

where $e^2Q\Delta q$ is the change in the quadrupolar coupling constant produced by an order-disorder flip, in the principal axis system, and λ_I, λ_2 and λ_3 are the direction cosines of H_o with respect to the crystal axes. The expression may be considerably simplified if a powder average is taken and also, as turns out to be the case in $NH_4C\ell$, if $\omega_o^2\tau^2 \ll I$. This gives

$$I/T_I = 48/5\left(\frac{e^2Q\Delta q}{\hbar^2}\right)^2\left(\frac{I-\eta^2}{4}\right)\tau \quad (6)$$

For $\eta = O\,(T>T_c)$ $I/T_I \propto \tau_0\left(\frac{T}{T-T_c}\right)$. This result predicts a different temperature dependence for T_I than that obtained by other workers (Bonera et al I97I and Blinc et al I97I) who used an approach involving dynamic susceptibility and the fluctuation dissipation theorem. An advantage of the present approach is that the order parameter η is explicitly incorporated in the final expressions which should be valid over a fairly wide temperature range. It must be admitted that within the molecular field approximation the order parameter critical exponent β, defined by $\eta \sim \varepsilon^\beta$ is predicted to be 0.5 and strictly speaking this result should be used in Eq (6). We shall assume that Eq(6) applies to a good approximation in situations where β departs from the molecular field predictions. Experiments are required to test this assumption. Very close to T_c the molecular field approach will fail and some other approach (e.g. dynamical scaling) must be used.

3. Results and Discussion . Fig. I shows log T_I vs IOOO/T for powder samples of $NH_4C\ell$ and $ND_4C\ell$. Digital signal averaging was used and temperature control was good to about 0.5K. Also shown are the recent results of Speight and Jeffrey (I973) for $NH_4C\ell$.

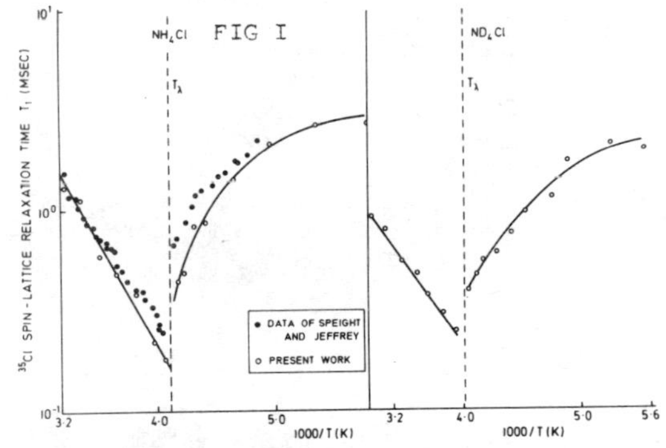

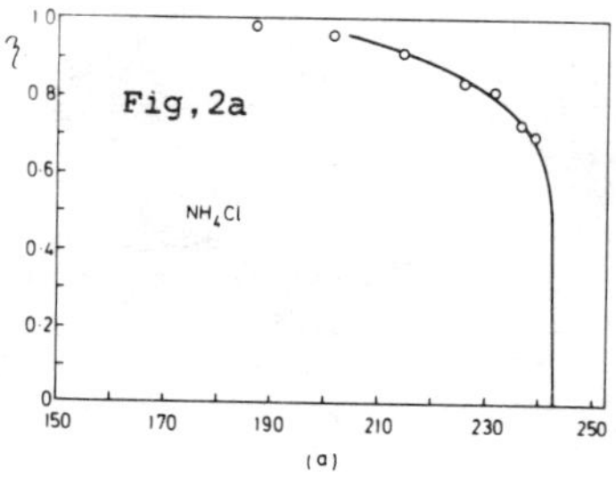

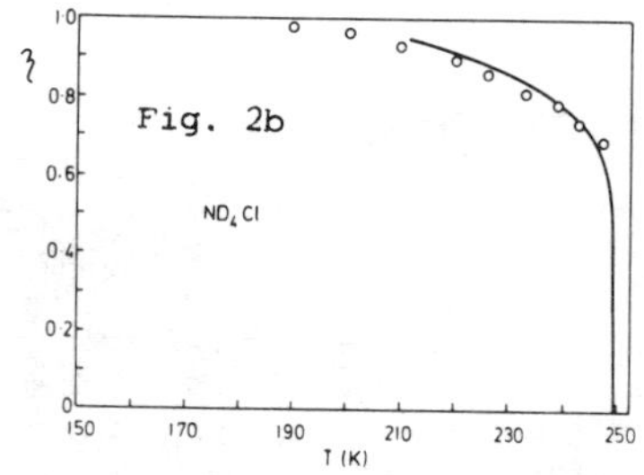

In phase II ($T > T_\lambda$) T_I follows a simple activation law behaviour with an activation energy of $\sim$ 2IOOK which is in good agreement with valuesobtained from p.m.r. results. Critical fluctuations do not appear to have any significant effect on $^{35}C\ell$ relaxation in this temperature interval. Further more precise measurements may reveal departures from Arrhenius behaviour very close to T_λ and such measurements are planned.

Below T_λ in phase III the T_I results reflect changes in the order parameter. The plotted points in Fig.2 are values of η for $NH_4C\ell$ and $ND_4C\ell$, as a function of T, obtained from the experimental data using Eq(6) and the modified Eq(I). The parameters in the theory $\left(\frac{e^2Q\Delta q}{2}\right)\tau_\infty$ and E are deduced from the phase II data. The curves in Figs. 2a and 2b both have the form $\eta = I.25\varepsilon^{0.I5}$ where ε is here given by (Slichter et al I97I) $\varepsilon = (T-T_o/T_o + n(V-V_o)/V_o^n)$. The parameter n and the quantities T_o and $\Delta V/V_o$ are taken from the work of Fredericks (I97I). (The volume dependent term provides only a small correction for the present results) The fit to the experimental points is quite acceptable. The value β=O.I5 may be deduced from the work of Slichter et al (I97I) and Fredericks (I97I). Yelon(I974) has also given β=O.I5 for $ND_4C\ell$ at zero pressure. Because of the various assumptions made in the present treatment a somewhat higher β value is not excluded by our results. However, the results do suggest that the β values are very similar for $NH_4C\ell$ and $ND_4C\ell$ at zero pressure and also that they are likely to be substantially lower than the 3-D Ising model prediction β = 5/I6.

References

Abragam, A ., I96I, The Principles of Nuclear Magnetism (O.U.P.)
Blinc, R.,Zumer,S and Lahajnar, G.,I970,Phys.Rev. BI, 4456-4463
Bonera, G., Borsa,F., and Rigamonti, A., I970, Phys.Rev. BI 2784-2795
Burgess , R.E., I958, Can.J.Phys., 36, I569-I58I
Fredericks, G.E., I97I, Phys.Rev., B4, 9II-9I9
Itoh,J,and Yamagata,Y.,I962,J.Phys.Soc.Japan,I7, 48I-507
Slichter,C.P.,Seidel,H,Schwartz,P and Fredericks,G,I97I,Phys.Rev,B4 907-9II
Speight,P.A,and Jeffrey,K.R.I973,J.Mag.Res,Io, I95-202
Suzuki,M, and Kubo, R, I968,J.Phys.Soc.Japan, 24, 5I-60
Yelon,W.B.I974,Anharmonic Lattices,Structural Transitions and Melting,255-274(Ed.Riste, T.)Noordhof.

SECULAR AND NON-SECULAR EPR RELAXATION NEAR THE $SrTiO_3$ PHASE TRANSITION

G.F. Reiter, W. Berlinger, T. Schneider, and K.A. Müller
IBM Zurich Research Laboratory, 8803 Rüschlikon, Switzerland.

Abstract. We show that the dependence of the EPR linewidth of the $Fe^{3+} - V_0$ center provides a means of separating the effect of secular and non-secular perturbations. The secular contributions show the effect of critical fluctuations, exhibit a cross-over from a "fast" (motionally narrowed) to "quasi-slow" regime, and dominate at T_c. The non-secular contributions are always "fast", and dominate at temperatures far above T_c.

1. Introduction. The $Fe^{3+} - V_0$ center in $SrTiO_3$ is well described (von Waldkirch et al., 1972) above T_c by the Hamiltonian $\mathcal{H} = D(S_z^2 - 1/4) + g\mu_B [H \sin(\theta)\, S_x + H\cos(\theta)\, S_z]$ with $g \simeq 2$, and $D = 1.4\ cm^{-1}$. $D >> g\mu_B H$ for the experiment we will discuss. Rotations of the quantization axis due to the motion of the oxygen octahedra in which the impurity is imhedded lead to a coupling to the lattice that can be described by the perturbation $\mathcal{H}_1 = \phi_x D(S_x S_z + S_z S_x) + \phi_y D(S_y S_z + S_z S_y) = \phi_x O_x^2 + \phi_y O_y^2$ where ϕ_x, ϕ_y give the rotational displacements of the oxygen octahedra, and are fluctuating variables. Systems of this type have been treated generally, and a theory developed (Reiter, 1974) that is adequate for both the "slow" and "fast" regimes. This distinction is determined by whether the coherence frequency Γ_L for the lattice fluctuations and the linewidth of the EPR line satisfy $\Delta\omega/\Gamma_L >> 1$ or $\Delta\omega/\Gamma_L << 1$, respectively. In $SrTiO_3$ near T_c, there are in fact two characteristic frequencies for the lattice fluctuations. Γ_c, characterizing the width of the central peak at $q = q_R$ is known from neutron scattering measurements (Shapiro et al., 1973) to satisfy $\Gamma_c \gtrsim 1\ cm^{-1}$. The short wavelengths are characterized by a much larger frequency, Γ_{nc}, which we take to be $25\ cm^{-1}$. As a consequence there are three regimes that can be distinguished, a "fast", "slow" and "quasi-slow" regime, the latter being determined by the condition $\Delta\omega/\Gamma_c >> 1 >> \Delta\omega/\Gamma_{nc}$. In the fast regime the linewidth due to $\mathcal{H}_1$ is

$$\Gamma(H,\theta) = \sum_{\alpha=x,y} \sum_k |\langle i|O_\alpha^2|k\rangle|^2\, J[\omega + (E_k - E_i)/\hbar] + |\langle j|O_\alpha^2|k\rangle|^2\, J[\omega + (E_j - E_k)/\hbar] \quad (1)$$

where ω is the resonance frequency for the transition from level i to level j, i.e., $\omega = (E_j - E_i)/\hbar$, and $J(\omega)$ is the Fourier transform of $\langle\langle \phi_x(t)\, \phi_x(0) \rangle\rangle$. In the experiment (Müller et al., 1974) $\hbar\omega = .648\ cm^{-1}$ and the levels i and j are the lowest lying pair. The secular contributions to the linewidth are defined to be those arising from the diagonal terms in (1) and are proportional to $J(0)$, the non-secular terms are the remainder. $J(0)$ is sensitive to the longest time scale in the lattice fluctuations, Γ_c. The separation between the energy levels is on the order of $5\ cm^{-1}$, and hence the non-secular terms depend upon values of $J(\omega)$ at $\omega/\Gamma_c >> 1$ and are insensitive to the presence of critical fluctuations. For frequencies in this range, the scale for significant variations of $J(\omega)$ is Γ_{nc}, and since $\Delta\omega/\Gamma_{nc} << 1$, these fluctuations are always in the fast regime. It happens that in this system, the angular dependence of the non-secular terms is sufficiently distinct that it provides a mean of separating the two contributions. The angular dependence of the non-secular terms depend to some extent upon the values of $J[(E_i - E_k)/\hbar]$, $J[(E_j - E_k)/\hbar]$. However, these frequencies are considerably smaller than the scale over which $J(\omega)$ varies in this range, Γ_{nc}, and it is reasonable to assume that these values are all the same. In fact, since the contribution from one of the terms actually dominates the non-secular contribution, these can be allowed to vary over a much smaller frequency range, on the order of $1\ cm^{-1}$, and the angular dependence will not change significantly. We have calculated this dependence assuming that the values of $J(\omega)$ are constant over the range of frequencies that enter the sum in (1). There is no ambiguity in the determination of the angular dependence for the secular terms. The angular variation of the linewidth, measured at constant frequency by scanning the field, is shown in Fig. 1, for both contributions.

2. Comparison with Experiment. The behavior of the angular variation of the linewidth for temperatures ranging from just above T_c ($\sim 106°$) to room temperature is shown in Fig. 2. It can be seen from a comparison of Figs. 2 and 1 that the linewidth at high temperature is due primarily to the secular term, that the contribution from this term decreases as $T \to T_c$, and that, in the immediate vicinity of T_c the secular term increases dramatically. The decrease in the intensity of the non-secular term with temperature is to be expected due to the reduction in the amplitude of the fluctuations. The increase in the intensity of the secular term occurs because Γ_c approaches zero,

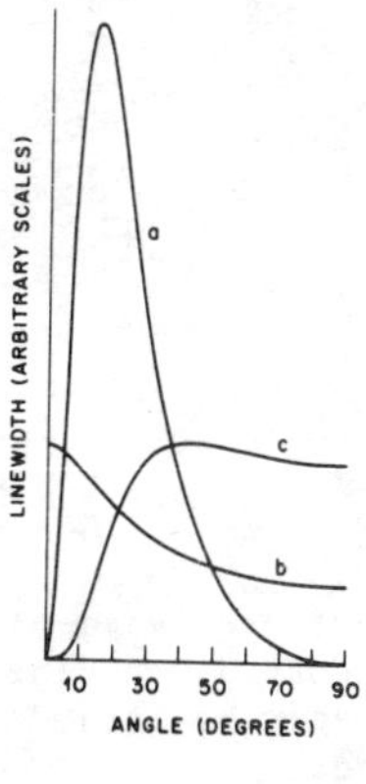

Fig. 1. Theoretical dependence of contributions to linewidth on angle between field and crystal axes. (a) $\mathcal{H}_1$-secular, (b) $\mathcal{H}_1$, $\mathcal{H}_2$-non-secular and (c) $\mathcal{H}_2$-secular(see section 2.)

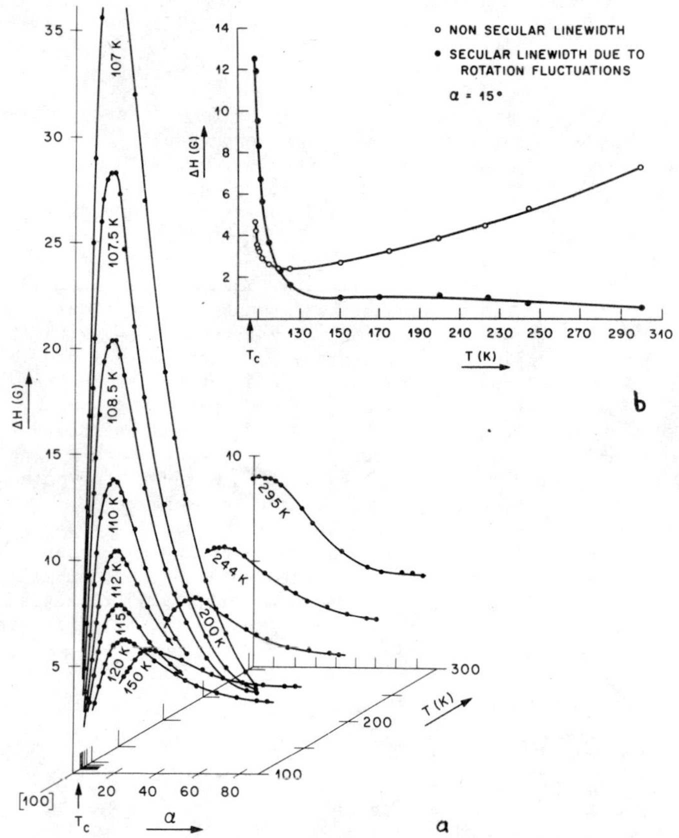

Fig. 2. a) Experimental linewidths as a function at angle and temperature.
b) Fit at $\alpha = 15°$

leading to a divergence of $J(0)$ (Müller et al., 1970). The actual linewidth does not diverge because Eq. (1) becomes inapplicable when $\Delta\omega/\Gamma_c \gtrsim 1$, and the system crosses over into the "quasi-slow" regime. When this occurs, the lineshape is no longer Lorentzian, the angular dependence of the linewidth changes, since whether one is in the "quasi-slow" regime or not depends upon the angle, and the linewidth saturates at a finite value. The data clearly shows all of these effects and indicates that the boundary of the fast regime should be taken to be at approximately 109°K. In order to obtain a quantitative comparison with the data, it is necessary to allow for the presence of an additional scource of linewidth other than that due to $\mathcal{H}_1$. We have considered two possibilities. Coupling to nuclei or to neighboring spins can yield a linewidth that is independent of angle, when measured at constant frequency. There is also an additional contribution due to the distortion of the oxygen octahedra, $\mathcal{H}_2 = \phi^2 E(S_x^2 - S_y^2)$. The non-secular terms due to this perturbation have essentially the same angular dependence as do those do to $\mathcal{H}_1$, while the secular variation is distinct, as can be seen from Fig. 1. Since the Fourier transform of $\langle\langle \phi_x^2(t)\ \phi_x^2(0) \rangle\rangle$ at zero frequency is not expected to diverge, there should be no critical effects due to this term. This is born out by the fit. Both forms can be used to fit the data equally well. The qualitative temperature dependence of the non-secular terms is the same in both cases, and the value for $J(0)$ essentially identical in the range $109° \leq T \leq 125°$. In Fig. 1 we have shown the contribution to the linewidth calculated using H_2. The fitted curves using both forms are shown in Fig. 3 for two temperatures. The lineshape for the data at 112° is very nearly Lorentzian even at 15°, indicating that the system is in the fast regime, at that temperature for all angles, and that the critical increase in linewidth cannot be due to impurities (Folk et al., to be published) which would yield a Gaussian lineshape.

3. Conclusion. The angular dependence of the linewidth provides a means of distinguishing between contributions to the linewidth corresponding to secular and non-secular perturbations. Since the non-secular terms are also non-critical, this is an effective means of subtracting the background contributions to the linewidth, and obtaining a less ambiguous measurement of $J(0)$. The angular dependence and lineshape data are inconsistent with the critical increase in linewidth at T_c being the effect of a static distribution of fluctuations.

4. References

Folk, R. and Schwabl, F., to be published.
Müller, K.A., Berlinger, W., West, C. and Heller, P., 1974, Phys. Rev. Lett. 32, 160.
Reiter, G.F., 1974, Phys. Rev. B 9, 3780.
Shapiro, S.M., Axe, J.D. and Shirane, G., 1973, Phys. Rev. B 6, 4332.
von Waldkirch, Th., Müller, K.A. and Berlinger, W., 1972, Phys. Rev. B 5, 4324.

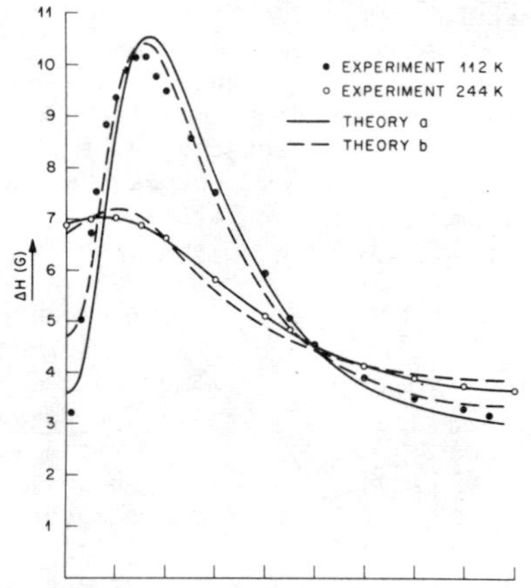

Fig. 3. Fit to the experimental linewidths. Theory a: secular terms due to $\mathcal{H}_2$. Theory b: includes constant background.

SECTION I

Metals, Alloys and Semiconductors

ELECTRON SPIN RESONANCE AND SUSCEPTIBILITY IN THE SUPERCONDUCTOR $La_xCe_{1-x}Ru_2$

U. Engel, K. Baberschke, and G. Koopmann
Fachbereich Physik, Freie Universität Berlin, 1 Berlin 33, Boltzmannstr. 20, Germany

Abstract. The ESR of Gd^{3+} in the solid solution $La_xCe_{1-x}Ru_2$ ($0\le x\le 1$) has been measured. The change of the negative g-shift and the thermal broadening of the linewidth with respect to x is attributed to a shift of a narrow d-band of the conduction-electrons at the Fermi surface.

1. Introduction. In a number of papers(1-4) the observation of the magnetic resonance of Gd^{3+} in the superconducting intermetallic compound $La_xCe_{1-x}Ru_2$ was reported. In this paper the dependence of the g-shift Δg and the thermal broadening b (linewidth DH=a+bT) as a function of the susceptibility χ of the undoped host $La_xCe_{1-x}Ru_2$ will be discussed. The g-shift Δg measures the exchange interaction parameter J(q) at q=0, while the thermal broadening b is associated with the square of J(q), summed over all wave vectors $|\vec{q}| = |\vec{k}-\vec{k}'|$ from 0 to $2k_F$.

$$\Delta g = J(0)\,\frac{\chi}{2\mu_B^2 L} \qquad (1)$$

$$b = \frac{\pi k_B}{2\mu_B}\,\langle J^2(q)\rangle \left(\frac{\chi}{2\mu_B^2 L}\right)^2 \qquad (2)$$

χ is the susceptibility and L the number of atoms per mole. With a rigid band model for a d-band metal the susceptibility may be taken as (5,6)

$$\chi = \chi_s^{spin} + \chi_d^{spin}(T) + \chi^{dia} + \chi^{orb} \qquad (3)$$

χ_s^{spin} and $\chi_d^{spin}(T)$ are the Pauli spin susceptibilities for the s- and d-bands. χ^{dia} holds for the diamagnetism of the ion cores and χ^{orb} is a temperature independent field induced orbital paramagnetism of Van Vleck type which can be appreciable in transition metals and its intermetallic compounds.

2. Experimental Results. Samples of about 100mg were prepared in an arc-furnace followed by annealing. The ESR was performed at X-band frequencies. The temperature was varied between liquid He and 30K. The g-value did not show a temperature dependence, while the linewidth increased linear with T. All experiments reported here were performed in the normal conducting phase. The Gd concentrations ranged from 80 to 3000 ppm. No dependence of Δg and b on the Gd-concentration could be detected, indicating that the systems are not bottlenecked. Fig.1 shows the g-shift Δg and the thermal broadening b as a function of La-concentration x. The susceptibility mwasurements were performed with bulk samples in a standard vibrating sample magnetometer. Fig. 2 shows the susceptibility χ of $La_xCe_{1-x}Ru_2$ as a function of x at T=10K, T=25K and T=300K. The curves were interpolated graphically. It can be seen that χ shows nearly no temperature dependence up to 25K which is in agreement with the fact that the g-value is temperature independent. At higher temperatur es a strong T-dependence of χ was observed. But because of increasing linewidth no ESR measurement could be performed in the high temperature region. The susceptibility values were corrected for the diamagnetism of the ion cores and for a small Curie-like contribution to the susceptibility at low temperatures (resulting mainly from about 160 ppm Fe impurities). The spin susceptibility χ_s^{spin} is very small(6) and was neglected. The values of the corrected susceptibility of our samples are higher than those reported in literature(6,7,8) for $LaRu_2$ and $CeRu_2$. Careful analysis showed that the susceptibility per mole increases if the homogeneity of the samples is improved. Samples with a large ESR residual linewidth, a low superconducting transition temperature T_c and a broader width of transition temperature δT_c yielded also smaller values for χ. This is attributed to La(Ce)-rich and Ru-rich phases in the peritectic solid solution $La_xCe_{1-x}Ru_2$ and the present values for χ are believed to be more accurate than the previous.

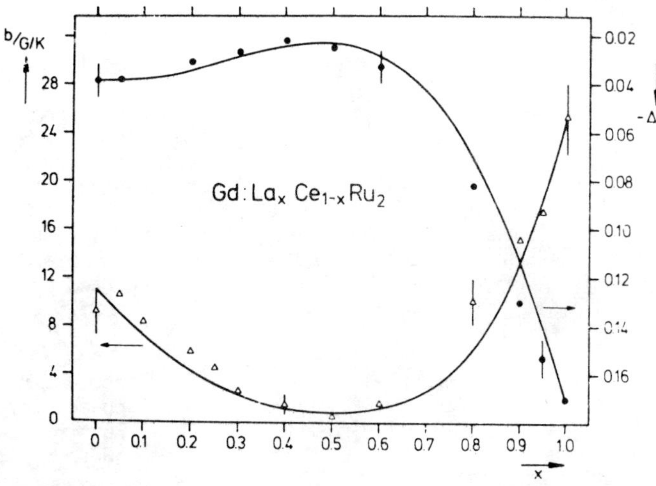

Fig.1: g-shift Δg and thermal broadening of linewidth b as a function of composition x. Circles and triangles are experimental velues, the drawn lines were calculated.

3. Discussion. In a first approximation it was tried to fit the experimental values for Δg and b shown in Fig.1 with the susceptiblity data from Fig.2, using a constant exchange parameter J(q) for all x. This produced no agreementwith experiment. The assumption of a J(q) scaling line-

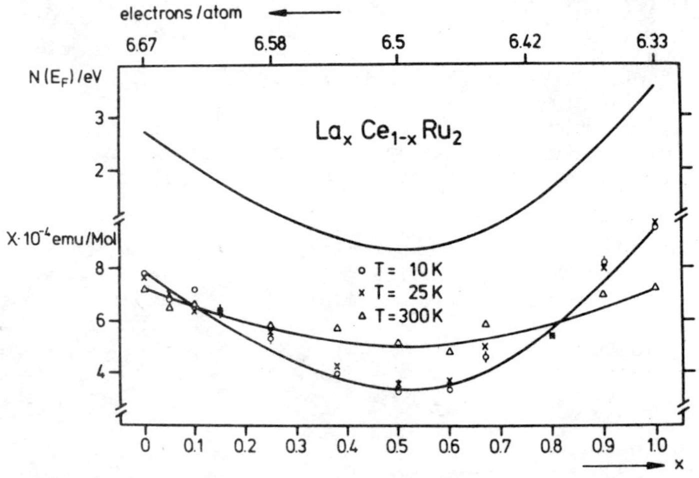

Fig.2: Measured susceptibility χ corrected for core diamagnetism. $N(E_F)$ was calculated from the low temperature susceptibility. This density of states is attributed mainly to the d-electrons.

arly with x between x=0 and x=1 gave improvement. The agreement shown in Fig.1 by drawn lines was achieved after subtracting a constant value of $1.8\cdot10^{-4}$ emu/mole from all susceptibility data, and variing J(0) and $<J^2(q)>^{1/2}$ respectively linearly between the fixed values at x=0 and x=1. This leads to the conclusion that an amount of $1.8\cdot10^{-4}$emu/mole from the total susceptibility does not contribute to the exchange polarization of the conduction electron seen by the Gd-ions and must be attributed to a temperature independent orbital susceptibility. Replacing χ in (1) and (2) by $\chi - 1.8\cdot10^{-4}$ emu/mole we derived J(0) = -12.6 meV and $<J^2(q)>^{1/2}$ = 7.3 meV for $CeRu_2$ and J(0) = -42.8 meV and $<J^2(q)>^{1/2}$=8.3 meV for $LaRu_2$ from the fitted curves for Δg and b respectively. The difference between J(0) and $<J^2(q)>^{1/2}$ is attributed to the q-dependence of the exchange integral.

From our Knightshift experiments on $LaRu_2$[9] we are able to give an estimate of about $2.5\cdot10^{-4}$ emu/ molefor χ^{orb} agreeing well with Shaltiels[6] number. This independent measurements of χ^{orb} substantiates the above estimate ($\chi^{orb} = 1.8\cdot10^{-4}$ emu/mole) needed to fit the ESR data. The density of states curve shown in Fig.2 was calculated from the low temperature susceptibility (shown below) after having subtracted χ^{orb}. $N(E_F)$ for $LaRu_2$ [10] and $CeRu_2$ [11] derived from specific heat data agrees very well with the values in Fig.2. This confirms the assumption of a large orbital contribution to χ.

Summarizing the fit of the ESR data as a function of composition leads to the conclusion that a susceptibility of approximately $1.8\cdot10^{-4}$ emu/mole does not contribute to Δg and b and is therefore attributed to a temperature independent orbital susceptibility. The variation of Δg and b with composition x may be caused by the shift of a d-band with respect to the Fermi surface.

4. Acknowledgement. We thank Prof. Hüfner for various helpful discussionsand his permanent interest in this work. Drs. M. Wilhelm and B. Hillenbrand are acknowledged for their help in the initial state of the project. This work was supported by Deutsche Forschungsgemeinschaft, SFB 161.

5. References.

1) U. Engel, K. Baberschke, G. Koopmann, S. Hüfner and M. Wilhelm, Sol.St.Comm. 12, 977 (1973)
2) C. Rettori, D. Davidov, P. Chaikin, and R. Orbach, Phys. Rev. Lett. 30, 437 (1973)
D. Davidov, C. Rettori, and H.M. Kim, Phys. Rev. B9, 147 (1974)
3) K. Baberschke, U. Engel, G. Koopmann, and S. Hüfner, AIP Conference Proceedings 18, Magnetism and Magnetic Materials, p. 984 (1974)
4) K. Baberschke, U. Engel, and S. Hüfner, accepted for publication in Sol.St. Comm. (1974)
5) A.M. Clogston and V. Jaccarino, Phys. Rev. 121, 1357 (1961)
6) D. Shaltiel in: Hyperfine Interactions; New York, London (1967) and ref. therein
7) P. Donzé, Archives des Sciences, Genève 22, 667 (1969)
8) B. Hillenbrand, K. Schuster, and M. Wilhelm, Z. Naturforsch. 26a, 1684 (1971)
9) K. Baberschke, U. Engel, and G. Clark, to be published
10) H.E.Hoenig, measured $\gamma = 51\pm2$ mJ/mole K^2 for our $LaRu_2$ samples with $T_c = 4.45\pm0.03$ K; private communication
11) R.R. Joseph, K.A. Gschneidner, and D.C. Koskimaki, Phys. Rev. B6, 3286 (1972)

COUPLED FERRO-PARAMAGNETIC RESONANCE IN COPPER-PERMALLOY DOUBLE LAYERS

A. Jánossy* and P. Monod**
*Central Research Institute for Physics, H-1525 Budapest P.O.B. 49, Hungary and
**Université Paris-Sud, Laboratoire de Physique des Solides, 91405 Orsay, France

Abstract The phase and amplitude of the transmission ESR spectrum of copper foils with a thin permalloy layer deposited onto it is strongly dependent on the difference of the paramagnetic and ferromagnetic resonance frequency of the separate layers. The g factor coincides with that of pure copper. The observations are explained within a phenomenological model.

At present very little is known about the nature of interfaces of ferromagnetic and paramagnetic layers. In a previous publication [1] we reported on striking effects in the transmission ESR (TESR) spectra of double layers of contacting ferromagnetic and paramagnetic metallic layers. We pointed out [1] that these effects are due to the coupling of magnetisations of the layers. In this paper we shall discuss the experimental results in the frame of a phenomenological model.

In the TESR experiment of a pure paramagnetic metal the sample is much thicker than the penetration depth of the eddy currents so that under off-resonance conditions no microwave power is transmitted through it. At resonance the conduction electrons polarized in the skin layer carry through the sample their magnetisation by diffusion. This leads to a small emission of radiation towards the receive cavity. The amplitude and the phase variation of the transmitted wave is measured as a function of the external field.

In the experiments [1] on double layers the thickness of the ferromagnetic layer (200-1000 Å permalloy) was much less than its skin depth while the paramagnetic layer (50 μ copper) was much thicker. Contrary to the paramagnetic case demagnetising effects give rise to a strong dependence of the ferromagnetic resonance frequency on the direction of the external static magnetic field with respect to the sample. The main features of the TESR spectra of ferromagnetic-paramagnetic double layers are the following [1]:

The g-factor is equal to that of the paramagnetic layer within experimental error. The intensity of the signal is strongly enhanced compared to that of pure copper at all static magnetic field directions. The enhancement is greatly increased when the FMR and conduction ESR frequencies are coinciding. The phase of the transmitted resonance changes sign when the FMR frequency is increased from a much lower to a much higher value than the CESR frequency. The transmission is independent of which side of the sample is excited.

To explain these features we suggest the following model. The motion of magnetisation may be described by a Bloch type of equation in both layers where in the ferromagnetic layer exchange interaction and in the paramagnetic layer spin diffusion is taken into account. At the interface we assume a short range coupling between magnetisations which formally may be described as a mixture of an exchange field and a spin diffusion type of terms. These assumptions lead to a set of coupled equations for the ferromagnetic and paramagnetic magnetisations very similar to Hasegawa's equations describing the motion of conduction and localized spin magnetisations in dilute alloys. We solved these equations making several non-essential approximations. It turns out that depending on the strength of the coupling one may distinguish between two regimes:

a. Strong coupling. One resonance is expected, the frequency and width of which are weighted averages of those of the separate layers. The spectrum is independent of the coupling constant in analogy with the "bottleneck" case for dilute alloys.

b. Weak coupling. Two resonances are expected, each having an amplitude proportional to the coupling constant $\Gamma = \lambda + i\Theta$, where λ is the strength of the exchange field type and Θ the diffusion type coupling. A shift and an additional relaxation proportional to λ and Θ respectively occurs compared to the spectra of the pure layers. In case the ferromagnetic layer is thinner than its skin depth the paramagnetic one is excited also directly giving rise to a contribution to the resonance not proportional to the coupling but still shifted and broadened. The resonance may be further broadened due to interface spin relaxation.

The obtained expressions describe well all the main features of the experiments on the copper-permalloy films. The observed signal is the paramagnetic branch of the weakly coupled spin resonance. The coupling is strong enough to make the resonance proportional to the coupling dominant. From the experimental data one may infer whether the coupling is dominantly of an exchange field or spin diffusion type. Although the absolute phase was not determined a comparison of the magnitude of the enhancement and the largest possible shift of the resonance compatible with experimental error indicates that the coupling is best described by a spin diffusion term only. The additional broadening obtained by this assumption is also in agreement with experiment taking into account that part of the observed linewidth is due to interface spin relaxation.

References

[1] A. Jánossy and P. Monod, Proceedings of the ICM Moscow, 1973 in press.

COMMENT ON THE CRYSTALLINE FIELD IN LaS:Er AND LaSe:Er INTERMETALLIC COMPOUNDS.

C. Rettori*, D. Davidov**, I. Amity**, L. J. Tao* and E. Bucher***.

* U.C.L.A., California.
** The Hebrew University of Jerusalem, Israel.
*** Bell Telephone Laboratories, Murray Hill, New Jersey.

Electron spin resonance (ESR) measurements yield Γ_7 crystalline field ground state for Er in both LaS and LaSe. This in contrast with previous measurements on the analogue intermetallic compounds LaSb:Er and LaBi:Er where $\Gamma_8^{(1)}$ ground state has been observed. This indicates an appreciable reduction in the crystalline field parameters ratio $A_4\langle r^4\rangle/A_6\langle r^6\rangle$ upon going from LaBi and LaSb to LaS and LaSe.

The ESR spectra of LaS (performed on single crystal containing 2000 ppm Er) exhibit slight anisotropy in both g value and line-width. The anisotropy of the g value, as well as the observed relatively large thermal broadening ($\Delta H/T \cong 14$ G/deg) in LaS, indicates that the first excited crystalline field level is a low lying $\Gamma_8^{(1)}$.

Taking the change of the lattice constants into consideration, we were not able to interpret the appreciable change in $A_4\langle r^4\rangle/A_6\langle r^6\rangle$ in terms of the point charge model. It seems, however, that the change in $A_4\langle r^4\rangle/A_6\langle r^6\rangle$ might be associated with the charge on the ligands, noting that both S and Se belong to the VIth column in the periodic table, whereas Sb and Bi belong to the Vth.

C.E.S.R. IN ULTRA PURE ALUMINIUM

L. Janssens, A. Stesmans and J. Witters
Laboratorium voor Vaste Stof-Fysika en Magnetisme, University of Leuven, Leuven (Belgium).

Abstract. The paper reports the observation of C.E.S.R. in pure bulk Aluminium samples by a reflection spectrometer at 21 GHz. The results are analysed in the framework of the theory by Fredkin and Freedman. The value of the many body parameter B disagrees considerably with the value reported by Lubzens et al. (1972).

1. Introduction. Measurements of C.E.S.R. in Aluminium have been reported twice by Schultz and co-workers (1966-1972) using the transmission technique. The results, presented here, are obtained by a reflection technique which allows a broader temperature range to be covered.

2. Experimental Technique. A homodyne spectrometer has been adapted for measurements of the electron paramagnetic resonance signal in metallic samples. A general description of the setup has been given at the previous Congress Ampere (Turku 1972, Janssens et al. 1973), one only wants to insist on a few improvements made since then. The cavity setup has been reconsidered. Modulation coils are wound through newsilver tubes chosen for their low spurious signal content. These tubes at the same time act as a sample holder. Vibrations due to Lorentz forces are kept beneath an acceptable level by tuning the modulation frequency into a broad minimum of the mechanical response. The samples were prepared from a 6N pure Aluminium single crystal. Small disks have been cut by spark erosion, then mechanically and electrolytically thinned resulting in platelets with highly brilliant surfaces. The smallest thickness obtainable with this method was about 30μm with a homogeneity, still acceptable, of about 25 %. In order to expose a large surface to the microwave field, a stack of 20 platelets was build up using a signal free plastic spray. The same spray was used to glue the sample in between the modulation coil. The microwave and modulation fields were both parallel to the sample surface as to minimize the losses in the sample. In the early measurements the g-value was evaluated from the knowledge of the magnetic field, which was determined by a calibrated Hall-effect device. Some uncertainty in this method due to time effects is reflected in the size of the error bars. This has now been reduced by using a suitable g-marker.

3. Results.

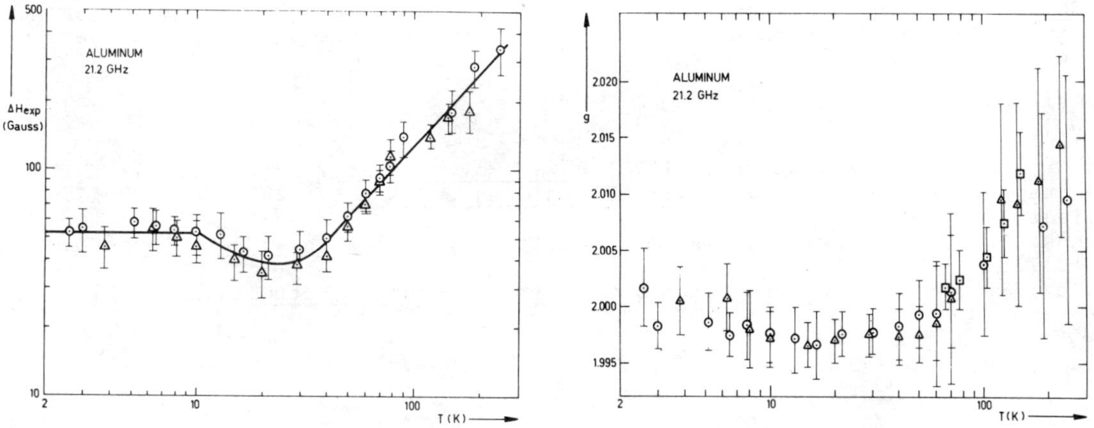

Fig. 1 Temperature dependence of the linewidth and the g-factor of C.E.S.R. in Aluminium.

Fig. 1 shows the linewidth and g-factor as a function of temperature between 2.7 K and room temperature. Up to 10 K a linewidth of 52 Gauss is measured and, within the experimental error found to be independent of the temperature. Before reaching the phonon dominated regime, where the slope indicates a T^1 behaviour, the linewidth decreases to a minimum of 38 Gauss at about 25 K. As could be expected from the results, at 35 GHz, of Lubzens et al. (1972) the g-value at 21 GHz, as a function of the temperature, shows a negative slope at low temperatures. At higher temperatures, where no other data are available to compare with, the g-value shows an unexpected increase. This has also been observed in samples with different characteristics. The sample dimensions mainly the thickness, strongly affect this slope.

4. Interpretation. The low temperature results are interpreted using the theory developed by Fredkin et al. (1972). Their formulas for the contribution to the linewidth and the g-factor by many body effects and anisotropy of the g-value over the Fermi surface are:

$$\Delta H_a = \Delta H_{max}\, 2X/(1 + X^2)$$

$$\Delta g = \Delta g_{max}\, X^2/(1 + X^2)$$

where:

$$\Delta H_{max} = (1 + B)^2 \sigma_g^2\, \omega/2g^2 B\gamma$$

$$\Delta g_{max} = (1 + B)^2 \sigma_g^2 \,/\, Bg$$

$$X = B\omega\tau/(1 + B)$$

B is a many body parameter, σ_g is an r.m.s. measure of the g-anisotropy and τ is the electron scattering time. From these expressions using the following experimental values:

$$\Delta g(21\ \text{GHz}) - \Delta g(9\ \text{GHz}) = 2.10^{-3}$$

$$\Delta H_a(21\ \text{GHz}) - \Delta H_a(9\ \text{GHz}) = 33\ \text{Gauss}$$

one can obtain values for B and σ_g as a function of τ. With a good estimation of τ, calculated from the resistivity value of the bulk material and the sample thickness, one finds:

$$0.13 \leqq B \leqq 0.18$$

$$0.051 \leqq \sigma_g/g \leqq 0.057$$

Comparing these values with those given by Lubzens et al. (1972) (B = 0.4 and σ_g/g = 0.04) one remarks a significant difference.

5. Conclusion. The difference (mainly in the value of B) together with the described temperature dependence of g at high temperatures justifies further theoretical and experimental work.

6. References.

Fredkin, D.R. and Freedman, R., (1972) Phys. Rev. Letters, 29, 1370.
Janssens, L. and Witters, J. (1973) XVIIth Congress Ampere, 285, North-Holland Publishing Co.
Lubzens, D. and Shanabarger, M.R. and Schultz, S. (1972) Phys. Rev. Letters, 29, 1387.
Schultz, S., Dunifer, D. and Latham, L., (1966), Physics Letters, 23, 192.

OBSERVATION OF CONDUCTION SPIN RESONANCE IN SMALL ALUMINIUM PARTICLES.

J.-L. Millet and R. Monot.
Laboratoire de physique expérimentale. Ecole polytechnique fédérale,
33, av. de Cour, 1007 Lausanne. Switzerland.

Abstract. A quantum size effect on the conduction electron spin resonance has been observed in the case of small Aluminium particles of sizes ranging between 10 and 200 Å. The width and shape of the resonance can be interpreted in terms of the theory of Kawabata and Kubo. However, the g value is not yet completely understood.

Introduction and Theory : in small metal particles of sizes of order of 10 Å, the localisation of the conduction electrons causes the conduction energy level to become discrete. As a result of this separation of the levels, some of the physical properties of the small particles can differ from those of the bulk (quantum size effect). The theory of the electronic properties of small metal particles has been principally developped by Kubo, Gorkov and Kawabata. In particular, for metals characterized by a spin relaxation which is due principally to the spin-orbit coupling Kawabata has shown that the quantum size effect on the conduction electron spin resonance (CESR) can be observed if the two following inegalities are satisfied :

$$\hbar\omega_z \ll \delta \qquad \hbar/\zeta \ll \delta$$

where δ is the energy splitting between two adjacent conduction electron levels and $\hbar\omega_z$ is the Zeeman energy. ζ is the spin relaxation time for the bulk resonance calculated according to the theory of Elliott but with the electron mean free path replaced by the particle size a ($\zeta = a/V_f\Delta g^2$ where V_f is the Fermi velocity and Δg the g shift of the bulk). When the two inequalities are satisfied, the normal spin relaxation process is blocked because the electrons cannot move between the discrete levels. For particles having exactly the same size, the CESR line is then the enveloppe of the contributions from individual electrons, each having a different g value (inhomogeneous broadening). The theory of Kubo and Kawabata predict that the modified resonance will have a width of order of $\hbar\omega_z/\zeta\delta$ and will be shifted towards lower frequency by the same amount. Experimental results have already been publised in the case of silver by Monot and al.. In this paper, we wish to report on some results obtained in the case of Aluminium. Assuming, as does Kubo that $\delta = 1/D_{ef}$ where D_{ef} is the density of state at Fermi level, both inequalities are satisfied for partices much smaller than 800 Å at X-band.

Experimental procedure : samples have been prepared using an atomic beam technique. Al and C_6H_6 or CO_2 were evaporated simultaneously under conventional vacuum (10^{-6} Torr) onto a surface maintained at 77°K. The concentration of Al could be modified by changing the rates of evaporation. After formation, the samples were annealed at various temperatures above 77°K to allow the metal atoms to diffuse and increase the size of the particles. Measurements were made at 77°K, at 450 MHz, 9 GHz, and 35 GH.

Results and discussion: for samples characterised by a concentration of 1 atom of Al for 10^6 molecules of C_6H_6 or CO_2 and before annealing, six sharp lines were observed which correspond to the hyperfine spectrum of isolated Al atoms (I = 5/2). In addition, the unannealed samples showed a resonance which is attributed to small particles and complicated structures which are attributed to agglomerates having a small number of atoms. After annealing at temperature ranging beween 90 and 120°K, only the resonance attributed to the small particles subsists. The width and position do not change much whereas the intensity increases slightly before decreasing. After annealing at 250°K this resonance disapears. If the concentration of Al in the matrix is increased, the intensities of atomic spectra and of the agglomerates spectra are relatively much smaller, even before annealing. Moreover, the width of the small particles resonance increases whilst its apparent g value becomes smaller. The principal characteristics of this resonance, corresponding to some samples, are presented in table 1.
Taking into account the size distribution, and assuming the Kawabata predictions, then the shape of CESR line is expected to be different from the shape of the size distribution. The experimental line shape could be fitted to the theory by a convolution calculation assuming a Gaussien distribution.

Sample	Matrix	Ann. T	35 GHz			9300 MHz			$\Delta H_1/\Delta H_2$	450 MHz			$\Delta H_2/\Delta H_3$
		°K	ΔH_1	g(app)	Δg_{so}	ΔH_2	g(app)	Δg_{so}	=3,76	ΔH_3	g(app)	Δg_{so}	=21
1	C_6H_6	116	81,5	2,0075	$1,87.10^{-2}$	31	2,007	$2,02.10^{-2}$	2,6	1,8	2,00	$2,01.10^{-2}$	17,2
2	CO_2	120	39,5	2,0091	$1,87.10^{-2}$	10,7	2,015	$1,92.10^{-2}$	3,64	–	–	–	–
3	C_6H_6	77	78	2,0089	$1,96.10^{-2}$	19,2	2,010	$1,93.10^{-2}$	4,1	1,0	2,01	$2,03.10^{-2}$	19,2
4	C_6H_6	77	99,5	2,0041	$1,77.10^{-2}$	37,4	2,003	$2,17.10^{-2}$	2,7	1,8	2,00	$1,60.10^{-2}$	20,7

Table 1

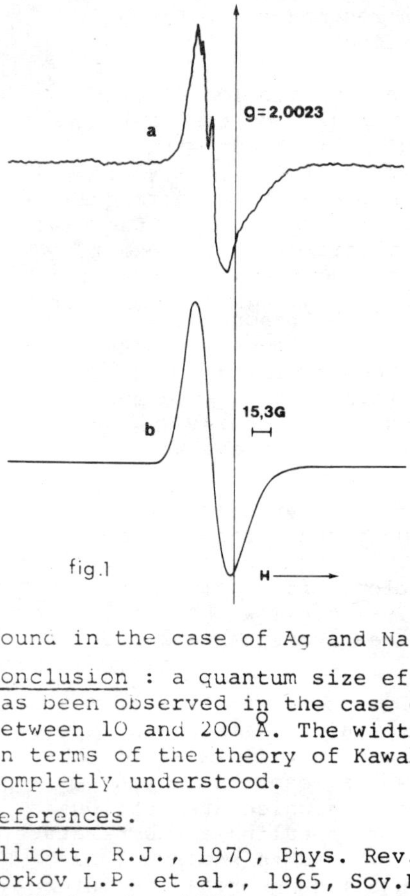

fig.1

Fig.1b shows for example the derivative of simulated signal calculated with a size distribution characterised by a central position value of 90 Å and a width of 30 Å. These datas were obtained from electron microscope observations; however, it is difficult to deduce the exact size distribution because it can be altered during the preparation of the grid. Fig. 1a shows the corresponding measured signal. Table 1 shows that the line width is roughly proportionnal to the measurement frequency as predicted by the theory and that g apparent and g bulk are different. Taking into account the theory and the size distribution, g bulk = g_{so} can then be calculated, since the resonance corresponding to an assembly of identical particles is shifted by an amount equal to its linewidth towards the higher values of the magnetic field if the measurement is made at constante frequency. g bulk was then found to be 2.02 ± 0.01 independently of the samples or the measurement frequencies. But, this value is not in agreement with the value of 1.997 given by Schultz and confirmed by Smithard. Problems alike are found in the case of Ag and Na and are not yet understood.

Conclusion : a quantum size effect on the conduction electron spin resonance has been observed in the case of small Aluminium particles of sizes ranging between 10 and 200 Å. The width and shape of the resonance can be interpreted in terms of the theory of Kawabata and Kubo. However, the g value is not yet completly understood.

References.

Elliott, R.J., 1970, Phys. Rev., 96,266.
Gorkov L.P. et al., 1965, Sov.Phys. JETP, 21, 940.
Kawabata, A., 1970, Journ Phys. Soc. Japan, 29, 902.
Kubo, R., 1 962, Journ. Phys. Soc. Japan, 17, 975.
Monot R. et al., 1974, Il Nuovo Cimento, 19 B, 253.
Scnultz S. at al, 1966, Physics Letters, 23 No 3, 192.
Smithard, M., 1970, Ph.D. Thesis, University of Exeter.

ELECTRON SPIN RESONANCE OF Gd S-STATE IONS IN La AND $LaAl_2$

G. Koopmann, K. Baberschke and U. Engel
Fachbereich Physik, Freie Universität Berlin, D-1000 Berlin 33, Boltzmannstr. 20, Germany

Abstract. Electron spin resonance (ESR) of Gd^{3+} in the cubic fcc La-lattice has been observed. The dependence of the g-value and of the thermal line broadening with the magnetic impurity concentration indicate the existence of a bottlenecked relaxation mechanism. This will be discussed in comparison with previous measurements on the $LaAl_2$-system.

1. Introduction. ESR-experiments of Gd^{3+} in metallic hosts allow to obtain information about the exchange interaction $J(q=0)$ and $<J^2(q)>$ of the localized moments with the conduction electrons (c.e.). The g-shift of Gd in various transition metals (Pd, Pt, Th) is negative (Davidov et al. 1972; Devine et al.1972) and seems to imply the dominance of a negative (covalence or d-band) contribution to the exchange interaction over the positive (s-band). But Gd^{3+} dissolved in Sc, Y and Lu (Harris et al. 1965) show a positive g-shift which can be attributed only to a dominant s-band contribution to J. In this paper we present new experimental data on Gd^{3+} ions in cubic fcc La-metal. The g-shift Δg and thermal broadening b (linewidth DH=a+bT) will be discussed in terms of the Hasegawa theory (Hasegawa 1959, Yafet 1968) and compared to previous measurements of Gd^{3+} in $LaAl_2$. One main conclusion to be drawn from the Hasegawa-equations is the appearence of bottlenecking whenever the conduction electron-impurity-ion relaxation rate, δ_{ei}, is large compared to the electron lattice relaxation rate, δ_{eL}. The ESR experiments in a bottlenecked system enable us to determine independently both rates δ_{ei} and δ_{eL}:

(1) $$g = g_{max} \cdot \left[\frac{\delta_{el}}{\delta_{el}+\delta_{ei}}\right]^2 ; \quad g_{max} = J_{sf}(0) \cdot N(E_f)$$

and (2) $$DH = DH_{max} \cdot \frac{\delta_{el}}{\delta_{el}+\delta_{ei}} ; \quad DH_{max} = \frac{\hbar}{2\mu_B} \delta_{ie}$$

with (3) $$\hbar\delta_{ie} = \pi <J_{sf}^2(q)> N(E_f)^2 k_B T$$

(4) $$\hbar\delta_{ei} = \frac{2\pi}{3} N(E_f) <J_{sf}^2(q)> S(S+1) \cdot C_{imp}$$

$N(E_f)$ is the density of states of the conduction electrons at the Fermi-level. It follows from equations (1) and (2) that δ_{ei} and δ_{el} turn out to be the controlling factors for the magnetic resonance bottleneck. The quantity δ_{ei} can be altered by variing the magnetic impurity concentration c. Addition of non-magnetic impurities or crystal-defects contribute to δ_{el}. That means the bottleneck can be broken by a decrease of exchange scattering (Koopmann et al. 1972) , or an increase of spin-orbit scattering (Gossard et al. 1967,1968; Yafet 1968; Taylor 1973). An interesting aspect of the knowledge of δ_{ei} lies in its correlation with superconductivity, because the pair breaking parameter of the Abrikosov-Gorkov theory (AG) is just δ_{ei}:

(5) $$\Delta T_c/\Delta c = \frac{3\hbar\pi}{16k_B} \Delta\delta_{ei}/dc$$

2. Experimental results and discussion. For Gd:La the measurements were performed on bulk samples at x-band frequencies and at temperatures $2K<T<25K$. We have been able to prepare the samples in an arc-furnace in the cubic fcc-phase (β-La). Careful x-ray analysis and measurements of the superconductivity transition temperature T_c and transition width ΔT_c gave evidence of the homogeneity of the samples and showed that they are at least 90% single phased. Any treatment after melting or keeping at room temperature for a day or longer disturbed the fcc-phase of the samples. Because of the low ESR-signal intensity up to now the smallest measured concentration on Gd_cLa_{1-c} was 880ppm. The results are shown in Fig.1. The concentration dependence of the superconducting transition temperature is determined as $\Delta T_c/\Delta c = 4K/\%$. The density of states at the Fermi-level $N(E_f)$ were calculated from the measured Pauli susceptibility of La $\chi_e = 1.40 \cdot 10^{-4}$ emu/mole (corrected for diamagnetism): $N(E_f) = 2eV^{-1}$ $spin^{-1}$. Calculation from specific heat data (Berman et al 1958) yields the same value. Also XPS-measurements indicate a high density of states (Baer et al. 1973). Therefore we neglect enhancement for the further calculations. ESR- and T_c-measurements on Eu_cLa_{1-c} are under current investigation. Preliminary results are

$$g_{max} \geq 0.05 \quad \text{and} \quad b_{max} \geq 25 \text{ G/K}.$$

From the experimental results the following relaxation rates were extracted: the conduction electron-lattice rate $\delta_{el}=5\cdot10^{11}$ sec^{-1} and the exchange scattering rate due to the Gd-impurities $\delta_{ei}=6\cdot10^{13} sec^{-1}\cdot C_{imp}$. With thesevalues the thermal broadening b and the g-shift Δg were computed as a function of the bottleneck factor $B = \frac{\delta_{el}}{\delta_{ei}} \propto \frac{1}{c}$. There is good agreement between the experimental data and the full curves (Fig.1), given by equations (1) and (2). This fit yields for the g-shift and thermal broadening the asymtotic values:

$$\Delta g_{max} = +0.12(1) \quad \text{and} \quad b_{max} = 55(5) \text{ G/K}.$$

The main fact to be drawn from theseresults is, the g-shift has a large positive value and in con-

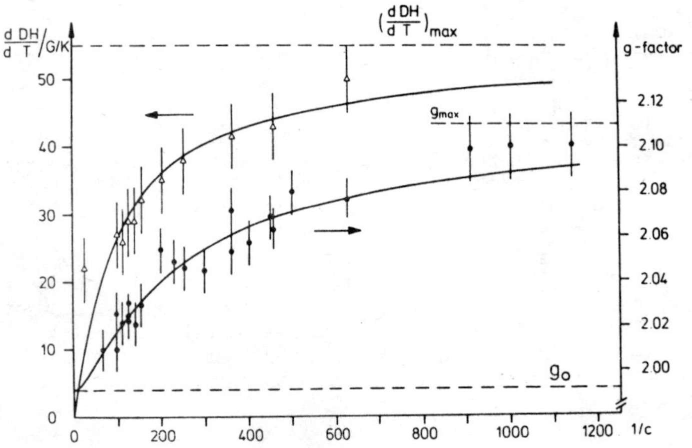

Fig. 1: g-value and thermal broadening of linewidth b as a function of reciprocal Gd-concentration c. The drawn lines were calculated.

sequence the exchange integral is positive J(0) = +0.06eV. A positive exchange interaction between the localized 4f-moment of the Gd and the c.e. can be explained, if the c.e. have s-character. In the same sense also the high density of states at the Fermi level (Baer,1973) should be attributed to s-electrons. Compared to results from the measurements at Gd:$LaAl_2$ (Davidov et al. 1973; Koopmann et al.1972) we get roughly the same values for the asymtotic g-shifts and the Korringa rates (for Gd:$LaAl_2$: $\Delta g=+0.10$,b=58G/K). As a consequence it is suggested that the exchange interaction is of short range. It seems that the Al in $LaAl_2$ have a negligible influence on the c.e. polarization at the La-site, in the contrast to other isomorphic systems, e.g. $LaRu_2$ (Engel et al. 1973; Rettori et al. 1973) and $LaOs_2$ (private communication Schrittenlacher et al. 1974) which both show a negative g-shift. Another fact, which supports the given argument is the similarity of the concentration dependence of the superconduction transition temperatures of $\Delta T_c/\Delta c$ = 4K/% in both systems. Summarizing the results, it has been shown by ESR-experiments that Gd in the dilute limit in La-metal has a positive g-shift and the bottleneck can be broken by decreasing the magnetic impurity concentration. This enables us to determine the spin-flip relaxation rates of the c.e. to the lattice and to the magnetic impurity. Both hosts La and $LaAl_2$ show roughly the same results.

La: $\delta_{ei} = 6 \times 10^{13}\ sec^{-1} \cdot c$; $\delta_{el} = 5 \times 10^{11}\ sec^{-1}$

$LaAl_2$: $\delta_{ei} = 14 \times 10^{13}\ sec^{-1} \cdot c$; $\delta_{el} = 1 \times 10^{11}\ sec^{-1}$

3. Acknowledgement. We thank Prof. Hüfner for various helpful discussions and his permanent interest in this work. This work was supported by Deutsche Forschungsgemeinschaft, SFB 161.

3. References.

Baer, Y. and Busch,G. (1973) Phys. Rev. Letters 31, 35

Berman, A., Zemansky, M.W. and Boorse, H.A. (1958), Phys. Rev. 109, 70

Davidov, D., Chelkowski, A., Rettori, C, Orbach R. and Maple, M.B., (1973), Phys. Rev. B 7, 1029

Davidov, D., Orbach, R., Rettori, C., Shaltiel, D., Tao, L.J. and Ricks, B., (1972), Sol. st. comm. 10, 451

Devine, R.A., Shaltiel, D., Moret, J.M., Ortelli, J., Zingg, W. and Peter, M. (1972), Sol. st. comm. 11, 525

Engel, U., Baberschke, K., Koopmann, G. and Hüfner, S., (1973) Sol. st. comm. 12, 997

Gossard, A.C., Kometany, T.Y. and Wernick, J.H., (1968) J. Appl. Phys. 39, 849

Harris, A.M., Popplewell, J. and Tebble, R.S., (1966) Proc. Phys. Soc. 88, 679

Hasegawa, H., (1959), Progr. Theo. Phys. 21, 483

Koopmann, G., Engel, U., Baberschke, K. and Hüfner, S., (1972) Sol. st. comm. 11, 1197

Rettori, C., Davidov, D., Chaikin, P. and Orbach, R., (1973) Phys. Rev. Lett. 30, 437

Taylor, R.H., (1973), J. Phy. F, 3 L 110

Yafet, Y., (1968), J. Appl. Phys. 39, 853

Gossard, A.C., Heeger, A.J. and Wernick, J.H., (1967) J. Appl. Phys. 38, 1251

SINGLE CRYSTAL NMR STUDIES OF TRANSITION METAL IMPURITIES IN ALUMINUM

J.A.R. Stiles, N. Kaplan*, and D. Ll. Williams
Department of Physics, University of British Columbia, Vancouver, Canada

Abstract. The satellites to the main aluminum NMR line in single crystals of AlCr and AlMn, as a function of magnetic field magnitude and direction, have been used to determine electric field gradient and Knight shift tensors for near neighbours to the impurity sites.

1. Introduction. The present work is a continuation of a program of single crystal NMR measurements in dilute aluminum alloys. First neighbour electric field gradients were first positively identified using the single crystal technique by Jorgensen et al. (1971). Additional field gradients for up to fifth neighbours to the impurity have been identified using this technique by Stiles and Williams (1974) for non magnetic impurities. Presented here are electric field gradients identified for up to fourth neighbours to the magnetic impurities in AlMn and AlCr. In addition, a Knight shift tensor has been determined for first neighbours. Some previous work in these alloys has been done on powdered samples. Berthier and Minier (1973) have measured several pure quadrupole resonances in these alloys using the Redfield technique. Isotropic differential Knight shifts for two near neighbours have been measured by Alloul et al. (1971) in AlMn. Janaussy and Grüner (1971) have measured the temperature dependence of some satellites in AlCr powder. The present data was obtained using cw NMR on single crystals of aluminum containing respectively 0.2 atomic percent manganese and 0.25 atomic percent chromium. The satellite structure out to about 400 kHz on either side of the main aluminum resonance was observed with the magnetic field along each of [100], [110] and [111]. The oscillator frequencies used were 8.950 kHz and 16.300 kHz for the manganese alloy and 9.010 kHz and 17.500 kHz for the chromium alloy. Fig. 1 shows a typical trace for AlMn. All measurements were performed at a temperature of 1.2 K.

2. Electric Field Gradients. In the AlMn alloy, electric field gradients were identified for first and third nearest neighbours. For the nearest neighbours, the pure quadrupole data of Berthier and Minier was used to calculate the satellite spectra with the magnetic field in the three main directions. The calculation involved the exact diagonalization of the Hamiltonian, and was done for the six choices of the principal axis directions for the electric field gradient tensor which are permitted by crystal symmetry. Of these six choices two are consistent with the observed satellite pattern and are indistinguishable from one another insofar as they predict the same satellite pattern. If one chooses a coordinate system so that [110] is along the radial direction from the impurity, then the largest principal value of the electric field gradient tensor has been determined to be along [110] or [1$\bar{1}$0] and the next largest along [100]. The choice of [110] for the largest value is intuitively more reasonable but this cannot be verified by this technique. The third neighbour satellite pattern was calculated using first order perturbation theory. It was found that the quadrupole frequency of 112 kHz with η=0 assigned to third neighbours by Berthier and Minier gave good agreement between the observed spectrum and the calculated third neighbour pattern. Although the third and fourth neighbour patterns have many satellites in common, the existence of a line at 182 kHz with the magnetic field along [111] which could be attributed only to third neighbours proved to be the deciding factor in favour of assigning the 112 kHz quadrupole frequency to the third neighbours. This line is shown in Fig. 1. It was necessary to obtain this spectrum at a relatively high magnetic field so as to Knight shift the more intense adjacent first neighbour satellite away from the line in question. Other less well defined lines closer to the main line also supported the choice of third neighbours over fourth. Because of the problem of overlapping lines and the difficulties of observing satellites close to the main line it was not possible to unambiguously

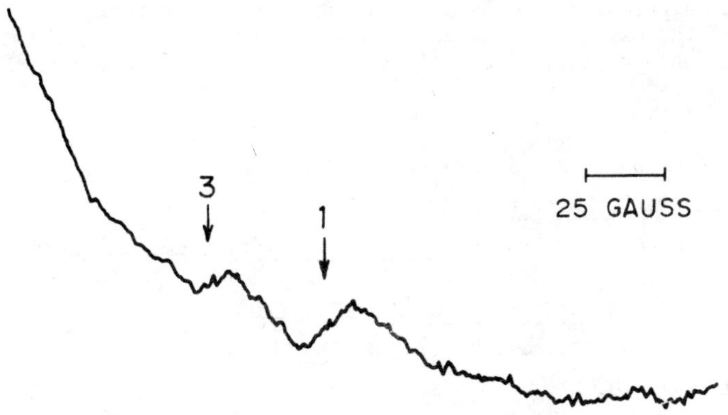

Fig.1 - Satellite structure in AlMn. A first and a third neighbour satellite are shown with the magnetic field along [1$\bar{1}\bar{1}$].

identify field gradients for additional neighbours to the impurity. A similar analysis was performed for the AlCr alloy. It was possible to identify electric field gradients for first, third and fourth nearest neighbours to the impurity. For first neighbours the observed lines were consistent with the same choice of principal axes chosen for the AlMn alloy. Also, a better fit to the data was obtained using an asymmetry parameter of .19±.01 rather than the value of .23±.03 quoted by Berthier and Minier. The third and fourth neighbour patterns were calculated by first order perturbation theory using the pure quadrupole frequencies observed by Berthier and Minier. A consistent correspondence between the calculated spectra and the observed satellites was obtained by choosing the quadrupole frequency of 70 kHz with $\eta = 0$ for the third neighbours which is also the assignment made by Berthier and Minier. For the fourth neighbours, however, the quadrupole frequency of 128 kHz with $\eta = 0$ was necessary to fit the data. In this case our results differ from those of Berthier and Minier, since they assigned this frequency to second neighbours. It is noted that both third and fourth neighbour satellites occur at 64 kHz or within a linewidth of this frequency. Thus a similar line observed in powders by Janaussy and Grüner, which they attributed to seventh or other shell of neighbours containing a large number of sites, because of the large intensity of the powder line, is in fact due to a combination of third and fourth neighbour lines. The same satellite was attributed to first nearest neighbours by Campbell and Williams (1972). It should be pointed out the satellite pattern for first and fourth neighbours are identical, given the same quadrupole frequency and asymmetry parameter, so that the reassignment of the structure to fourth neighbours is not inconsistent with their data. They did not, however, see the additional structure which identifies the third neighbours.

3. Knight Shifts. In many cases the quadrupole satellites for first neighbours are attributable to only two inversion related sites. For these lines one can measure the difference between their expected position calculated on the basis of the known quadrupole parameters and their measured positions. The difference can be attributed to a differential Knight shift. This interpretation can be established by examining the field dependence of the shift and verifying a linear relationship between the applied field and the measured shift. The Knight shift measured is characterized by the angles between a coordinate system referred to the crystallographic axes and the magnetic field direction. By examining several satellites and by varying the orientation of the magnetic field one can map out the orientation dependence of the Knight shift. This was done for the first neighbours in AlMn. The results are listed in Table 1. An anisotropy is observed and appears not to be consistent with that associated with a second rank tensor. The results appear to be slightly lower than the isotropic differential Knight shift of -109% of the aluminum Knight shift measured by Alloul et al. in a powdered sample. For the third neighbour we found no measurable differential Knight shift; and so a second differential Knight shift observed by Alloul et al. must be due to second, or to the fourth or further neighbours to the impurity.

TABLE 1

Differential Knight shifts as a function of magnetic field direction in AlMn. The polar angle θ is measured from the axis joining a first neighbour to the impurity. The azimuthal angle ϕ is measured from the perpendicular direction $[1\bar{1}0]$.

θ	ϕ	ΔK. (percent)
90	90	-.140 ± .006
35	90	-.156 ± .006
90	35	-.156 ± .006
90	0	-.135 ± .006
45	0	-.178 ± .006
0	-	-.165 ± .020

In AlCr a differential Knight shift of -.09±.01% was observed for the first neighbours. Within the present accuracy of the experiment, no anisotropy was observed. Clearly the experiments should be repeated at higher magnetic field to search for any such anisotropy.

4. References.

Alloul, H., Bernier, P., Launois, H., and Pouget, J.P., J. Phys. Soc. Japan 30 (1971) 101.

Berthier, C., and Minier, M., J. Phys. F. 3 (1973) 1169.

Campbell, G.R., and Williams D. Ll., Proc. XVII Congress Ampere, Editor V. Hovi, North Holland, Amsterdam (1972) 299.

Janaussy, A., and Grüner, G., Sol. Stat. Comm. 9 (1971) 1503.

Jorgensen, L., Nevald, R., and Williams, D. Ll., J. Phys. F. 1 (1971) 972.

Stiles, J.A.R., and Williams, D. Ll., J. Phys. F. (to be published).

* Present address: Hebrew University of Jerusalem, Israel.

MAGIC ANGLE ROTATION AND KNIGHT SHIFT DETERMINATION FOR ALUMINIUM AND CADMIUM

E. R. Andrew, W. S. Hinshaw and R. S. Tiffen
Department of Physics, University of Nottingham, University Park, Nottingham NG7 2RD, ENGLAND.

Abstract. The isotropic Knight shifts of ^{27}Al, ^{111}Cd and ^{113}Cd in pure aluminium and cadmium have been determined with much improved precision using the fast magic-angle rotation technique to narrow the NMR lines; the values found are 1640 ± 1, 4321 ± 7 and 4324 ± 7 ppm respectively. The specimens of aluminium used, produced by a metallization process, were devoid of the anomalous excess second moment usually found hitherto. Cadmium provides an excellent example of the removal of anisotropy of Knight shift by magic-angle rotation.

Introduction. The Knight shift of a metal, namely its NMR shift relative to non-metallic material, provides a valuable source of information concerning the conduction electron density in the metal (Knight, 1956). For comparison with theory the Knight shift must be determined as precisely as possible, but in practice the accuracy of its measurement has hitherto been limited to several per cent by the problem of precise location of the centre of the resonance line whose width is usually comparable with the shift to be determined. We have overcome this limitation by using the high-speed magic-angle rotation method (Andrew et al., 1958, 1959, Lowe, 1969) to reduce the linewidth by typically a factor of twenty, so enabling the isotropic Knight shift to be measured with more than an order of magnitude greater precision than before.

The application of this technique to ^{63}Cu and ^{65}Cu has been reported previously (Andrew et al., 1971a,b, Andrew and Hinshaw, 1973). In this paper we report the extension of the method to the precise determination of the isotropic Knight shifts of ^{27}Al in aluminium metal and of ^{111}Cd and ^{113}Cd in cadmium metal. Aluminium is cubic and its NMR linewidth of about 9 kHz is generated mainly by nuclear magnetic dipolar broadening, which is removed by the high-speed magic-angle rotation; any residual quadrupolar broadening caused by defects in the metal is also removed (Andrew, 1971, 1973). The two cadmium isotopes have spin ½ and therefore have no quadrupolar interactions. However the structure of cadmium metal is hexagonal, and in addition to the magnetic dipolar broadening there is a substantial broadening arising from anisotropy of the Knight shift. Cadmium therefore provides a good test of the removal of both sources of broadening by the magic-angle rotation method.

Materials. The specimens were prepared from wire supplied by Johnson Matthey Chemicals Ltd. The purity of aluminium was at least 99.995% and of the cadmium at least 99.999%. The wire was fed into a metallizing gun which produced a fine spray of molten metal. This was collected in distilled water and then sieved to limit the size to 44 μ. Viewed under a microscope the particles were spherical or spheroidal in shape. They were annealed under high vacuum for two hours at 250°C.

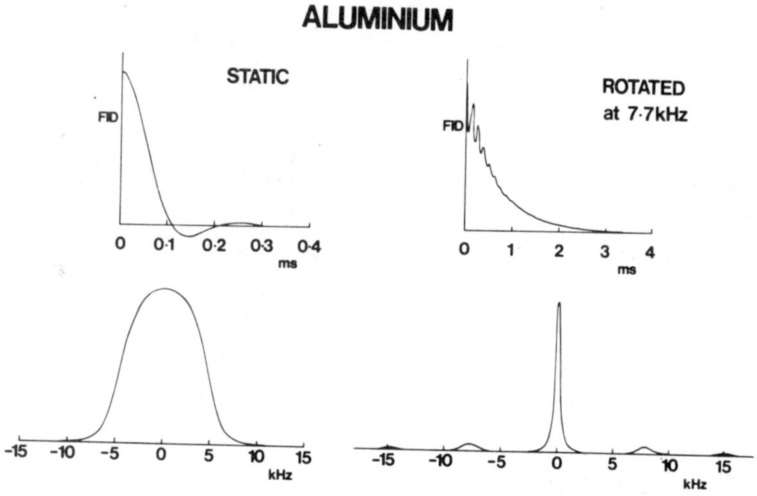

Figure 1. ^{27}Al free induction decays and Fourier-transformed NMR spectra of polycrystalline aluminium. The left hand diagrams refer to the static specimen. The right hand diagrams refer to the specimen rotated at 7.7 kHz about an axis inclined at 54° 44' to the direction of the laboratory magnetic field.

The second moment of the ^{27}Al NMR spectrum of the aluminium specimens was 7.7 ± 0.4 G^2 in good agreement with the theoretical dipolar second moment of 7.5 G^2. Many previous workers have reported consistently higher values of second moment in the range 11.0 ± 1.5 G^2. This has been attributed to quadrupolar broadening introduced by processes of filing or rolling in the preparation of the specimens used earlier, and apparently absent in the present metallized specimens (Andrew et al., 1974).

The reference material for ^{27}Al was a saturated aqueous solution of pure $AlCl_3$. The reference solution for ^{111}Cd and ^{113}Cd was a pure aqueous solution of cadmium nitrate with concentration 0.06 mole $Cd(NO_3)_2$ per mole H_2O.

Results for Aluminium. High-speed magic-angle rotation narrowed the ^{27}Al NMR spectrum dramatically. In Figure 1 is shown the FID and the Fourier-transformed spectrum for the static specimen (on the left) and for the specimen rotated at 7.7 kHz about an axis inclined at 54° 44' to the magnetic field direction (on the right). The linewidth at half maximum is reduced from 9 kHz to 400 Hz.

This reduction in resonance linewidth of aluminium metal by a factor of more than twenty has made it possible for us to determine the centre of the line, and thus the isotropic Knight shift, with a corresponding increase of accuracy. The Knight shift, defined as $(\nu_m - \nu_r)/\nu_r$, was determined from many measurements of the resonant frequency ν_m for the metal and ν_r for the reference material, repeated on different days and with different specimens using a double coil probe box slid on a rigidly mounted track, the exchange being made rapidly and without stopping the turbine. The mean value of the isotropic Knight shift of ^{27}Al relative to $AlCl_3$ solution at 298 ± 2 K was found to be 1640 ± 1 ppm. Earlier values have ranged from 1500 to 1680 ppm.

Results for Cadmium. A series of Fourier-transformed spectra for ^{111}Cd in cadmium metal are shown in Figure 2; the spectra for ^{113}Cd are very similar. The spectrum of the specimen when static shows the asymmetric profile characteristically generated by an axially-symmetric Knight shift. The spectrum has been fitted to the ideal powder spectrum for an axially-symmetric Knight shift tensor, yielding the difference $(K_{\parallel} - K_{\perp})$ between principal values of 480 ± 20 ppm for both isotopes.

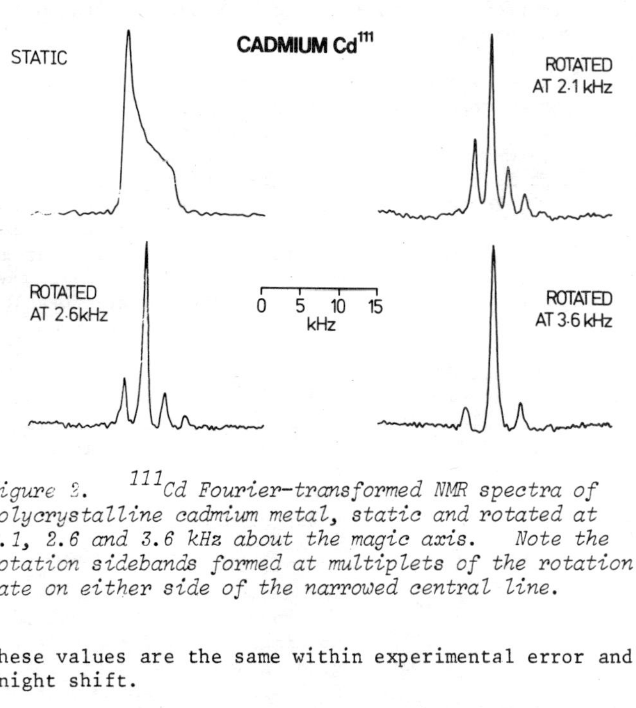

Figure 2. ^{111}Cd Fourier-transformed NMR spectra of polycrystalline cadmium metal, static and rotated at 2.1, 2.6 and 3.6 kHz about the magic axis. Note the rotation sidebands formed at multiplets of the rotation rate on either side of the narrowed central line.

The three other spectra in Figure 2 at different rates of rotation of the specimen about the magic axis provide a good illustration of the ability of the rapid specimen rotation technique to remove anisotropic shift broadening. As the rate of rotation is increased the spinning sidebands move away from the central component and by 3.6 kHz are quite separate. At this speed the dipolar, pseudo-dipolar and Knight shift anisotropy contributions have been removed from the central line whose residual shape is determined by the Ruderman-Kittel indirect scalar nuclear interaction and by lifetime broadening.

With the removal of the anisotropic sources of broadening, especially that due to the large anisotropy of Knight shift, it became possible to measure the isotropic shift with greater accuracy. With rapid interchange of spinning sample and reference sample we have obtained values of 4321 ± 7 and 4324 ± 7 ppm for ^{111}Cd and ^{113}Cd respectively relative to cadmium nitrate solution at 285 ± 2 K. These values are the same within experimental error and show no evidence of isotope dependence of Knight shift.

As with aluminium the results just quoted for cadmium represent an order of magnitude improvement in precision on the results of previous workers, who have reported values from 3900 to 4300 ppm relative to solutions of cadmium chloride or nitrate. The improved precision obtained underlines the importance of a careful choice of reference material (Krüger et al., 1974).

References

Andrew, E. R., 1971, Prog. NMR Spectroscopy, 8, 1.
Andrew, E. R., 1973, 17th Ampere Congress, Turku, 1972, North-Holland, 18.
Andrew, E. R., Bradbury, A. and Eades, R. G., 1958, Nature, 182, 1659.
Andrew, E. R., Bradbury, A. and Eades, R. G., 1959, Nature, 183, 1802.
Andrew, E. R., Carolan, J. L. and Randall, P. J., 1971a, Phys. Letters, 35A, 435.
Andrew, E. R., Carolan, J. L. and Randall, P. J., 1971b, Phys. Letters, 37A, 125.
Andrew, E. R. and Hinshaw, W. S., 1973, Phys. Letters, 43A, 113.
Andrew, E. R., Hinshaw, W. S. and Tiffen, R. S., 1974, J. Phys. F: Metal Physics, in press.
Knight, W. D., 1956, Solid State Physics, 2, 93.
Krüger, H., Lutz, O., Schwenk, A. and Stricker, G., 1974, Z. Phys., 266, 233.
Lowe, I. J., 1959, Phys. Rev. Letters, 2, 285.

DETECTION OF VACANCY-INDUCED SELF-DIFFUSION BY ROTATING-FRAME SPIN-LATTICE RELAXATION IN ALUMINIUM

R. Messer, S. Dais, D. Wolf
Max-Planck-Institut für Metallforschung, Institut für Physik, Stuttgart, Germany.

Abstract. The spin-lattice relaxation times T_1 and $T_{1\rho}$ have been measured in Al in a wide temperature and field range(90 - 930 K, 1.4-15 G). A pertubation approach and a modified Slichter-Ailion (SA) theory were used to relate the data to the coefficient of self-diffusion. The transition from the SA-region to the high field region is discussed.

I. Introduction. In recent theoretical papers by Wolf [1,2] the effect of point-defect diffusion on the temperature and the field dependence of the spin-lattice relaxation time $T_{1\rho}$ in the rotating frame has been investigated in both the high-field region ($H_1 \gg H_{L\rho}$) and in the Slichter-Ailion (SA) region ($H_1 \approx H_{L\rho}$). Here H_1 denotes the rotating field amplitude and $H_{L\rho}$ the local field in the rotating frame. The two main results of these investigations are: (i) shape and width of the $T_{1\rho}$-minimum as a function of temperature depend on the diffusion mechanism. (ii) the extrapolation of $T_{1\rho}$ derived with the framework of the SA-theory to high fields reproduces the results deduced from perturbation theory.

To check these predictions experimentally NMR studies on Al by Sun [3] and Fradin [4] have been used. It was found, however, that these measurements are unsatisfying in that they are not quite consistent with each other and that the temperature and field range over which the data were taken is not wide enough. Therefore both $T_{1\rho}$ for seven different values of H_1, ranging from 1.4 Gauss ($H_1<H_{L\rho}$) to 15 Gauss ($H_1 \gg H_{L\rho}$), and T_1 have been measured on Al from 90 K up to the melting point at 930 K.

II. Experimental Results and Discussion. The sample used in our experiments consisted of a few hundred layers of Al foil (99.999 purity, 27-30μ thickness) which were insulated from each-other by mica sheets. At the larmor frequency of 19 MHz used, the skin depth was calculated to be greater than about 20μ above room temperature.

II.1. Spin-Lattice Relaxation in the Laboratory Frame. Our T_1-measurements agree with the Korriga-relation with a slightly temperature dependent Knight-shift. This is seen, e.g., from the following products of $T_{1e} \cdot T$ for different temperatures: $T_{1e} \cdot T$ = 1.835 at 130K, $T_{1e} \cdot T$ = 1.880 at room temperature, $T_{1e} \cdot T$ = 1.886 at 750K. Within the accuracy of our measurements (better than 1%) only this conduction electron contribution T_{1e} to T_1 could be found. This is in contradiction to the results of El Hanany and Zamir [5] who found a small quadrupolar contribution to T_1 at temperatures just below the melting point.

II.2. Spin-Lattice Relaxation in the Rotating Frame. From measurements of $T_{1\rho}$ both a dipolar diffusion contribution $(T_{1\rho})_d$ and a conduction electron contribution $(T_{1\rho})_e$ are found. No quadrupolar effects could be detected. The interpretation of the H_1-dependence of $(T_{1\rho})_e$ in terms of the relationship (see, e.g.[6]

$$1/(T_{1\rho})_e = 1/T_{1e} \cdot (H_1^2 + \alpha H_{L\rho}^2) \;/\; (H_1^2 + H_{L\rho}^2) \qquad (1)$$

with a temperature independent parameter $\alpha (2<\alpha<3)$ is not possible for the entire temperature range. Also, the transition from the rigid-lattice region to the motionally narrowed region cannot be described well by a simple decrease of $H_{L\rho}$. Recent theoretical studies by Jung and Wolf [7] suggest that $H_{L\rho}^2$ in the numerator of (1) should be replaced by a temperature dependent "effective" local field while in the denominator the full rigid-lattice value of $H_{L\rho}$ should enter.

The temperature and field dependence of the dipolar diffusion contribution to $(T_{1\rho})_d$ cannot be interpreted in terms of a single theoretical approach. For $H_1 \gg H_{L\rho}$ (high-field region) perturbation theory is applicable in the entire temperature range. Therefore our data taken at H_1 = 15 G and 8 G, respectively, were analysed in terms of the so-called encounter model by Wolf [1] which takes into account the effect of vacancy-induced self-diffusion on $T_{1\rho}$. From a computer fit of shape and width of the $T_{1\rho}$-minimum as a function of temperature good agreement with the theoretical predictions is found. For $H_1 \approx H_{L\rho}$ (low-field region) the Slichter-Ailion theory has been extended by Wolf [2] to account for the temperature region in which $\tau_v < T_2^{rl} < \tau$ (τ_v denotes the mean time between consecutive vacancy jumps), i.e., where no spin temperature is established between vacancy jumps but between consecutive encounters. In the limiting case $H_1 \gg H_{L\rho}$ this theory reproduces the result obtained from Wolf's perturbation treatment [1] if $H_{L\rho}^2$ is replaced by $3/4 \cdot H_{L\rho}^2$ as first suggested by Slichter and Ailion [6]. In the intermediate region non of the two approaches is applicable. Fig.1 shows the theoretical predictions of Wolf's SA-approach both for low fields and with the modification for high fields mentioned above. They are plotted in units which are independent of experimental parameters like, e.g., activation energy, lattice constant, and nuclear parameters γ, I, etc. They only depend on the diffusion mechanism and the lattice structure. In general, on the low-temperature side of the $(T_{1\rho})_d$-minimum the following dependence of $(T_{1\rho})_d$ on $\omega_1 = \gamma \cdot H_1$ and τ holds:

$$1/(T_{1\rho})_d = 1/(\omega_1^2 \cdot \tau) \cdot KON \cdot f(H_1'^2) \quad (2)$$

where the constant KON is given by $3/2\gamma^4 h^2\ I(I+1)\ a_o^{-6}$ and H_1' is related to H_1 by [2]

$$H_1'^2 = H_1^2 \cdot a_o^6/(1/4\gamma^2 h^2\ I(I+1)) \quad , \quad (3)$$

$2a_o$ denotes the cube edge of a unit cell. $f(H_1'^2)$ accounts for the non-quadratic H_1-dependence of $(T_{1\rho})_d$ in the SA-region and may easily be obtained from [2]. In the perturbation region f is independent of $H_1'^2$ and equal to 0.1592 for vacancy-induced self-diffusion in an fcc-lattice [1]. Fig.1 shows the function $f(H_1'^2)$ as obtained for the different theoretical approaches. The experimental points shown in this figure were obtained by determining f from (2) with the experimental values of $(T_{1\rho})_d$, τ, ω_1^2, and calculating $H_1'^2$ from (3).

III. Self-Diffusion in Aluminium. The theoretical approaches discussed in II were used to determine τ from the measured $(T_{1\rho})_d$ values. The coefficient of self-diffusion, D^{SD}, is related to τ through the Einstein-relationship, while the tracer diffusion coefficient D^T is obtained from D^{SD} by multiplication with the correlation factor f (geometrical correlation factor).

Fig.2 shows D^T determined in the way described above as a function of 1/T together with tracer measurements of Lundy, Murdock [8] and Beyeler, Adda [9]. On the low temperature side of the $T_{1\rho}$-minimum 1.28 eV was obtained for the activation energy E_A and 0.137 cm^2/s for the preexponential factor D_o (earlier NMR-results: Fradin: E_A = 1.25 eV, D_o = 0.035 cm^2/s; Sun: E_A = 1.32 eV, D_o = 0.143 cm^2/s). The same values of E_A and D_o are in agreement with our data on the high-temperature side of the $T_{1\rho}$-minimum. However, the shape of the $T_{1\rho}$-minimum is in good agreement with the theoretical shape for the migration of single vacansies, but within the experimental accuracy a slight discrepancy is found. Therefore it seems to be necessary to include divacancies. This is also seen from the slight curvature of the plot of all experimental D^T-values versus 1/T including tracer measurements in Fig.2 (dashed line: D^T=0.137 exp(1.28eV/KT).

Acknowledgment: We like to thank Prof. A. Seeger and P. Jung for many discussions.

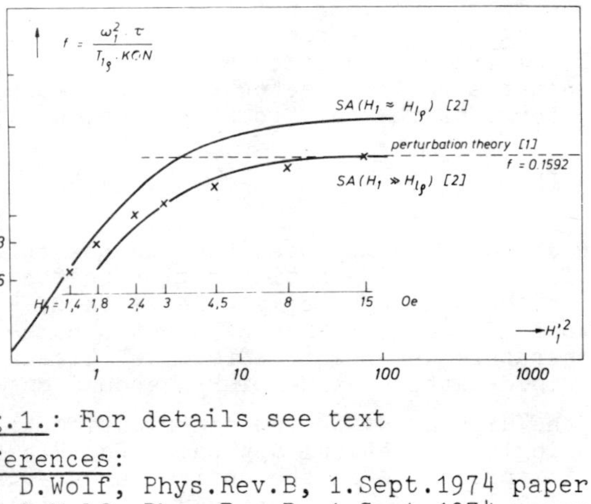

Fig.1.: For details see text

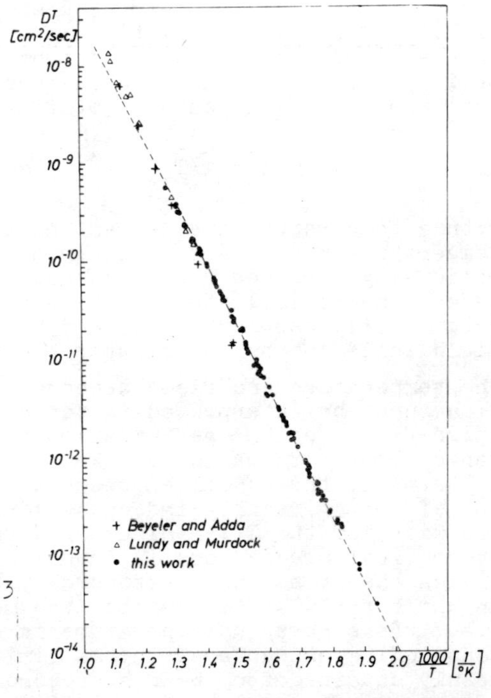

Fig.2.: For details see text

References:

[1] D.Wolf, Phys.Rev.B, 1.Sept.1974 paper 1
[2] D.Wolf, Phys.Rev.B, 1.Sept.1974 paper 2
[3] Ch.Y.Sun,Thesis,University of Illinois (1971)
[4] F.Y.Fradin,Thesis,University of Illinois (1967)
[5] El.Hanany,D.Zamir,Solid State Comm.10,1223 (1972)
[6] Ailion-Slichter,Phys.Rev.135A,1099,(1964)
[7] P.Jung,D.Wolf to be published
[8] Lundy,Murdock,J.Appl.Phys..33,1671(1962)
[9] Beyeler,Adda,J.Phys. 29,345 (1968)

THE KNIGHT SHIFT IN LIQUID Na-Li ALLOYS.

P.D. Feitsma, G.K. Slagter and W. van der Lugt
Solid State Physics Laboratory, Melkweg 1, Groningen, The Netherlands

Abstract. The Knight shifts of Li^7 and Na^{23} in liquid binary Na-Li alloys have been determined. The Knight shift depends linearly on concentration. The nuclear contact density is distinctly dependent on concentration.

1. Introduction. The measurements in Na-Li alloys form part of a systematic investigation of the physical properties of alkali metals (see e.g., Dekker and van der Lugt, 1973). Recently, van Hemmen et al. (1974) have reviewed the Knight shift results in alkali alloy systems not containing lithium. Using their notation, the experimental results can be summed up as follows.

a) If b refers to the heavier component of a binary alloy, the relation between the Knight shift K and the atomic concentration c_b is almost perfectly linear.

b) The shifts of the two components are approximately related by

$$K_a^{-1}\,(dK_a/dc_b) = K_b^{-1}\,(dK_b/dc_b) \qquad (1)$$

c) $(dK_a/dc_b) > 0 \qquad (dK_b/dc_b) > 0 \qquad (2)$

Discussing the results in terms of the fundamental formula

$$K = \frac{8\pi}{3}\,\chi_p\,\Omega_o\,P_F \qquad (3)$$

(where χ_p is the Pauli susceptibility, Ω_o is the average atomic volume and $P_F = \langle|\psi(o)|^2\rangle_F$ is the nuclear contact density of the conduction electrons at the Fermi level), Ω_o appears to be responsible for the major part of the variation of the Knight shift on alloying.

The linear relation between K and c can be explained in terms of two models. One is based on a conjecture by Kaeck (1968), that P_F be independent of concentration. In this model, the susceptibilities of K, Rb and Cs appear to be nearly equal. Then, for binary alloys of these three components, the linear relation follows as a consequence of the assumed additivity of atomic volumes: $\Omega_o = c_a\,\Omega_a + c_b\,\Omega_b$. This approach fails for sodium alloys. Indeed, measurements in ternary Na-Rb-Cs alloys have clearly demonstrated that P_F is dependent on concentration in these alloys.

According to the second model (the "Single APW model") developed by van Hemmen et al. (1969) the susceptibility is taken to be proportional to the free-electron susceptibility and P_F varies as a function of c_b as a consequence of wave function normalization. This approach is applicable also to sodium alloys and reproduces the linear K(c) relations quite accurately. Indeed, the assumption of free-electron like susceptibilities still lacks sufficient theoretical or experimental justification.

For the study of metallic properties, the alkali metals have the advantage of their relatively simple electronic structure. At the other hand, they are untypical in one respect: the average conduction electron density is much smaller than it is for most other metals, and it varies from Li to Cs by as much as a factor of 6. Lithium takes a special position among the alkali metals. The sodium-lithium system exhibits a liquid miscibility gap up to approximately 300°C. Lithium easily forms nitrides and it destroys almost all non-metallic refractories. From the large size difference between sodium and lithium ions one expects a large Nordheim resistivity, which, according to our measurements is in fact almost negligible (Feitsma and van der Lugt, 1974). Fortunately, the spin susceptibilities of Na and Li are known fairly well from C.E.S.R. experiments.

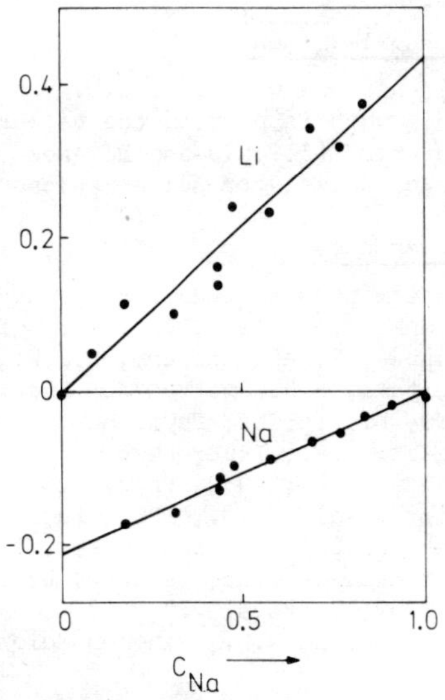

Fig. 1. $(K-K_o)/K_o$ as a function of $c_b = c_{Na}$. K and K_o are defined in the text.

2. Results and discussion. As a consequence of the liquid immiscibility, the measurements were performed at 400 °C. A dewar, of the type designed by Clark (1969), was constructed to fit the Varian VF 16 probe head. The Knight shift results are shown in figure 1, where $\Delta K/K_o = (K-K_o)/K_o$ is plotted for each of the components (K_o is the Knight shift for the pure component at 400°C). Li-Na alloys share with the other alkali systems the linear K(c) as well as the validity of eqs. 2. The values of $(1/K_o)dK/dc_b$ are 0.44 for Li and 0.21 for Na. However, P_F depends rather strongly on c_b. This can be shown in the following way, without making use of the experimental values of χ_p. In the limit of $c_{Na}\to 0$, we find $K_{Na}(0)/K_{Li}(0) = \chi_{P,Li}\Omega_{o,Li}P_{F,Na}(0)/\chi_{P,Li}\Omega_{o,Li}P_{F,Li}(0) = P_{F,Na}(0)/P_{F,Li}(0) = 3.74$.

where the argument of P_F and K is $c_b = c_{Na}$. In the same way, for $c_{Li} \to 0$, one finds:

$P_{F,Na}(1)/P_{F,Li}(1) = 3.31$, so $P_{F,Na}/P_{F,Li}$ depends on c_b.

Quantitatively, substituting known atomic volumes and susceptibilities in eq. 3, one calculates

$P_{F,Na}(1)/P_{F,Na}(0) = 1.26$ and $P_{F,Li}(1)/P_{F,Li}(0) = 1.43$.

The observed behavior of P_F is similar to that of Na in the ternary alloys and forms an indication that the conjecture of constant P_F does not apply to systems with atomic volumes smaller than that of potassium. This behavior is also incompatible with the APW model, which demands $P_{F,Na}/P_{F,Li}$ to be independent of concentration.

There is still another reason, why an evaluation of the results in terms of the APW model is unfeasible: the ratio of the spin susceptibilities of Li and Na deviates strongly from its free-electron value. For free electrons, χ_{Li}/χ_{Na} should be 1.22, whereas it is 1.83 in reality. In this connection, it is interesting to note that the linearity of K(c) appears to hold also where the conditions for its theoretical justification seem to fail.

For Li, χ_P amounts to approximately 2.6 times its free electron value, probably due to an increased density of states. Indeed, calculations by Ham (1962) and Rudge (1969) as well as specific heat measurements, properly corrected for electron-phonon and electron-electron interactions (Ashcroft, 1965), result in an effective mass ratio of, say, 1.7. For sodium, m^*/m is generally believed to be close to 1. The behavior of m^*/m in the Na-Li alloys is still uncertain and its strong variation from Li to Na precludes any meaningful interpolation. The exchange enhancement factor is approximately 1.5 for Li and it increases slowly and smoothly as a function of Ω_o (Dupree and Geldart, 1971).

From the large number of data now available for alkali alloys we have calculated the factor $\xi = P_F/P_A$ for all alkali metals dissolved in sodium. The results are shown in figure 2, and apply to the limit of vanishing solute concentration. The free atom contact densities P_A are derived from optical pumping experiments. The ξ derived in this way is defined by substituting $P_F = \xi P_A$ in eq. (3) and so is an effective ξ including possible corrections for core polarization.

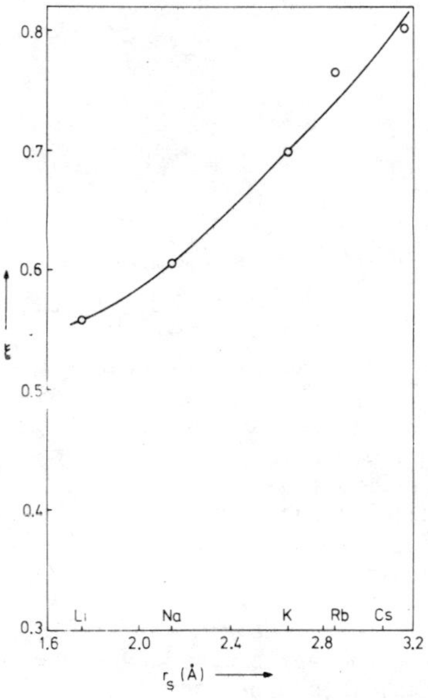

Fig. 2. Values of $\xi = P_F/P_A$ for different alkali solutes in sodium.

Acknowledgements

Discussions with drs. J.L. van Hemmen have essentially contributed to this paper.
This work is part of the research program of the "Stichting voor Fundamenteel Onderzoek der Materie" (F.O.M.) and has been made possible by financial support from the "Nederlandse Organisatie voor Zuiver Wetenschappelijk Onderzoek" (Z.W.O).

References

Ashcroft, N.W., 1965, Phys. Rev. 140A, 935
Clark, L.D., 1969, Rev. Sc. Instr. 40, 1498
Dupree, R. and Geldart, D.J.W., 1971, Sol. State Comm. 9, 145
Feitsma, P.D. and van der Lugt, W., 1974, to be published
Ham, F.S., 1962, Phys. Rev. 128, 2524
Kaeck, J.A., 1968, Phys. Rev. 175, 897
Rudge, W.E., 1969, Phys. Rev. 181, 1024
Silverstein, S.D., 1963, Phys. Rev. 130, 1703
Van der Lugt, W. and Dekker, A.J., 1973, Physics 69, 148
Van Hemmen, J.L., Caspers, W.J., van der Molen, S.B., van der Lugt, W., and van de Braak, H.P., 1969, Z. Phys. 222, 253
Van Hemmen, J.L., van der Molen, S.B. and van der Lugt, W., 1974, Phil. Mag. 29, 493.

THEORY OF NUCLEAR QUADRUPOLE RELAXATION IN LIQUID METALS AND ALLOYS

H. Gabriel and W. Schirmacher
Institut für Atom- und Festkörperphysik, Freie Universität Berlin

Abstract. The nuclear quadrupole relaxation rate in liquid metals is expressed in terms of an effective ion-ion potential and the van Hove function S(k,0). Numerical calculations are based on the assumption that ion diffusion is the dominant contribution to the relaxation rate in many cases. Effective-potential calculations of Appapillai and Williams were utilized. The examples considered are liquid In and the liquid alloy SbIn.

Within the last 10 years the quadrupolar part of the nuclear relaxation time in liquid metals has been the subject of various experimental and theoretical investigations (see e.g. Sholl (1974) for references) since it provides information about the electric field gradients (efg) as well as the atomic motions in the liquid metal. Most experiments indicate that the dynamic quadrupole interaction is caused by diffusive motion of the ions in the liquid, an interpretation first presented by Sholl (1967). Further theoretical studies relating the quadrupolar relaxation rate to the van Hove neutron scattering law have been given by Warren (1971) Gabriel (1974) and Sholl (1974). The aim of this contribution is to calculate within a simple model the temperature dependence of the quadrupolar relaxation rate R_{1Q} in monatomic liquid metals and its concentration dependence on binary alloys.

We start from the approximate formula for R_{1Q} proposed by one of the above authors which is also a special case of a more general expression derived by Sholl (1974):

$$R_{1Q} = \text{const.} \sum_{q=-2}^{+2} \int |\tilde{u}(\underline{k})|^2 \, S(\underline{k},0)\, d^3\underline{k} \quad (1) \qquad \tilde{u}_q(\underline{r}) = u_q(\underline{r})\alpha(\underline{r}) \quad (2)$$

The relaxation rate is defined in terms of the van Hove function $S(\underline{k},\omega)$ (thus including two- and three particle correlations), the efg $u_q(\underline{r})$ and a hard sphere step function $\alpha(r) = 0$ (1) for $r<r_o$ ($r>r_0$) accounting for the fact that the relaxating ion excludes a certain volume for any neighbouring ion. Values of r_0 may be simply identified with hard-sphere radii for model calculations or taken from the measured radial distribution function g(r) e.g. by the lowest value of r for which g(r) = 1 holds.

In the following we derive the efg from an effective ion-ion potential $\phi_{eff}(r)$

$$\phi_{eff}(r) = \phi_{dir}(r) + \phi_{ind}(r) = Z^+e/r\,\{1-\frac{2}{\pi}\int_0^\infty F_N(k)\,\frac{\sin kr}{k}\,dk\} \quad (3)$$

which includes the direct coulomb interaction and an indirect interaction via the conduction electrons. Values of the effective charge Z^+ and tables of the normalized energy-wavenumber characteristics $F_N(k)$ are available from model- or pseudopotential calculations (e.g. Appapillai and Williams 1973).

The efg entering the numerical calculations was taken as

$$u_q(\underline{k}) = -(1-\gamma_\infty)\frac{4}{3}\sqrt{\frac{\pi}{5}}\,Y_{2q}(\theta_{\underline{k}},\phi_{\underline{k}})k^2\,\{\phi_{dir}(k) + \varepsilon\phi_{ind}(k)\} \quad (4)$$

where $1-\gamma_\infty$ is a common antishielding factor and Y_{2q} a second order spherical harmonic. The parameter $0<\varepsilon<1$ allows one to give different weights to the direct and indirect interaction. Since no theory of shielding effects with respect to the indirect interaction is available at present, we merely consider ε to be a means of studying the effect it has on the numerical calculation.

The relaxation rate is calculated from

$$R_{1Q} = \left(\frac{Z^+e^2Q(1-\gamma_\infty)}{\hbar}\right)^2 \frac{4\pi(2I+3)}{15I^2(2I-1)}\,\rho\int_0^\infty k^2F(k)^2\,S(k,0)dk$$

$$F(k) = 3j_1(kr_0)/kr_0+\varepsilon\{r_0^3\,\frac{2}{\pi}\int_0^\infty k'^2F_N(k)\,\chi(kr_0,k'r_0)dk'-F_N(k)\}$$

$$\chi(a,b) = \frac{1}{2ab}\left(\frac{\sin(a-b)}{a-b}-\frac{\sin(a+b)}{a+b} - 6j_1(a)j_1(b)\right)$$

$$j_1(a) = (\sin a- a\cos a)/a^2 \quad (5)$$

which depends on the effective ion-ion interaction and on the "quasi-elastic" peak S(k,o) as a function of k. For simplicity we have chosen the self-diffusion approximation $S(k,o) = S(k)/2\pi Dk^2$ in our calculation, which then yields a D^{-1} dependence of R_{1Q}, but, of course, one may also introduce either experimental data or appropriate interpolation formulae for S(k,o) if available.

The generalization to binary alloys of composition c_α (α=1,2) is straightforward yielding

$$R_{1Q} = \left(\frac{e^2Q}{\hbar}\right)^2 \frac{4\pi(2I+3)}{15I^2(2I-1)}\,\bar{\rho}\sum_{\alpha=1}^{2}\sum_{\beta=1}^{2} Z_\alpha^+Z_\beta^+\,(1-\gamma_\infty^{(\alpha)})(1-\gamma_\infty^{(\beta)})\int k^2F_\alpha(k)F_\beta(k)c_\alpha S_{\alpha\beta}(k,o)dk \quad (6)$$

and only introduces partial quantities (like the liquid structure factors

$S_{\alpha\beta}(k)=\delta_{\alpha\beta}+c_\beta\Gamma_{\alpha\beta}(k)$ instead of $S(k)$) in an obvious way. An average density is defined in terms of the pure liquid densities $\rho_o^{(\alpha)}$ by $\bar{\rho} = [\rho_o^{(\alpha)}/c_\alpha + \rho_o^{(\beta)}/c_\beta]^{-1}$. An implicit concentration dependence enters (6) via the $\Gamma_{\alpha\beta}(k)$.
The influence of the parameters r_o and ε on R_{1Q} is shown in Table 1 for In as an example. In the calculation we have chosen I = 9/2, Q = 0.834 b; $1 - \gamma_\infty = 25.9$, T = 573K, ρ = 0.037 $Å^{-3}$, D = 3.3 $\cdot$ 10^{-5} cm^2/s, in agreement with Sholl (1974). We observe from the numerical data that increasing r_o at constant values of ε decreases R_{1Q} systematically. This is due to the fact that the efg decreases very rapidly with distance in the range of interest. The overlap with the step function $\alpha(r)$ is thus reduced with increasing r_o. A much smaller variation of r_o than presented in the table arises if the temperature is varied. According to Ocken and Wagner (1972) and Edwards and Jarzynski (1972) r_o changes from 2.87 Å to 2.83 Å between 430K and 930K. Recent measurements of Bräuer et al. (1974) of $R_{1Q}(T)$ for Sn^{115} in liquid In show some slight deviation from the 1/D(T) behaviour which could be due to the temperature variation of r_o. However, we find that the resulting changes of $R_{1Q}(T)$ are small and well within the experimental error.

Table 1 also displays a strong influence of ε on R_{1Q} for any value of r_o. Reducing the relative weight of the efg due to the indirect interaction, via modified antishielding, as one may call it, tends to enhance R_{1Q} to its maximum value for ε = 0. If the uncertain parameter $(1-\gamma_\infty)$ were exact, values of R_{1Q} that are close to the experimental one ($R_{1Q}(573)=2100s^{-1}$) result only for particular pairs (ε, r_o) in the range $\varepsilon \leq 0.4$, $2.4 \leq r_o \leq 3.2$. (Smaller or larger values of r_o are ruled out for physical reasons). Notice that e.g. an increase of $(1-\gamma_\infty)$ would shift ε to larger values. Also given in Table 1 is I_2/I_1, the ratio of integrals (I_2, I_1) related to three- and two-particle correlations, resp.. It is interesting that values close to -0.97 (the value deduced by Claridge et al. (1972) from R_{1Q} values for alloys to get reasonable agreement between experiment and Sholl's first approach) appear just in the regime where we get "realistic" values of R_{1Q}.

We now turn to the concentration dependence of binary alloys. Using the hard-sphere model of Ashcroft and Langreth (1966) one can, in principle, calculate the partial structure factors $\Gamma_{\alpha\beta}(k)$ as a function of composition, the hard-sphere diameter ratio σ_1/σ_2, and $\bar{\rho}$. We have attempted to fit the InSb alloy data of Claridge et al. by setting $\sigma_1 = \sigma_2$ = 2.83 Å (which then gives $\Gamma_{11}=\Gamma_{12}=\Gamma_{22}$) and by varying ε and $\gamma = (1-\gamma(Sb))/(1-\gamma(In))$. R_{1Q} was adjusted to the pure Sb value. No improvement was achieved by including the concentration dependence of the $\Gamma_{\alpha\beta}$ within the Ashcroft-Langreth model. The experimental points together with the theoretical curves for some ε and γ are displayed in Fig. 1. Again values of ε much lower than one lead to better agreement with the experiment except in the high-concentration regime.

r_o \ ε	0.0	0.2	0.4	0.6	0.8	1.0	
2.4	3174	2214	1435	837	418	180	R_{1Q}
	.974	.973	.968	.957	.910	.709	$-I_2/I_1$
2.8	2116	1446	911	509	241	106	R_{1Q}
	.979	.977	.973	.963	.916	.127	$-I_2/I_1$
3.2	1181	792	487	267	133	83	R_{1Q}
	.985	.984	.981	.975	.944	.851	$-I_2/I_1$

Table 1

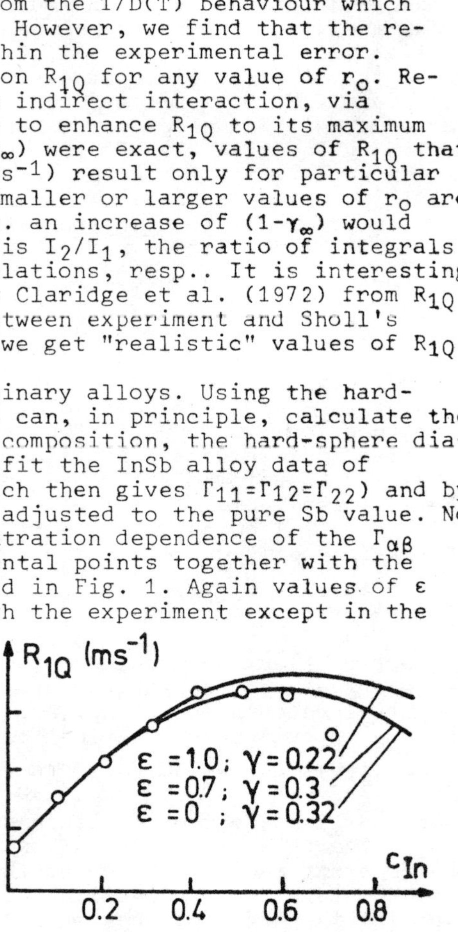

Fig. 1 Quadrupolar relaxation rate in the InSb alloy as a function of composition

References
Appapillai, M. and Willimas, A.R., 1973, J. Phys. F.: Met.Phys. 3, 759
Ashcroft, N.W. and Langreth, D.C., 1967, Phys. Rev. 156, 685
Bräuer, N., Dimmling, F., v.Hartrott, M., Kornrumpf, Th., Nishiyama, K., Quitmann, D. and Riegel, D., 1974, in the Int. Conf. on Hyperfine Interactions, Uppsala, p.254
Claridge, E., Moore, D.S., Seymour, E.F.W., and Sholl, C.A., 1972, J. Phys. F.: Met. Phys. 2, 1162
Edwards, D.J. and Jarzynski, J., 1972, J. Phys. C.: Sol. St. Phys., 5, 1745
Gabriel, H., 1974, Phys. Stat. Sol., vol. 64
Ocken, H. and Wagner, C.H., 1966, Phys. Rev. 149, 122
Sholl, C.A., 1967, Proc. Phys. Soc. 91, 130; 1974 to be published in J. Phys. F.: Met. Phys.
Warren, W.W., 1971, Bull. Am. Phys. Soc. 16, 338

This work is part of the research program of SFB 161 supported by Deutsche Forschungsgemeinschaft.

NUCLEAR RELAXATION IN LOW MOBILITY METALLIC SOLIDS

D. P. Tunstall
Department of Physics, The University of St Andrews, St Andrews, Scotland, KY16 9SS

Abstract A diffusive regime of conduction in metals, where the mean free path L is of the order of the interatomic distance, often occurs close to metal-insulator transitions. Strong exchange and correlation effects can be expected so close to the transition, but when the conduction is diffusive, independent T_1^{-1} enhancements can also occur. Data from different materials are correlated to clarify the situation.

1. Introduction. The large changes in conductivity that occur in some materials on variation of such parameters as pressure, doping concentration or temperature, can be driven by real band overlap effects, interactions between electrons (Hubbard band overlap) and by disorder, Mott (1974). All of these mechanisms can have strong effects on the nuclear resonance properties K, the Knight shift, and T_1, the spin-lattice relaxation time.

Real band overlap effects where, on variation of an external parameter, two or more bands are displaced with respect to each other, can lead to a change in the character of the wave-function at the Fermi surface. Such a change in character leads to a different magnitude of hyperfine field acting on the nuclei.

Inter-electron correlation effects can split a band, causing a metal-insulator transition, Hubbard (1964). Strong static and dynamic magnetic effects are predicted close to the transition on the metal side, Brinkman and Rice (1970).

Mott (1974) has argued that a minimum metallic conductivity exists for all materials, in the absence of percolation. The predicted value for particular materials varies with the density of the electron donors; in doped semiconductors with high dielectric constants the donors are far apart and the critical minimum metallic conductivity is $\sim 50(\Omega\text{cm})^{-1}$ whereas where the donors are $\sim 3\text{Å}$ apart predictions are ~ 200 to $300(\Omega\text{cm})^{-1}$. It is also clear that the standard nearly free electron (N.F.E) treatment of metallic conduction breaks down well before such small conductivities are reached, most obviously because the mean free path L for the electrons gets reduced down to an inter-donor distance and so $k_F L \sim 1$. This breakdown, where the description based on extended k-states ceases to have meaning, occurs at a conductivity value, again varying with density, of $\sigma \sim 4000(\Omega\text{cm})^{-1}$ for 3Å donor separations. The quite large range of conductivity values between 200 and $4000(\Omega\text{cm})^{-1}$ (or between 50 and $700(\Omega\text{cm})^{-1}$ for doped semiconductors) is labelled the diffusive regime of conduction.

Warren (1971) has pointed out in his work on liquid semiconductors that, in this diffusive mode of electron motion, T_1 depends on the non-thermally activated hopping time τ_e for electron motion. Since σ also depends on τ_e, a link between T_1 and σ is established; as σ decreases towards the metal-insulator transition T_1^{-1} enhancements are predicted.

2. Methods of Interpretation. It is important to realize that many metal-insulator transitions are driven by complex amalgams of band-overlap and electron-electron effects, and so all of the three resonance responses, outlined in the introduction, could be present in any particular system, with some relative weighting factor. It is fortunate that various constraints can be imposed on any assessment of n.m.r. data; in particular, the Korringa (1950) relation is, for non-interacting, nearly free, s-electrons,

$$K^2 T_1 T = \frac{\hbar}{4\pi k_B}\left(\frac{\gamma_e}{\gamma_n}\right)^2 \qquad (1)$$

where γ_e, γ_n are the electronic and nuclear gyromagnetic ratios respectively. Two points are worth emphasizing; (i) this is a N.F.E, one-electron, expression, dependent on k being a good quantum number at the Fermi surface, (ii) in situations where K is independent of temperature T_1T is a constant, the signature of metallic relaxation in many cases.

Consider briefly the various modifications to equation (1) that can arise in practical situations.

(a) Through band overlap effects, the relative s- and d- character of the wave-function at the Fermi surface might change. Both K and T_1, and therefore necessarily K^2T_1T, can be reduced in such cases.

(b) Electron-electron interactions tend to enhance both K and T_1^{-1}, but with a net enhancement of K^2T_1T, Narath and Weaver (1968).

(c) In the diffusive regime, as the conductivity worsens and the electron motion slows up, an increased spectral density of electronic magnetic fluctuations at the Larmor frequency is produced, and so T_1^{-1} is enhanced by a factor η over the value that would correspond to that obtained from equation (1) and the measured Knight shift. Warren (1971) predicts that the product $(\sigma\eta)$ remains constant.

In the diffusive regime, the reduced electronic motion leads to a reduced Korringa product and reduced T_1, whereas electron-electron interactions increase the Korringa product but reduce T_1. Where real band overlap considerations are important, the Korringa product may well be reduced.

3. Comparison. Figures 1(a), 1(b) and 1(c) show plots of T_1 through the transition for three systems. The estimated positions of the boundaries between the conduction regimes are represented by vertical dashed lines.

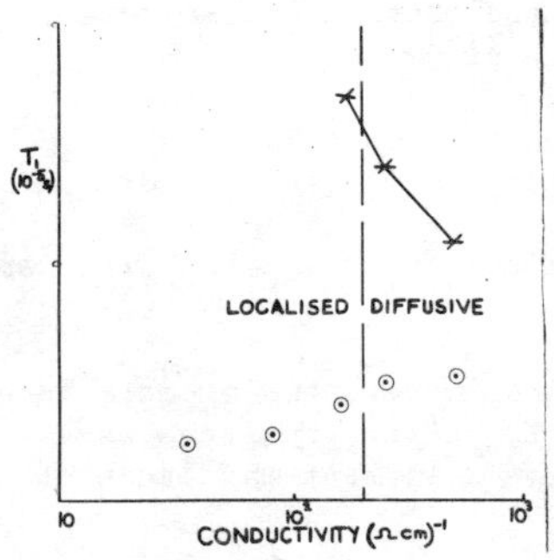

Fig 1(a) Liquid Semiconductor Ga_2Te_3. Band-overlap driven by temperature variation causes the transition. The interpretation, Warren (1971), relies on slow electronic motion.

x - deduced T_1 from measured K plus equation (1).

⊙ - Ga^{71} T_1

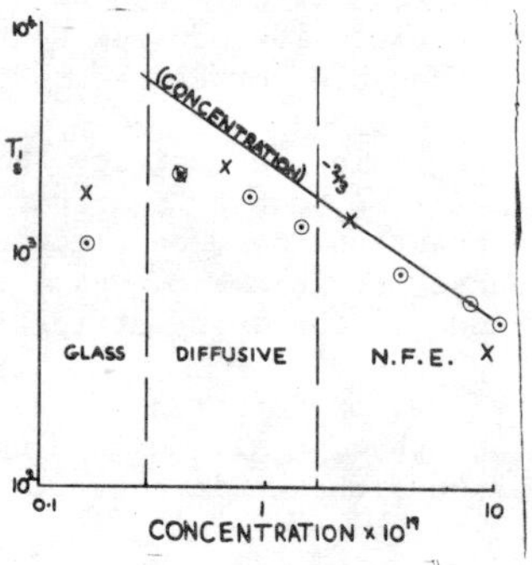

Fig 1(b) Doped Semiconductor Si:P, ⊙ -Sundfors (1964) 1.6K, x - Sasaki (1974) A free electron analysis predicts $T_1 \propto n^{-2/3}$. Hall concentrations for Sundfors data have been obtained by transposing Irvin curve concentrations.

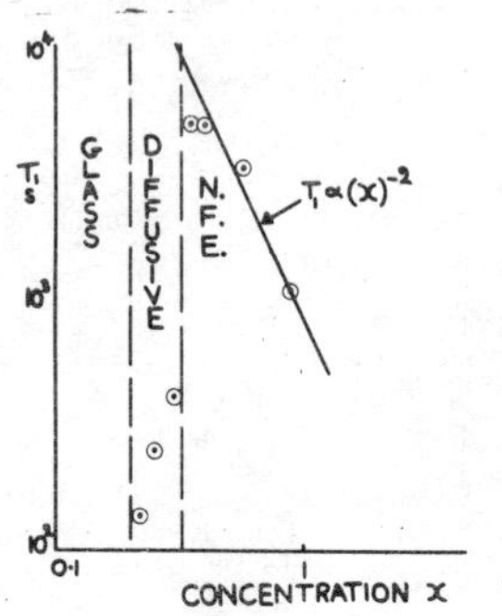

Fig 1(c) Alloy, Na_xWO_3. 4.2K .89 value from Fromhold and Narath (1964). Rest from Tunstall (1975).

The $T_1 \propto x^{-2}$ dependence in the NFE regime fits in with specific heat measurements.

The three sets of data all indicate some tendency for T_1 to decrease below the full lines, which in all cases represent extrapolations into the diffusive regime of N.F.E behaviour. The effect is most marked in the sodium tungsten bronze data. In fact, in both Ga_2Te_3 and Na_xWO_3, electron dynamic effects in the diffusive regime seem to dominate the relaxation process; for both systems the conductivity σ and the relaxation enhancement have approximately a constant product in this region.

The doped semiconductor Si:P exhibits only relatively small departures from N.F.E behaviour, and these could be the result of electron-electron or electron dynamic effects. From the form of the data it looks as though, on decreasing concentration, departures from N.F.E behaviour do not develop until the diffusive regime is reached; however, the most recent Knight shift measurements, Sasaki et al. (1974), taken with these T_1 results, show the presence of electron-electron enhancements of the Korringa product. The absence of marked electronic dynamic effects in the diffusive regime in Si:P is curious.

Two other points about Si:P deserve consideration. Firstly, the criterion for a transition from diffusive to N.F.E regimes leads to a concentration close to that accepted for the merging of the impurity and conduction bands (right-hand dashed vertical line in Fig 1(b)). Secondly, the relaxation process immediately below the Anderson transition at 3.0 x 10^{18} might be unusual. Mott (1974) has predicted a strong resonance line with zero shift at such densities. The T_1 for this line should presumably be very long.

4. References

Brinkman, W.F., and T.M. Rice, 1970, Phys.Rev. B2, 4302.
Fromhold, A.T., and A. Narath, 1964, Phys. Rev. 136, A487.
Hubbard, J., 1964, Proc. Roy. Soc. A277, 237.
Korringa, J., 1950, Physica 16, 601.
Mott, N.F., 1974, 'Metal-Insulator Transitions', Taylor and Francis, London.
Narath, A., and H. T. Weaver, 1968, Phys. Rev. 175, 373.
Sasaki, W., S. Ikehata and S. Kobayashi, 1974, J. Phys. Soc. Japan 36, 1377.
Sundfors, R.K., and D.F. Holcomb, 1964, Phys. Rev. 136, A810.
Tunstall, D.P., 1975, (to be published).
Warren, W.W., 1971, Phys. Rev.3, 3708.

NMR STUDY OF $La_3X_{1-y}X'_y$ AND La_3XC TERNARY ALLOYS (X = Al,Ga,In,Tl,Sn,Pb).

P.Descouts, B.Perrin and A.Dupanloup
Département de physique de la matière condensée,Université de Genève,GENEVE(CH)

Abstract. In $La_3X_{1-y}X'_y$ the Knight-shift of ^{115}In and ^{205}Tl is strongly temperature and composition dependent and the supraconducting properties are very close to those of corresponding binary compounds. In the La_3XCperovskite carbides the X-site Knight-shift is temperature independent and the electronic specific heat coefficient γ and supraconducting temperature T_c are strongly reduced.

1. Introduction. In arecent paper ,Heiniger et al(1973) have reported some results on superconducting and other electronic properties of La_3In,La_3Tl and some related phases like $La_3X_{1-y}X'_y$ and La_3XC. These authors have shown that La_3In and La_3Tl both crystallizing in Cu_3Au structure,have remarkable properties which suggest that these compounds belong to the class of strong-coupling superconductors and are closely relatedto the behavior of fcc La under pressure. The present work is an extension of our NMR study of these binary compounds to the corresponding ternary alloys.

2. Experimental results. All the alloys have been prepared in an argon arc furnace from the elements. The powders of these compounds are extremely pyrophoric in air and they have been prepared by filing under argon atmosphere and sealed in quartz capsules under helium gas for NMR experiments.
NMR measurements were done with cw-method using a Varian 16-B wide-line spectrometer modified to work at low temperatures.
We have extensively investigated the $La_3In_{1-y}Tl_y$ system and the results of Knight-shift measurements for the ^{115}In and ^{205}Tl nuclei are summerized on figures I and II. We can see on these plots the

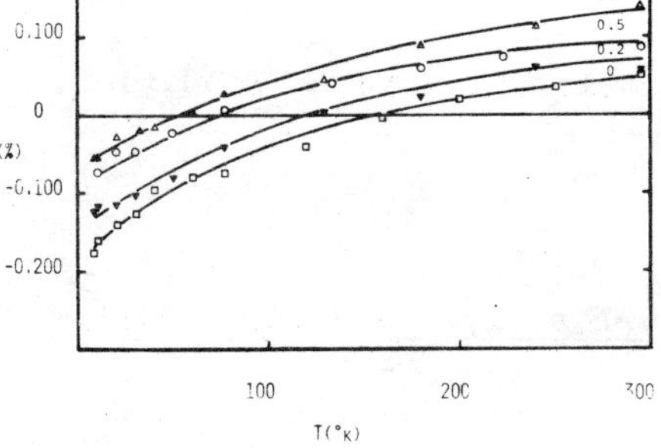

FIG I - Knight shift of ^{115}In vs T in $La_3In_{1-y}Tl_y$

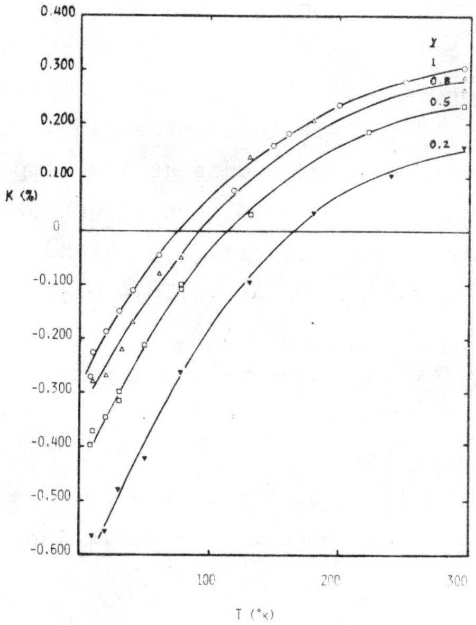

FIG II - Knight shift of ^{205}Tl vs T in $La_3In_{1-y}Tl_y$

strong dependence of K_{In} and K_{Tl} with temperature and composition.
On the tables 1. and 2. we have reported some results of Knigth-shift measurements in other ternary alloys. In $La_3X_{1-y}X'_y$, K_X is also strongly temperature dependent and of the same order as in La_3X compounds but in La_3XC, K_X is much more positive at room temperature and temperature independent.

NMR results in $La_3X_{1-y}X'_y$ and La_3XC ternary alloys

1 - Knight shift K(%) of ^{115}In

T (°K)	295	77	8
La_3 In	+0.052 (5)	-0.077(5)	-0.177(10)
La_3 $In_{0.7}Al_{0.3}$	+0.060 (5)		
La_3 $In_{0.7}$ $Ga_{0.3}$	+0.041 (5)	-0.065(5)	
La_3 $In_{0.8}$ $Sn_{0.2}$	+0.086(10)	+0.002(10)	
La_3 $In_{0.5}$ $Tl_{0.5}$	+0.085 (5)	+0.011 (5)	-0.080(10)
La_3 In C	+0.210(10)		

2 - Knight shift K(%) of ^{205}Tl

T (°K)	295	77	8,3	4,2
La_3 Tl	+0.305(5)	-0.005(5)	-0.273(10)	
$La_3Tl_{0.5}In_{0.5}$	+0.260(5)	-0.105(5)	-0.396(10)	
$La_3Tl_{0.8}Pb_{0.2}$	+0.274(5)	0.000(50)	-0.080(20)	
La_3 Tl C	+0.755(5)			+0.740(10)

3. Analysis of results. Both strong variation of X-site Knight-shift with temperature and also the X-rays patterns indicate that $La_3In_{1-y}Tl_y$ alloys exhibit long range order. We therefore analyse these results using the same schematic band structure model as for binary compounds (Descouts, 1973). This is formed essentially by a wide free-electron like s-band (La-6s) which overlaps a narrow pd-band (La-5d and In-5p or Tl-6p) near the Fermi level.
A plot of X-Knight-shift vs magnetic susceptibility gives us the core-polarization hyperfine fields H^{cp}_{hfs} which give the same values asfound previously in the binary compounds ($-0.59\ 10^5$ Oe μ_B^{-1} for ^{115}In and $-1.83\ 10^5$ Oe μ_B^{-1} for ^{205}Tl). If we suppose that s- and orbital contributions are unchanged by substituting Tl to In in the alloys we find that the variation of X-site Knight-shift by alloying is only due to the core-polarization contribution and seems to depend on the variation of the density of states on the X-site. A 80% substitution on the X-site gives 30% variation of K_X^{cp} whithout any change in other electronic and superconducting properties.
These results suggest that the superconducting properties of La_3X compounds are essentially due to the La 5d-electrons and support the idea of aclose relation between La_3X and fcc La under pressure.
For La_3XC perovskite carbides, if we compare the Knight-shift K_X to that of corresponding binary compounds at low temperature we find that the core-polarization contribution seems to be negligible in these alloys. If weassociate this result with the strong reduction of γ and T_c (Heiniger et al, 1973), we may suppose that La_3XC are true ternary alloys without relation with the corresponding binary alloys.
This work was financially supported by the Swiss National Science Foudation.

4. References.

Descouts, P., Thesis, University of Geneva (1973), unpublished.
Heiniger, F., Bucher, E., Maita, J.P., and Descouts, P., Phys.Rev., B8, 3194 (1973)

NMR IN DILUTE TERNARY Al-Me_1-Me_2 AND Al-Me-v SYSTEMS

M.A.Adawi[+], C.Hargitai[+], E.Kovács-Csetényi[++], K.Tompa[+]
+ Central Research Institute for Physics, Budapest, Hungary
++ Research Institute for Nonferrous Metals,Budapest,Hungary

Abstract. Slowly cooled and quenched Al-based dilute binary alloys, and slowly cooled Al-based ternary dilute alloy samples were investigated by NMR method. The quadrupole perturbation of the quenched-in vacancies was detected in binary alloy samples and the addition of quadrupole effects was found in ternary systems.

1. Introduction. NMR measurements were made on Al-based dilute alloys containing two kinds of substitutional impurities /Al-Me_1-Me_2/ and a substitutional impurity and quenched-in vacancies /Al-Me-v / respectively. The long range first order quadrupole perturbation were investigated in order to see the syncronous effects originated from two kinds of impurities.

2. Experimental. We have investigated the following alloys: Al-Mg /300...... 3000 ppm Mg/, Al-Si /50...3700 ppm Si/, Al- Cu /120...1800 ppm Cu/, Al-Fe /2.. 200 ppm Fe/, Al-Mg-Si /500, 1000, 2000 ppm Si and Mg respectively/, Al-Fe-Si /500, 1000, 2000, 3000 ppm Si and 200 ppm Fe/. The samples were annealed at 600 C^o /Al-Mg 450 C^o/ for an hour, then they were slowly cooled or quenched into water. Foil-sandwich samples /Tompa, et al 1963/ were used, and the measurements were done on a conventional c.w. NMR spectrometer at room temperature.

3. Experimental results. The amplitude of the NMR spectrum and the "wipe - out" number characteristic of the first order quadrupole effect were measured /Abragam, 1961; Tompa,1972/. Typical results are shown on the 1a. and 1b.figures.

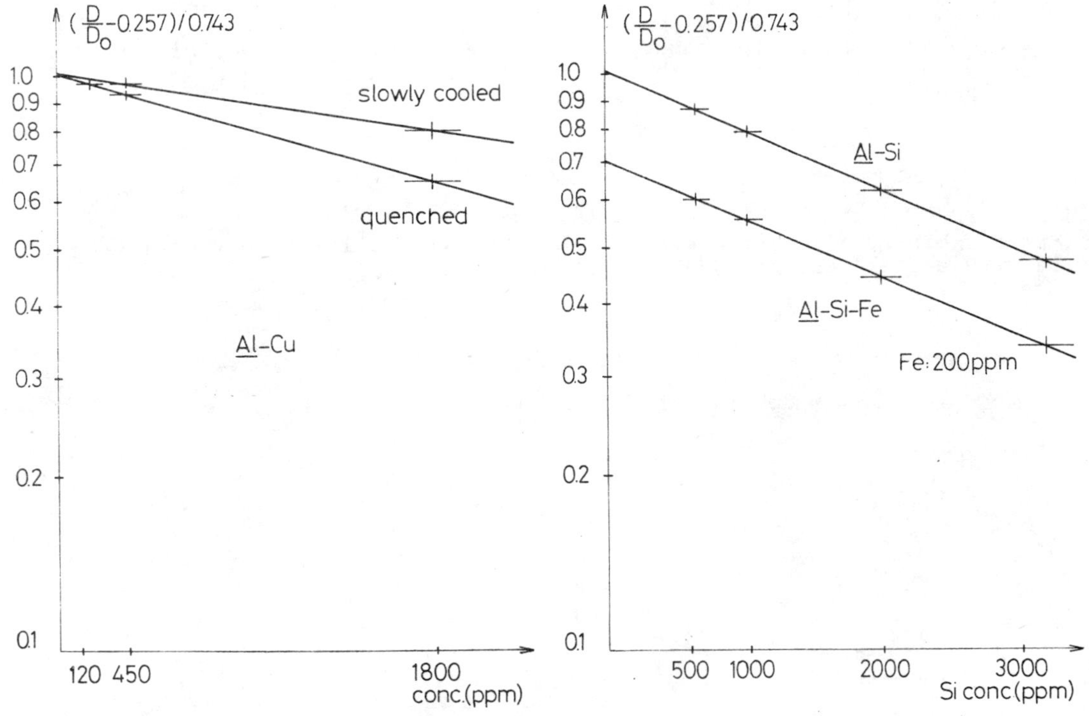

Fig.1a.Effect of quenching

Fig.1b. The change of NMR spectrum on ternary alloys

The orther alloys show similar results.

Quenching changes the "wipe-out" numbers characteristic of the first order quadrupole effects. The results are summarized in the table I.

Table I. Wipe-out numbers on binary alloys

"wipe-out" number*	Al-Mg	Al-Si	Al-Cu	Al-Fe
$n_{1_{SL}}$	124	180	235	1170
n_{1_Q}	150	250	310	1570

* standard error ± 10 %

In a ternary alloy there is a paralell shift in the amplitude-concentration curve as the effect of second alloyant element, but the slope of this straight line is unchanged in this representation /Fig.1b./

4. Comments. We describe the NMR amplitude of the ideal ternary alloys /containing non-interacting impurities/ by the following formula

$$\frac{D}{D_o} = /I-c_I/^{n_{I1}} /I-c_2/^{n_{I2}}$$

where c_I = concentration of impurities, n_{Ii} wipe-out numbers characteristic of binary systems. In this dilute ternary alloys the second moment of the NMR spectrum is a linear function of the corresponding parameters measured on binary alloys.

The excess "wipe-out" numbers of the quenched binary alloys $\Delta n_1 = n_{1Q} - n_{1_{SL}}$ are presumably brought about by the excess quenched-in vacancies. Moreover it seems reasonably to suppose that the impurity-vacancy binding energy can be scaled by the wipe-out number $n_{1_{SL}}$, therefore the number of quenched in vacancies should be

$$\Delta n_1 = a \cdot e^{b\, n_{1SL}}$$

For our binary alloys the experimental results follow rather well this formula with a=51±14 and b= 1,9±0,2/.10^{-3} /see Fig.2./,

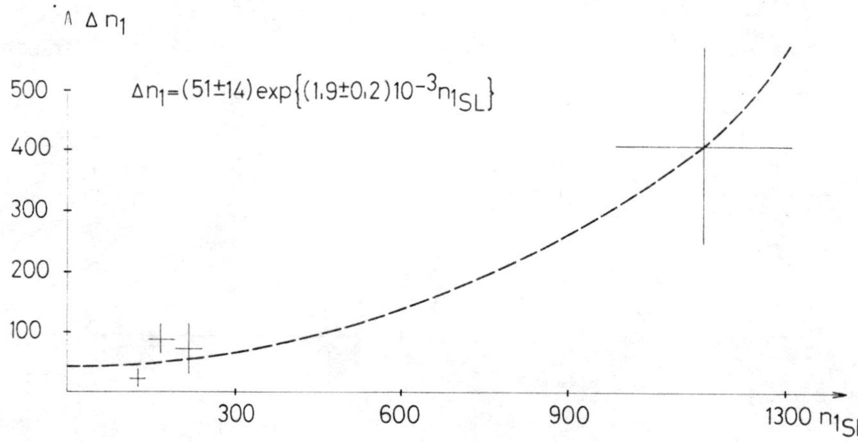

Fig.2. Excess "wipe-out" numbers in quenched Al-based dilute alloys

it suggests that the excess wipe-out numbers are due to quenched-in vacancies and the impurity-vacancy binding energy is proportional to the amplitude of the asymptotic perturbation around the impurities.

5. References

Tompa,K., Tóth F., 1963.Phys.Status Solidi 3, 2051

Abragam,A., Principles of Nuclear Magnetism, Oxford University Press,1961.

Tompa,K.,1972. J.Phys.Chem.Solids 33, 163.

DIFFUSION OF HYDROGEN IN VANADIUM HYDRIDE AS STUDIED BY NMR TECHNIQUES

G.J. Krüger and R. Weiss
Laboratory of Magnetic Resonance, Physics Division, EURATOM CCR, 21020 Ispra, Italy

Abstract. The diffusion coefficient D and the spin lattice relaxation time T_1 of protons in $VH_{0.3}$ have been measured by NMR pulse technqiues. D exhibits an activation energy of (0.11 ± 0.04) eV in the α phase whereas the jump time obtained from T_1 shows (0.5 ± 0.1) eV in the $\alpha+\beta$ phase at low temperatures.

1. Introduction. Though protons are very suitable for NMR studies the diffusion of hydrogen in transition metal hydrides has only been investigated via relaxation time measurements (Cotts, 1972). The direct measurement of the hydrogen diffusion coefficient D by spin echo techniques has not been reported so far. In this paper we report the results of a spin echo study of D in $VH_{0.3}$ at a proton resonance frequency of 48 MHz. We used the 90°-180° echo with pulsed field gradient (Abragam, 1961, Ch. III, Stejskal et al., 1965). In addition the spin lattice relaxation time T_1 of the protons was measured in the same sample using standard pulse techniques.

2. Diffusion coefficient. In our measurements the sample of $VH_{0.3}$ was ground down to particles of about 0.1 mm diameter. This is a compromise between the need to have particles sufficiently large for the measurement of D and sufficiently small to remain within the skin depth at the measuring frequency. The measurements were made by first heating the sample to the highest temperature so that the measuring temperature was reached from above. The results are shown in fig.1. By assuming an Arrhenius type of activation law the activation energy turns out to be (0.11 ± 0.04) eV and $D_0 = 6.8\times10^{-4}$ cm²/s. This is in fairly good agreement with results obtained by neutron scattering in similar systems (Gissler, 1972). Our measuring accuracy for the diffusion coefficient is at present ±30%. In addition we have a systematical error due to the particle size. With diffusion coefficients of nearly 10^{-4} cm²/s this error turns out to be approximately 10% if we assume spherical particles (Murday et al., 1971, Krüger et al., 1971). The diffusion coefficients given in fig. 1 are therefore about 10% too small.

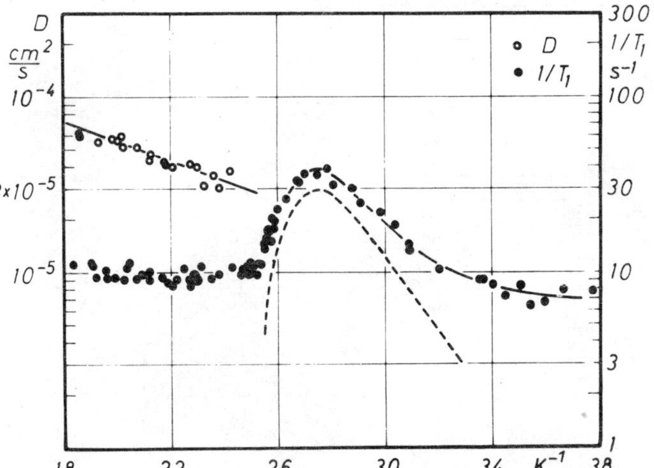

Fig. 1. Diffusion coefficient and spin lattice relaxation rate of protons in $VH_{0.3}$ vs. reciprocal temperature. The broken curve corresponds to $1/T_{1d}$ (see text).

From the phase diagram (Maeland, 1964) it is obvious that the results are for D in the α phase of the system. At lower temperatures in the mixed $\alpha+\beta$ phase D could not be measured because it was too small for our experimental set up. Therefore T_1 measurements were done over a larger temperature range.

3. Spin lattice relaxation time Fig. 1 shows that there are two contributions to T_1. The first one is not very temperature dependent and we think that it is due to interaction of the protons with conduction electrons of the form $T_{1e}\times T$ = const. (Abragam, 1961, Ch. IX). The second contribution exhibits a maximum relaxation rate at $10^3/T = 2.75$. This part after subtraction of $1/T_{1e}$ is the broken curve in fig. 1. We assume that it is mainly caused by dipolar interaction of protons and ^{51}V modulated by the hydrogen diffusion and call it $1/T_{1d}$. Furthermore we identify T_{1d} with T_{1II} in the two spin system of ^1H and ^{51}V (Abragam, 1961, Ch. VIII). Thus we have, neglecting the proton proton interaction, and assuming cubic symmetry:

$$(1)\quad \frac{1}{T_{1d}} = \frac{1}{T_1} - \frac{1}{T_{1e}} = \gamma_I^2 \gamma_S^2 \hbar^2 S(S+1)\left\{\frac{1}{2} J_1(\omega_I-\omega_S)+\frac{3}{2} J_1(\omega_I)+3J_1(\omega_I+\omega_S)\right\}$$

where γ_I and γ_S are the gyromagnetic ratios of ^1H and ^{51}V respectively and S is the spin of ^{51}V. The intensity function we take for a first approach from Torrey's theory of jump diffusion in a cubic lattice (Abragam, 1961, Ch. X., Resing et al., 1963) and obtain:

(2) $$J_1(\omega) = \left[16\pi n_S \tau/(45\ k^3 l^3)\right] G(k,y)$$

and

(3) $$G(k,y) = \int_0^\infty J^2_{3/2}(kx)\ \frac{1-(\sin x)/x}{[1-(\sin x)/x]^2 + y^2}\ \frac{dx}{x}$$

where $J_{3/2}(kx)$ is a Besselfunction and $y = \omega\tau$ (because we assumed the vanadium spins to be stationary). n_S is the number of ^{51}V spins per cm^3, τ is the jump time and l the jump length of the protons. k is a normalization constant which was calculated to be $k_{oct} = 0.7372$ for octahedral and $k_{tetr} = 1.0896$ for tetrahedral interstitials of the protons in a bcc vanadium lattice. Our model is not quite correct, since we obtained T_{1d} only in the $\alpha+\beta$ phase where the β phase exhibits a bct structure (Maeland, 1964).

The jump times obtained with this theory range from about 10^{-10}s to 4×10^{-8}s in our temperature range. At low temperatures they exhibit an activation energy of (0.5 ± 0.1) eV for both octahedral and tetrahedral interstitials. A more precise interpretation should include the temperature dependence of the amount of α and β phase in the mixed $\alpha+\beta$ phase and then the relaxation theory of the two phase system should be used (Zimmerman et al., 1957).

The quantum theory of diffusion of hydrogen in the α phase of Nb (Sussmann et al., 1972) has not been used so far for the calculation of intensity functions of nuclear magnetic relaxation.

4. Acknowledgement. The authors would like to thank Dr. G.A. Helcké for revision of the manuscript.

5. References.

Abragam, A., 1961 The Principles of Nuclear Magnetism, Oxford University Press, London
Cotts, R.M., 1972 Ber. Bunsenges. 76, 760
Gissler, W., 1972 Ber. Bunsenges. 76, 770
Krüger, G.J., W. Müller-Warmuth and A. Klemm, 1971, Z. Naturforsch. 26a, 94
Maeland, A.J., 1964, J. Phys. Chem. 68, 2197
Murday, J.S., and R.M. Cotts, 1971, Z. Naturforsch. 26a, 85
Stejskal, E.O., and J.E. Tanner, 1965, J. Chem. Phys. 42, 288
Sussmann, J.A., and Y. Weissman, 1972, phys. stat. sol. (b) 53, 419
Resing, H.A., and H.C. Torrey, 1963, Phys. Rev. 131, 1102
Zimmerman, J.R., and W.E. Brittin, 1957, J. Phys. Chem. 61, 1328.

PROTON SPIN LATTICE RELAXATION IN TITANIUM HYDRIDE

C. Korn
Department of Physics, Ben-Gurion University of the Negev, Beer-Sheva, Israel

Abstract. The spin lattice relaxation time T_1 of hydrogen in titanium hydride was measured as a function of temperature and hydrogen concentration in the temperature range where the overriding relaxation mechanism is the interaction with the conduction electrons. The variation of $(T_1T)^{-\frac{1}{2}}$ with these parameters is presented in order to obtain information about the electronic structure and density of states.

Introduction. It is well known that transition metals absorb large amounts of hydrogen (Mueller et al., 1968) and considerable controversy exists as to how this effects the electronic band structure (Stalinski, 1972). Models have been proposed wherein the hydrogen simply contributes its electrons to a rigid band (Mott et al. 1958; Merriam et al. 1963), or that the hydrogen removes electrons from a rigid band (Bos et al. 1967), or that a rigid band model is inapplicable (Switendick, 1970).

Nuclear magnetic resonance parameters are sensitive to the electronic structure (Narath,1967) and can be used as probes to help its determination. The spin lattice relaxation time of hydrogen was thus obtained as a function of hydrogen concentration and temperature in the temperature range where the primary relaxation mechanism is due to the conduction electrons.

Experimental Procedure. Sample preparation method was described elsewhere (Korn et al. 1970). The spin lattice relaxation time T_1 of hydrogen was measured at 24 MHz using a Bruker B-KR 321s pulsed spectrometer with temperature varying from 100°K to 350°K. The low temperature limit was determined by our present instrumentation and the high value was chosen to overlap the region where a cubic-tetragonal transformation takes place (Yakel, 1958) but does not yet contain a significant diffusional relaxation contribution. T_1 was obtained by flipping the spins with an initial 180° pulse and measuring the signal amplitude following a second 90° sampling pulse as a function of time between the two pulses. H_1 was approximately 50 gauss which is about four times the dipolar line width. A box car integrator was used to determine signal amplitude.

Experimental Results. T_1 was obtained as a function of temperature and hydrogen concentration. The diffusional contribution to the relaxation rate was known from a previous study (Korn et al. 1970) and found to be negligible for most of the samples over most of the temperature range. Where necessary, this small contribution was subtracted from the total relaxation rate giving the net electronic contribution $(T_1)_e$. Since $[(T_1)_eT]^{-\frac{1}{2}}$ is approximately proportional to the density of states (Narath, 1967), it is this parameter that has been plotted in figures 1 and 2 as a function of temperature and x, the ratio of the number of hydrogen to titanium atoms. The broken lines of Fig. 1 are simply drawn through the experimental points while the solid lines are probably closer to the true behavior (see magnetic susceptibility results of Trzebiatowsky et al.1953).

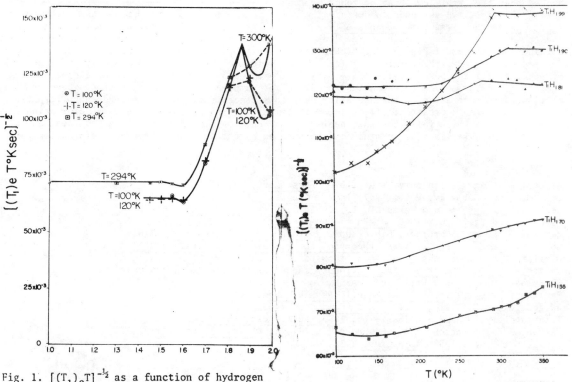

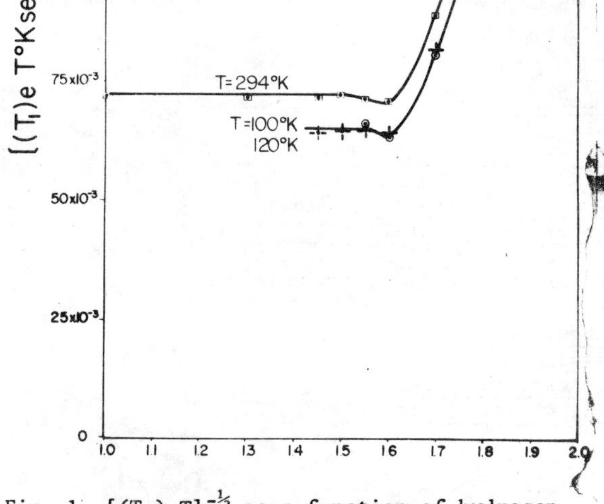

Fig. 1. $[(T_1)_eT]^{-\frac{1}{2}}$ as a function of hydrogen concentration $x = \frac{H}{Ti}$

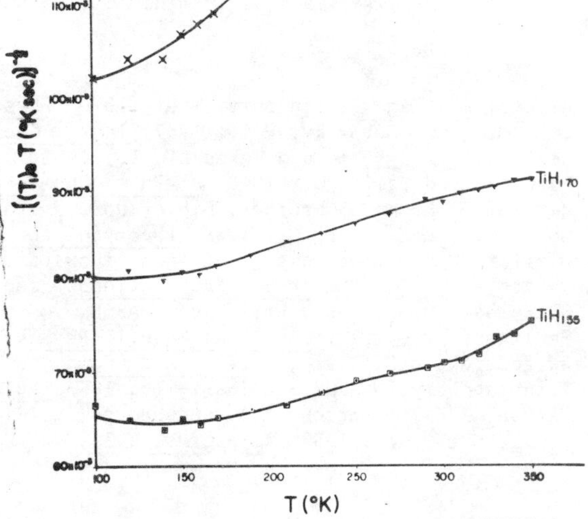

Fig. 2. Temperature Dependence of $[(T_1)_eT]^{-\frac{1}{2}}$

$[(T_1)_eT]^{-\frac{1}{2}}$ is constant below x = 1.5 since this is a mixed phase region containing pure Ti and $TiH_{1.5}$. Upon increasing x, $[(T_1)_eT]^{-\frac{1}{2}}$ dips slightly, rises to a cuspoidal break, dips and rises again.

Fig. 2 shows the temperature dependence of $[(T_1)_eT]^{-\frac{1}{2}}$. The most outstanding features are: (a) the break in slope near 300°K for samples having hydrogen concentrations above x = 1.8 and the absence of this break for lower concentrations and (b) the strong rise of $[(T_1)_eT]^{-\frac{1}{2}}$ with temperature for the high concentration sample, intersecting the lower concentration curves.

<u>Discussion.</u> In interpreting the data we contend that $[(T_1)_eT]^{-\frac{1}{2}}$ closely follows the density of states at the Fermi level. As x is increased, electrons are contributed to the conduction band so that the Fermi level is swept through the density of state curve for increasing x. Schreiber (1965) has put forward the theory that the cubic symmetry of metal hydrides causes the d band to split into two subbands, representing a doublet and a triplet, analogously to d_γ - d_ε crystal field splitting with d_γ at the lower energy. Since the d_γ subband can hold four electrons, metals having an outer structure of d^n electrons can accommodate an extra 4-n electrons in the d_γ subband and thus should tend to absorb a maximum of 4-n hydrogen atoms for each metal atom. This is generally the case. Our T_1 results, susceptibility measurements (Trzebiatowsky, 1953) and transport properties (Gesi et al. 1963; Bickel et al. 1970) show however that when approaching the maximum hydrogen concentration we are not approaching the end of a split subband. Thus we modify the Schreiber theory by assuming that for our case the subbands overlap. Now in the stoichiometric hydride each titanium atom is surrounded by four hydrogen atoms located at the corners of a cube, with the d_γ type orbitals pointing towards the cube faces and d_ε towards the cube edges. It is thus reasonable to assume that the two types of subbands have different bonding characteristics with respect to the hydrogen. If one pursues this further, it can be argued that the Fermi level may be found by considering the first two subbands to contain the original two titanium d electrons and by counting states under the first two subbands only as if each hydrogen atom contributes one electron. Applying these ideas to the rigid band fcc density of states calculations of Watson et al.(1971), one obtains the results of Fig. 3. The broken curves are the d subbands and the solid curve their sum. The second row shows the Fermi level for various values of x using the previously mentioned scheme. We see that for the pure hydride phase between x = 1.5 and x = 2 the N(E) curve closely resembles that of $[(T_1)_eT]^{-\frac{1}{2}}$. The tetragonal deformation for samples having concentrations above x = 1.8 is associated with a Van Hove singularity near this concentration, causing the sharp cusplike shape. The strong increase of $[(T_1)_eT]^{-\frac{1}{2}}$ with temperature shows that there is a deformation of the higher subbands which fills the cusp near x = 1.8, until a temperature is reached (near 300°K) where the cusplike behavior is minimized to such an extent that no tetragonal deformation takes place.

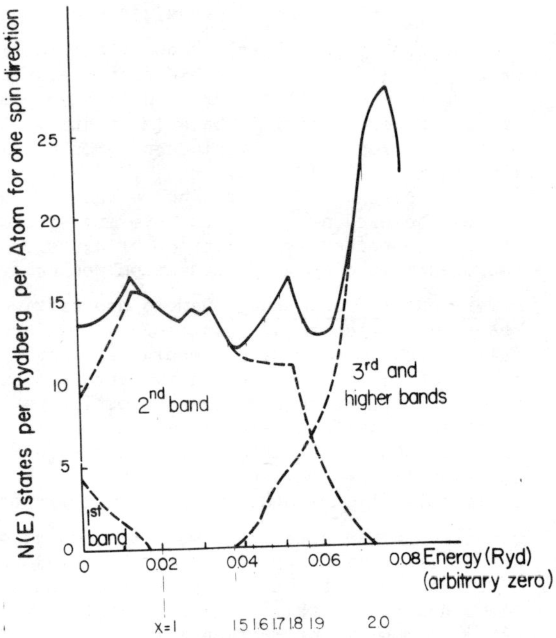

Fig. 3. Density of states and Fermi level for various hydrogen concentrations, according to theory in text.

References.

Bickel, P.W. and Berlincourt, T.G., 1970, Phys. Rev., B2, 4807.
Bos, W.G. and Gutowsky, H.S., 1967, Inorg. Chem., 6, 552.
Gesi, K., Takagi, Y. and Takeuchi, T., 1963, J. Phys. Soc. Japan, 18, 306.
Korn, C. and Zamir, D., 1970, J. Phys. Chem. Solids, 31, 489.
Merriam, M.F. and Schreiber, D.S., 1963, J. Phys. Chem. Solids, 24, 1375.
Mott, N.F. and Jones, H., 1958, Theory of the Properties of Metals and Alloys, Dover.
Mueller, W.M., Blackledge, J.P. and Libowitz, G.G., 1968, Metal Hydrides, Academic Press.
Narath, A., 1967, Hyperfine Interactions, Freeman and Frankel ed., Academic Press.
Schreiber, D.S., 1965, Phys. Rev., A860.
Stalinski, B., 1972, Ber. Bun. Gesell. Physik. Chem., 76, 724.
Switendick, A.C., 1970, Solid State Comm., 8, 1463.
Trzebiatowsky, W. and Stalinski, B., 1953, Bull. Acad. Polon Sci., 1, 131.
Watson, R.E., Misetich, A.A. and Hodges, L., 1971, J. Phys. Chem. Solids, 32, 709.
Yakel, H.L., Jr., 1958, Acta Cryst., 11, 46.

COMPARISON OF THE H, D AND V^{51} SPIN-LATTICE RELAXATION TIMES IN V-D CONTAINING A FEW PERCENT HYDROGEN

Arons, R.R., Bohn, H.G., and Lütgemeier, H.
Institut für Festkörperforschung, KFA Jülich, D 517 Jülich, Germany.

Abstract: The relaxation times $(T_1)^H$ and $(T_1)^D$ in the α'- and δ-phases of V-D show, that only in the α'-phase quadrupole coupling contributes to $(T_1)^D$. As the relaxation in α' and the line splitting in δ lead to the same value for ν_Q, the same sites must be occupied by the D-atoms in the two phases.

1. Introduction. From NMR (Arons et al. 1972, 1974) and neutron diffraction experiments (Asano and Hirabayashi, 1973), it has been shown that the phase diagram of the vanadium-deuterium (V-D) system is very complex and completely different from that for V-H. For D/V ratios above 0.6, an orthorhombic δ-phase appears, in which the deuterium NMR line is found to be split with a quadrupole frequency ν_Q = 55 kHz. The tetragonal β-phase known from V-H, on the other hand, is only found in a region near D/V = 0.4 and gives ν_Q = 100 kHz for the D-line. Also the behaviour of the V-line was shown to be completely different in the two phases. In the various phases the jump frequency $1/\tau$ of H and D was investigated by measurements of the spin-lattice relaxation times, T_1, of V, H and D. Whereas in the β-phase a clear T_1 minimum at about 250 K is observed, indicating that $1/\tau$ is about 10^8 Hz, the minimum in the δ-phase appears suspiciously near the α'-δ phase transition. In order to determine the behaviour at the α'-δ transition more accurately, a V-D sample containing a few percent hydrogen was prepared.

2. Experimental details. In contrast to the previous experiments a superconducting magnet was used, allowing measurements of V, H and D at the same frequency up to 46 MHz. In the δ-phase, the split D line was completely saturated by a comb of 90° pulses.

3. Experimental results. Both the D and the V resonances show the same characteristics as observed previously for the pure V-D samples. Particularly from the value of ν_Q = 55 kHz obtained for the D resonance at low temperatures, it follows that here the δ-phase of V-D appears. Fig. 1 shows T_1 of V, H and D for 33 MHz as a function of 1/T. In the α'-phase, protons and deuterons show the

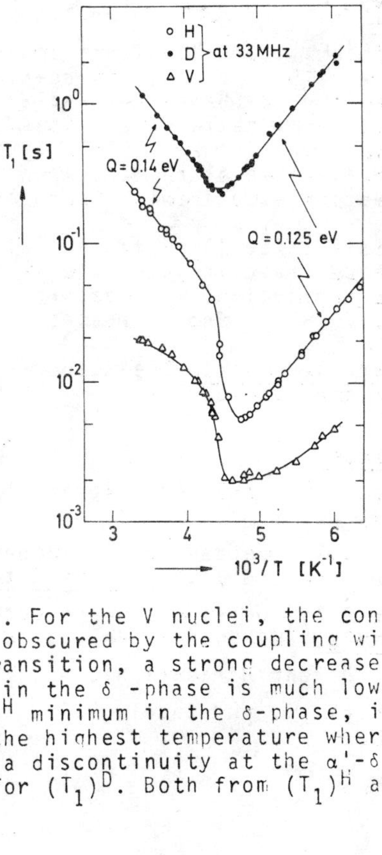

Fig. 1: T_1 vs 1/T for H, D and V^{51} at 33 MHz in a V-D sample containing a few percent hydrogen.

The activation energies for the diffusion of H and D in the α'- and δ-phases, deduced from the slopes in the high- and low-temperature regions, respectively, are indicated. The steep fall in $(T_1)^H$ at about 224 K represents the α'- δ transition.

same activation energy for diffusion Q = 0.14 eV. For the V nuclei, the contribution of the diffusing H and D atoms is rather obscured by the coupling with the conduction electrons. At the α'- δ phase transition, a strong decrease in the proton T_1 is observed, indicating that $1/\tau$ in the δ-phase is much lower than in the α'-phase. From the absence of a $(T_1)^H$ minimum in the δ-phase, it follows that $1/\tau$ is lower than 2π x 33 MHz at the highest temperature where the δ-phase is stable. Whereas also $(T_1)^V$ shows a discontinuity at the α'-δ transition, a quite normal minimum is obtained for $(T_1)^D$. Both from $(T_1)^H$ and

$(T_1)^D$, the activation energy for the diffusion in the δ-phase is found to be Q = 0.125 eV.

4. Discussion. Due to their diffusion, the H and D atoms are subjected to fluctuating dipolar fields produced by the neighbouring V-atoms. As is readily calculated, the H-H, H-D, and D-D dipolar coupling can be neglected. The fluctuating fields give rise to spin-lattice relaxation, and a T_{1d} minimum occurs for $\omega_H \tau_H = \omega_D \tau_D \approx 1$ (Abragam, 1961). At the higher and lower temperature sides of the T_{1d} minimum, respectively, it holds that

$$(T_{1d})^D/(T_{1d})^H = (\gamma_H/\gamma_D)^2 \quad \tau_H/\tau_D = 42.5 \qquad \text{for} \quad \omega\tau << 1$$

and

$$(T_{1d})^D/(T_{1d})^H = 1.48\ (\gamma_H/\gamma_D)^2 \quad \tau_D/\tau_H = 62.8 \qquad \text{for} \quad \omega\tau >> 1.$$

The two calculated values are valid under the condition that $\tau_H = \tau_D$ in the V-D-H mixture (see e.g. Arons et al. 1967).

Additionally for D the coupling of the quadrupole moment with fluctuating electric field-gradients (EFG) must be taken into account.

In the δ-phase, we observe that $(T_1)^D/(T_1)^H = 65$, indicating that the D relaxation rate is completely produced by the dipolar coupling with the V-atoms. This means that the D-atoms only occupy those sites for which the EFG has the same orientation and the same magnitude. This agrees with the fact that in spite of the large mobility at 200 K, the value of γ_Q is equal to that observed at 4 K, where the D-atoms are at fixed positions (Arons et al. 1974).

In the α'-phase, where the static quadrupole coupling is averaged to zero, the ratio $(T_1)^D/(T_1)^H$ is measured to be 6, instead of 42.5 expected from the dipolar coupling. This means that the rate produced by quadrupole coupling, $(1/T_{1Q})^D$, represents more than 85 % of the total rate $(1/T_1)^D$. From this, γ_Q can be calculated. Taking into account, that D has spin 1 and V spin 7/2, we have for $\omega\tau << 1$ (Abragam, 1961)

$$(1/T_{1d})^D = (63/3)\gamma_D^2\ \gamma_V^2\ \hbar^2 (\sum_k r_k^{-6})\tau$$

$$(1/T_{1Q})^D = (2/3)\pi^2\ \gamma_Q^2\ \tau.$$

Here $\sum_k r_k^{-6}$ is known to be 146.5 $/a^6$, where a is the lattice parameter (Stalinski and Zogal, 1965). With a equal to 3.15 Å (Asano and Hirabayashi, 1973), we find γ_Q = 52 kHz. This is nearly the same value as deduced from the line splitting in the δ-phase. This value was previously concluded to correspond to the occupation of tetrahedral sites, whereas for the octahedral sites γ_Q is about 100 kHz (Arons et al. 1974). Thus, in the α'-phase the D-atoms occupy the tetrahedral sites at random, in agreement with the conclusion drawn from neutron diffraction experiments (Asano and Hirabayashi, 1973).

Apparently, the $(T_1)^D$ minimum at about the α'- δ transition arises accidently, as the decrease in jump rate is compensated by the freezing out of the quadrupole contribution. $T_{1\rho}$ measurements are planned for comparing the absolute 1/τ values in α'- and δ-phases.

5. Acknowledgement. The authors gratefully acknowledge Mr. K.H. Klatt for preparing the sample.

References

Abragam, A., 1961, Principles of Nuclear Magnetism, Oxford University Press, London, Ch. 8.

Arons, R.R., Bouman, J., Wijzenbeek, M., Klaase, P.T.A., Tuyn, C., Leferink, G., and de Vries, G., 1971, Acta Met. 15, 144-147.

Arons, R.R., Bohn, H.G., and Lütgemeier, H., 1972, Jülich Conference on Hydrogen in Metals, Jül-Conf-6, Vol. I, 272-279; 1974, J. Phys. Chem. Solids 35, 207-214.

Asano, H. and Hirabayashi, M., Phys. Status Solidi (a) 15, 267-279.

Stalinski, B. and Zogal, O.J., 1965, Bull. Acad. Polon. Sci. Ser. Sci. Chim. XIII, 397-402.

SPIN-LATTICE RELAXATION TIME OF DEUTERIUM IN TITANIUM DEUTERIDE

Drora Kedem and D. Zamir
Soreq Nuclear Research Centre - Yavne, Israel.

Abstract. The spin-lattice relaxation time of deuterium was measured in several deuterides TiD_x as a function of temperature. The relaxation time due to diffusion of the deuterium atoms was obtained and compared to that of hydrogen in the similar hydrides TiH_x.

1. Introduction. Titanium, like other transition metals, can absorb large amounts of hydrogen or deuterium, thus forming hydrides and deuterides. The properties of the TiH_x system were extensively investigated (Coogan and Gutowsky, 1962, Stalinski et al., 1961, Korn and Zamir,1970). Additional information on the hydrides can be obtained from measurements on deuterium, as the structure of the deuterides is identical to that of the hydrides in atomic aspects. The deuterium, having a quadrupole moment, will experience a quadrupole interaction with its environment, which is sensitive to the location of the deuterium atoms in the lattice and to the electronic structure of the system. The diffusion of the deuterium in the lattice causes the quadrupole interaction to be time-dependent, and as a result, it affects the spin-lattice relaxation time. In the present work we study the quadrupole interaction of deuterium in the TiD_x system by means of spin-lattice relaxation time measurements.

The maximum atomic ratio x in the TiH_x and TiD_x systems is 2. The phase diagram of the hydrides (Lenning et al., 1954) shows that for x>1.6 the fcc γ phase exists, while for x<1.6 we find a mixture of γ phase $TiH_{1.6}$ and hcp α phase of Ti, with a negligible amount of hydrogen dissolved in the metal.

In the γ phase, the hydrogen atoms occupy the tetrahedral sites in the titanium fcc lattice. In the stoichiometric TiD_2 compound, the deuterium atoms form a simple cubic sublattice, with a lattice parameter being half that of the titanium lattice. As the local symmetry of the deuterium atoms in this compound is cubic, we do not expect any quadrupole interaction with the environment. For x<2, the vacancies in the deuterium sublattice will produce a deviation from the cubic symmetry and a quadrupole interaction is expected.

2. Results and discussion. Several samples with different atomic ratios x were investigated in a pulsed NMR System. The spin-lattice relaxation time T_1 was measured at temperatures ranging from 20° to 500°C. The two most effective mechanisms for relaxation are the interaction of the nucleus with the conduction electrons and the time-dependent interactions (dipole and quadrupole), caused by the deuterium diffusion. The measured relaxation time T_1 can be described by

$$\frac{1}{T_1} = \frac{1}{(T_1)_e} + \frac{1}{(T_1)_d} \qquad (1)$$

$(T_1)_e$, the electronic relaxation time, yields the Korringa relation

$$(T_1)_e\, T = \text{const.} \qquad (2)$$

The value of $(T_1)_e T$ can be measured at temperatures low enough so that the contribution of the diffusion can be neglected. The ratio of the electronic relaxation times $(T_1)_e$ of the deuterium and hydrogen is γ_H^2/γ_D^2 as the electronic structure of the two systems is similar.

The relaxation time due to diffusion, $(T_1)_d$, is described by

$$\frac{1}{(T_1)_d} = <\Delta\omega^2> \left\{ \frac{\tau}{1+\omega^2\tau^2} + \frac{4\tau}{1+4\omega^2\tau^2} \right\} \qquad (3)$$

(Abragam, 1961), where ω is the Larmor frequency and τ is the correlation time for the diffusion. A minimum value of the diffusion relaxation time $(T_1)_d^{min}$ obtained for $\omega\tau=0.616$ is:

$$\frac{1}{(T_1)_d^{min}} = \frac{1.425}{\omega} <\Delta\omega^2> \qquad (4)$$

In Fig. 1 the values of $(T_1)_d^{min}$ are shown as a function of 2-x. For comparison, the values of $(T_1)_d^{min}$ of hydrogen in the TiH_x system are plotted (Korn and Zamir, 1970). Those values are multiplied by the ratio of $\gamma^4 I(I+1)$ of the two isotopes and the ratio of the frequencies at which the two systems were investigated. It is clearly seen that for the nearly stoichiometric $TiD_{1.98}$, the main contribution to the interaction is dipolar, and as the atomic ratio x decreases, a steep drop in $(T_1)_d^{min}$ is observed.

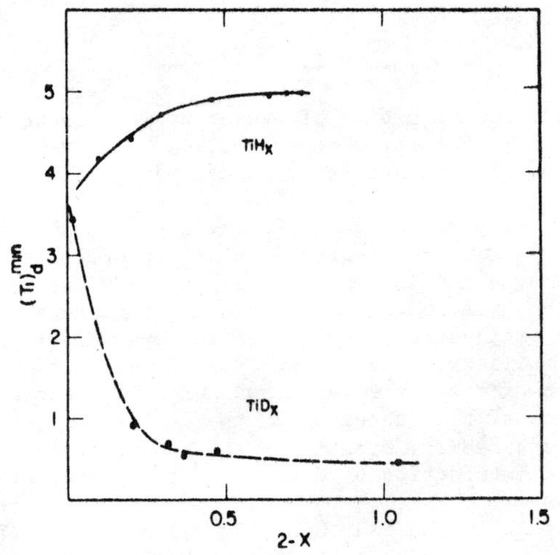

Fig. 1

$(T_1)_d^{min}$ of deuterium and of hydrogen in the TiD_x and TiH_x systems

From these data, the contribution obtained for the quadrupole interaction $< \Delta\omega^2 >$ rises from about 6×10^7 sec^{-2} for x=1.8 to 11×10^7 sec^{-2} for x=1.0. These data are consistent with measurements of line widths of deuterium magnetic resonance spectra (Nakamura, 1974, Kedem and Zamir, to be published).

3. References

Abragam, A., 1961, The Principles of Nuclear Magnetism, Oxford University Press, London
Coogan, C.K. and Gutowsky, H.S., 1962, J. Chem. Phys. 36, 110.
Korn, C. and Zamir, D., 1970, J. Phys. Chem. 31, 489.
Lenning, G.A. et al., 1954, Trans. AIME 200, 367.
Nakamura, K., 1974, J. Magn. Resonance 14, 31.
Stalinski, B. et al., 1961, J. Chem. Phys. 34, 1191.

ESR EXPERIMENTS ON Gd IN THE SUPERCONDUCTING STATE OF $CeRu_2$ and $LaOn_2$.

K. Baberschke, W. Schrittenlacher and U. Engel.
Fadbereid Physik, Freie Universitat, Berlin, I Berlin 33, Boltzmannstr.2O.
W.Germany.

Abstract. The ESR of Gd in the superconducting state of $CeRu_2$ and $LaOn_2$ was measured at X-, K- and Q-band frequencies. The dependence of g-values on the different frequencies is attributed to a change of spatial internal field variation.

The localised resonance of Gd in the superconducting state was reported previously (I,2,3). These experiments were performed at X-band frequencies ($H_o \sim 3.5$ kG). A change of g-value as a function of temperature ($T<T_c$) was observed and attributed to a change of the spin susceptibility (4). Another effect which shifts the field of resonance has to be taken into consideration. This is the spatial field distribution in the vortex state of a superconductor which already has been used to discuss NMR-experiments (5). The resonance field increases (g-value decreases) for Gd spins which do not see the field at the vortex center but a reduced one between the vortices. This can be proved using different microwave frequencies according to different external fields, as we showed for $CeRu_2$ recently (6). Here we report additional results of K-band (H_o = 8.5 kG) measurements for $CeRu_2$. The field-dependent shift of resonance in the superconducting state is confirmed by results in the isomorphic intermetallic compound $LaOn_2$. It has a superconducting transition temperature $T_c(H_o=0)$ = 9K and an upper critical field $H_{c2}(T=0)$=30 kG. The results are shown in the table:

System	$T>T_c$: g_n	T=I.5K: gs^X	gs^K	gs^Q
$CeRu_2$	(I) I.952(5)	(6) I.930(5)	I.952(IO)	(6) I.96I(5)
$LaOs_2$	I.930(5)	I.870(IO)	I.9I5(IO)	I.930(IO)

g_n is the temperature independent g-value in the normal conducting state. In both systems, no dependence on Gd-concentrations between 80-2000 ppm could be detected. In the superconducting state, g_s roughly changes linearly between T_c and T=I.5K. The superscripts X,K,Q, indicate the different microwave frequencies, while subscript s denotes superconductivity.
Discussion of the results for $CeRu_2$ show that for an external field of H_o=3.5 kG the field distribution effect gives the major contribution while for H_o=I3 kG (Q-band) the decrease of spin susceptibility dominates. At the intermediate field H_o=8.5 kG (K-band) both effects compensate. For $LaOs_2$ it can be seen that the field distribution effect at H_o=3.5 kG is stronger than for $CeRu_2$. This is confirmed by the fact that at H_o=I3 kG both effects compensate, while for K-band measurements a slight decrease of the g-value was observed. The g-factors given in the table were calculated by the ratio of external field H_o and the microwave frequency. However, to determine the "true" g-shift one has to take into account that the local field at the vortex core, H_v, is even higher than H_o. The results in $CeRu_2$ and $LaOs_2$ give evidence that the field distribution effect plays an important rôle and has to be eliminated before discussing the change of local susceptibility and homogeneous linewidth. Both quantities are important to study the static and dynamic properties of a localised moment in the superconducting state.

We acknowledge Profs. R. Orbach and B.J. Kochalayev for helpful discussion. Furthermore we thank Prof. B. Elschner, Darmstadt, for providing the K-band spectrometer. This work was supported by Deutsche Forschungsgemeinschaft, SFB I6I.

References

1. Engel, U., Baberschke, G., Koopman, G., Hufner, S. and Wilhelm, M., Sol.St.Comm., 12,977 (1973).
2. Rettori, C., Davidov, D., Chaikin, P. and Orbach, R., Phys.Rev.Lett., 30, 437 (1973).
3. Alekseevski, N. E., Garufullin, I. A., Kochalayev, B. I. and Kharakhash'yan, Zh ETF Pis.Red., 18, No. 5, 323-326 (5 September, 1973) and B. I. Kochalayev at this conference.
4. Anderson, P. W., Phys.Rev.Lett., 30, 437 (1973).
5. Fite, W. and Redfield, A. G., Phys.Rev., 162, 358 (1967); Redfield, A. G., ibid, p. 367.
6. Baberschke, K., Engel, U. and Hüfner, S., accepted for publ. in Sol.St.Comm. (1974).

ON THE QUADRUPOLE INTERACTION IN METALS

D. Quitmann, K. Nishiyama and D. Riegel
Institut für Atom- und Festkörperphysik (A)
Freie Universität Berlin, D1000 Berlin 33, Boltzmannstr. 20, Fed. Rep. Germany

Abstract: The lattice sums for the electric field gradient (efg) in Zn, Cd, In, and Sn is calculated using the Coulomb potential of the ions and three approximations for the shielding by the conduction electrons. To obtain agreement with the experimental temperature dependence we invoke the phonons and take $\langle x^2 \rangle$ from X-ray scattering in a Taylor expansion of the efg.

The quadrupole interaction energy in noncubic metals has been determined for many pure and substitutional systems by classical NMR techniques and, more recently, by using radioactive nuclear states. As to the electric field gradient (efg) q, its sign has been determined in some cases; the absolute magnitude is usually rather uncertain; but the temperature dependence is known well for a few systems, and the present note concentrates on it.

The charges exterior to the nucleus in question may be separated as: (a) ions, i.e. nuclei and non-conduction electrons, in other lattice cells, (b) inner (i.e. non conduction) electrons in the reference lattice cell. Contributions a) and b) are as in ionic crystals. The corresponding lattice sum has been calculated by de Wette (deW61). Its dependence on the lattice constants (c,a) is generally much too small to explain the observed decrease of q with T (see e.g. Fig. 1). (c) In the noncubic metals the conduction electrons in and outside the reference cell contribute due to the noncubic part of their density distribution. Though this is the overwhelming part in the heavy metals, it is rather uncertain to calculate since it consists of large compensating contributions from the various parts of the bands (MohS72). The band structure calculations appear at present good enough only to reproduce the sign and approximate magnitude.-It is, therefore, not uncommon to ascribe the deviation of sign, magnitude and temperature dependence between observed efg and the lattice sum as a whole to "the conduction electron contribution".

Let us try to trace the response of the conduction electrons somewhat more precisely in the following model: a lattice of point ions (charge Ze position $\vec{r}_i$) which would produce potentials $v(\vec{r}_i)$ at the reference point r=0; all the conduction electrons form a shielding medium with dielectric function ε. The Sternheimer effect in the reference lattice cell is described by an effective shielding factor ξ. We denote the screened potential by $V(r_i)$. Then

(1) $q_0 = \Sigma_i V_2(r_i)$ with (2) $V_2(r_i) = r\,\partial/\partial r\left[r^{-1}\partial/\partial r V(r_i)\right]\cdot\left[3\cos^2\vartheta_i - 1\right]$

is the z-component of the efg without Sternheimer correction (axial symmetry assumed). Then

(3) $q = \xi q_0$

is the efg effective at r=0. We consider the unscreened Coulomb potential

(4) $V^c = Zer_i^{-1}$ and three approximations for it after screening:

(5) $V^t = Zer_i^{-1}\exp(-\kappa r_i)$ in Thomas Fermi approximation with shielding length 1/κ.

Larger changes of q_0 might be expected from potentials which fluctuate, like

(6) $V^f = 2Ze/(\pi a_0)\cdot\varepsilon^{-2}(2k_F)\cdot\cos(2k_F r_i)\cdot(2k_F r_i)^{-3}$,

the Friedel potential which includes an expansion valid only for large r. Finally we consider the static Hartree dielectric function which leads to the potential

(7) $V^h = \frac{4\pi\Omega}{(2\pi)^3}\int \frac{\tilde{V}^c(q)}{\varepsilon(q)}\frac{\sin qr}{r}\,q\,dq$ (8) $\varepsilon(q) = 1 + \frac{4\pi e^2}{q^2}\frac{n(E_F)}{2}\left[1 + \frac{1-\eta^2}{2}\ln\left|\frac{1+\eta}{1-\eta}\right|\right]$

(9) $\tilde{V}^c = 4\pi Ze/q^2$; $\eta = q/2k_F$; for details see e.g. (Har 70).

The lattice sums were performed in r-space out to r = 100 Å or 200 Å and converged rapidly ; for the shielded potentials, the relative error in q_0 is $< 10^{-3}$. The results for Zn, Cd, In and Sn are presented in table 1 below.

We add the following remarks:

1) Although the signs of q_0 remain unchanged from the Coulomb potential calculation in the cases presented, sign changes occur for some other lattices, e.g. Hg.

2) The efg q is known experimentally to be q > 0 for CdCd, CdSn (KaR74a); the effective Sternheimer correction ξ of eq. (3) is thus negative at least in Cd. This is observed in many calculations for metals (e.g. (WatG65), (MohS72)).

3) The relative change with temperature is for In essentially independent of the choice for V and much too small. For CdSn, experiment shows a decrease of q with T (HeuH74), while all calculated potentials give an increase (for SnSn). The same applies to Zn.

As the additional mechanism which reduces the efg at higher temperatures, the phonon modes havebeen discussed for ionic crystals by Bayer (Bay 51) and by Kushida et al. (KusB56). Redistribution of the conduction electrons at the Fermi surface was considered by Watson et al. (WatG65); however, due to rather large uncertainties it appears undecided whether this mechanism is responsible for the gross variation. We therefore return to the phonons which may make a large contribution because a) the metals under study have low Debye temperatures Θ and b) the nearest neighbour bonds are weaker here than in the ionic crystals, while contributing strongly to the efg.

We calculate the efg as if unchanged conduction electron clouds were carried adiabatically with the ion cores. This is only to say that we assume all effects, especially the electron redistribution, to be still proportional to the changes of the neighbour ion positions, even though

the total effect on the efg may be large. The same assumption is used by Sternheimer and in (WatG65). Contrary to these analyses, we are not concerned here with the mechanisms which determine the proportionality constants ξ of eq. (3) and q" of eq. (12). These are considered temperature independent. In a Taylor expansion for the displacements $x_{\mu i}$ (μ=1,2,3) of the neighbours i
(10) $q = \xi(q_0 + q_1 + q_2 + \ldots)$ replaces eq. (3) with the linear term averaging to zero and
(11) $q_2 = \langle \Sigma_{\mu\nu}\Sigma_{ij}(\partial^2 q_0/\partial x_{\mu i}\partial x_{\nu j})dx_{\mu i}\, dx_{\nu j}\rangle$. Now we make the drastic assumption
(12) $q_2 \approx q''\cdot\langle x^2\rangle$ and take (13) $\langle x^2\rangle = 3\hbar^2 T/(mk_B\Theta^2)\cdot(\Phi(y) + y/4)$
as in the X-ray studies; $y = \Theta/T$. The temperature dependence of $\langle x^2\rangle$ is thus described in the Debye approximation. We want to stress that this treatment of q_2 takes into account the collective nature of the atomic motions which change q.

Fits have so far been produced for In, see fig. 1. Of the three fit parameters at our disposal, ξ, q", and Θ, we take Θ (isotropic) from the literature, since the fits are not very sensitive to its value. The effective shielding factor turns out to be $|\xi| = 128$. Lattice sums for the second derivatives eq. (11) were studied in some detail for V^c. They are by no means small and the fit parameter $q''/q_0 = 7.3$ Å^{-2} is, therefore, considered reasonable.

The temperature dependence is reproduced rather well for In, but not perfectly. It remains open whether this is due to some other mechanism not comprised in the model (e.g. thermal electron redistribution at E_F) or to the approximations made in the evaluation of q_2. The apparently general $T^{3/2}$-dependence (HeuH74) is not obtainable with the complicated approximations used here.

The fact that different impurities (Cd,Ta) in the host lattice Ti show different behaviour of q vs. T (KauR74b) would in the picutre presented here be related to localized vibrations.

We are obliged to the Deutsche Forschungsgemeinschaft for financial support, and to the Hahn-Meitner-Institut Berlin for use of the Siemens 4004.

Table 1

	q_0	V^c	V^t	V^f	V^h	Experimental q_{exp}		References
Zn q(300)	a	-4.964	-0.959	-2.042	-1.485	^{67m}ZnZn +320 i	^{111m}CdZn +240 j	i BerR74
q(630)/q(300)	b	1.09	1.07	1.05	1.04	0.84 i	0.86 j	j RagR 74
Cd q(300)	c	-4.017	-0.644	-1.699	-1.108		^{111m}CdCd +234 k	k KaR74a
q(470)/q(807	d	1.05	1.02	1.03	1.03		0.84 l	l RagR71
In q(4)	e	-0.581	-0.092	-0.860	-0.631	^{115}InIn +73 m	^{111m}CdIn +45 n	m HewT62
q(400)/q(4)	f	0.72	0.72	0.78	0.78	0.47 m	0.49 n	n HeuH74
Sn q(4)	g	-0.332	-0.128	-1.521	-1.249		^{111m}CdSn +78 p	p KaR74a
q(485)/q(4)	h	2.27	1.32	1.08	1.13		0.69 q	q HeuH74

-Charges used: Zn(2),Cd(2),In(3),Sn(4).

Calculated and experimental efgs, q_0 and q_{exp}, in 10^{13} esu cgs units; q(300) is q(T=300K). To derive q_{exp} from the measured interaction frequencies, estimated values of the quadrupole moments had to be chosen from the references: Q(^{67m}Zn) = 0.2b; Q(^{115}In) = 0.86b; Q(^{115m}Cd) = 0.77b. The lattice constants were taken from: a) B. Ancker: Ann. d. Physik 12 (53) 121, b) H.M. Gilder et al.: Phys. Rev. 182 (69) 771, c) D. Edwards et al.: J. Am. Chem. Soc. 74 (52) 5256, d) J.C. McLennan et al.: Trans. Roy. Soc. Canada III23 (29) 255, e) C.S. Barrett: Adv. Xray Anal. 5 (62) 33, f) J. Graham et al.: J. Int. Met. 84 (55) 86, g) J. Rayne et al.: Phys. Rev. 120 (60) 1658, h) D. Solomon et al.: Phil Mag. 11 (31) 1090.

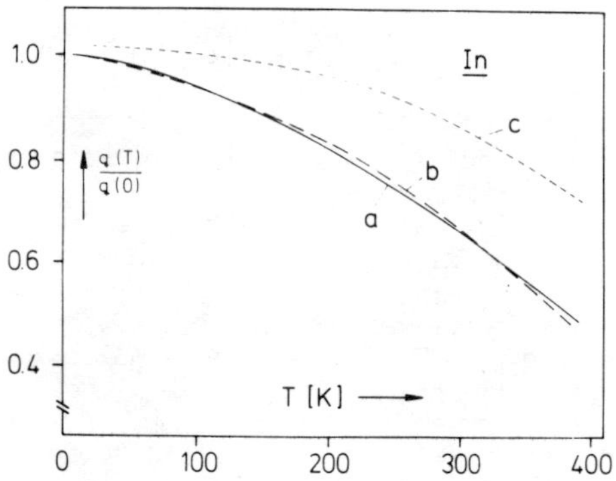

Fig. 1 Temperature dependence of the electric field gradient for In a) experimental data of (HewT62) b) fit curve using Θ =116K c) is the pure lattice sum q_o from the Coulomb potential using the lattice constants vs. T from (SmiS64).

References

Bay 51 H.Bayer: Z. Physik 130 (51) 227
BerR74 H.Bertschat et al.: Phys. Rev. Lett. 32 (74) 18
deW61 F.W.de Wette: Phys. Rev. 123 (61) 103
Har70 W.A.Harrison: Solid State Theory, New York 1970, chap. III, 4.3
HeuH74 P. Heubes et al.: Proc. Int. Conf. Hyperfine Interactions, Uppsala 1974, p. 208
HewT62 R.R.Hewitt and T.T.Taylor: Phys. Rev. 125 (62) 524
KaR74a E.N.Kaufmann et al.: Proc. Int. Conf. Hyperfine Interactions, Uppsala 1974, p. 200
KaR74b E.N.Kaufmann et al.: Phys. stat. sol. b63 (74) 719
KusB56 T.Kushida et al.: Phys. Rev. 104 (56) 1364
MohS72 N.C.Mohapatra et al.: Phys. Rev. Lett. 29 (72) 456
RagR71 R.S.Raghavan and P. Raghavan: Phys. Lett. 36A (71) 313 and Phys. Rev. Lett. 27 (71) 724; J. Bleck et al.: Phys. Rev. Lett. 29 (72) 1371
RagR74 P.Raghavan et al.: J. Phys. F4 (74) L 80
SmiS64 Smith and Schneider: J. Less. Com. Met. 7 (64) 17
WatG65 R.E.Watson et al.: Phys. Rev. A140 (65) 375

IMPURITY NMR IN PARAMAGNETIC NICKEL

Y. Chabre, V. Jaccarino* and P. Segransan
Laboratoire de Spectrométrie Physique†, Université Scientifique et Médicale de Grenoble, BP 53 Centre de tri, 38041 Grenoble Cédex, France.

* On leave from Physics Department, UCSB, Santa Barbara, Calif. 93106, U.S.A.

† Laboratoire associé au C.N.R.S.

Abstract. We report measurements of the V^{51} Knight Shift and relaxation rate $1/T_1$ in Ni alloys with 4 and 1.4 at % V, above T_c. The Knight Shift, as well as the susceptibility exhibit a Curie-Weiss like behavior in the 600 K to 1400 K region. The smallish values of $dK/d\chi$ suggest the local susceptibility and exchange enhancement at the V impurity to be diminished relative to that of the host. $1/T_1$ exhibits an extremely rapid increase with temperature, going from $330 s^{-1}$ at 600 K to $10^4 s^{-1}$ at 1400 K. Such a behavior might be consistent with a transition from itinerant to localized character for the electron gas -- at least in the vicinity of the impurity.

1. Introduction. Much has been done from both a theoretical and experimental point of view to understand the ferromagnetic properties of Fe, Co and Ni. The paramagnetic state of these metals has received considerably less attention experimentally, particularly because of the high transition temperatures T_c of Fe and Co. Ni, which would otherwise be a good candidate for an NMR study, unfortunately has only one isotope, Ni^{61}, in very low natural abundance and with small gyromagnetic ratio. The addition of V to Ni decrease T_c drastically and the relatively high concentration V-Ni alloys have been studied using NMR (Panissod, 1973) in the 300 K and below region, as a result. We have elected to study the very dilute (1-4 at. % V) region above T_c in the hope of using the NMR as a static and dynamic probe of the paramagnetic spin susceptibility. We were, in part, motivated by the relatively recent work of Doniach (1968) and Moriya et al (1973) who have attempted to calculate the dynamical susceptibility of itinerant ferromagnets and relate this to the nuclear spin lattice relaxation rate.

2. Experimental Procedure and Results. The alloys were prepared from 5N Nickel and 3N Vanadium by high frequency melting on a cooled silver plate, ground with alumina paper and sieved to 40 µ size. After mixing with thorium oxide powder to allow both better r.f. penetration and to avoid sintering at high temperatures, each sample was inserted into a silica tube and kept under vacuum during experimental NMR runs. The NMR measurements were performed at 10 MHz with a coherent pulsed spectrometer, using a boxcar for Fourier transformation in the Knight Shift studies. Fig. 1 shows the Knight Shift vs. temperature results for the 4 and 1.4 at. % samples. The solid curves are the measured susceptibilities that were made on parts of the same ingots used for the NMR studies. All of the Knight Shift data may be represented in Curie-Weiss like fashion : $K = K_\infty + A_K/(T-T_c)$; with $K_\infty = (0.52 \pm 0.02)$ % and $A_K = -(0.92 \pm 0.1)$ [% K], $T_c = (425 \pm 30)$ K for the 4 at. % and $K_\infty = (0.50 \pm 0.02)$ % and $A_K = -(0.69 \pm 0.07)$ [% K], $T_c = (635 \pm 40)$ K for the 1.4 at. % sample. In Fig. 2 the spin-lattice relaxation rates $1/T_1$ vs. temperature for the 4 at. % samples are plotted. Although only a few $1/T_1$ measurements were made on the 1.4 at. % samples they were in agreement with the 4 at. % results. Perhaps the most striking feature of the data is the very rapid increase of $1/T_1$ with T reaching $10^4 s^{-1}$ at 1400 K.

At low temperatures $1/T_1$ is asymptotic to a Korringa-like value of 1.8 s.K. In fact, the entire $1/T_1$ data may be expressed as the sum of a Korringa rate and one associated with a thermally activated process, with the latter having an activation energy $\Delta E = 0.6 \pm 0.1$ eV.

3. Discussion. A two band (s and d) model has been widely used to interpret and relate the various contributions to the Knight Shift (K) and susceptibility (χ) of the pure transition metals (Clogston et al, 1964) :

$$K = K_s + K_d(T) + K_{orb} \qquad \chi = \chi_s + \chi_d(T) + \chi_{orb} + \chi_{dia}$$

with only the d spin paramagnetism χ_d (T) assumed to be temperature dependent. Thus a K vs. χ plot, with T the implicit parameter, should be linear, with the slope a measure of the hyperfine field per spin of the d electrons. Obviously, in the present case, where one compares the Knight Shift of the impurity K_{imp} with the total susceptibility χ_t, a plot of $K_{imp}(T)$ vs. $\chi_t(T)$ need not be linear. Indeed, a departure from linearity is observed at the lower temperatures. Nor can one interpret the slope $dK_{imp}/d\chi_t$ in the simple manner as is done for the pure metals, since both the local d-density of states varies in the vicinity of the impurity as does the exchange enhancement. Nevertheless the values of $dK_{imp}/d\chi_t$ found from the linear portion of the plots ($H_{hf} = -45 \pm 5$ and -35 ± 5 kG/spin for the 4 and 1.4 at. %, respectively) are of some interest. They indicate a strong decrease in the local spin susceptibility of the impurity relative to that of the host since the characteristic h.f. fields, due to core polarization, associated with V are of the order of $H_{hf} = -250$ kG/spin.

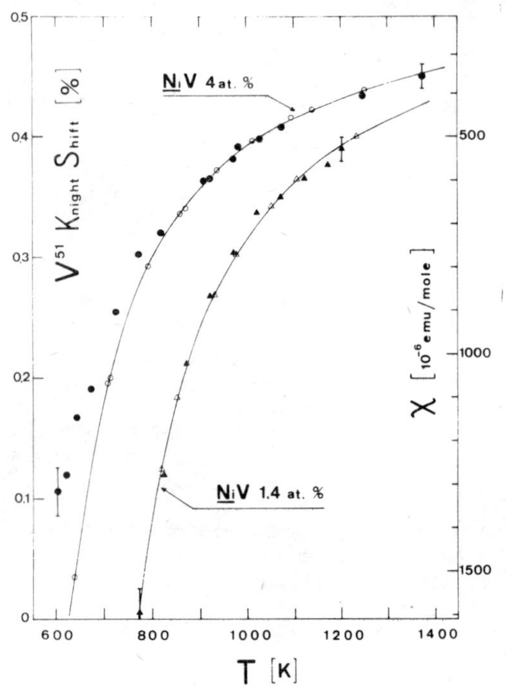

Fig. 1 : V^{51} Knight Shift vs. temperature in Ni V 1.4 at. % (▲) and 4 at. % (●). The solid curves are the respective χ data (Δ, ○) fitted with Curie-Weiss laws.

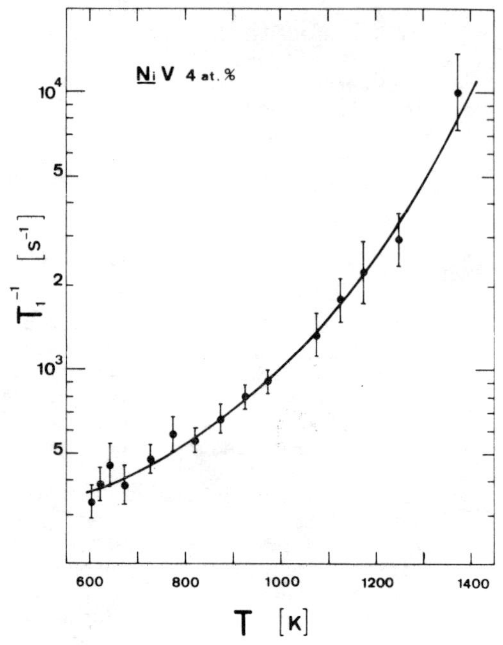

Fig. 2 : Relaxation rate vs. temperature in a Ni V 4 at. % sample.

As to $1/T_1$, the exponential dependence on temperature might, at first, suggest a quadrupolar induced mechanism caused by diffusing impurities such as oxygen or nitrogen. However, such a contribution would exhibit a maximum when the "jump" and Zeeman frequencies are of the same order. According to the diffusion data available on interstitial migration this maximum would appear below 1000 K, as is observed for pure V (Segransan et al). If we are to interpret the relaxation in terms of the dynamical susceptibility $\chi(q, \omega)$ we know that for the pure metals

$$T_1^{-1} = \gamma^2 \, kT \sum_q A_q^2 \, \frac{\chi''_\perp(q, \omega)}{\omega}$$

The fact that $1/T_1$ increases exponentially with T suggests that we are observing a transition from a region where mainly only the low q components of $\chi(q, \omega)$ are strongly exchange enhanced to one where the large q ones are as well. This might come about if either localization manifests itself in going from degenerate to non-degenerate character in the electron gas or insofar as a true localized state, associated with the impurity level, exists some half an eV or so above E_F. Clearly further experiments on this and related systems are necessary to explain this intriguing result.

4. Acknowledgements. We would like to thank Dr. P. Averbuch for many stimulating ideas and discussions. One of the authors (V.J.) wishes to acknowledge the support of a Guggenheim Memorial Foundation Fellowship.

5. References.

Clogston, A.M., Jaccarino, V., Yafet, Y., 1964, Phys. Rev., 134 A, 650.

Doniach, S., 1968, J. Appl. Phys., 39, 483.

Moriya, T., Kawabata, A., 1973, J. Phys. Soc. Jap., 34, 639 and 35, 669.

Moriya, T., Ueda, K., J. Phys. Soc. Jap., to appear.

Panissod, P., 1973, Solid State Com., 13, 575.

Segransan, P., Chabre, Y., to be published.

NUCLEAR MAGNETIC RESONANCE IN A SERIES OF SUPERCONDUCTING $V_{1-x}Pt_x$ COMPOUNDS

L.A.G.M. Wulffers and N.J. Poulis
Kamerlingh Onnes Laboratorium der Rijksuniversiteit, Leiden, The Netherlands

Abstract. Measurements of the Knight-shift of ^{51}V and ^{195}Pt in a series of $V_{1-x}Pt_x$ compounds show that the $N(E_F)$ depends on the concentration x of the Pt. This agrees with the observed behaviour of the T_c as a function of the concentration.

1. Introduction. The Knight-shift of ^{51}V and ^{195}Pt were measured in a series of $V_{1-x}Pt_x$ compounds, where x varies from 20% to 30%. Measurements of the superconducting transition temperature T_c show that the T_c is strongly dependent on the Pt concentration x. (Flükiger, 1972). This can be explained by assuming that the concentration of Pt influences the density of states at the Fermi level $N(E_F)$. When the concentration of Pt decreases from 30% to 25% the number of Pt atoms on the V chains will decrease, which will raise the $N(E_F)$ (Labbé and van Reuth, 1970). As is known the $N(E_F)$ has a maximum when all the V atoms are well-ordered on the chains. Going from x = 25% to 20% the T_c decreases which indicate that the $N(E_F)$ also decreases. Probably the V chains of the A-15 structure loose their one dimensional character which will lower the $N(E_F)$. Because the $N(E_F)$ will also influence the NMR properties we measured the ^{51}V and ^{195}Pt Knight-shift and nuclear spin-lattice relaxation time T_1 as a function of the Pt concentration.

2. Knight-shift measurements.

2a. Vanadium Knight-shift. The Knight-shift of the V can be written as a sum of the following terms: K_s, K_d, K_{orb} and K_{dip} (Ancher, 1971). The K_s and K_d are isotropic; the others are anisotropic. The Knight-shift of V can thus be written as $K_V = K_{iso} + K_{ax}$. The K_{iso} is the sum of the isotropic terms K_s and K_d and the isotropic parts of K_{orb} and K_{dip}. K_{ax} is the sum of the anisotropic parts of K_{orb} and K_{dip}. Our measurements show that the K_{ax} remains constant over the concentration range and has a value of K_{ax} = 0.02%. It is known that the K_d has a negative sign and is a function of the $N(E_F)$ (Ancher, 1971).

The K_{iso} decreases going from x = 20% to 25%, from K_{iso} = 0.55% to K_{iso} = 0.52% and then increases again to K_{iso} = 0.55% going from x = 25% to x = 30%. This is in agreement with the T_c measurements, as an increasing $N(E_F)$ will cause an increasing K_d. As K_d has a negative sign, this will result in a decreasing K_{iso} when $N(E_F)$ increases.

2b. Platinum Knight-shift. The K_{Pt} shows a similar behaviour. From x = 20% to x = 25% the K_{Pt} decreases from K = +0.08% to K = -0.90% and increases then to K = -0.80% when x goes to x = 30%. As the K_{Pt} is mostly negative it is assumed that the K_d is the most important term in the K_{Pt}. This term shows the same behaviour as the K_d term of K_V. However the change in K_{Pt} from x = 20% to x = 25% is very strong compared to the change in K_V in the same concentration range. The dependence of $N(E_F)$ on the concentration as found in the Knight-shift measurements of ^{51}V and ^{195}Pt is also reflected in the behaviour of the spin-lattice relaxation time as a function of the concentration.

3. Superconducting state. The temperature dependence of the ^{51}V and ^{195}Pt Knight-shift of the samples V_3Pt and $V_{.76}Pt_{.24}$ is measured in the superconducting state. After correction for the effects of the demagnetization caused by the Meissner effect, the K_V shows a slight temperature dependence in the superconducting state. As a result of the theories of the Knight-shift in the superconducting state (Ancher, 1971) this leads us to the conclusion that the terms K_s and K_d, which are the terms that are expected to be influenced by the superconducting state, are nearly equal in magnitude. The K_{Pt} changes from -0.9% at T_c to -0.8% at T = 0 what is much less than expected. Probably spin-orbit effects will diminish the expected disappearing of the negative K_d term.

4. Effects of heattreatment. As is known from the work of Flükiger (Flükiger, 1972) it is possible that samples with x < 25% are not total homogeneous. This agrees with the observed weaker additional Pt lines in these samples as the K_{Pt} is very sensitive for changes in x in that concentration range. So all samples were annealed at a temperature of 900° C in the hope that the samples became better ordered and more homogeneous. The results of our K_{Pt} measurements showed that this failed. A second fase ($AuCu_3$ fase) was formed in the samples. The remaining part with A-15 structure of each sample turns out to have the stoichiometric ratio, cause now the main Pt resonance line of the samples has a K_{Pt} = -0.9%. Besides some additional Pt resonance lines were observed. T_c measurements on the annealed samples are in agreement with this result, cause the samples $V_{.78}Pt_{.22}$ $V_{.76}Pt_{.24}$ and V_3Pt have now the same T_c = 3.1 K.

5. References

Ancher, L.J., 1971, thesis, University of Leiden.
Flükiger, R., 1972, thesis, University of Geneva.
Labbé, J. and Van Reuth, E.C., 1970, Phys. Rev. Letters 24, 1232.

NMR AND MOSSBAUER STUDY OF ELECTRON AND PHONON SPECTRA IN THE SUPERCONDUCTING V_3X COMPOUNDS*

F. Y. Fradin and C. W. Kimball
Argonne National Laboratory, Argonne, Ill. 60439 and Northern Illinois Univ., DeKalb, Ill. 60115

Abstract. Results of ^{51}V nmr measurements of spin-lattice relaxation rates and nuclear-electric-quadrupole interactions, and results of ^{119}Sn Mössbauer-effect measurements of the recoil-free fraction and thermal shift in vanadium-based A-15 compounds are discussed in terms of the electron and phonon spectra.

From the measured normal-state values of the ^{51}V spin-lattice relaxation rates and nuclear electric-quadrupole interactions in the pseudobinary compounds $V_3Ga_{1-x}Si_x$ and $V_3Ga_{1-x}Sn_x$, the bare electronic density of states at the Fermi level N(0) has been calculated using a tight-binding formalism (Fradin and Zamir, 1973; Fradin and Williamson, 1974). In the A-15 compounds with high-superconducting transition temperature T_c, N(0) is dominated by the π-symmetry d-subband, in qualitative agreement with the band structure calculations of Weger and Goldberg (1973).

In the $V_3Ga_{1-x}Sn_x$ compounds, the ^{119}Sn thermal shift $\delta(T)$ and recoil-free fraction $f(T)$ yield direct measurements of the thermal averages over the phonon spectrum $\langle\omega\rangle$ and $\langle\omega^{-1}\rangle$, respectively, in the harmonic approximation. In the compositions of low x, T_c approaches 15°K and there is evidence of phonon softening at low temperatures (Kimball et al., 1974). The temperature dependencies of $\delta(T)$ and $f(T)$ have been analyzed in a quasiharmonic approximation with a two Debye model for the phonon spectrum; the results are consistent with the sum rule on $\langle\omega\rangle$. The concentration dependence of $\langle\omega\rangle/\langle\omega^{-1}\rangle$ indicates a weaker dependence of the phonon only part of the electron-phonon coupling parameter λ than that derived from McMillan's (1968) equation for T_c.

It has been found (Fradin and Williamson, 1974) that the superconducting transition temperature normalized by the average Debye temperature T_c/θ_D is a universal function of N(0). T_c can be fit to the McMillan (1968) equation with λ given by

$$\lambda = \frac{\lambda_o}{1 - BN(0)} \tag{1}$$

and $\lambda_o = 0.36$, B = 0.015 Ry V-spin.

A model density of states has been constructed that agrees with the shape of $N(\varepsilon)$ in the vicinity of ε_F for V_3Ga (Weger and Goldberg, 1973) and is in agreement with the nmr results. The density of states is given by

$$N(\varepsilon) = \frac{2}{3} N_0\, e^{-(\varepsilon - \varepsilon_o)^2/(200°K)^2} + \frac{1}{3} N_0 \quad \text{for } \varepsilon \geq \varepsilon_o \tag{2}$$

$$N(\varepsilon) = N_0 \quad \text{for } \varepsilon \leq \varepsilon_o$$

where $\varepsilon_F = \varepsilon_o$ for V_3Ga. This density of states has been used to calculate the temperature dependence of the chemical potential μ and in the constant matrix element approximation the temperature dependence of T_1T, the spin susceptibility χ, and the electronic specific heat coefficient

$$\gamma = \frac{1}{T}\frac{\partial}{\partial T}\int_0^\infty \varepsilon N(\varepsilon)\, f(\varepsilon,\mu,T)\, d\varepsilon. \tag{3}$$

See Fig. 1. The temperature dependencies of χ and T_1T are in agreement with experiment for $T > T_c$. The temperature dependence of γ is consistent with high-temperature specific-heat results (Fradin et al. 1974). The increase of μ with T, $\mu(300°K) - \mu(0°K) = 124°K$, predicts a small positive increase of the isomer shift with temperature.

The model density-of-states calculations suggest that the phonon renormalization due to electron screening should yield a temperature-dependent phonon self-energy, yielding temperature-dependent phonon frequencies. This is in agreement with the Mössbauer measurements. According to McMillan (1968), the electron mass enhancement due to electron-phonon coupling is given by

$$\lambda = \frac{N(0)\langle I^2\rangle}{M\langle\omega\rangle_o/\langle\omega^{-1}\rangle_o} \,. \tag{4}$$

where $\langle I^2\rangle$ is the double average over the Fermi surface of the electron-scattering matrix element. The form of Eq. (1) follows from Eq. (4) if on average the phonon moments are given by

$$\frac{\langle\omega^{-1}\rangle_o}{\langle\omega\rangle_o} \approx \frac{1}{\langle\omega^2\rangle_o} \approx \frac{1}{\langle\Omega^2\rangle[1 - BN(0)]} \tag{5}$$

where $\langle\Omega^2\rangle$ is the average of the square of the bare phonon frequencies and we have used a simplified form for the average T = 0° phonon renormalization.

*Based on work performed under the auspices of the U. S. Atomic Energy Commission and the National Science Foundation.

References

Fradin, F. Y., Knapp, G. S., and Kimball, C. W., 1974, Solid State Commun. 14, 89.

Fradin, F. Y. and Williamson, J. D., 1974, Phys. Rev. B, to be published.

Fradin, F. Y. and Zamir, D., 1973, Phys. Rev. B, 7, 4861.

Kimball, C. W., Taneja, S. P., Weber, L., and Fradin, F. Y., 1974, Mössbauer Methodology, Vol. 9, (ed. I. Gruverman and C. Seidel, Plenum Press, N. Y.) to be published.

McMillan, W. L., 1968, Phys. Rev. 167, 331.

Weger, M. and Goldberg, I. B., 1973, in Solid State Physics (ed. H. Ehrenreich, F. Seitz, and D. Turnbull, Academic Press, N. Y.) 28, 1.

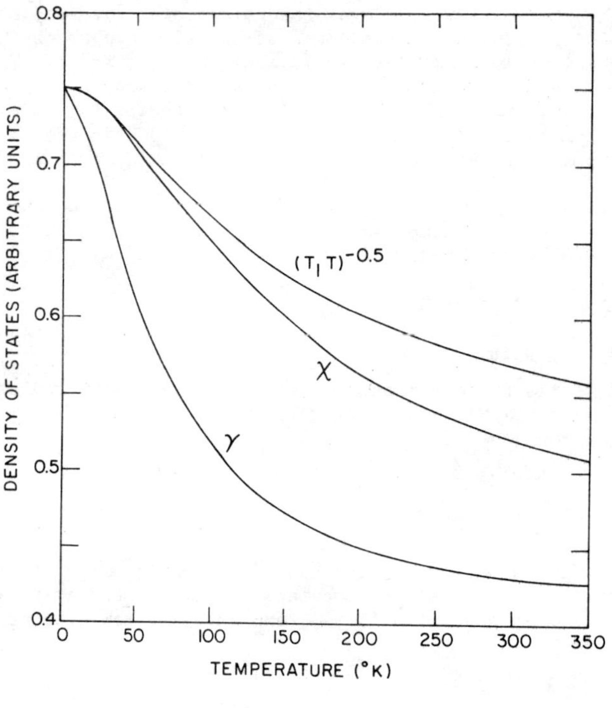

Figure 1. Calculated temperature dependencies of the electronic specific-heat coefficient γ, the spin susceptibility χ, and the spin lattice relaxation time T_1 for V_3Ga.

NUCLEAR MAGNETIC RESONANCE IN DILUTE ALLOYS OF VANADIUM IN URANIUM

L.E. Drain & W.A. Hines*
Materials Physics Division, A.E.R.E., Harwell, Didcot, Oxon. OX11 ORA, England.

Abstract The observation of the nuclear magnetic resonance of vanadium in solid solution in uranium is reported. The line width and structure of the resonance is attributed to a combination of quadrupole interaction and Knight shift anisotropy and the observation of the resonance at 4.2°K indicates that a transition to a magnetically ordered state has not occurred.

1. Introduction There is considerable evidence for a transition of some kind in α-uranium at about 43°K. Anomalies have been observed in a number of physical properties such as expansion coefficients (Marples, 1970), elastic constants (Fisher & Dever, 1968) and specific heat (Crangle & Temporal, 1973). As there is apparently no change in crystal structure, the transition is presumably electronic and the possibility of antiferromagnetic ordering has been suggested. The transition does indeed show some similarity to the Néel point of chromium at 311°K. But neutron diffraction measurements (Lander & Mueller, 1970) have given no evidence for antiferromagnetism. Magnetic resonance techniques are more sensitive to small magnetic moments and could be helpful in establishing whether or not magnetic ordering does occur. In a polycrystalline sample, static internal magnetic fields produced by ordered magnetic moments are randomly oriented with respect to an externally applied field. This gives a spread in field values at the resonant nuclei, line broadening and consequent loss in signal strength. Internal fields of a few gauss would be sufficient to produce an appreciable effect on a normal nuclear resonance signal.

Uranium itself is not very suitable for magnetic resonance investigations as the ^{238}U isotope has zero spin and the ^{235}U resonance has not been observed owing presumably to a large quadrupole interaction. The use of the nucleus of another element in solid solution was therefore considered. An element having appreciable solid solubility and a strong nuclear resonance is required. A U-Pt alloy was tried without success but it was found possible to observe the ^{51}V resonance from a solid solution of vanadium in uranium and the results are presented here. It was hoped that the concentrations of impurity used, 1.0, 0.5 and 0.2 atomic %, were sufficiently low that the transition of uranium would not be inhibited. Similar measurements (Drain, 1962) (Barnes & Graham, 1965) on dilute Cr-V alloys show a disappearance of the ^{51}V resonance below the Neel point.

2. Experimental Measurements were made at room temperature, liquid nitrogen and helium temperatures with a crossed coil spectrometer suitable for low temperature work described elsewhere (Drain, 1972). To achieve penetration of the sample by the radio-frequency field, foil samples were used. The alloy ingots were homogenized at the 900°C for 2 days, quenched and rolled to 50μm thickness. The foils were annealled at 650°C cut and stacked with 12 μm thick Mylar or Melinex film as insulation. About 100 layers were needed to fill the sample space. A powder sample of one alloy (1.0 at.%V) was also prepared but did not prove satisactory. Filing the sample was tedious as this had to be done in an inert atmosphere to prevent oxidation and the sample became contaminated with ferromagnetic impurity. Although the ^{51}V was observed in this sample, it was very broad and weak.

To observe the weakest signals satisfactorily, the technique of signal averaging was employed. A Laben multichannel analyser was adapted for this purpose using suitable analogue to digital conversion of the signal.

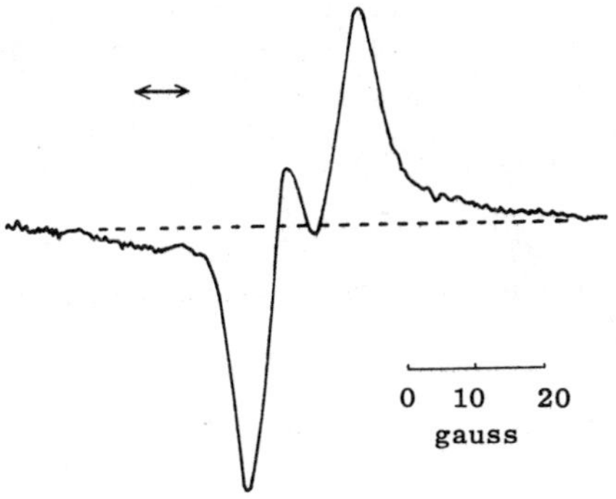

Figure 1 Trace of resonance derivative from U-1.0 at % alloy at 12.2 MHz, 4.2°K

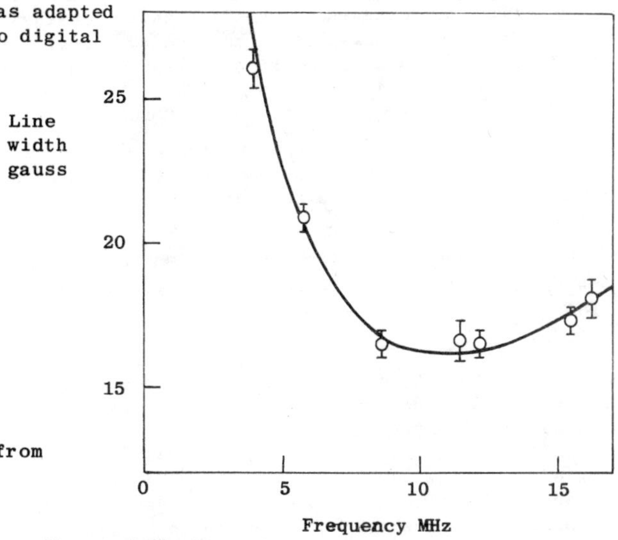

Figure 2 The dependence of line width on frequency ^{51}V resonance U-1.0 at% V alloy 4.2°K

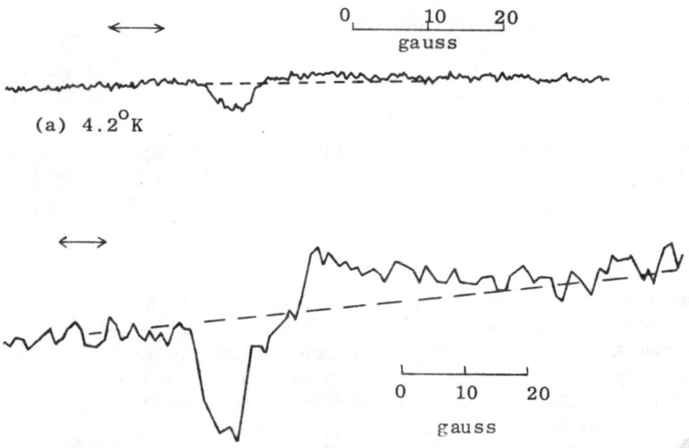

Figure 3 ^{51}V resonance derivative signals from a U-0.2 at.% V alloy at a frequency of 12.2 MHz. Curve (b) was signal averaged for 11 hours.

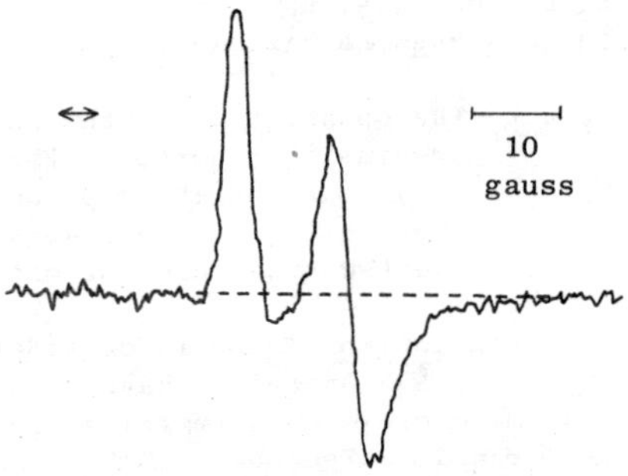

Figure 4 ^{51}V resonance from a U-0.5 at % V alloy, 4.2°K 12.2 MHz.

3. Results The best signals were obtained from the alloy containing the highest concentration of vanadium (1.0 at.%). The resonance from this alloy was studied in the most detail. As will be seen from Figure 1, the resonance shows a structure suggesting the powder pattern due to second order quadrupole perturbation. The variation of line width with field (figure 2) is however unexpectedly small. It was necessary to reduce the field below 5 kilogauss before an appreciable increase in line width was observed. The behaviour can be explained by a combination of quadrupole interaction and anisotropic shift and the curve drawn in Figure 2 is a theoretical one using the analysis of Jones et al. (1963). The assumption of axial symmetry has been made. Though this is not justified by crystal structure considerations, judging by the experimental results it does seem to be a reasonable approximation. We deduce a quadrupole coupling paramter ν_Q = 187 KHz. (e^2qQ/h = 2.62 MHz). and a Knight shift anisotropy K_{ax} = -0.029% ($K_\perp - K_\parallel$ = 0.088%). The isotropic Knight shift relative to $NaVO_3$ solution was 0.57± 0.01%.

The resonance was also observed at 77°K and 300°K at 15.5 MHz and appeared to have the same form as at 4.2°K though the structure could not be so well resolved owing to the expected reduction in signal to noise ratio. It is clear that no transition involving magnetic ordering occurred in the temperature range studied. However it can be argued that the addition of 1% vanadium has significantly affected the behaviour of uranium and in attempt to meet this objection, the observation of the resonance in more dilute alloys was attempted. Good signals were obtained from a 0.5 at.% alloy at 4.2°K (figure 4). The resonance shapes were similar to those from the 1% alloy with a somewhat better resolution of the structure. The signals from the 0.2 at.% alloy at 4.2°K were however disappointing, (Figure 3a), though resonances observed at 77°K and 300°K by signal averaging had about the expected amplitudes. The problem at 4.2°K seems to be due to the skin effect. It was noticed that the noise level was also reduced in these measurements apparently due to a loss of Q in the receiving circuit. If we assume that the sample has as small a residual resistivity as low as that observed in some samples of nominally pure uranium (0.7μΩcm) (Arajo & Colvin, 1964), the skin depth would be 14μm. and clearly the foil used would be too thick. Unfortunately, attempts to roll a foil of about 25μm thickness were unsuccessful. If thinner foils could be produced, it should be possible to detect the ^{51}V resonance at a concentration of 0.1 at.% or even less.

4. Conclusion The observation of the ^{51}V resonance at 4.2°K in uranium containing 0.2 to 1.0 at.% vanadium shows that magnetic ordering does not occur in these alloys. It seems probable that the same is true of pure uranium and the transition at 43°K must have some other explanation.

5. Acknowledgements The authors wish to thank Mr B.C. Moss for help with the construction and operation of equipment and the Argonne National Laboratory for the preparation of some samples.

6. References

Arajo S. & Colvin, R.V. 1964, J. Less Common Metals 7 54-65
Barnes, R.G., & Graham, T.P., 1965, J. Appl. Phys. 36, 938-9
Crangle J. & Temporal J., 1973, J. Phys. F. 3, 1097-1105
Drain, L.E. 1962 J. Phys. Rad. 23 745-9
Drain, L.E. 1972 Rev. Sci. Inst. 43 1648-50
Fisher, E.S. & Dever, D., 1968 Phys. Rev. 170 607-13
Lander, G.H. & Mueller, M.H. 1970 Acta Crystallographica B26 129-36
Marples, J.A.C. 1970 J. Phys. Chem. Solids. 31 2421-30

*Now at Physics Department, The University of Connecticut, Storrs, Connecticut, USA.

CADMIUM ALLOYS BAND STRUCTURE PECULIARITIES FROM THE NMR DATA

V.V. ZHUKOV, I.V. SVECHKAREV, Physico-Technical Institute of Low Temperatures, Kharkov, 310086,USSR, R.V. KASOWSKI, E.I. Du Pont de Nemours, Wilmington, Del., USA, H.E. SCHONE, Phys. Dep. College of William and Mary, Williamsburg, Va., 23185, USA.

The Knight shift components in Cd-Hg and Cd-Mg alloys have been investigated in a wide range of concentrations and temperatures on the nuclei of the matrix (Cd^{113}, Cd^{111}) and imputity (Hg^{199}). The NMR frequency shift high sensitivity to various influences in the hexagonal α-phase of Cadmium alloys is the specific result of their band structure, in the liquid phase of these alloys the concentration and temperature dependences being absent; in the tetragonal ω -phase these dependences are weaker; the effects on different types of nucleis are quite similar. The results of the band structure and Knight shift calculation in pure Cd with model nonlocal pseudopotential were used for examining shift behaviour in alloys. The empirical analysis gives a satisfactory agreement with the experiment if the spectrum smearing due to the disordered impurities distribution is considered.

SPIN-ECHO MEASUREMENTS OF CONDUCTION ELECTRON SELF-DIFFUSION COEFFICIENT IN METALLIC LITHIUM.

Cherkasov F.G.,Kessel A.R.,Kharakhashyan E.G.,Zhikharev V.A.,Medvedev L.I. and Yudanov V.F.*

Physical-technical Institute, Kazan, 420029, USSR
*Institute of Chemical Kinetics and Combustion, Novosibirsk, 630090 , USSR.

ABSTRACT Measurements of conduction electron spin-echo signal decay in linear gradient of magnetic field was made in metallic lithium plate. The magnitude of conduction electron diffusion coefficient was extracted. The dependence of the signal damping upon the angle between gradient direction and plate plane was considered.

1. Introduction. It is well knoun that the orbital motion of the conduction electron during the time intervals considerably surpassing the "transport relaxation time" may be looked upon as diffusion with the characteristic self-diffusion coefficient $\mathcal{D} = 1/3\, v_F \lambda$ where v_F -the Fermi velocity, λ - the mean free path. The spin-echo method (Hahn, 1950) is considered to be one of the most effectiv methods of diffusion coefficient measurements. For this purpose the additional spin-echo signal damping in magnetic field gradient is investigated. In the present article there are given the results of the first - as far as the authors know - application the conduction electron spin-echo method (CESE) for direct measurements of diffusion coefficient in metal.

2. The experimental results. In the CESE experiments on the small spheroidal particles of lithium metal carried out earlier (Zhikharev et al.,1973) the influence of diffusion was limited by the small particle size (the diameter was equal to 10^{-4} cm.). In present work the new technik, based on the high-temperature electrolysis in solid alkali-haloid monocrystalline (Kurchatov,1932) was used for preparation of the lithium samples of extreme purity. The application of such technik leads to the formation of large ($\sim 10^{-2}$ cm.) connection of metal, "closed" in airproof container made of its pure salt. In preliminary stressed monocrystalline LiF it was possible to get the samples in which the metal forms flat plate. The measurements was made on similar lithium plate (the area ~ 0.1 cm^2, the thickness $\sim 1.5 \cdot 10^{-3}$cm.) at room temperature with the help X-band relaxometer using two pulse method. To extracte the magnitude of diffusion coefficient the stationary linear gradient G = (15 $\pm$ 3) gauss/cm. was put on the sample in the direction of the static magnetic field. The measurements was taken at different angles between the gradient direction and the plate plane. The experimental data are plotted on Fig. 1.

3. Discussion. The theoretical description of CESE in the present work is based on the Bloch equation with stochastic parameters (Zhikharev,Kessel, 1974) When averaging the solution of Bloch equation over the trajectories of conduction electron one should take into account the finitness of the sample sizes (Robertson,1966). In the case of angle θ between gradient direction and

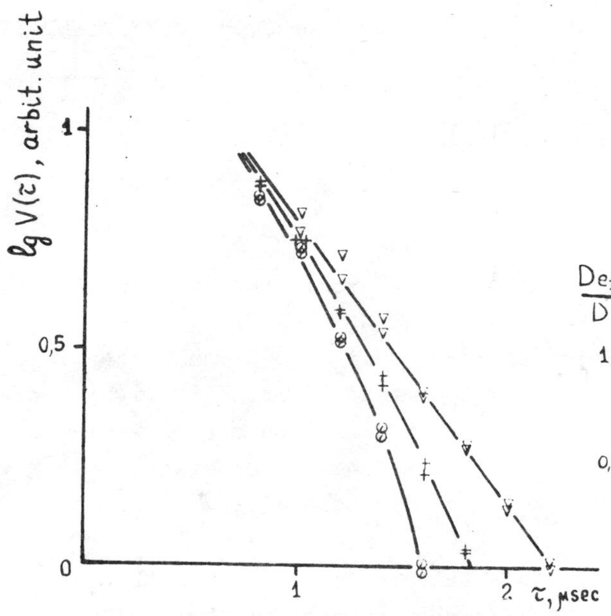

Fig. 1. Dependence of CESE signal upon the interval between pulses τ.
G = 0 ∇ - for all θ.
G = 17 gauss/cm ⊘ - θ = 0°
\+ - θ = 45°
∇ - θ = 75°

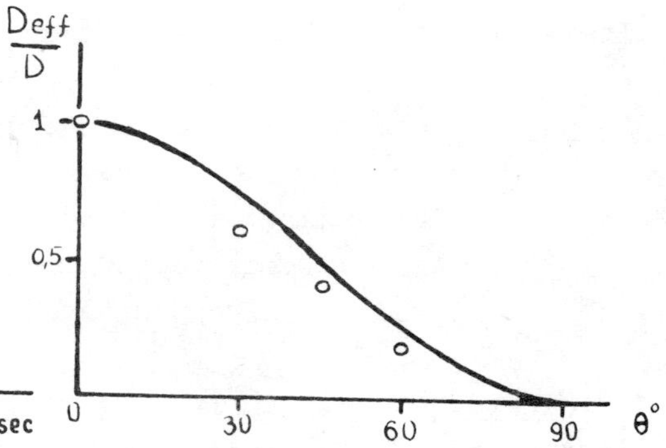

Fig. 2. $\mathcal{D}_{eff}$ as function of angle $\theta°$
o - experimental results
solid line - $\cos^2\theta$ -function.

plate plane one obtaines the expression for the amplitude of echo signal:

$$V(2\tau) \propto \exp\{-2\tau/T_v - 2\gamma^2 G^2 [A(\tau,d_1)\sin^2\theta + A(\tau,d_2)\cos^2\theta]\}$$

$$A(\tau,d) = \sum_{n=0}^{\infty} \frac{8d^4}{\pi^6(2n+1)^6}\left[2\tau - \frac{d^2}{\mathcal{D}\pi^2(2n+1)^2}\left(3 - 4e^{-\frac{\pi^2\mathcal{D}}{d^2}(2n+1)^2\tau} + e^{-\frac{\pi^2\mathcal{D}}{d^2}(2n+1)^2 2\tau}\right)\right] \quad (1)$$

where d_1 and d_2 are the thickness and linear size of the plate, T_v - bulk relaxation time, τ is the interval between pulses. In our case $\frac{d_2^2}{\pi^2\mathcal{D}} \gg \tau \gg \frac{d_1^2}{\pi^2\mathcal{D}}$. Then

$$A(\tau,d_1) = \frac{1}{60}\frac{d_1^2\gamma^2G^2}{\mathcal{D}}\tau \ ; \quad A(\tau,d_2) = \frac{1}{12}\gamma^2G^2\mathcal{D}\tau^3 \quad (2)$$

The linear upon τ term is in our case small and may be omitted in (1). The experimental results plotted on Fig.1. may be phenomenologically described by $V(2\tau) = \exp\{2\tau/T_v + 2/3\,\gamma^2G^2\mathcal{D}_{eff}\tau^3\}$. The angular dependence of $\mathcal{D}_{eff}$ is shown on Fig.2 and it coincides with $\cos^2\theta$ -function, wich predicted by Ex.(2). With Ex. (1) one may obtaines the magnitude of conduction electron diffusion coefficient in lithium at room temperature: $\mathcal{D}$ = (17 ± 5) cm^2/ sec.

Thus the principal possibility of using of the spin-echo method for measurements of conduction electron orbital motion parameters is demonstrated on the example of lithium metal. In other metals the low temperature is required to observe the CESE signal. Note: under these conditions the diffusive character of electron motion may be broken and the echo damping dependes upon $\vec{v}_F$.

4. References.

Hahn, E.L. 1950, Phys. Rev., 80, 580,
Kurchatov,I.V., K.D.Sinelnikov,A.K.Valter,O.N.Trapeznikova 1932,Phys.Z.Sov,1,337
Robertson,B. 1966, Phys.Rev. 151, 273.
Zhikharev V.A.,A.R.Kessel,E.G.Kharakhashyan,F.G.Cherkasov,1973,Zh.Eksp.Teor.Fiz 64, 1356.; Zhikharev V.A.,A.R.Kessel,1974, Zh.Eksp.Teor.Fiz, to be published.

COPPER NUCLEI RELAXATION IN DILUTE CuMn ALLOYS DOWN TO 25 mK.

D. Bloyet[†], P. Piéjus[†], E.J.A. Varoquaux[†] and H. Alloul[*].

[†] Institut d'Electronique Fondamentale, Université Paris Sud, 91405 Orsay, France.
[*] Laboratoire de Physique des Solides, Université Paris Sud, 91405 Orsay, France.

Abstract. The shape of the magnetization recovery of copper nuclei in dilute CuMn alloys has been found to vary from an exponential form at high temperature to an $\exp-(t/\tau)^{1/2}$ form at very low temperature, characteristic of a regime where the spin diffusion is blocked.

1. Introduction. We have studied experimentally the longitudinal relaxation rate of the host nuclei of dilute Cu Mn alloys with 12,20 and 40 ppm of manganese in a temperature range extending from 2 K down to 25 mK. The magnetization recovery in time is observed by means of a series of $\frac{\pi}{2}, \frac{\pi}{2}$ pulse sequences. The measurement are performed at a frequency of 1.5 MHz ($H_o \simeq 1330$ Oe) using a highly stable NMR spectrometer developped for thermometry purposes (D. Bloyet et al. 1972). Heating effects during the sequence at the lowest temperatures have been checked to be negligible.

At temperatures much higher than the temperature T_i at which the impurities fall into an ordered magnetic state, our results confirm those obtained by Alloul et al (1974) in a different experimental set-up, on the same samples. At lower temperatures, i.e. for values of the ratio T/c smaller than 10 (T in mK, c in ppm), the shape of the magnetization recovery with time departs from the exponential behaviour $\exp-(t/T_{1ob})$ to reach a limiting functional form $\exp-(t/\tau)^{1/2}$ below T_i.

2. High temperature regime. In the high temperature regime, the observed relaxation is exponential. It arises from the combined actions of the conduction electrons, which yield a Korringa constant $T_{1K}\,T = 1.25$ sec.K for Cu^{63} in pure copper (with a small admixture of Cu^{65} in the signal) and of the impurities

$$(1)\qquad \frac{1}{T_{1ob}} = \frac{1}{T_{1imp}} + \frac{1}{T_{1K}}$$

The impurity contribution $(T_{1imp})^{-1}$ is found to be proportional to the impurity concentration c and to be essentially independent of temperature which is, with $H_o = 1.33$ kOe,

$$(2)\qquad \frac{T}{H_o}\,\frac{c}{(T_{1imp})^{-1}} = (40 \pm 4) \quad T \quad \left[\frac{\text{K.sec.ppm}}{\text{kOe}}\right]$$

in agreement with the work of Alloul et al (1974). This brings further support to the conclusion of these authors, based on the magnitude of the effect and on the lack of temperature dependence, that the relaxation occurs by a coupling of the copper spins through the RKKY electronic density oscillations as proposed by BGS (Benoit et al 1963). The corresponding value of the transverse electronic relaxation time is

$$(3)\qquad \tau_2\,T = \quad (g) \quad 10^{-12}\ [\text{sec.K}]$$

3. In the ordered state. The ordering temperatures among the magnetic impurities is given in low fields by $T_i \simeq (1.87\,c + 6.6)$ (T_i in mK, c in ppm) according to Hirschkoff et al (1971). At higher fields - 1330 Oe in the present work - the transition is expected to be broadened and the value of T_i slightly raised.

In the ordered state, we have observed a magnetization recovery going, for long times, as $\exp-(t/\tau)^{1/2}$. More precisely, the shape of the recovery has been fitted to better than 0.5%, using a computer, to the expression proposed by Mc Henry et al (1972)

$$(4)\qquad P(t) = \exp\{n_o c - n_o c \exp\left[-t/\pi\tau(n_o c)^2\right] - (t/\tau)^{1/2}\,\mathrm{erf}\left[t/\pi\tau\,(n_o c)^2\right]^{1/2}\}$$

where n_o is an instrumental wipe-out number. This behaviour reveals that :

1) the spin diffusion between the host nuclei is quenched by the magnetic field gradients fostered by the impurities ;

2) the impurity induced relaxation goes as $\mathcal{C}/r^6$ about each impurity, observation which is consistent with the BGS mechanism already responsible for the relaxation at high temperature. This, apart from other things, rules out the GH mechanism which would yield, according to Mc Henry et al (1972), a law in $\exp-(t/\tau)^{3/5}$, as well as the longitudinal dipolar mechanism which is fully saturated at 1.3 kOe and below 100 mK.

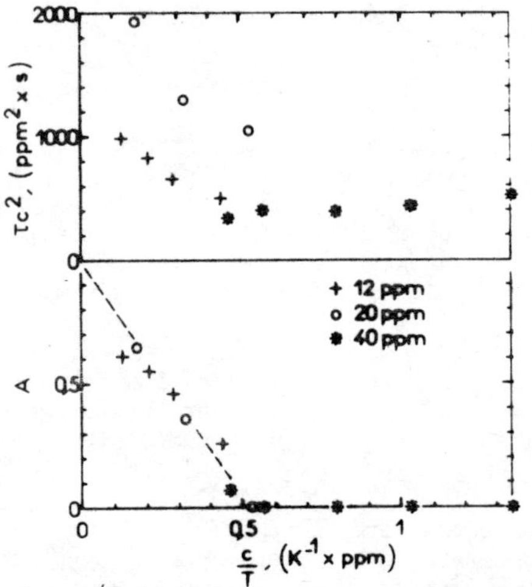

Fig. 1 : c/T dependence of two parameters of the calculated form, A in the lower part, τc^2 in the upper part.

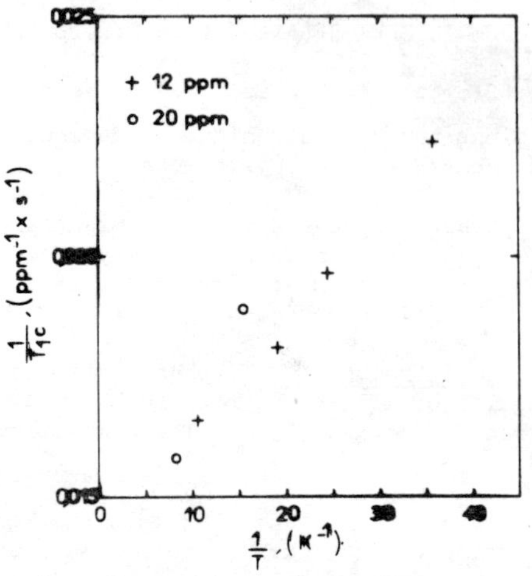

Fig. 2 : Temperature dependence of $(T_{1imp})^{-1}_{1ppm}$ parameter of the calculated form at very low temperature.

4. <u>In the intermediate regime</u>. At temperature such that $2 < \frac{T}{c} < 10$ in mK/ppm, the relaxation lies in between the two limiting behaviours. Because of the random nature of the impurity spatial distribution, the onset of interactions between impurities occurs gradually, leaving a fraction A of the host nuclei free to exchange magnetization by a spin diffusion process and a fraction (1-A) where this process is blocked by local field gradients. According to this simplified picture, the magnetization recovery can be expressed by

$$(5) \quad \frac{M(t)}{M(\infty)} = 1 - \{A \exp -\left(\frac{t}{T_{1imp}}\right) + (1-A)\,P(t)\,\} \exp -\left(\frac{t}{T_{1K}}\right)$$

We have adjusted the four parameters n_o, τ, T_{1imp}, and A appearing in expression (5) to fit the experimental shape. The Korringa relaxation time T_{1K} has been taken to be that of pure copper for both the spin diffusive and the diffusionless parts, although the Korringa process might be significantly altered for the latter part. The residual errors in the fitting procedure are smaller than 0.5% on most of the data points which have been analysed.

Parameter A has been found to vary linearly with c/T from 1 at high temperature to zero for a critical ratio T/c, as shown on fig. 1. This linear dependence is consistent with the assumption of a random impurity distribution. The critical value of T/c turns out to be 1.88 (mK/ppm) , very close to the figure of Hirschkoff et al (1971) obtained by static susceptibility measurement in low fields. We feel that both these features of the variation of parameter A and the accuracy with which expression (5) can be fitted to the experimental data establish the plausibility of the model that we have devised to describe the observed facts.

The variation of $(T_{1imp})^{-1}/c$ due to impurities in the intermediate temperature range is shown on fig. 2 and departs from the constant value (2) obtained at high temperature. The diffusionless relaxation time τ, normalized to c^2 is plotted versus c/T in the top part of fig. 1. It must be noted that the longitudinal relaxation rate as described by these results increases as the impurities order magnetically, showing that fairly large fluctuations of the transverse part of the localized electronic susceptibility are still taking place at high values of H/T.

<u>References</u>

Alloul, H., Bernier, P., to be published.

Benoit, H., De Gennes, P.G., Silhouette, D., 1963, <u>C.R.A.S</u> 256, 3841-4.

Bloyet, D., Ghozlan, A., Piéjus, P., Varoquaux, E.J.A., August 1972, Proc. of LT 13, Boulder, 503-507.

Hirschkoff, E.C., Symko, O.G., Wheatley, J.C., 1971, J. of Low Temp.Physics 5, 155-76.

Mc Henry, M.R., Silbernagel, B., Wernick, J.H., 1972, Phys. Rev. B, 5, 2958-72.

NUCLEAR MAGNETIC RESONANCE IN DILUTE Cu-BASED ALLOYS:SPIN POLARIZATION AND KONDO EFFECT.

H. ALLOUL*, J. DARVILLE+** and P. BERNIER*

*Laboratoire de Physique des Solides, Université Paris-Sud, Centre d'Orsay, 91405 - Orsay (France)

**Institut de Physique, Université de Liège, Sart-Tilman, 4000 par Liège I (Belgium)

Abstract++. ^{63}Cu NMR linewidth measurements have been performed on dilute Cu.Mn, Cu.Fe and Cu.Cr alloys. The existence of interactions between impurities is well shown and the hypothesis of a progressive change from a magnetic to a non magnetic Hartree-Fock state is proved to be unvalid.

^{63}Cu NMR linewidth measurements have been performed on dilute Cu.Mn, Cu.Fe and Cu.Cr alloys, from 0.34 to 150°K and between 1 and 50kG. The data have been compared with magnetization measurements and it has been shown that the existence of interactions between impurities is reflected by the NMR results even for very dilute Cu.Mn and Cu.Fe samples (a few tens of ppm).

From the single impurity contributions to the NMR width, the sd effective exchange interactions have been deduced and are found to scale as expected from the Kondo temperature. The amplitude of the asymptotic spin polarization, as deduced from the present results, have been compared with its values near the impurity given by the satellite Knight shift data of Boyce and Slichter in Cu.Fe (1974). It is shown that the observed slight deviations with respect to the RKKY asymptotic limit can be explained with results deduced from the Hartree-Fock magnetic limit of the Anderson model, which gives a rough estimate of $\sim$ 1 eV for the virtual bound state widths. This analysis also shows that the Kondo effect cannot be represented as a gradual change with temperature from a magnetic to a non magnetic Hartree-Fock state. It rather shows that a single Hartree-Fock picture allows to correlate the measured magnetization and the spin polarization, in which the phase shifts stay those calculated from the high temperature value of the moment, though the mean values of the moment vanishes below T_K.

These results emphasize that the Kondo condensation is really a dynamic effect, well localized on the impurity, which requires a picture such as the one developed by Anderson et al. (1970) for which the statistical orientation of the total spin localized on the impurity changes with temperature, while its coupling with the conduction electrons is not modified.

References

P.W. ANDERSON, G. YUVAL and D.R. HAMANN, 1970, Phys. Rev. B, 1, 4464
J.B. BOYCE and C.P. SLICHTER, 1974, Phys. Rev. Letters, 32, 61

+ Aspirant au F.N.R.S. (Bruxelles-Belgium)

++ The article giving the results and developing their interpretation cited in this communication will be published in J. Phys. F (Metal Physics) by the end of 1974

THE Ga AND P NUCLEAR RELAXATION IN GaP:Te

A. Rogerson and D.P. Tunstall
Department of Physics, University of St Andrews, North Haugh, St Andrews, Fife KY16 9SS.

Abstract. We report T_1 measurements of the Ga^{69}, Ga^{71} and P^{31} nuclei in a series of GaP samples doped with Te concentrations between $2 \cdot 10^{17}$/cc and 10^{19}/cc. The results reflect the changing nature of the electron system due to the increasing Te concentration.

1. Introduction. Much current interest is directed towards understanding the nature of the metallic transition in doped semiconductors (Mott, 1972). Nmr has proved particularly useful for studying this phenomenon as the nuclear spin-lattice relaxation time of the host nuclei is very sensitive to changes in the electron system in the intermediate doping region preceding the transition. GaP samples have become available which are of interest in the nmr study of this doping region and nuclear spin diffusion processes, as all three isotopes are magnetic and abundant, with the two Ga isotopes having different quadrupole moments.

2. Results. We report the results of measurements of the T_1's of Ga^{69}, Ga^{71} and P^{31} nuclei in seven LEC grown, gallium phosphide samples doped with tellurium which, at infinite dilution forms a donor level 93 meV below the conduction band. The samples had room temperature carrier concentrations between $1 \cdot 10^{17}$/cc and $9 \cdot 10^{18}$/cc and analysis of Hall and Schottky barrier data allowed estimates of N, the total uncompensated donor concentration, which were between $2 \cdot 10^{17}$/cc and 10^{19}/cc, to be made. T_1 measurements were made at 6.5 MHz between 1.8 K and 300 K using the pulse train method. The results are presented for each isotope in Figs 1, 2 and 3.

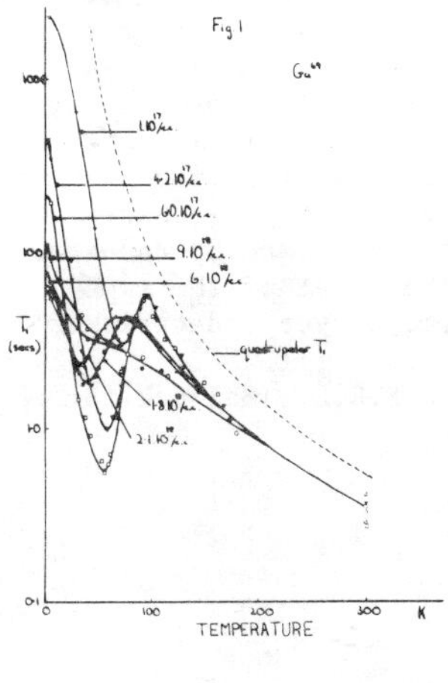

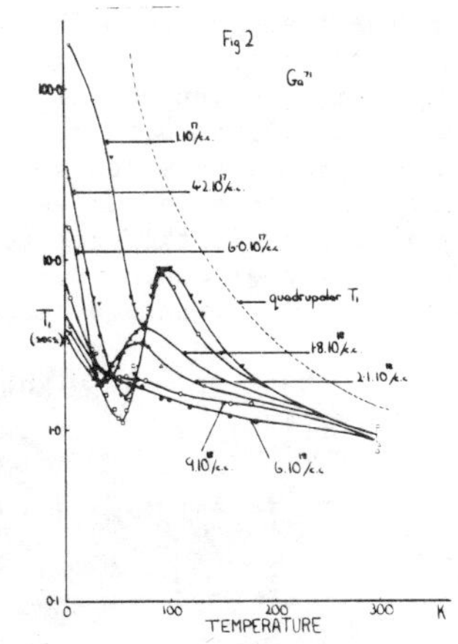

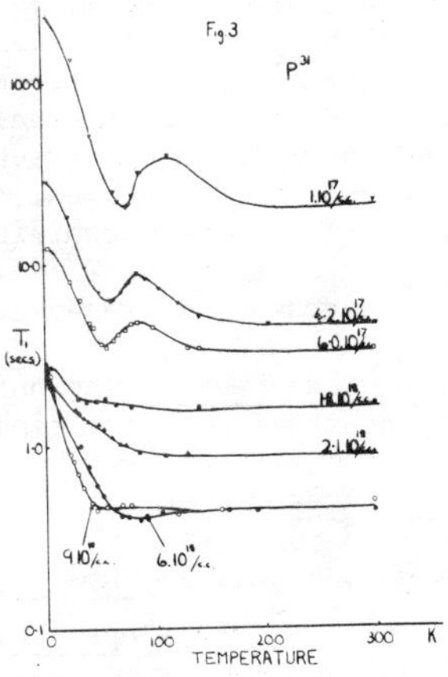

Each curve is characterised by a minimum, which broadens and moves to lower temperatures as the doping density increases and which is absent in the most heavily doped samples. The depth of this minimum and the concentration at which it disappears differs for the P and two Ga isotopes.

3. Discussion. In this discussion we outline the reasons for the variation of the depth and position of the T_1 minimum and interpret them in terms of the changes which occur in the Te donor system with density. After identifying the relaxation processes for the three isotopes in the lightly doped samples, we present calculations of the nuclear spin diffusion coefficients of the two Ga isotopes in GaP. Ga^{69} and Ga^{71} both have spin 3/2 and have quadrupole moments of $0.23 \cdot 10^{-21}$ cm^2 and $0.15 \cdot 10^{-24}$ cm^2 respectively. Superimposed on Figs 1 and 2 we have the theoretical T_1 curves due to quadrupolar relaxation only and, in any quantitative analysis concerning the relaxation mechanisms responsible for the additional, low temperature, structure this quadrupolar component has to be subtracted beforehand. P^{31} has spin 1/2 and no quadrupole moment.

Bloembergen (1949) has shown that nuclear relaxation in diamagnetic crystals containing small quantities of paramagnetic impurities is dominated by spin diffusion to, and relaxation by, these impurities. The coupling between the impurity and the nucleus is through the dipolar Hamiltonian linking their magnetic moments. Such a mechanism, with nuclear relaxation induced by the fluctuating magnetic field of the Te donors, is likely to be dominant in our samples at low temperatures. The observation of short-time nonexponentialities in the magnetization recovery after saturation of the two Ga resonances around the minimum is evidence of diffusion limited relaxation for which $1/T_1 = 8.5\, N\, C^{1/4}\, D^{3/4}$ where C is a coefficient relating to the transition probability of a nucleus a distance r away from an impurity, and D is the nuclear spin diffusion coefficient. C is proportional to $\tau/(1 + \omega^2\tau^2)$ where ω is the nuclear angular Larmor frequency, and τ is the electron correlation time which, at low doping densities, is just the bound electron spin-lattice relaxation time. A resonance occurs when $\omega\tau = 1$ with C having a maximum and T_1 a minimum. Using the orientational average values of C at the minimum for the two Ga isotopes, estimates of D were made from graphs of $1/T_1$ at the minimum against N for the three most lightly doped samples. The values obtained for Ga^{69} and Ga^{71} were $2.8 \cdot 10^{-12}$ cm^2/sec and $1.4 \cdot 10^{-12}$ cm^2/sec respectively which are in reasonable agreement with the values obtained from Khutsishvili's (1966) simple expression for a face-centred cubic lattice, $D = a^2/13.5\ T_2$. No such non-exponentials were observed in the P^{31} T_1's indicating that the rapid diffusion case, as described by Blumberg (1960) is dominant for which $1/T_1 = 4\pi/3\ N\ C\ b^{-3}$, where b is the spin diffusion barrier radius. We believe the rapid diffusion relaxation in the P^{31} system results from the enlarged b for P^{31} nuclei around each impurity due to the preferential build-up of the electron wavefunction on P sites, as predicted by Morgan (1968). Because b varies little over the temperature range in which the T_1 minima for the lightly doped samples occur, a straight line plot of $1/T_1$ minimum against N gives a value of b = 13Å. The theoretical value is 10Å. The theoretical pseudopotential radius ρ is only 6Å and so, with b > ρ, rapid diffusion is to be expected. In comparison, the values of b and ρ for the two Ga isotopes are approximately equal with b = ρ = 5Å and hence diffusion limited relaxation is expected.

Esr linewidth measurements in GaP:Te suggest electron spin-lattice relaxation via an Orbach process (Thomson, 1970) for which $\tau \propto e^{\Delta/kT}$, where Δ is the donor ionization energy. A reduction in Δ with increasing donor concentration would account for the temperature variation in the position of the T_1 minimum. However Hall measurements (Montgomery, 1968) indicate that at a donor density of $9 \cdot 10^{18}$/cc, the ionization energy is approximately 40 meV and, as the Debye temperature, T_D, in GaP is 400 K, the usual condition that $\Delta/k \ll T_D$ does not appear to hold. Fortunately there are alternate mechanisms, such as a spin exchange process with conduction electrons, which would produce a similar dependence of τ on T and Δ.

The reduction in depth of the T_1 minimum and its rapid movement to lower temperatures in our samples is, we believe, evidence of the onset of impurity conduction and a weakening of the effectiveness of nuclear relaxation to Te centres. The disappearance of the minimum is evidence of strong electron interactions. A theoretical calculation of the critical donor density for metallic conduction gives a value of between 2.3 and $4.5 \cdot 10^{19}$/cc. It is unlikely therefore that the reduction in donor ionization energy has resulted solely from wavefunction overlap. Random potentials due to ionized centres and other impurities, of which there are many in LEC material, are likely to contribute significantly to the observed reduction and broadening. The disappearance of the P^{31} T_1 minimum at a lower doping density than that of the two Ga isotopes suggests the presence of an additional relaxation mechanism, as yet undetermined.

4. Acknowledgements. The authors wish to thank Dr D.R. Wight of S.E.R.L., Baldock for the samples and the S.R.C. for financial assistance.

5. References.

Bloembergen, N., 1949, Physica, 15, 386.

Blumberg, W.E., 1960, Phys. Rev. 119, 79.

Khutsishvili, G.R., 1966, Soviet Physics Uspekhi, 8, 743.

Montgomery, H.C., 1968, J. Appl. Phys., 39, 2002.

Morgan, T.N., 1968, Phys. Rev. Lett., 21, 819.

Mott, N.F., 1972, Adv. Phys., 21, 785.

Thomson, F. et al. 1970, BRSG Conference, not published.

A NUCLEAR MAGNETIC RESONANCE STUDY OF SINGLE CRYSTALS OF $NbSe_2$*

M.I. Valic, K. Abdolall and D. Ll. Williams
Department of Physics, University of British Columbia, Vancouver, B.C., Canada, V6T 1W5

Abstract. Pure $NbSe_2$ samples exhibit fluctuations in the electric field gradients above a phase transition at 26K, and a distribution of Knight shifts is observed below. An anomalous feature appears at low temperatures. Less pure samples show no effect. The behaviour is suggestive of the establishment of a charge density wave.

1. Introduction. The properties of the layered structure transition metal dichalcogenides have received a great deal of attention (Wilson and Yoffe (1969)) and the recent suggestion (Wilson et al (1974)); Williams et al (1974) that the properties of some of the tantalum dichalcogenides may be interpreted in terms of charge density waves has intensified interest in this field. Huntley and Frindt (1974) have recently studied the transport properties of single crystals of the two layer (2H) modification of $NbSe_2$ and have shown that the change in sign in the Hall coefficient observed by Lee et al (1969) is sample dependent, varying from a reversal at 27K for the purent specimen to no temperature dependence for the samples having the lowest residual resistance ratio. We have studied the nuclear magnetic resonance (NMR) spectrum of Nb^{93} in two batches of their specimens and find a direct correlation with the Hall coefficient. One batch showing no temperature dependence in the Hall coefficient showed similarly no temperature dependence in its NMR spectrum, whereas the batch showing a Hall reversal at 26K showed behaviour characteristic of a phase transition at 26K in its NMR spectrum.

2. High Temperature Phase. In the hexagonal high temperature phase it is necessary from symmetry that the c axis of the crystal is also the Z principal axis of the electric field gradient tensor. Measurements were made with the magnetic field H_0 parallel to the c axis to determine one component of the Knight shift tensor from the $(\frac{1}{2}\leftrightarrow-\frac{1}{2})$ transition frequency and subsequently to determine the quadrupole coupling constant $\frac{e^2qQ}{h}$ from another transition frequency such as the $(\frac{1}{2}\leftrightarrow{}^3/_2)$ line. A measurement with the magnetic field in the basal plane then determined the remaining component of the Knight shift tensor. These values are given in Table 1 for 77K and 300K using the convention $K(\theta)=K_{iso} + K_{ax}(3\cos^2\theta - 1)$, where θ is the angle between H_0 and c. They are sensibly temperature independent below 77K and the same values within experimental error were obtained for both batches of specimens. The very small isotropic Knight shift at 77K together with the very large anisotropic shift implies negligible s-character in the electron wavefunctions at the Fermi Surface and is consistent with the d-character predicted by band structure calculations (e.g., Mattheis 1973).

Table 1.

T(K)	K_{iso} (%)	K_{ax} (%)	$\frac{e^2qQ}{h}$ (MHz)
300	0.06±0.02	0.19±0.02	59.94±0.06
77	0.00±0.02	0.19±0.01	61.86±0.03

As the temperature is lowered, the behaviour of the two batches of specimens differs. In the batch exhibiting a phase transition we observe that all the resonance lines whose position depends upon the quadrupolar interaction broaden and eventually disappear into the noise. The remaining line, the $(\frac{1}{2}\leftrightarrow-\frac{1}{2})$ transition whose position depends only upon the magnetic interaction, is totally unaffected. Furthermore, over the observable temperature range of $(T-T_c) > 8K$, the excess broadening of the quadrupolar influenced lines is well represented by a $(T-T_c)^{-1}$ dependence, where $T_c = 26K$. Clearly the broadening is a consequence of fluctuations in the electric field gradients throughout the sample brought about by the onset of a phase transition. Since the $(\frac{1}{2}\leftrightarrow-\frac{1}{2})$ line is unaffected, relaxation time broadening may be ruled out and a magnetic disturbance such as a spin density wave is not possible. However, a charge density wave could well be responsible.

3. Low Temperature Phase. Below 26K we were only able to observe the $(\frac{1}{2}\leftrightarrow-\frac{1}{2})$ transition and in figure 1 we present the peak-to-peak linewidth as a function of temperature at two magnetic field strengths. The low temperature line widths shown, together with data at two other fields (K. Abdolall (1974)) show a linear dependence upon field strength and extrapolate at zero field to the line width above the transition. We, therefore, conclude that this line broadening results from a distribution of Knight shifts which, in principle, one may determine by a deconvolution procedure. In fact this is not unambiguous but we can assert that the line shape does not result from just two non-equivalent sites as suggested by Ehrenfreund et al (1971) on the basis of a powder NMR spectrum. And while work at higher magnetic fields is required to determine the distribution unambiguously, we may remark that any non-commensurate (i.e. having no simple relation to the lattice periodicity) periodic perturbation would result in a doubly peaked Knight shift distribution with a flat variation in between. Such a distribution is compatible with the present data and could explain the earlier interpretation of the powder data. This again suggests a charge density wave.

4. Anomalous Features. Our analysis of the existing data was complicated by the occurence of a novel feature in the spectrum at low temperatures. As shown in figure 2, on lowering the temperature a hump centred at the NMR frequency appears superimposed on the 'normal' spectrum. We have established that relative to the 'normal' spectrum the amplitude of the hump is proportional to T^{-3} and to $H_1{}^2$ where H_1 is the rf amplitude. It also increases with increasing magnetic field. The hump always occurs at the Larmor frequency and shows an angular dependence characteristic of an NMR line in the presence of a quadrupolar interaction of the same magnitude as that observed at 77K. An example of this feature under conditions such that it dominates the 'normal' spectrum is shown in figure 3. Since we are observing a derivative spectrum, it appears that there is an onset of some additional absorption at fields lower than the resonance field. Since this onset always occurs at the NMR condition we conclude that this represents a new coupling mechanism between the rf field and the nuclei. However, the form of this additional absorption is difficult to understand.

5. Conclusion. We have demonstrated the existence of a low temperature phase in pure $NbSe_2$ which is characterised by a distribution of non-equivalent sites. This together with the fluctuations above T_c in the quadrupolar interaction are suggestive of a charge density wave. The sample dependence of the results make it desirable to characterise the specimens by their Hall coefficient behaviour. Further work is necessary at higher magnetic fields and with a wider range of specimens to determine the distribution of non-equivalent sites and to study the origin of the anomalous low temperature feature.
We are grateful to R.F. Frindt and D.J. Huntley for their encouragement and cooperation.

6. References.
Abdolall, K., 1974, M.Sc. Thesis, U.B.C. (unpublished).
Ehrenfreund, E., Gossard, A.C., Gamble, F.R. and Geballe, T.H., 1971, J. Appl. Phys. 42 1491.
Huntley, D.J., and Frindt, R.F., 1974, Can. J. Phys. 52 861.
Lee, H.N.S., McKinzie, H., Tannhauser, D.S. and Wold, A., 1969, J. Appl. Phys. 40 602.
Wilson, J.A., DiSalvo, F.J. and Mahajan, S., 1974, Phys. Rev. Lett. 32 882.
Wilson, J.A. and Yoffe, A.D. 1969, Adv. Phys. 18 193.
Williams, P.M., Parry, G.S. and Scruby, C.B., 1974, Phil Mag. 29 695.

*Research supported by the National Research Council of Canada.

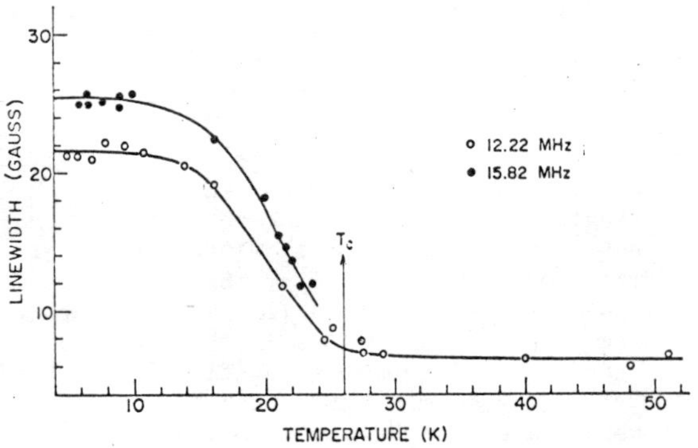

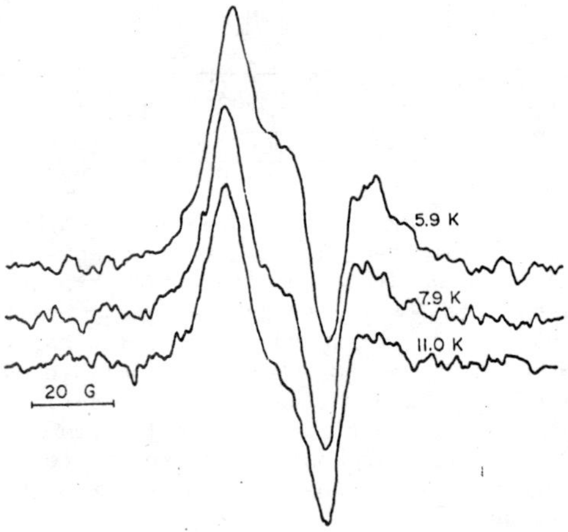

Figure 1. Peak-to-peak linewidth of ($\frac{1}{2}\leftrightarrow-\frac{1}{2}$) transition at two magnetic fields.
Figure 2. Temperature dependence of the ($\frac{1}{2}\leftrightarrow-\frac{1}{2}$) transition at constant H_1 and 12.22 MHz.

Figure 3. The ($\frac{1}{2}\leftrightarrow-\frac{1}{2}$) transition at 5K and15.82 MHz with a larger H_1 to enhance the anomaly.

SECTION J

Motion in Molecular Solids

THERMALLY INDUCED DIPOLAR POLARIZATION IN SAMPLES OF γ-PICOLINE CONTAINING IMPURITIES OF HELIUM AND BENZENE

J. Haupt
Laboratory of Magnetic Resonance, Physics Division, Euratom, CCR, 21020 Ispra,Italy

Abstract. The enhancement of the dipolar polarization obtained after a rapid change of temperature is given for samples of γ-picoline containing helium and benzene impurities. These impurities cause a stronger relaxation whilst the polarization rate is not increased. The relaxation and polarization behaviour is qualitatively discussed.

1. Introduction. Since the discovery of the temperature induced dipolar polarization (Haupt, 1972) in γ-picoline some systematic measurements with this substance have been made (Haupt, 1973) and the typical parameters were interpreted with the help of a simple phenomenological model. Up to now no other substance has been found giving an enhanced dipolar signal which is larger than the Zeeman signal. This last condition is necessary for an easy experimental study, because it is difficult to measure a much smaller dipolar signal in the presence of a large Zeeman signal.

The particular properties of γ-picoline ly in the very low hindering barrier of the methyl-group-rotator. Such a behaviour was indicated by the temperature dependence of T_1. Recent T_1-measurements over a larger temperature range (Zweers et al., 1974), including the T_1-minimum, confirmed this behaviour. We have therefore the unique situation that the hindering barrier of the methyl group in the solid is still smaller than for the free molecule. In addition the "efficiency" of the relaxation (i.e. the value of $1/T_1$ at the maximum as compared to the maximum value of a classical hindered rotator) is much lower than that of other materials containing methyl groups like toluene, xylene etc.

The qualitative explanation of the effect was given on the basis of a quantum mechanical treatment with harmonic oscillator wave functions (Haupt, 1971, 1972) by an extrapolation to low barriers. Because of the extremely low barrier one can eventually improve the explanation by starting the calculation with free rotator wave-functions and extrapolating to low barriers (Haupt, in preparation). Another attempt of an explanation based on less rigorous assumptions was given by Clough (1973).

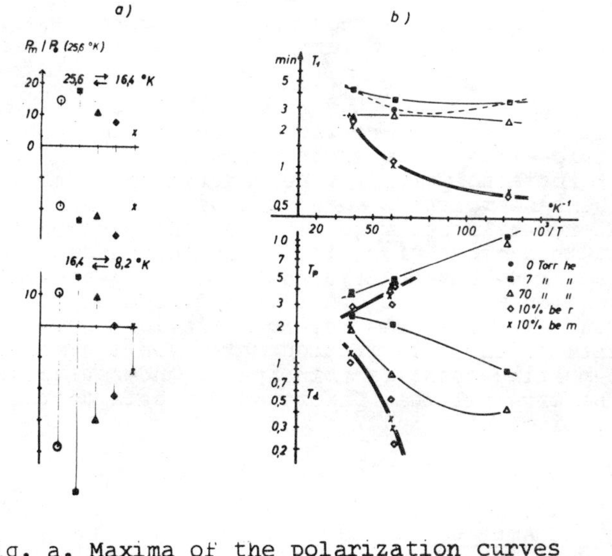

fig. a. Maxima of the polarization curves for temperature jumps 25,6 to 16,4°K, 16,4 to 8,2°K and in the opposite directions
fig. b. Temperature dependence of T_1, T_p and T_d

2. Measurements. In this paper experimental results with modified samples of γ-picoline are presented. A pure sample of γ-picoline is compared with one containing 7 Torr and another with 70 Torr helium gas pressure in the sample tube. A fourth sample contains 10 weight % benzene with 1 Torr of helium. The freezing speed of the latter is varied in the region of the melting point from about 0,5°K/s (index r in the fig.) to about 0,05°K/s (index m). The samples were prepared by degasing with repeated freezing and melting under vacuum and vacuum distillation.

The dipolar polarization has been measured with a 15 MHz pulse-spectrometer after temperature jumps from 25,6 to 16,4°K and from 16,4 to 8,2°K and in the inverse directions in dependence of the time. The maximum amplitudes of the polarization curves are given in fig.a. $P_m/P_0(25,6°K)$ is the maximum dipolar signal as compared to a Zeeman signal at 25,6°K. The shape of the polarization curves is analysed with the help of a simple phenomenological model as described earlier(Haupt 1973) in terms of the polarization time T_p and the dipolar relaxation time T_d. T_p and T_d are defined by the differential equation $dp/dt = -p/T_d \pm K \exp(-t/T_p)$. The solution for $T_p > T_d$ is given by

$$p(t) = \pm K.(T_p.T_d/(T_p-T_d))\cdot(\exp(-t/T_p)-\exp(-t/T_d)).$$

K is a constant depending on the temperature and the jump amplitude. In the lower part of fig.b T_p and T_d are given. In the upper part the measured longitudinal relaxation times T_1 are plotted. The polarization curves for the pure sample have not been analysed because a too slow increase after the temperature jump indicated an insufficient heat conductivity which is accompanied with a deformation of the whole curve.

3. Discussion of the results.

i. By adding sufficient helium to the sample the maximum of the polarization is reduced. This can be seen by comparing the maxima for 7 and 70 Torr helium pressure. An exception is made for the pure sample because of insufficient heat conductivity. A rather small quantity of helium garantees already a sufficient thermal contact between crystallites in the polycrystalline material. In earlier measurements we stored simply the sealed samples for about 1 week in helium gas.

If the amount of helium is large the diffusing helium atoms contribute to a perturbation and modulation of the methyl group barrier. Therefore the longitudinal relaxation and still more the dipolar relaxation is increased. The polarization rate is practically not increased as one sees in the figure. This behaviour might be understood in the following way. A diffusing helium atom which passes close at a methyl-group disturbs the barrier so much that the conditions for a polarization process are no more fulfilled whilst it causes still relaxation. With a larger barrier the relaxation rate (especially the relaxation efficiency) is larger. The increased dipolar relaxation leads therefore to a smaller polarization maximum. The polarization maxima for opposite temperature jumps differ quite a lot especially for jumps between 16,4 and 8,2°K. This is due to the different ratio T_p/T_d at the initial and final temperature; and it's the final temperature which counts.

ii. The addition of 10% benzene increases especially at low temperatures the relaxation considerably as one can see from the T_1-measurements and the T_d-Curves. This leads to a very low maximum of the dipolar polarization for a temperature jump from 16,4 to 8,2°K.

In an earlier paper (Haupt, 1971) a proposal for the relaxation mechanism of e.g. benzene in vitreous o-terphenyl at low temperature was given. A quantum mechanical type of motion even for molecules like benzene was proposed in the cavities of the glass. Although γ-picoline forms a polycrystalline matrix some similar arguments hold for that part of the benzene which is more or less unregularly distributed in between the γ-picoline crystallites. Together with the perturbed potential barriers of the methyl-group this might give rise to an increased relaxation. Whilst T_1 decreases only for a factor 4 between 25,6 and 8,2°K, T_d decreases so strongly that the dipolar polarization is in the order of the Zeeman polarization at 8,2°K. For these short times and small signals the polarization curves could no more be analysed. The dependence of the polarization maxima on the freezing procedure in the vicinity of the melting point is principally understandable in terms of the different size of the crystallites. Unfortunately data of the crystal structure are not existing for γ-picoline.

4. References.

Clough, S., 1973, Physics Letters 42A, 5, 371
Haupt, J., 1971, Proc. 16th Congress Ampere p. 630
Haupt, J., 1971, Z. Naturforsch. 26a, 1578
Haupt, J., 1972, Physics Letters, 38A, 6, 389
Haupt, J., 1973, Z. Naturforsch. 28a, 98
Zweers, A.E., H.B. Brom and W.J. Huiskamp, 1974, Physics Letters 47A, 4, 347

SYMMETRY-RESTRICTED SPIN DIFFUSION AND NONEXPONENTIAL SPIN-LATTICE RELAXATION OF RAPIDLY REORIENTING OR TUNNELING METHYL GROUPS

S. Emid and R.A. Wind
Laboratorium voor Technische Fysica, Technische Hogeschool Delft, Lorentzweg 1, Delft
The Netherlands

Abstract. Spin diffusion is symmetry-restricted in rapidly reorienting or tunneling methyl groups. As a result the populations of the energy levels can not be described by a single spin temperature. Magnetic relaxation is nonexponential due to a coupling to other dynamical variables of the system.

1. Introduction. We have shown previously (Emid and Wind, 1974) that in rapidly reorienting methyl groups, satisfying the condition of motional narrowing, fast spin diffusion involves only flip-flops which do not change the symmetry of the spin states as classified according to the irreducible A, E^a and E^b representations of C_3. This so-called symmetry-restricted spin diffusion (SRSD) model predicts a nonexponential magnetic relaxation due to its coupling to other dynamical variables of the system. The original model and several modifications have been worked out in more detail (Emid et al., 1974 a) and the results explain the observed nonexponential magnetic relaxations reported in the literature. The SRSD model has also been used (Emid et al., 1974 b) to predict the so-called rotational polarization and thermally induced dipolar polarization (Haupt, 1972). The purpose of this contribution is to give a brief review of our results and to extend the previous treatment to the case of rapidly tunneling methyl groups. (Nonexponential relaxation of tunneling methyl groups has also been predicted by Punkkinen and Clough (1974) using a phenomenological model.)

2. Symmetry-restricted spin diffusion. The energy levels and spin states of rapidly reorienting or tunneling methyl groups are as shown in figure 1, where ω_0 is the Larmor frequency, d the intramethyl dipolar shift, b the intermethyl broadening and ω_t the tunneling frequency, which is zero for the classical reorientation. In this paper we only consider the case d ≃ b.

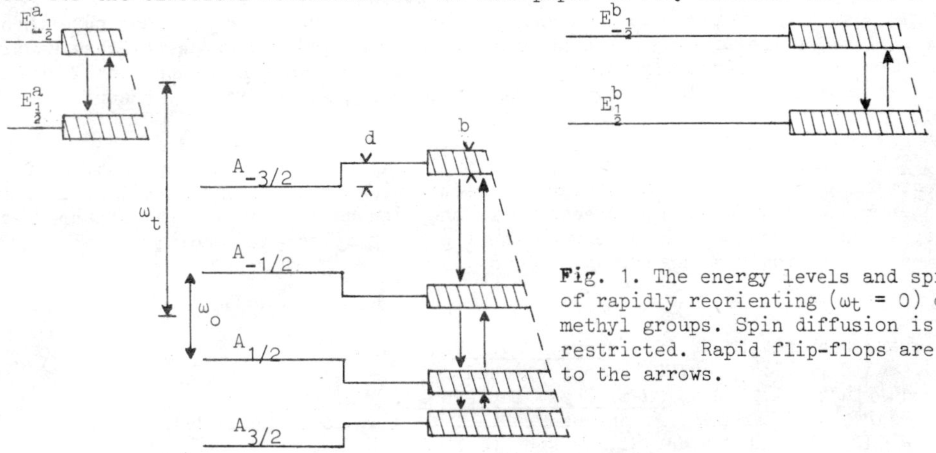

Fig. 1. The energy levels and spin states of rapidly reorienting (ω_t = 0) or tunneling methyl groups. Spin diffusion is symmetry-restricted. Rapid flip-flops are according to the arrows.

Symmetry-restricted spin diffusion means that rapid flip-flops with probability of order b occur only when a methyl group flips from X_m to X_{m+1} and another metyl group from $Y_{m'}$ to $Y_{m'-1}$(X, Y = A, E^a or E^b), m is the magnetic quantum number. As a result rapid spin diffusion can only restore a Boltzmann distribution within a symmetry, but cannot change the total populations of different symmetries, so the populations of all levels can not be described by a single spin temperature.

3. Relaxation of rapidly reorienting or tunneling methyl groups. Magnetic relaxation of rapidly reorienting or tunneling methyl groups can be treated in two ways: 1. By solving the rate equations for the populations (Emid and Wind, 1974, Emid et al., 1974 a). 2. By introducing new parameters (spin temperatures) to describe the populations after rapid spin diffusion (Emid et al., 1974 b). We first consider a single crystal with a unique direction of the methyl rotation axes. Following the second method we label the A, E^a and E^b species by the parameter μ_i= 0, 1 -1, respectively. In the high temperature approximation the populations of the levels are given by

$$p_i = N^{-1}\{1 + \beta_z m_i + \beta_d(\mu_i^2 - 1)(m_i^2 - 5/4) + \beta_r\mu_i + \beta_t(1 - 2\mu_i^2)\}, \quad (1)$$

where β_z is a measure for the Zeeman energy, β_d for the dipolar energy, β_r for the population difference between E^a and E^b and β_t for the population difference between A and E. Introducing intramethyl relaxation transition probabilities W_{ij} one obtains

$$\frac{dp_i}{dt} = -\sum_j W_{ij}\{p_i - p_j - (p_i - p_j)^{eq}\} \qquad (2)$$

from which coupled equations of motion for the β's follow. Due to the mutual couplings the relaxation of e.g. β_z is in general a sum of four exponentials. For classical reorientation

$$W(m_i\ \mu_i, m_j\ \mu_j) = W(-m_i\ -\mu_i, -m_j\ -\mu_j)\ , \qquad (3)$$

it follows (Emid et al., 1974 b) that β_z is coupeld only to β_r, and β_d to β_t. In the tunneling case the W_{ij}'s generally do not satisfy eq.(3). Elaborate calculations for the general case will be given elsewhere. In the extreme narrowing limit or in the case $\omega_t >> \omega_0$ eq.(3) again holds and as a consequence the only nonzero couplings among the β's are the above mentioned. The coupling between β_z and β_r accounts for nonexponential Zeeman relaxation (governed by two time constants) and rotational polarization. The coupling between β_d and β_t accounts for thermally induced dipolar polarization (Haupt, 1972), as also shown by Clough and Hill (1974). Concerning the Zeeman relaxation, for the case of classical reorientation the relaxation has been calculated explicitly for single crystals and polycrystalline samples. For polycrystalline samples two distinct situations may occur. 1. If spin diffusion is effective only within a crystallite, the isotropic averaged Zeeman relaxation remains nonexponential. The rate is slightly faster than for isolated methyl groups (Hilt and Hubbard, 1964), but is just in better agreement with the experiments (Baud and Hubbard, 1968). Further references to nonexponential relaxations in single crystals and polycrystalline samples are given in our previous papers. 2. If spin diffusion, though still symmetry-restricted, is effective throughout the whole sample, or if the methyl groups are oriented in different directions, the relaxation becomes exponential, with exactly the same rate as according to the usual spin temperature theory. Hence the SRSD model explains both the observed nonexponential and exponential relaxation behaviour. The model predicts also nonexponential Zeeman relaxation in the tunneling case. In the limits considered it differs only in a lower relaxation efficiency (Haupt, 1971) compared to the classical case.

4. Conclusion. Symmetry-restricted spin diffusion is essential for the explanation of nonexponential magnetic relaxation of rapidly reorienting or tunneling methyl groups. It offers interesting experimental perspectives in the high as well as low temperature region, by exploiting the mutual couplings between the β's under different conditions, as exemplified in the previous papers.

5. References.

Baud, M.F. and Hubbard, P.S., 1968, Phys. Rev. 170, 384.
Clough, S. and Hill, J.R., 1974, to be published.
Emid, S. and Wind, R.A., 1974, Chem. Phys. Letters 27, 312.
Emid, S., Baarda, R.J., Smidt, J. and Wind, R.A., 1974 a, to be published.
Emid, S., Wind, R.A. and Clough, S., 1974 b, to be published.
Haupt, J., 1971, Z. Naturforsch. 26a, 1578.
Haupt, J., 1972, Phys. Letters 38A, 389.
Hilt, R.L. and Hubbard, P.S., 1964, Phys. Rev. 134, A392.
Punkkinen, M. and Clough, S., 1974, to be published.
References to nonexponential relaxations are given in our paper: 1974, Chem.Phys. Letters 27, 312.

THE APPARENT TEMPERATURE DEPENDENCE OF THE METHYL GROUP TUNNELLING SPLITTING AS OBSERVED IN NUCLEAR RESONANCE LINE SHAPES

P. S. Allen and D. G. Taylor
Department of Physics, University of Nottingham, University Park, Nottingham NG7 2RD, ENGLAND.

Abstract. Temperature dependent line shapes are calculated for tunnelling methyl groups hindered by barriers intermediate between 14 kJ mole^{-1} and 25 kJ mole^{-1}, in order to illustrate the possibility of temperature dependent line shapes without the presence of thermally activated reorientation.

Johnson and co-workers (1973) have recently established that the nuclear resonance line shape of the protons in a tunnelling methyl group may be temperature dependent. One possible interpretation of this data is to suppose that the temperature dependence of the line shape reflects a temperature dependent tunnelling splitting Δ of the methyl group torsional ground state. For example as an initial expedient Johnson has compared his line shapes with the ground line shape calculations of Apaydin and Clough (1968), which are Δ dependent, and so obtained an effective Δ for each temperature. The temperature dependence of this effective tunnelling splitting can be explained (Johnson and Mottley, 1973; Allen, 1974) in terms of the random jumping of the torsional oscillator between Boltzmann populated torsional energy states containing tunnelling splittings of differing magnitudes and signs, as established by the solution of the Schrodinger equation for a threefold torsional oscillator. We denote, in what follows, the magnitudes of the ground and first excited state tunnelling splittings by Δ_o and Δ_1 respectively.

This approach to the explanation of the line shape is not entirely correct. It assumes that for an identical magnitude of the tunnelling splitting, the line shape derived from the ground state energy level structure will be identical to that derived from the excited state structure. Such an assumption is not in fact valid (Taylor, 1974), except in the limits of $\Delta_o,\Delta_1 >> \Gamma$ and $\Delta_o,\Delta_1 << \Gamma$, where the line shape is insensitive to Δ. (Γ is a measure of the dipolar interaction strength and is given by $\Gamma = \gamma^2\hbar^2/r^3$, where γ is the nuclear gyromagnetic ratio and r is the interproton distance.) Thus a simple averaging of the tunnelling splitting will not suffice and a more rigorous approach than the one outlined above must therefore be adopted.

Using a Hamiltonian (Allen, 1968) composed of a rotational term $\mathscr{H}_R$, a Zeeman term $\mathscr{H}_Z$ and a secular term $\mathscr{H}_D$, i.e.

$$\mathscr{H} = \mathscr{H}_R + \mathscr{H}_Z + \mathscr{H}_D$$

together with zeroth order spin-rotor product wave functions, each factor of which transforms according to an irreducible representation of the group C_3, the energy matrix was obtained for both the ground and first excited torsional states of the methyl group. This matrix was diagonalized to give both the energy eigenvalues and eigenfunctions in terms of Δ_o or Δ_1. From the eigenstate data, the individual component transitions were generated, thus enabling the total line shape to be calculated for both the ground and first excited states.

When $\Delta_o,\Delta_1 > 5\Gamma$, the line shape for both torsional states is the same and identical to that of a rapidly rotating triangle. When $\Delta_o,\Delta_1 < 0.5\Gamma$ each torsional state exhibits a line shape identical with that of a static triangle of spins $\frac{1}{2}$. However, in the intermediate region both line shapes are functions of their respective tunnelling splittings but, as pointed out earlier, they are not identical functions of Δ.

In order to compute the actual line shape, a stochastic averaging of the ground and first excited state line shapes was employed. The mechanics of doing this was to decompose each line into its respective components, divide each component into a large number of slices and then average each slice of the ground state with its equivalent slice from the first excited state, before reassembling the components and subsequently the whole line, from the resultant slice averages. The method of evaluating the slice averages was that due to Anderson (1954), which assumes a random jumping between two discrete frequencies and can be adapted to take account of the Boltzmann population difference between the two torsional states and the jump rate between them, each of which is temperature dependent, thus giving rise to a temperature dependent line shape.

It must now be emphasised that for a given barrier height Δ_o and Δ_1 are not the same, $\Delta_1 \sim 70\Delta_o$. Thus as Figure 1 shows only outside the barrier range 14 kJ mole^{-1} to 25 kJ mole^{-1} will both Δ_o and Δ_1 satisfy the aforementioned criteria and give the same line shape. Below this barrier range the resultant line shape is identical to that of a rapidly rotating group and is temperature independent. Above this interval the line shape is the same as for a rigid triangle and remains so until such temperatures as random thermally activated rotational jumps impart a time dependence to the dipolar Hamiltonian at the time scale of the inverse line width. Within the above barrier range the line shape is temperature dependent, even without thermally activated reorientation. Figure 2 illustrates the temperature variation of the spectrum for three different barrier heights within the given range and at temperatures sufficiently low for the spectra to be unaffected by thermally activated rotational jumps. The overall resultant line shape so produced does not necessarily correspond to the shape of any ground state resonance line. As a consequence the comparison of experimental results with computed ground state line shapes in order to extract a temperature dependent Δ is invalid. The stochastic averaging of the ground and first excited state spectra predicts a temperature dependent line shape whose shoulder moves in more rapidly with increasing temperature, than that predicted by the procedure adopted by Johnson (1973). This makes it possible to fit the data to Johnson over a much wider temperature range than he was able to do himself.

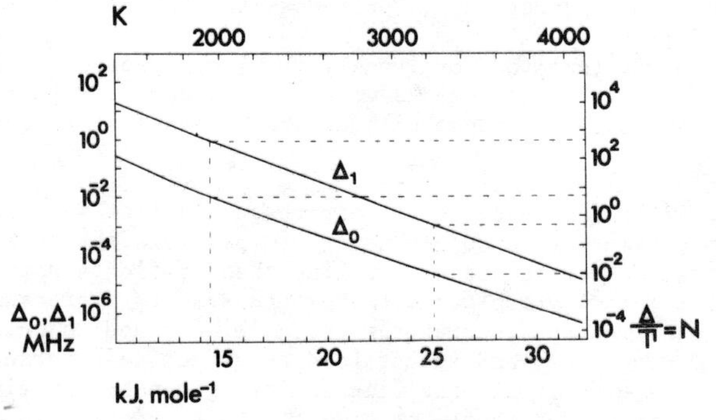

FIGURE 1

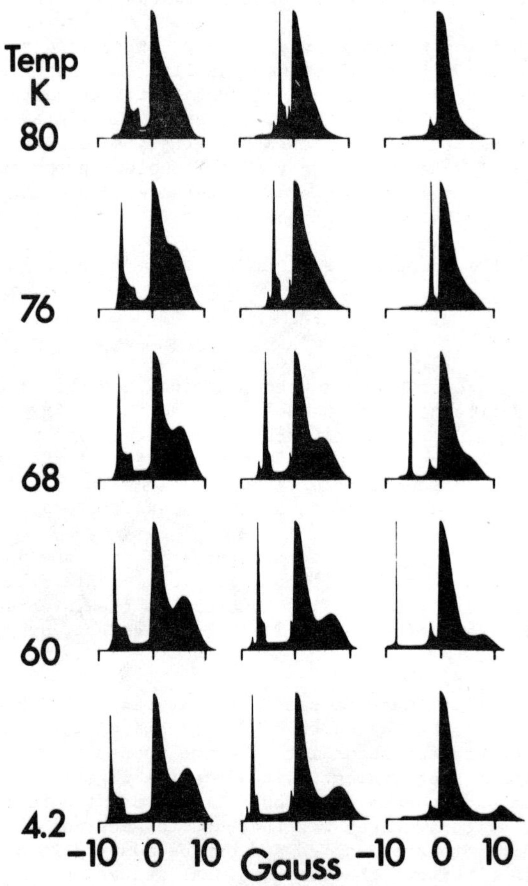

FIGURE 2

References

Allen, P. S. (1968), J. Chem. Phys., 48, 3032-3036.
Allen, P. S. (1974), J. Phys. C: Solid State Phys., 7, L22-5.
Anderson, P. W. (1954), J. Phys. Soc. Japan, 9, 316-39.
Apaydin, F. and Clough, S. (1968), J. Phys. C: Solid State Phys., 1, 932-9.
Johnson, C. S. and Mottley, C. (1973), Chem. Phys. Letters, 22, 430-2.
Taylor, D. G. (1974). Thesis, University of Nottingham, unpublished.

TUNNELLING MOTION IN β- AND γ-PICOLINE AND SOME PICOLINE COMPOUNDS, STUDIED BY PROTON SPIN LATTICE RELAXATION MEASUREMENTS.

H.B. Brom, A.E. Zweers and W.J. Huiskamp
Kamerlingh Onnes Laboratory of the State University of Leiden, Leiden, the Netherlands

Abstract. Proton spin-lattice relaxation data, obtained in γ-picoline, β-picoline, $Zn(\gamma\text{-picoline})_2Cl_2$ and $Cd(\gamma\text{-picoline})_4Cl_2$ are presented and analysed. For γ-picoline, the only compound, which shows dipolar polarization, the highest activation energy was found.

1. Introduction. As reported by Haupt (Haupt, 1972) in γ-picoline (abbreviated γ-pic) a jump in lattice temperature in the 20 K region leads to appreciable polarization of the proton dipolar spin system. In our cw NMR experiments a temperature jump from 4 K to 20 K resulted in a dipolar signal 14 times larger than the equilibrium Zeeman signal at 4 K, corresponding to an increase in dipolar polarization of about a factor 10^3. This phenomenon is primarily connected with the rotation properties of the methyl group, linked to the pyridine ring. Also the difference in energy splitting between the various Zeeman levels of the proton spins in the I = 3/2 ground state (of A symmetry), caused by the dipolar interaction, is of essential importance (Haupt (1972), Clough (1973)). However, both theories differ in other aspects. Our aim was a further elucidation of the kind of mechanisms involved, for which purpose more knowledge about various parameters, as tunnelling splitting, relaxation times, are needed.

2. Proton-Relaxation Times and Activation Energies, Theory. Let us denote by $\hbar\Lambda_o$ the tunnelling splitting between the ground and first excited torsional oscillator state. Nuclear relaxation is effectuated by transitions between the eigenstates of the rotator system, without spin flips. Clough (1971) assumes tunnelling rotation and spin states to be independent; Haupt (1971) supposes correlation between rotation and spin states. Following Haupt lifetime broadening of the rotational levels, dependent on the transition rate between these levels, gives the necessary energy conservation to make nuclear spin flips possible. In the limit of large tunnelling splitting Clough's relaxation formula for the nuclear spin lattice relaxation time (T_{1n}) gets the familiar form,

$$T_{1n}^{-1} \propto \tau/(1 + \omega^2\tau^2) + 4\tau/(1 + 4\omega^2\tau^2) \qquad (1)$$

with ω = the nuclear Larmor frequency. Haupt's theory gives

$$T_{1n}^{-1} \propto \tau/(1 + \Lambda_o^2\tau^2) \ . \qquad (2)$$

In both equations the correlation time $\tau = \tau_o \exp(E_a/kT)$. We assume that at low enough temperatures the activation energy E_a becomes equal to the tunnelling splitting $\hbar\Lambda_o$. Then, if $\omega_o\tau > 1$ or $\Lambda_o\tau > 1$, T_{1n} is proportional to $\exp(\hbar\Lambda_o/kT)$, and from a $\ln T_{1n}$ vs 1/T plot $\hbar\Lambda_o$ can be calculated.

3. Proton Relaxation Times and Activation Energies, Experiment. T_{1n} was measured as a function of temperature in β-pic, γ-pic, $Zn(\gamma\text{-pic})_2Cl_2$ and $Cd(\gamma\text{-pic})_4Cl_2$, the results of which will be the subject of this contribution. In fig. 1 T_{1n} data of β-pic are shown, together with the earlier published γ-pic results. The β-pic plot is typical for the other compounds. Straight lines result in a log T_{1n} vs 1/T plot. No minima are found. The

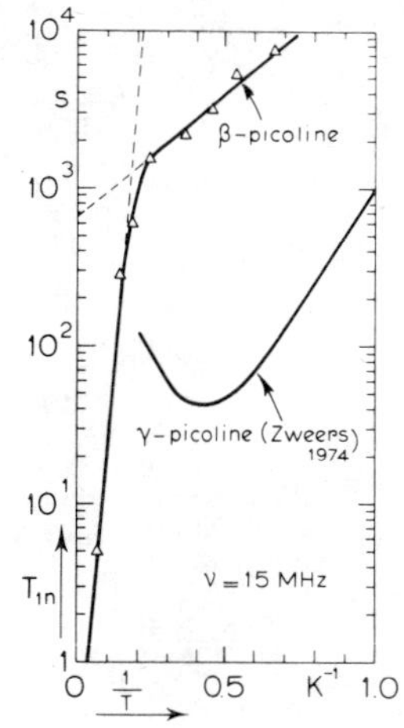

Fig. 1 The nuclear relaxation time (T_{1n}) as a function of 1/T in β-picoline. The supposed exponential form of the correlation time is confirmed by the linear dependence of log T_{1n} on 1/T. The experimental data, connected by the drawn line, can be fitted by the sum of the two dotted lines, corresponding to two activation energies. The β-picoline plot is typical for the other compounds.

activation energies become
β-pic : E_{a1} = 3.6 K and E_{a2} ≈ 50 K,
Zn(γ-pic)$_2$Cl$_2$: E_{a1} = 0.85 K and E_{a2} ≈ 22 K,
Cd(γ-pic)$_4$Cl$_2$: E_{a1} = 3.9 K and E_{a2} ≈ 15 K.
In γ-pic we, Zweers et al. (1974), found E_a = 6.9 K, while from microwave spectroscopy data Rudolph et al.(1967) derive a potential barrier of 6.8 K for the gaseous state.

In γ-pic no field dependence was found. In the other compounds $T_{1n} \propto H^n$ with $n \lesssim 1$ for 5 MHz $\lesssim \nu \lesssim$ 30 MHz.

4. Concluding remarks. a) The field independence of T_{1n} in γ-pic is in agreement with Haupt's model. The slight field dependence in the other compounds is possibly caused by intermolecular proton-proton interaction. b) Conclusions about the height of the barrier are hard to make. At given tunnelling splitting the barrier for sixfold symmetry is much larger than for threefold symmetry. c) We have not observed any effect of the direction of the magnetic field with respect to the sample on the magnitude or sign of the dipolar polarization, which seems to be in contradiction with Clough's model. d) In heat capacity measurements in our laboratory by Klaaysen et al. no energy splitting of a few Kelvin is seen in β-pic and γ-pic. This may be explained by very fast conversion between E and A states.

5. References

Clough, S., 1971, J. Phys. C4, 2180.
Clough, S., 1973, Phys. Lett. 42A, 371.
Haupt, J., 1971, Z. Naturforsch. 26a, 1578.
Haupt, J., 1972, Phys. Lett. 38A, 389.
Rudolph, H.D. , Dreizler, H. and Seiler, H., 1967, Z. Naturforsch. 22a, 1738.
Zweers, A.E. , Brom, H.B. and Huiskamp, W.J., 1974, Phys. Lett. 47A, 347.

THE INDUCTION OF DIPOLAR POLARIZATION BY SPIN SYMMETRY CONVERSION OF TUNNELLING METHYL GROUPS

S. Clough and J. R. Hill
Department of Physics, University of Nottingham, University Park, Nottingham NG7 2RD, England.

Abstract. It is shown that when the populations of nuclear spin symmetry species of tunnelling methyl groups depart from thermal equilibrium, then the subsequent relaxation will induce dipolar order due to coupling of the corresponding differential equations. The model assumes only that all transition probabilities contributing to the relaxation are proportional to squares of matrix elements of intra-methyl group dipole-dipole interactions. For a powder sample it is found that the reciprocal dipolar temperature departs from equilibrium with a time dependence $-C(\exp(-at) - \exp(-bt))$ where C is a constant dependent on the initial departure of spin symmetry populations from equilibrium and $b/a = 5.7$. It is suggested that this mechanism may account for the transient dipolar polarization observed by Haupt after a sudden temperature change in γ-picoline.

Introduction. In an experiment which has attracted a lot of interest, Haupt (1972, 1973) has demonstrated that a sudden change in the temperature of a sample of γ-picoline (4-methylpyridine) may cause a very large transient dipolar polarization. In other words, during the recovery towards equilibrium which follows the thermal disturbance the dipolar spin temperature falls to a low value (either positive or negative depending on the sign of the temperature change). An understanding of this effect is of importance because of possible application to nuclear polarization techniques and also because it relates to other experiments on solids containing molecular groups undergoing tunnelling rotation and a unified explanation for these related phenomena is sought. There is little doubt that the peculiar properties of γ-picoline are due to the quantum tunnelling rotation of the 4-methyl group which at low temperature causes the molecular ground state to split into an orbital singlet (A) and a doublet (E), the splitting $\hbar\omega_t$ defining the tunnelling frequency ω_t which is simply related to the rate of tunnelling between the three equivalent orientations of the hindered methyl group. When the nuclear spin of the protons is taken into account the singlet A is a nuclear spin quartet and the E states are nuclear spin doublets. In the presence of a magnetic field the energy level diagram is shown in Figure 1. The intra-methyl group dipole-dipole interactions shift the A($\pm$ 3/2) levels and the A($\pm \frac{1}{2}$) levels in opposite directions by an amount $\hbar\omega_d(1 - 3\cos^2\phi)$ which depends on the angle ϕ between the methyl group symmetry axis and the applied magnetic field. In Figure 1 $\omega_t \gg \omega_o \gg \omega_d$. Haupt (1972) has given an explanation of the effect which depends on the properties of excited torsional oscillator states of the methyl group. In this paper we suggest an explanation which involves only assumptions which are conventional at higher temperatures, namely the existence of a dipole-dipole spin temperature and the occurrence of random rotational hops through $\pm$ 120° with a correlation time τ. The effect of the latter on the tunnelling states of Figure 1 is to cause a random phase modulation of the E states. Some dipole-dipole matrix elements connecting pairs of levels of Figure 1 are randomly modulated and this gives rise to transition probabilities between these levels, governing the way the system returns to equilibrium following a disturbance. We shall be particularly concerned with those matrix elements of the intra-methyl group interaction which are modulated. These connect A and E states and so cause spin symmetry conversion and also change the dipolar energy since the A levels are shifted by the intra-methyl group dipole-dipole interaction while the E levels are not. Interactions between the methyl group and neighbouring protons are relatively unimportant in this respect though they may be most important for Zeeman relaxation. Transition probabilities C, D, 3C, B, are shown in Figure 1; the sum of those from the A($\pm$ 3/2) levels is aX and the sum from A($\pm \frac{1}{2}$) levels is aY where a is a number dependent on ω_d, ω_t, and X and Y give the angular dependence.

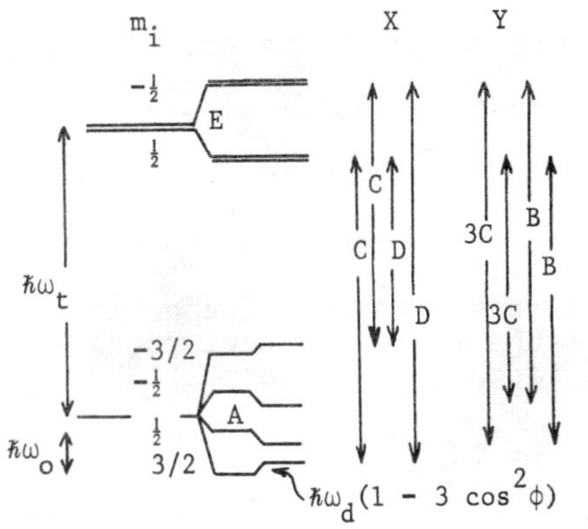

Figure 1. Energy levels and transition probabilities for a tunnelling methyl group.

$$X = 12(1 + 3\cos^2\phi) \quad ; \quad Y = 36(1 - \cos^2\phi) \quad . \qquad (1)$$

Induced Dipolar Polarization. The physical idea of the model may be simply understood from Figure 1 and Equation (1). If the sample temperature is suddenly increased, the relative populations of A and E levels are no longer in thermal equilibrium and spin conversion transitions from A to E levels occur. For methyl groups with $\phi = 0$ we see from (1) that $X > Y$ while for $\phi = 90°$ we have $Y > X$. For both cases though it is the transitions from those A levels whose dipolar shift is negative which occur most rapidly in the early stages of the relaxation.

The relative depletion of the downward dipolar shifted levels implies the existence of a negative dipolar temperature.

The energy levels of Figure 1 may be written

$$E_i = -\hbar(\omega_t \nu_i + \omega_o m_i + \omega_d f_i) \quad (2)$$

where m_i is the magnetic quantum number for the three protons, ν_i is ½ for A levels and −½ for E levels and

$$f_i = -(1 - 3\cos^2\phi)(2\nu_i + 1)(m_i^2 - (5/4))/2. \quad (3)$$

For an isotropic distribution of methyl group orientations, common Zeeman and dipolar temperatures and an angularly dependent temperature θ_t, we can use the high temperature approximation to write the populations $p_i(\phi)d\phi$ of methyl groups of orientation ϕ in level i

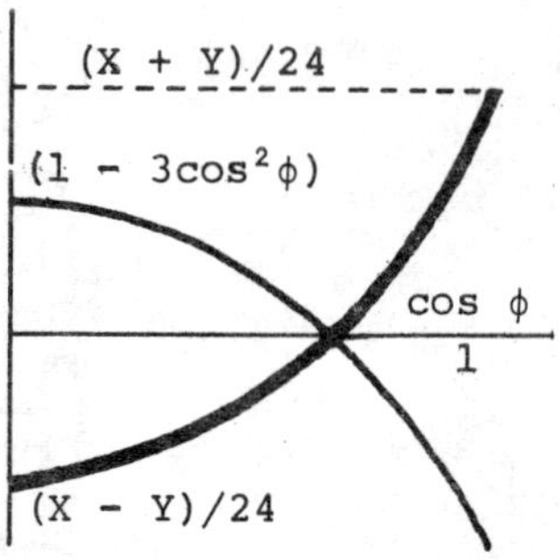

Figure 2. Dependence on cos ϕ of (X $\pm$ Y) and dipolar shift.

$$p_i(\phi)d\phi = (N/8)(1 + \beta_z m_i + \beta_t \nu_i + \beta_d f_i)\sin\phi\, d\phi \quad (4)$$

where β_z, β_d, β_t, are respectively $\hbar\omega_o/k\theta_Z$, $\hbar\omega_d/k\theta_d$, $\hbar\omega_t/k\theta_t$, θ_Z and θ_d being the Zeeman and dipolar spin temperatures. From (4) one easily obtains

$$\beta_t = \sum_i \nu_i p_i(\phi))/(N \sin\phi); \qquad \beta_d = 5(\int_0^{\pi/2} \sum_i f_i p_i(\phi)d\phi)/(2N) . \quad (5)$$

Differentiating these with respect to time and using

$$dp_i(\phi)/dt = \sum_j W_{ij}(p_j(\phi) - p_i(\phi)) - (p_j(\phi) - p_i(\phi))_{equilibrium}) \quad (6)$$

and (4) to eliminate $p_i(\phi)$ results in coupled equations for β_d' and β_t' where the primes indicate deviations from equilibrium ($\beta_d' = \beta_d - \hbar\omega_d/k\theta_L$; $\beta_t' = \beta_t - \hbar\omega_t/k\theta_L$ with θ_L the lattice temperature). The W_{ij} are transition probabilities.

$$\text{Defining } S(t) = -(5a/16)\int_0^{\pi/2} (X - Y)(1 - 3\cos^2\phi)\beta_t' \sin\phi\, d\phi . \quad (7)$$

The coupled equations are (Clough and Hill, 1974)

$$d\beta_d'/dt = -12a\beta_d' - S(t) \quad ; \quad dS(t)/dt = -24a\, S(t) - (864/7)a^2\beta_d' \quad (8)$$

which are easily solved with the above initial conditions to give

$$\beta_d' = -0.22\beta_o(\exp(-5.4at) - \exp(-30.6at)) \quad (9)$$

indicating a maximum change in the reciprocal dipolar temperature of $0.1\omega_t/\omega_d$ times the change in reciprocal lattice temperature in the thermal disturbance. This factor may be about 10^5.

References

Clough, S. and Hill, J. R., 1974, Physics Letters, to be published.

Haupt, J., 1972, Physics Letters, 38A, 389.

Haupt, J., 1973, Zeitschrift für Naturforschung, 28a, 98.

RELAXATION OF FOUR SPIN ½ GROUPS AT LOW TEMPERATURES

R. HALLSWORTH and M. M. PINTAR

Department of Physics, University of Waterloo, Waterloo, CANADA.

Proton spin lattice relaxation times of the NH_4 (or CH_4) groups in various lattices are reported in the temperature range 300 to 1.4°K. Data on $(NH_4)_2Ce(NO_3)_6$, NH_4SnCl_3 and a CH_4 clathrate are compared with the published results on several other NH_4 and CH_4 compounds. The hindrance of the group reorientation in these lattices varies from 5000 to 50 cal/mole. A discussion of the low temperature tunnelling relaxation based on the Clough model will be given. Spin rotational relaxation of the almost free rotor CH_4 will be analysed.

EXPERIMENTAL METHOD DEPENDENT, NON-EXPONENTIAL ZEEMAN RELAXATION IN SYSTEMS CONTAINING TUNNELLING METHYL GROUPS

M. Punkkinen*, D. G. Taylor and P. S. Allen
Department of Physics, University of Nottingham, University Park, Nottingham NG7 2RD, ENGLAND
*On leave from Wihuri Physical Laboratory, University of Turku, Turku, FINLAND

Abstract. The proton zeeman spin-lattice relaxation was observed in solid acetone between 20 and 77 K by using (a) $180^o - \tau - 90^o$ and (b) saturation $- \tau - 90^o$ resonant rf pulse sequences at 24.5 MH_z. Below 50 K the results (a) can be described in terms of two exponentials with time constants T_{1s} and T_{1f} which refer to the slow and the fast relaxing components respectively. T_{1s} and T_{1f} differ by a factor of about 40. The (b) results are largely exponential with a time constant equal to T_{1s}. All the results are explained by a model which assumes a common spin temperature for the A and E species of the methyl groups and a tunnelling splitting comparable with the zeeman splitting. An estimate of 50 MH_z is obtained for the tunnelling splitting at 32 K.

1. Experimental. Nuclear magnetic resonance and relaxation of samples containing tunnelling methyl groups have been the object of many investigations in recent years (see for instance Allen and Snell, 1973, and references therein). To pursue these studies further, the zeeman spin-lattice relaxation of solid acetone was observed at 24.5 MH_z in the temperature range 20 - 77 K. A degassed sample of spectrophotometric grade acetone (99.9%) obtained from Aldrich Chemical Company, USA was used and two different pulse sequences were employed to study the relaxation. First Method (a), in which the thermal equilibrium magnetization was initially inverted by a 180° pulse and then, after a variable time, had its recovery monitored by a 90° read pulse; and secondly sequence (b), in which magnetization was saturated by a number (10 to 20) of equidistant 90° pulses and its recovery then observed by a 90° pulse.

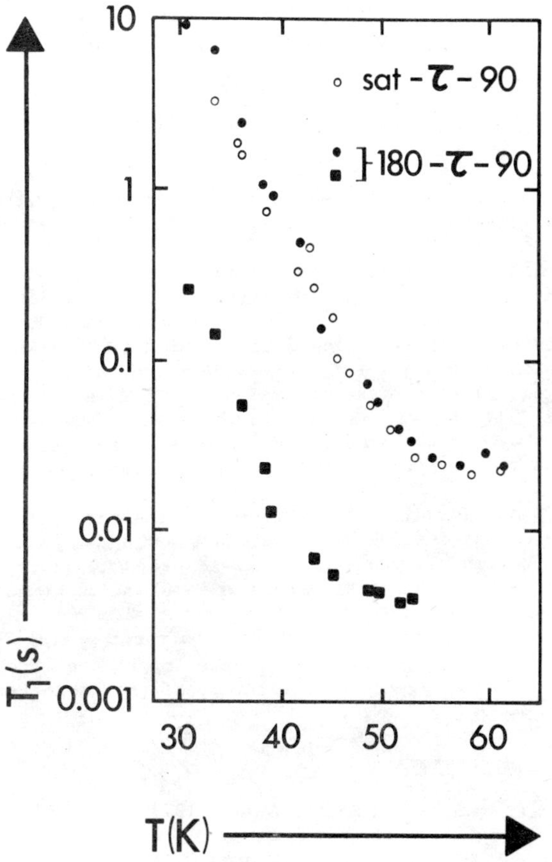

Below 55 K the non-exponentiality of the relaxation as observed by method (a) is very pronounced and the recovery can be reasonably well described as a sum of two exponentials. The temperature dependance of the time constants T_{1s} and T_{1f} for the slow and fast relaxing components respectively, is given in Fig. 1. Since the fast component did not relax rigorously exponentially, T_{1f} was defined as the half amplitude time $t_{A/2}$ (fast) divided by ln2. In this temperature range the ratio $R_T = T_{1f}/T_{1s}$ is in the region of 0.025 and the relative amplitude of the fast relaxing component w_f, as shown in Fig. 2, increases from 0.20 to 0.47 with decreasing temperature. No reliable data for R_T and w_f could be obtained above 55 K, because T_{1f} and T_{1s} differ to little from each other.

According to the results obtained by the saturation method (b) however, the relaxation proceeds largely exponentially with a relaxation time roughly equal to the T_{1s} obtained by the method (a). The best exponentiality for the recovery was obtained when the separation between the successive saturation pulses was about T_{1f}. With the exception of the point at 42.5 K, the method (b) data shown in Fig. 1 were obtained by saturation sequences with only a 50 μs time separation between individual pulses. So a small amount of non-exponentiality remained which caused the time constants (defined as $t_{A/2}/\ln 2$) to be consistantly shorter than T_{1s} of method (a).

2. Discussion. Below 55 K the tunnelling splitting of the methyl group torsional ground state, is believed to have a major effect on the relaxation. A qualitative explanation of the present data is provided by a model (Punkkinen and Clough), which suggests that the relaxations of the excess magnetization $\Delta M_z = M_z - M_0$ and of the excess E species concentration $\Delta \mathcal{N}_E^o = \mathcal{N}_E^o - \mathcal{N}_{E_o}^o$ are

coupled. A common label E is used for the E^a and E^b species of the methyl groups, $\mathcal{N}_E$ is the concentration of the E species and $\mathcal{N}_{E_0}$ its value at the thermal equilibrium. The theory assumes a common zeeman spin temperature for the A and E species in order to derive coupled differential equations for ΔM_z and $\Delta \mathcal{N}_E$. The solution of these equations for a powder sample is given by Punkkinen and Clough as:-

$$\Delta M_z = \left\{ \frac{k_{12}}{S} \cdot \gamma\hbar \cdot \Delta \mathcal{N}_{Ei} + \frac{-k_1 + k_2 + 2S}{4S} \cdot \Delta M_{zi} \right\} \exp(-t/T_{1s}) + \left\{ -\frac{k_{12}}{S} \cdot \gamma\hbar \cdot \Delta \mathcal{N}_{Ei} + \frac{k_1 - k_2 - 2S}{4S} \cdot \Delta M_{zi} \right\} \exp(-t/T_{1f})$$

$$\Delta \mathcal{N}_E = \left\{ \frac{k_1 - k_2 + 2S}{4S} \cdot \Delta \mathcal{N}_{Ei} + \frac{k_{12}}{3S} \cdot \frac{\Delta M_{zi}}{\gamma\hbar} \right\} \exp(-t/T_{1s}) + \left\{ \frac{-k_1 + k_2 + 2S}{4S} \cdot \Delta \mathcal{N}_{Ei} - \frac{k_{12}}{3S} \cdot \frac{\Delta M_{zi}}{\gamma\hbar} \right\} \exp(-t/T_{1f})$$

where

$$k_1 = 4(R_{-1} + R_{-2} + R_1 + R_2)/3$$

$$k_2 = 4R_{-1} + R_{-2} + 4R_1 + R_2 + 2R_0$$

$$k_{12} = -2R_{-1} - R_{-2} + 2R_1 + R_2$$

$$R_{\pm n} = \left(\frac{27}{160}\right) \cdot a_n \cdot \left(\frac{\gamma^4 \hbar^2}{r^6}\right) \cdot \frac{\tau}{\{1 + (\bar{\Delta}_t \pm n\omega_0)^2 \tau^2\}}$$

$$a_0 = 2 \;,\; a_1 = 1 \;,\; a_2 = 4$$

$$S = \sqrt{(k_1 - k_2)^2/4 + 4k_{12}^2/3}$$

$$T_{1s}^{-1} = (k_1 + k_2 - 2S)/2 \;;\; T_{1f}^{-1} = (k_1 + k_2 + 2S)/2$$

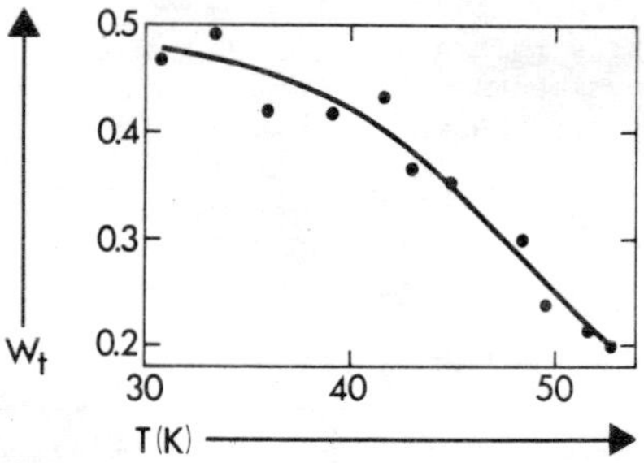

Here r is the intramethyl proton separation, τ the correlation time for the methyl group reorientations, ω_0 the larmor frequency and $\bar{\Delta}_t$ the average tunnelling splitting.

Immediately after a 180° pulse the initial values in equations (1) are $\Delta M_{zi} = -2M_0$ and $\Delta \mathcal{N}_{Ei} = 0$. So according to equations (1) the magnetization recovery can be described as a sum of two exponentials, in agreement with the experimental data (a). For a narrow tunnelling splitting distribution in a slow motion region $(\bar{\Delta}_t \pm n\omega_0)^2\tau^2 \gg 1$ one can plot (Punkkinen and Clough) the ratio $R_T = (k_1+k_2-2S)/(k_1+k_2+2S)$ and the relative weight $w_f = (k_1-k_2+2S)/4S$ as functions of $\bar{\Delta}_t/\omega_0$. The experimental value $R_T \sim 0.025$ is consistent with the theory only if $\bar{\Delta}_t$ is in the vicinity of 0, ω_0 or $2\omega_0$. On the other hand the experimental result of 0.47 for w_f at 32 K implies that $\bar{\Delta}_t \sim 2\omega_0$. So we get a rough estimate 50 MH_z for $\bar{\Delta}_t$ at 32 K. The temperature dependence of w_f is no doubt affected by the temperature dependence of $\bar{\Delta}_t$ (Allen, 1973, and Clough and Hill, 1973).

If equation (1) is considered for the saturation method (b), then after the first saturating 90° pulse the initial values are $\Delta M_{zi} = -M_0$ and $\Delta \mathcal{N}_{Ei} = 0$. At a time t_2 of the magnitude of T_{1f} we have approximately $\Delta \mathcal{N}_E(t_2) \sim -\frac{M_0}{\gamma\hbar} \cdot \frac{k_{12}}{3S} \cdot \exp(-t/T_{1s})$. If at that time the second saturating pulse is applied, the initial values are $-M_0$ and $\Delta \mathcal{N}_E(t_2)$. It is easy to realize that equations (1) now leads to an increased weight of the slow component and to a decreased weight of the fast component in comparison with those after the first saturating pulse. By applying many saturating pulses at intervals of T_{1f}, the importance of the fast component decreases continuously so that finally an exponential recovery of the magnetization is observed, in accordance with the experiments (b)

3. References

Allen, P. S. and Snell, A. J., 1973, J. Physics C: Solid state Physics 6, 3478-90

Allen, P. S., 1974, J. Physics C: Solid State Physics 1, L22-5

Clough, S. and Hill, J. R., 1974, J. Physics C: Solid State Physics 1, L20-1

Punkkinen, M, and Clough, S., J. Physics C: Solid State Physics., in the press.

ELECTRON-ELECTRON DOUBLE RESONANCE STUDIES OF A METHYL GROUP UNDERGOING TUNNELLING ROTATION

Carolyn Mottley, Lowell Kispert and Pu Sen Wang
Department of Chemistry, The University of Alabama, Tuscaloosa, Alabama, USA 35486

Abstract. Electron-electron double resonance (eldor) spectroscopy was used to study the tunnelling rotation undergone by the unstable $CH_3\dot{C}HR$ radical in irradiated ℓ-alanine crystals.

1. Introduction. Magnetic resonance techniques have been used extensively to study the quantum mechanical tunnelling of protons in methyl groups. Several models have been proposed to explain the change in the spectral lineshape from one determined by a quantum mechanical tunnelling motion to one characteristic of classical rotation about the symmetry axis of the methyl group (Gamble et al., 1968; Davidson and Miyagawa, 1972; Clough et al., 1970; Clough and Poldy, 1969; Johnson and Mottley, 1973; Allen, 1974). It was thought that the use of eldor spectroscopy could further elucidate the mechanism responsible for this spectral change.

The low temperature esr spectrum of the unstable radical $CH_3\dot{C}HR$ in ℓ-alanine was first studied in 1970 (Davidson and Miyagawa, 1970). The temperature dependence of the esr lineshape was explained in terms of a Brownian mixing of all the $M_I=+1/2$ (or $-1/2$) states with A and E rotational symmetry which failed to explain the relative line intensities above 15°K (Davidson and Miyagawa, 1972). This failure can be corrected by assuming that the change is brought about by the onset of spin uncorrelated motion (classical rotation over the top of the potential barrier) which has no effect on the A levels since they are symmetric under rotation but which causes a selective mixing of the E levels (Clough and Poldy, 1969 and 1973). The classical rotation model predicts that the A symmetry lines will not decrease below their low temperature intensity values while the E symmetry lines will decrease with increasing temperature. This behavior has been observed (Davidson and Miyagawa, 1972).

In the present study eldor techniques indicate that there is little cross relaxation between A and E lines at 15°K and that classical rotation is responsible for the change in esr lineshape.

2. Experimental. All spectra were run on a standard Varian E-800 eldor accessory coupled to a Varian E-12 esr spectrometer. An Air Products Heli-Tran LTD-3-110 refrigeration system was used and the temperatures were measured using a calibrated carbon resistor. The unstable $CH_3\dot{C}HR$ radical was prepared by irradiating ℓ-alanine crystals at 77°K, warming to 170°K and recooling to 77°K (Davidson and Miyagawa, 1972 and 1970). All eldor spectra were obtained along the c crystal axis with a modulation frequency of 100 kHz and an observing power of 8 uw. The maximum pump power was 336 mw.

3. Results and Discussion. The temperature dependent esr spectra of the $CH_3\dot{C}HR$ radical in ℓ-alanine are shown in Figure 1. The nine line pattern at 15°K is the overlapping of two seven line tunnelling spectra due to the splitting of the alpha proton which has no effect on the motion of the methyl group (Davidson and Miyagawa, 1970). With the increase in temperature the E lines disappear from the spectrum and the $M_I=\pm 1/2$ A symmetry lines grow in intensity approaching the 1:4:6:4:1 intensity ratio expected for the rapidly rotating methyl group split by the alpha proton. The intensity ratio for the low temperature spectrum is not the 1:1:2:3:2:3:2:1:1 predicted due to a slight misalignment of the crystal.

The frequency swept eldor spectra were obtained with lines 9,8 and 7 (see Fig. 1) respectively as the observed line while the rest of the esr spectrum was pumped. Figure 2 shows the temperature dependence of the eldor spectrum obtained with the observing position set at the crossing point of line 9 (high field A line, $M_I=-3/2$). No lines were observable at frequencies higher than 80 MHz. At 15°K two strong, forbidden lines at 21 and 47.5 MHz and a weak, allowed line at 60.5 MHz are observed. The intensities of these lines are directly proportional to the pump power. As the temperature is raised no real drastic change is observed in the eldor spectrum of line 9. This behavior is to be expected, since the high field A line plays no part in the motional averaging.

The eldor spectra obtained with the observing position set at the crossing point of line 8 (high field E line, $M_I=-1/2$) are shown in Figure 3. At 15°K the eldor spectrum of line 8 has major lines at 15 MHz (ν_H), 62 MHz (a_H), and 126 MHz ($2a_H$) and the associated spin flip lines. The intensities of these lines are inversely proportional to the pump power, leading to the conclusion that the A and E esr lines have different saturation properties and thus are not communicating. This behavior is to be expected if the tunnelling splitting is large since the transition probabilities between A and E lines will be small. At 22°K lines have appeared at 30.5 MHz ($a_H/2$) and 95 MHz ($3a_H/2$), corresponding to pumping lines 7 and 5 respectively. These are allowed lines when the E states are mixing and collapsing onto lines 5 and 7. The spectrum at 24°K is very weak since line 8 (the observing line) has almost disappeared from the ESR spectrum.

The ELDOR spectra obtained with line 7 (A symmetry, $M_I=-1/2$) as the observing line are shown in Figure 4. At 15°K the spectrum is essentially the same as that obtained observing line 9 and has the same pump power dependence as line 9. At 22°K there is a prominent line at 65 MHz (a_H) and 94 MHz ($3/2a_H$) corresponding to pumping lines 5 and 4 respectively. The observation of the 94 MHz line at higher temperature is due to the collapsing of the E lines onto the A line (7) not to a mixing of A and E levels. The 94 MHz line disappears at higher temperatures corresponding to the disappearance of line 4 (E symmetry) from the ESR spectrum.

The difference in behavior with respect to pump power of the A and E symmetry lines, the fact that at low temperature (15°K) there is little cross relaxation between the A and E lines as demonstrated by the eldor spectra, and the observation by Davidson and Miyagawa that the intensities of the A lines never dropped below their low temperature values all strongly support the classical rotation model. This model has been previously used to explain the temperature dependence in radicals with low barriers to rotation (Clough and Poldy, 1969 and 1973) but does not work for radicals such as $CH_3\dot{C}COOH$ in the methyl malonic acid where the rotation barrier is higher and where the temperature dependence of the tunnelling frequency (separation between A and E energy levels) can be directly observed. It appears that the coupling between the methyl group and the lattice must be considered in order to arrive at a model which can be generally applied (Clough and Hill, 1974).

Eldor techniques can be used to study quantum mechanical tunnelling in methyl groups and potentially may provide information not now available. Investigations are currently being carried out on other methyl containing radicals. This research was supported by the U.S. Atomic Energy Commission. Dr. I. Miyagawa is acknowledged for the loan of the ℓ-alanine crystals.

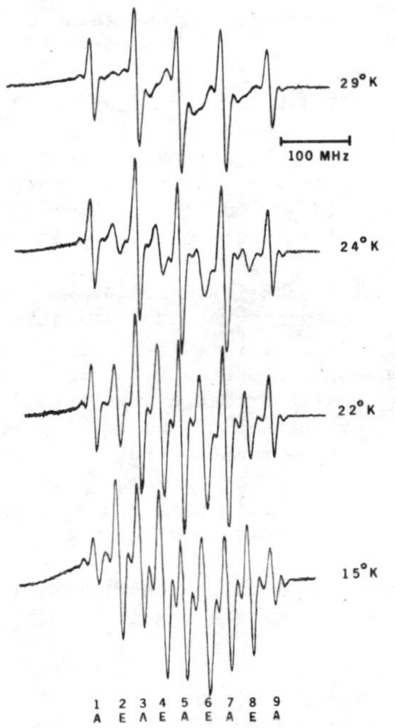

Figure 1. Temperature dependent ESR.

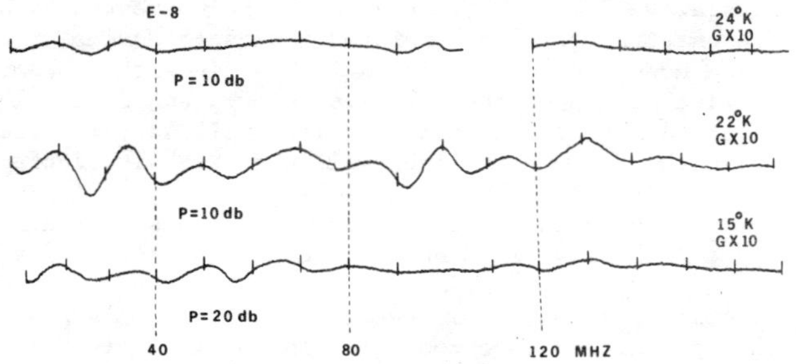

Figure 3. Eldor spectra using the crossing point of line 8 as the observing position. P=Pump Power

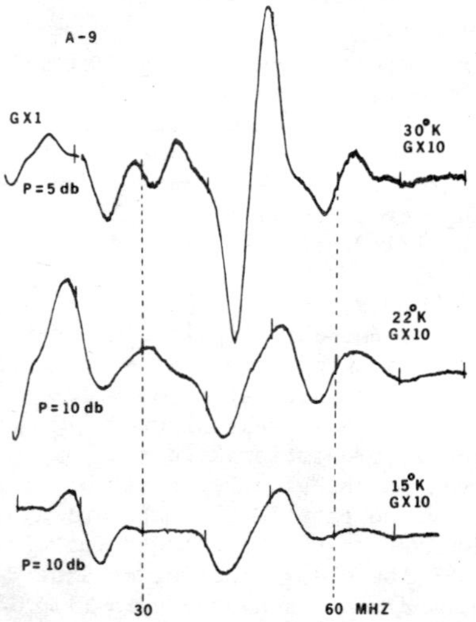

Figure 2. Eldor spectra using the crossing point of line 9 as the observing position. P=Pump Power

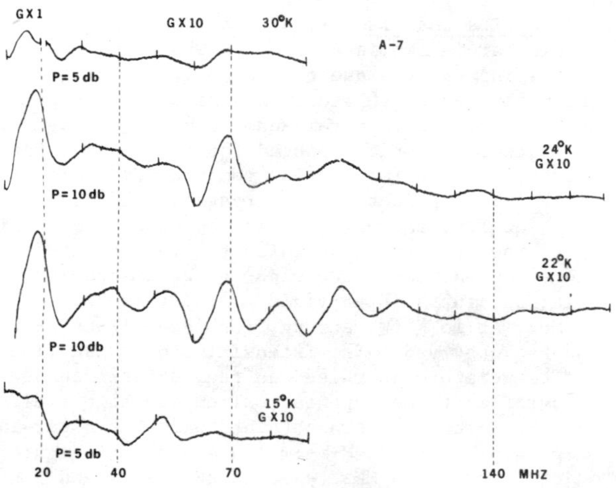

Figure 4. Eldor spectra using the crossing point of line 7 as the observing position. P=Pump Power

4. References

Allen, P.S., 1974, J. Phys. C.; Solid State Physics, 7, L22.

Clough, S. and Hill, J.R., 1974, J. Phys. C.; Solid State Physics, 7, L20.

Clough, S. and Poldy, F., 1969, J. Chem. Phys., 51, 2076.

Clough, S. and Poldy, F., 1973, J. Phys. C.; Solid State Physics, 6, 1953.

Clough, S. and Starr, M. and McMillan, N.D., 1970, Phys. Rev. Letters, 25, 839.

Davidson, R.B. and Miyagawa, I., 1970, J. Chem. Phys., 52, 1727.

Davidson, R.B. and Miyagawa, I., 1972, J. Chem. Phys., 57, 1815.

Gamble, W.L., Miyagawa, I. and Hartman, R.B., 1968, Phys. Rev. Letter, 20, 415.

Johnson, C.S. Jr. and Mottley, C., 1973, Chem. Phys. Letters, 22, 430.

TUNNELLING MAGNETIC RESONANCES

S. Clough and T. Hobson
Department of Physics, University of Nottingham, University Park, Nottingham NG7 2RD, ENGLAND.

Abstract. Quantum tunnelling rotation of molecular groups in solids at low temperature can be studied by observing the effects of resonant relaxation processes involving paramagnetic impurities. The processes involve a flip of the electron spin and a simultaneous change in the motional state of the molecular group and occur when the rotational tunnelling frequency and the electronic Larmor frequency are close. Since the latter is field dependent and the former is not, the effects show up as resonant phenomena in the field dependence of relaxation studies. Some of the transitions involve a simultaneous flip of a nuclear spin, leading to an induced nuclear polarization which serves as the most sensitive way of detecting the resonances. Experiments are described which demonstrate the inhomogeneous broadening of the resonances due to a spread of tunnelling frequencies, the very long life times of tunnelling states and the slow rate of tunnelling diffusion (the analogue of spin diffusion in which molecules exchange rotational rather than magnetic energy).

Introduction. At low temperature the ground state of a molecule containing a hindered methyl group undergoing tunnelling rotation about its symmetry axis is split into an orbital (motional) singlet (A) and a doublet (E), the splitting $\hbar\omega_t$ depending on the frequency of rotational tunnelling between the three equivalent orientations of the group. The state of the proton spins is constrained by the exclusion principle to be a quartet state (total nuclear spin of the three protons = 3/2) when the motional state is A and a doublet (total nuclear spin of the three protons = $\frac{1}{2}$) when the motional state is E. Transitions between these levels (spin symmetry conversion) require a symmetry change in both motional and proton spin functions. Effectively this demands an socillating field gradient of frequency ω_t, only the difference in the field at the three proton sites being effective in causing the transition. At low temperatures conversion may be very slow and the populations N_A and N_E are not in thermal equilibrium with the lattice. It is convenient to define a tunnelling temperature θ_t through the equation

$$N_E/N_A = \exp(-\hbar\omega_t/k\theta_t) \quad .$$

When a sample is cooled to a low temperature θ_L, we expect to find $\theta_t = \theta_o > \theta_L$ where θ_o is the temperature at which the conversion rate becomes very slow.

Resonant Nuclear Spin Symmetry Conversion. Paramagnetic impurities in solids in an external field behave on account of the Larmor precession as localized sources of microwave field with a well defined frequency ω_s which is simply proportional to the field. Because of the r^{-3} radial dependence the local field gradient is also large. The interaction of this oscillating field gradient with the proton spins of a neighbouring tunnelling methyl group may induce spin symmetry conversion of the latter if $\omega_s \sim \omega_t$ (Glattli et al., 1972, Clough and Mulady, 1973, Clough et al., 1973). Since ω_s is field dependent, it is possible to switch on and off by adjustment of the magnetic field, a process which allows an interchange of methyl group rotation energy and electronic magnetic energy. Since the electron spin-lattice relaxation time is short (< 1 s) it provides a cross-relaxation route for the equilibration of spin symmetry populations with the lattice, $\theta_t \rightarrow \theta_L$. However, the range of the effect from the electron is rather limited because the electron field gradient falls off as r^{-4} and the conversion rate as r^{-8}. Thus when the cross-relaxation process is switched on θ_t may fall to θ_L for methyl groups near the electron but stay at θ_o for more distant groups. A relatively small dispersion of tunnelling frequencies would ensure that tunnelling diffusion (analogous to spin diffusion) is ineffective in eliminating spatial inhomogeneity in θ_t. A distribution of tunnelling frequencies $G(\omega_t)$ means also that we must regard θ_t as a function of ω_t as well as having a spatial variation. Figure 1 illustrates the form of the surface $\theta_t(\omega_t,r)$ at a moment during an experiment in which the sample is first cooled and then the magnetic field is slowly swept down from an initially high value. As ω_s is varied, successive sets of methyl groups become resonant and θ_t falls to θ_L for small r. In Figure 1, groups with low values of ω_t are yet unaffected and for them $\theta_t = \theta_o$ for all r.

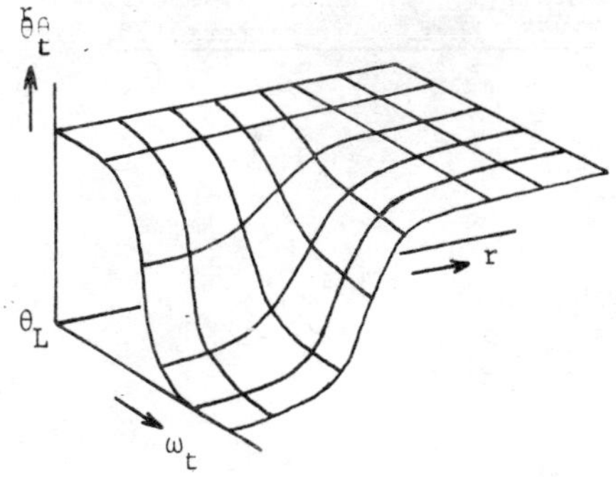

Figure 1. The surface $\theta_t(\omega_t,r)$ during initial field sweep

Induced Nuclear Polarization. The dipole-dipole interaction between electron and methyl group contains the operators S_+I_-, S_+I_+, S_+I_z and Hermitian conjugates. The first two involve nuclear spin flips and are responsible for proton polarization which is induced during the resonant spin symmetry conversion. Transitions due to S_+I_- are resonant when $\omega_s + \omega_n = \omega_t$ (ω_n is the nuclear Larmor frequency); those due to S_+I_+ when $\omega_s - \omega_n = \omega_t$. The former induce magnetization antiparallel and the latter parallel to the external field. The rate at which those methyl groups distant r from an electron and having tunnelling frequency ω_t generate nuclear polarization is

$$dM/dt = G(\omega_t)d\omega_t r^2 dr(\theta_L^{-1} - \theta_t^{-1})r^{-8}X(\omega_t - \omega_s) \quad (1)$$

where $G(\omega_t)d\omega_t r^2 dr$ is proportional to the number of such methyl groups, $(\theta_L^{-1} - \theta_t^{-1})$ is proportional to the population difference driving the transitions, $X(\omega_t - \omega_s)$ is a response function with a positive peak at $\omega_t - \omega_s = -\omega_n$ and a negative peak at $+\omega_n$, while r^{-8} is the radial dependence. (The angular dependence is averaged out.) The evolution of θ_t is given by

$$d(\theta_L^{-1} - \theta_t^{-1})/dt = -(\theta_L^{-1} - \theta_t^{-1})r^{-8}Y(\omega_t - \omega_s) \quad (2)$$

where the response function Y has positive peaks at $\omega_t - \omega_s = \pm\,\omega_n$ and 0.

After cooling (2) is integrated over the history of ω_s variation so that the surface $\theta_L^{-1} - \theta_t^{-1}$ is known at any time. This can then be used in (1) which after integration successively over t, r, ω_t gives the nuclear polarization induced in any desired time interval.

Experiments. The nuclear polarization induced in 1 minute is measured as the amplitude of a free induction decay following a 90° pulse. Measurements are made at a high field far from the resonance and at 4.2 K where the nuclear spin-lattice relaxation time is much longer than 1 minute. The procedure is first to reduce the nuclear magnetization to zero by means of a series of pulses, then to reduce the magnetic field to a value near resonance for a period of 1 minute and finally to return to the high field and measure the magnetization which was induced at the low field. Successive experiments using a sequence of values of ω_s (i.e. field) during the polarization period provide the data to check the model. One remarkable feature of such measurements is shown in Figure 2. Results taken with a freshly cooled sample and a sequence of values of ω_s decreasing form a bell-shaped curve. A second sequence taken with ω_s increasing then traces out a completely different curve which is approximately proportional to the negative derivative of the bell-shaped curve.

The explanation is to be found in (1). Because X is an odd function, the integral of (1) over ω_t is zero unless either G or $(\theta_L^{-1} - \theta_t^{-1})$ varies across the relatively narrow width of X. Figure 1 shows that there is such a sharp variation in θ_t during the first sequence of measurements and this variation, being caused by the measurements, progresses with the measurements. The integral over (1) is therefore proportional to $G(\omega_s)$. During the second sequence, the variation of θ_t with ω_t is very much diminished because methyl groups near to the electron are no longer important. Now only the variation in G prevents the integral of (1) being almost zero and the result is that this integral is proportional to the (negative) derivative of G. Detailed calculations (Clough and Hobson, 1974) confirm this prediction in agreement with the experiments.

References

Clough, S., Hinshaw, W. S. and Hobson, T., 1973, Phys. Rev. Letters, 31, 1375.

Clough, S. and Mulady, B. J., 1973, Phys. Rev. Letters, 30, 161.

Clough, S. and Hobson, T., 1974, J. Phys. C: Solid State Phys., in the press.

Glattli, W., Sentz, A. and Eisenkremer, M., 1972, Phys. Rev. Letters, 28, 871.

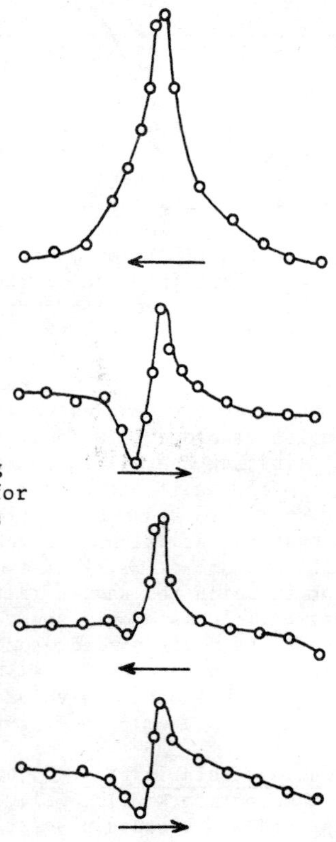

Figure 2. Tunnelling resonance lineshapes for four successive passes through the resonance.

THE MEASUREMENT OF METHYL GROUP TUNNELLING FREQUENCIES BY ENDOR

S. Clough, J. R. Hill and M. Punkkinen
Department of Physics, University of Nottingham, University Park, Nottingham NG7 2RD, ENGLAND.

Abstract. The frequencies of tunnelling methyl groups in free radicals in solids at low temperatures can be found from the difference in ENDOR frequencies observed when saturating the extreme high and low field lines of the ESR hyperfine structure. The method is suitable for both single crystals and powder samples.

Introduction. The quantum mechanical features of methyl group hindered rotation are exhibited at low temperatures by a splitting of the ground molecular state into a low-lying singlet (A) and a doublet (E). The magnitude of splitting $h\nu_t$ depends on the height of the hindering barrier, and ν_t which we refer to as the tunnelling frequency is related to the rate at which the methyl group tunnels through the barrier by means of a rotation of $2\pi/3$ about its symmetry axis. The measurement of ν_t in free radicals of the type $R\dot{C}\text{-}CH_3$ exploits the dependence of the hyperfine interaction, between the unpaired electron spin and the methyl group, on both the spin and space coordinates of the methyl protons. The motion of the methyl group then has an influence on the energy levels of the system to which the ESR spectrum is sensitive only if ν_t is of the order of the hyperfine coupling frequency (ν_c) (Clough et al., 1972). However when (i) $\nu_t >> \nu_c$ or (ii) $\nu_t << \nu_c$ the ENDOR frequencies are still dependent on ν_t. In particular it is the difference in ENDOR frequencies obtained when saturating the extreme high and low field lines of the ESR spectrum which is the most sensitive quantity. ENDOR then allows the measurement of ν_t from values of a few tens of kHz up to several GHz dependent on the ENDOR linewidth. This paper outlines the application of the ENDOR technique in both the limits (i) and (ii).

ENDOR Frequencies in the Slow Tunnelling Limit. We consider a free radical containing a single methyl group. In the absence of tunnelling the protons occupy well-defined positions with characteristic hyperfine coupling tensors with the unpaired electron spin. The Hamiltonian is

$$\mathscr{H}/h = \nu_s S_Z - \nu_n \sum_i I_Z^{(i)} + \underline{S}.\sum_i \underline{\underline{T}}.\underline{I}^{(i)} . \qquad (1)$$

The hyperfine coupling tensors are nearly isotropic with magnitudes dependent on the location of proton i. The frequencies ν_s and ν_n are the electron and proton Larmor frequencies. The tunnelling behaviour may now be included by inserting the effective Hamiltonian $\mathscr{H}_r$.

$$\mathscr{H}_r/h = -\nu_t(P + P^{-1})/3 \qquad (2)$$

where P is the operator which cyclicly permutes the methyl group spin states. With $\nu_t << T_{ZZ}^{(i)}$ the Hamiltonian (1) is diagonal to sufficient accuracy using the simple states $|m_s m_1 m_2 m_3\rangle$ where $m_s = \pm \frac{1}{2}$ and $m_i = \pm \frac{1}{2}$ are the electron and proton spin quantum numbers. The ESR spectrum then consists of eight equally intense lines at

$$\nu = \nu_s \pm \tfrac{1}{2}T_{ZZ}^{(1)} \pm \tfrac{1}{2}T_{ZZ}^{(2)} \pm \tfrac{1}{2}T_{ZZ}^{(3)} \qquad (3)$$

and is indistinguishable from the static case ($\nu_t = 0$). The difference in ENDOR frequencies measured while saturating the extreme high and low field lines of the ESR spectrum is

$$\Delta\nu_e = \Delta\nu_n - 4\nu_t/3 \qquad (4)$$

where $\Delta\nu_n$ allows for the difference in nuclear Larmor frequency across the ESR spectrum (Clough et al., I, 1974).

ENDOR Frequencies in the Fast Tunnelling Limit. When $\nu_t >> T_{ZZ}^{(i)}$ we must choose state functions which diagonalise $\mathscr{H}_r$. This is achieved by using functions which form a basis for the irreducible representations of the C_3 group. The ESR spectrum then consists of eight equally intense lines at (Clough and Poldy, 1969)

$$\nu = \nu_s = m(T_{ZZ}^A \pm \lambda T_{ZZ}^E)/3 \qquad (5)$$

where $\lambda = 0$ for $m = \pm 3/2, \pm \frac{1}{2}$ and $\lambda = 1$ for $m = \pm \frac{1}{2}$ with

$$\underline{\underline{T}}^A = \underline{\underline{T}}^{(1)} + \underline{\underline{T}}^{(2)} + \underline{\underline{T}}^{(3)}$$

and

$$\underline{\underline{T}}^E = 2(\underline{\underline{T}}^{(1)}\cos(2\phi) + \underline{\underline{T}}^{(2)}\cos(2\phi - 2\pi/3) + \underline{\underline{T}}^{(3)}\cos(2\phi + 2\pi/3))$$

where ϕ is the equilibrium orientation of the first methyl proton. Again, then, although it indicates a methyl group which is rapidly tunnelling, the ESR spectrum gives no precise information about ν_t. However, in second order, the difference in the extreme high and low field ENDOR frequencies is still sensitive of ν_t (Clough and Poldy, 1969)

$$\Delta\nu_e = \Delta\nu_n = (t^A)^2/9\nu_s + (t^E)^2\nu_s/(36\nu_t(\nu_t + \nu_s)) \qquad (6)$$

where t^A and t^E are one-third the traces of T^A and T^E both of which are measurable from the ESR spectrum (5).

Experimental Results

(a) α-Aminoisobutyric Acid. γ-irradiated crystals of α-aminoisobutyric acid yield the free radical $(CH_3)_2CCOOH$ which contains two methyl groups. Figure 1 shows the ENDOR frequencies obtained in a field of 0.33 T for various orientations of the magnetic field in a single crystal plane. Only the region 38 - 50 MHz and one of the non-equivalent crystal sites (Clough et al., I, 1974) is chosen for discussion. ENDOR obtained while saturating the extreme low field ESR line is labelled ν_L while the high field extreme is labelled ν_H. There is clearly a shift $\Delta\nu_e = \nu_H - \nu_L$ which is positive for line F (fast) indicating a rapid rotating methyl group. The magnitude of the shift only permits us to state that $\nu_t \gtrsim 4$ GHz since the shift is of the order of the ENDOR linewidth and we are in the high frequency limit of the ENDOR method. The shift in line S (slow), however, is negative and indicates a small ν_t. Use of equation (4) yields $\nu_t = 1.4 \pm 0.1$ MHz and does not verify the assumption of Wells and Box (1967) that $\nu_t = 0$ and their hyperfine coupling tensors must be corrected accordingly.

(b) Magnesium Lactate. Figure 2 shows the ENDOR lineshapes obtained in γ-irradiated powdered magnesium lactate. It is seen that the lineshapes are broad but because of the axially symmetric form of the hyperfine coupling tensors the powder average exhibits a well-defined edge. Since the shift in frequency due to a finite ν_t is isotropic the shape of the powder average spectrum is unaltered but is shifted in frequency only (Clough et al., II, 1974). Looking at the ENDOR lineshapes obtained while saturating the extreme high (ν_H) and low (ν_L) field ESR lines we see the lineshapes are the same with a sharp edge near 38.5 MHz. The frequency shift is positive and use of (6) gives $\nu_t = 1.8 \pm 0.3$ GHz.

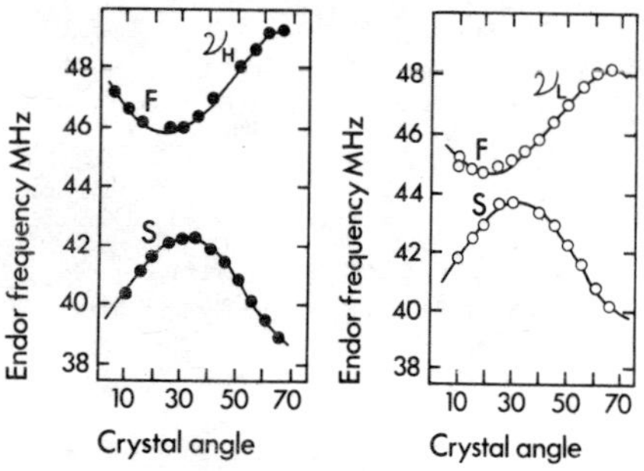

Figure 1. Angular dependence of the ENDOR frequencies in γ-irradiated α-aminoisobutyric acid.

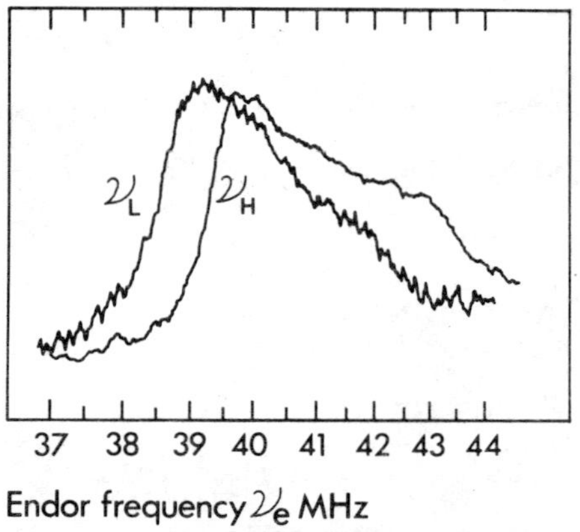

Figure 2. Lineshapes in γ-irradiated powdered magnesium lactate.

References

Clough, S., Hill, J. R. and Poldy, F., 1972, J. Phys. C: Solid State Phys., 5, 1739-44.
Clough, S., Hill, J. R. and Punkkinen, M., I, 1974, to be published.
Clough, S., Hill, J. R. and Punkkinen, M., II, 1974, to be published.
Clough, S. and Poldy, F., 1969, J. Chem. Phys., 51, 2076-84.
Wells, J. W. and Box, H. C., 1967, J. Chem. Phys., 46, 2935-8.

QUADRUPOLE SPIN-LATTICE RELAXATION OF ^{59}Co IN A SINGLE CRYSTAL OF $K_3Co(CN)_6$

M.I. Gordon*, M.J.R. Hoch+ and J.A.J. Lourens*
+ Department of Physics, University of the Witwatersrand, Johannesburg, South Africa.
* Department of Physics, University of South Africa, Pretoria, South Africa.

Abstract. The n.m.r. spin-lattice relaxation time for ^{59}Co in $K_3Co(CN)_6$ has been measured between 80 and 304 K. Relaxation proceeds via a quadrupolar mechanism involving internal optical modes of the $Co(CN)_6$ octahedron with wavenumber ~ 412 cm^{-1}. A calculation which attempts to explain this result is outlined.

1. Introduction. N.m.r. relaxation via a quadrupolar mechanism for nuclei incorporated in complex ions may be expected to be dominated by optical modes particularly if the nucleus involved occupies a central position in the complex. For example Armstrong and Jeffrey (1971) have shown that in $R_2MC\ell_6$ compounds rotary lattice modes with wavenumbers ~60 cm^{-1} dominate in determining the relaxation time behaviour for the Cℓ nuclei.

We have studied the ^{59}Co spin-lattice relaxation time (T_1) in a single crystal of K_3 $Co(CN)_6$. A somewhat surprising result has been the extent to which one, or possibly two, intermediate frequency optical modes of the $Co(CN)_6$ octahedron dominate in the relaxation process.

2. Measurements. ^{59}Co has I = 7/2 and largely due to a slight distortion of the $Co(CN)_6$ octahedron there is a static electric field gradient (e.f.g.) at the Co site (Lourens and Reynhardt 1974). T_1 measurements were made on the central component of the spectrum using a single crystal oriented with its c-axis perpendicular to the magnetic field (10.4 kG).

Since the energy level spacings are unequal the spin-temperature concept does not apply and the relaxation will not be governed by a single relaxation time. However, in our experiments which were made using a coherent pulsed spectrometer with digital signal averaging, no departures from linear relaxation could be detected. Fig. 1 shows a plot of log T_1 against log T(K) where T_1 denotes the initial relaxation time.

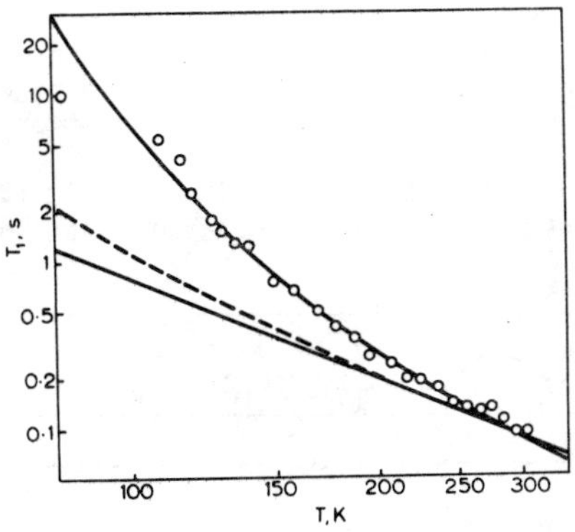

Fig. 1. Log T_1 vs log T for ^{59}Co in $K_3Co(CN)_6$. Plotted points from experiment. The curves are described in the text. (T_1 here denotes the initial relaxation time.)

3. Discussion. The three types of lattice vibrations which could play a role in relaxation are the acoustic modes, the rotary lattice mode of wavenumber ~ 25 cm^{-1} and the internal modes of the octahedron. The Debye temperature is Θ_D = 290 K and the departure of the data from a T^2 law at temperatures T > 200 K rules out the acoustic modes. The lowest curve in Fig. 1 represents a T^2 law normalized to pass through a data point near 300 K. Similarly the rotary lattice mode, which should give the temperature dependence shown by the dashed curve in Fig. 1, clearly does not fit the results. The best fit curve through the experimental points has the form $T_1 = 0.06 \sinh^2 (\Theta_E/kT)$ with Θ_E = 600 K corresponding to a mode of wavenumber ~412 cm^{-1} and is based on the Einstein lattice model. This empirical fit implies that a single internal mode, or perhaps more than one mode of roughly the same frequency, dominates in relaxation. In order to understand this it is necessary to consider the contributions of these modes in detail.

The basic theory of quadrupolar relaxation due to lattice vibrations (Van Kranendonk 1954; Kochelaev 1960) may be readily extended to cover the internal modes of a complex ion (Armstrong and Jeffrey 1971). The transition rate between spin states m and m + μ is given by perturbation theory as

$$W_{m,m+\mu} = \frac{2\pi}{\hbar} \sum_{\ell\ell'} \sum_{\underline{k}\underline{k}'} \left| \left\langle m+\mu, n_{\underline{k}'\ell'}+1, n_{\underline{k}\ell}-1 \left| H_Q \right| m, n_{\underline{k}'\ell'}, n_{\underline{k}\ell} \right\rangle \right|^2 \delta(E_{m+\mu} + E_{\underline{k}'\ell'} - E_m - E_{\underline{k}\ell}) \quad (1)$$

where $H_Q = \sum_{\mu=-2}^{2} F^{\mu} Q^{-\mu}$ and the F^{μ} and Q^{μ} are lattice and spin operators respectively. $n_{\underline{k}\ell}$ denotes the number of phonons with wave vector $\underline{k}$ in branch ℓ and $E_{\underline{k}\ell} = \hbar\omega_\ell(\underline{k})$ is the phonon energy. Following Van Kranendonk (1954) the F^μ are expanded in terms of the lattice displacements and only the quadratic term is retained. The relative displacements $\underline{u}_i = \underline{r}_i - \underline{r}_o$ of two atoms i and o may in turn be expanded in terms of the normal modes and the α'th component of the relative displacement is

$$u_{i\alpha} = (\hbar/2N)^{\frac{1}{2}} \sum_{\underline{k}\ell} [e_{i\alpha}(\underline{k},\ell)/m_i^{\frac{1}{2}} - e_{o\alpha}(\underline{k},\ell)/m_o^{\frac{1}{2}}]\{a(\underline{k},\ell) + a^+(-\underline{k},\ell)\}\omega_\ell(\underline{k})^{-\frac{1}{2}} \qquad (2)$$

where $e_{i\alpha}(\underline{k},\ell)$ denotes an eigenvector component for atom i (mass m_i). N is the number of elementary cells per unit volume and a and a^+ are phonon annihilation and creation operators.

In the expansion of the F^μ the second order term may be written as $\sum_{i\alpha\beta} A^\mu_{ii} u_{i\alpha} u_{i\beta}$ where the elements of the tensor A^μ_{ii} have been tabulated by Van Kranendonk. Substituting into Eq.(1) and assuming that each branch may be considered separately ($\ell = \ell'$) and approximating the first Brillouin zone by a sphere gives

$$W_{m,m+\mu} = \pi^3\hbar/40|\langle m+\mu|Q^\mu|m\rangle|^2 \sum_\ell |\sum_i \chi_i^\mu(\ell)|^2 \Big/ \Delta\omega_\ell\omega_{\ell o}{}^2 \sinh^2(\beta\hbar\omega_{\ell o}/2) \qquad (3)$$

where $\chi_i^\mu(\ell) = \sum_{\alpha,\beta} A^\mu_{ii}[e_{i\alpha}(\ell)/m_i^{\frac{1}{2}} - e_{o\alpha}(\ell)/m_o^{\frac{1}{2}}][e_{i\beta}(\ell)/m_i^{\frac{1}{2}} - e_{o\beta}(\ell)/m_o^{\frac{1}{2}}]$.
In integrating over $\underline{k},\underline{k}'$ a linear dispersion relation has been used $[\omega_\ell(\underline{k}) = \omega_{\ell o} - \Delta\omega_\ell(k/km)]$ and it has been assumed that the eigenvectors are independent of $\underline{k}$.

By writing down the rate equations for spin state populations it may be inferred that $\frac{1}{T_1} \sim W_{-3/2,-1/2} + W_{-3/2,1/2} + W_{-5/2,-1/2}$. Thus by computing the $\chi_i^\mu(\ell)$ we can compare the contributions of the various modes to T_1.

Degeneracies reduce the 33 normal modes of the $Co(CN)_6$ ion to 13. (O_h symmetry is assumed and the small distortion mentioned above is ignored in this treatment.) Jones (1962) has given G and F matrix elements for $Co(CN)_6$ and using standard procedures we have calculated the eigenfrequencies and eigenvectors $\underline{e}_i$. Table 1 gives the eigenfrequencies and calculated $|\sum_i \chi_i^1(\ell)|^2$ values; $|\sum_i \chi_i^2(\ell)|^2 = 1/8|\sum_i \chi_i^1(\ell)|^2$. The summation over i includes only the 6 nearest neighbour C atoms since previous work (Lourens and Reynhardt 1974) has shown that these will provide the major e.f.g. contribution. (A multiplying constant has been omitted in the tabulation.)

Table : 1.

Representation (Mode)	A_{1g} (Q_2)	E_g (Q_4)	F_{1g} (Q_5)	F_{1u} (Q_7)	F_{1u} (Q_8)	F_{1u} (Q_9)	F_{2g} (Q_{10})	F_{2g} (Q_{11})	F_{2u} (Q_{12})	F_{2u} (Q_{13})
Eigenfrequency $\omega_\ell(cm^{-1})$	408	342	358	564	416	(?)	480	98	440(?)	72
$\lvert\sum_i \chi_i^1(\ell)\rvert^2$	0	0	11	21	41	472	21	1	21	2

Two dominant modes are seen to be $F_{1u}(Q_9)$ and, to a smaller extent, $F_{1u}(Q_8)$. (The three very high frequency modes have been omitted.) The factor $1/\omega_{\ell o}{}^2 \sinh^2(\beta\hbar\omega_{\ell o}/2)$ should enhance the importance of the low frequency 98 and 72 cm^{-1} modes and make them comparable with Q_8 but this may be offset by larger $\Delta\omega_\ell$ values, (Eq.3), due to greater coupling to the rest of the lattice. Jones (1962) and other workers have somewhat arbitrarily assigned a frequency of $\sim 90\ cm^{-1}$ to $F_{1u}(Q_9)$ based predominantly on C-Co-C bending and ignoring Co-C stretching. Centre of mass considerations, however, show that there must be a combination of bond bending and stretching. This will result in a higher frequency similar to that of $F_{1u}(Q_8)$. This conclusion is supported by our results. To our knowledge this is the first time that relatively high frequency modes have been found to be of dominant importance in relaxation. Their dominance is due to pronounced bond bending motions. It is necessary to assume an anti-shielding factor of about 10^2 in order to obtain quantitative agreement between theory and experiment.

4. References.

Armstrong, R.L. and Jeffrey, K.R., 1971, Can. J. Phys. 49, 49-53.
Jones, L.H., 1962, J. Mol. Spectroscopy, 8, 105-120.
Kochelaev, B.I., 1960, Sov. Phys. - JETP, 10, 171-175.
Lourens, J.A.J. and Reynhardt, E.C., 1974, Advances in N.Q.R., 1, 315-321, (Ed. J.A.S. Smith) Heyden Press, London.
Van Kranendonk, J., 1954, Physica, 20, 781-800.

SOFT MODES IN THE LATTICE VIBRATIONS OF TRANSITION METAL HEXAMMINE HALIDES

I.L.A. Crick, A.R. Bates, R.O. Davies
Department of Physics, University College, Cardiff

Abstract A theoretical study has been made of the lattice vibrations of the hexammine nickel halide lattices, in order to investigate the phase transitions which occur in these crystals. Instabilities and soft lattice modes are predicted which support the previous theoretical calculations and which agree with available experimental data.

1. Introduction. It has been established that the transition metal II hexammine complexes undergo phase transitions at critical temperatures T_c. These phase transitions were first reported by Palma-Vitorelli et al (1960) for the nickel hexammine halides (chemical formula $Ni(NH_3)_6\{Halide\}_2$) who showed that the EPR linewidth suddenly increased at T_c: for the chloride $T_c = 76K$, and for the iodide $T_c = 19.5K$. Trapp and Chin-I Shyr (1971) showed that, above T_c, the ligand field at the metal ion was cubic, but that below T_c there was a trigonal component. This agreed with the theoretical predictions of Bates and Stevens (1969) and Bates (1970), in which it was proposed that the transitions were caused by a freezing out of NH_3 group rotations within the crystal. Above T_c the NH_3 groups would be able to tunnel or rotate between different static low-energy positions at a frequency greater than the measuring frequency, so as to exhibit a cubic field at the metal ion. Below T_c, the NH_3 groups would only be allowed to tunnel between equivalent low-energy positions which possess trigonal symmetry, and hence give a trigonal component to the field. It was proposed by Bates and Stevens, on the basis of a static electrostatic model, that the freezing of the NH_3 groups was due to the interaction between neighbouring hexammine clusters.
On the basis of this static model, no crystal structure change is expected at T_c. However, it has recently been shown that the chloride lattice suffers a trigonal distortion at T_c of approximately 10% (J.I. Langford, M.W. Thomas, private communications), whereas the iodide lattice appears to remain undistorted through T_c (M.W. Thomas, private communication). In order to develop the understanding of these phase transitions a dynamic model of the hexammine lattice has been studied in detail, with the object of investigating whether any instabilities or soft lattice modes occur. The following discussion and numerical results are appropriate to the nickel hexammine halides.

2. Theoretical model. The face centred cubic hexammine lattice contains one $Ni(NH_3)_6$ cluster and two halogen atoms per primitive cell, and, with no restriction on the motion of the atoms, possesses 81 degrees of freedom. This is reduced to 18 by assuming that the $Ni(NH_3)_6$ cluster moves as a rigid body, maintaining the arrangement of the six nitrogen atoms in three mutually perpendicular directions. The only internal motion allowed in the cluster is the rotation of the six H_3 triangles about their principal symmetry axes in planes at right angles to each Ni-N direction. This motion is included because of its suspected importance in the phase transitions. The frequencies of all internal modes are greater than those of the lattice modes (Sacconi et al, 1964), and coupling between the internal torsional and lattice modes is expected to be small. To a reasonable approximation, therefore, the triangles can be assumed to have no torsional motion and allowed only to tunnel between equivalent low-energy positions. Thus it is necessary to consider only 12 degrees of freedom made up of 9 translational (the translational motion of the cluster and the two halogen atoms) and 3 librational (the rigid body rotation of the cluster).

As stated in §1, well above T_c the triangles of charges are considered to be freely rotating, whereas below T_c they are considered to be tunneling between equivalent low energy positions. In order to investigate the behaviour of the modes near T_c, each tunneling H_3 triangle is regarded as a ring of charge q_r superimposed upon three charges q_t at the apices of the H_3 triangle. The total charge $q = q_r + 3q_t$ is considered constant. Well above T_c, $q_t \to 0$, $q_r \to q$ and the triangle structure no longer exists; a fall in temperature causes a reduction in the tunneling frequency and a consequent increase in q_t, until below T_c $q_t \to q/3$ and $q_r \to 0$.

3. The normal modes of the hexammine lattice. On the basis of the model outlined above the energy of interaction between neighbouring ions has been determined using a multipole expansion. In general, it is found that the $Ni(NH_3)_6$ cluster possesses all multipoles from and including the octopole, but that all poles greater than 16-pole give negligible contributions. The force constants have been determined from an expansion of the lattice energy up to second order. The Coulomb contributions have been evaluated using the rigid ion approximation of Kellermann (1940) and the short-range repulsive terms governing translational motion have been estimated from far infra-red data (Sacconi et al) and elastic constant measurements (Haussühl, 1963). No experimental data is available for the librational or the principal Raman-active frequencies, so realistic values for these frequencies have been chosen. The details of these calculations are to be published elsewhere.
The normal modes of the $Ni(NH_3)_6Cl_2$ lattice have been computed along high-symmetry directions within the first Brillouin zone. Figure 1 gives these modes well above T_c when $q_t = 0$. In figure 1, the symmetry species of each mode is labelled away from the zone centre. The nature is specified within brackets: longitudinal or transverse nature is specified by L or T respectively, translational, librational, or coupled motion is specified by t, r, or tr respectively.

At k = o, the modes separate into 9 pure translational modes (two triplets, a doublet, and a singlet) and a pure librational triplet. Away from k = o there is coupling between translational and librational modes, a possibility which seems to have been excluded by O'Leary and Wheeler (1970). Along [1 1 0] the mixed B_1 and B_2 modes drop rapidly to zero (and subsequently to an imaginary value) indicating a possible lattice instability. These modes are found to be very sensitive to the magnitude of the cluster-cluster interaction; variation of other parameters have much smaller effects. Since the chloride lattice is known to be stable above T_c a screening effect is assumed which reduces the cluster-cluster interaction by a factor of 2 which is enough to leave the modes stable throughout the zone (figure 2). For the $Ni(NH_3)_6I_2$ lattice, however, no such instability occurs, and this can be attributed to the smaller cluster-cluster interaction due to the larger lattice constant. For consistency, the dispersion diagram for the iodide lattice is shown with this same reduction (figure 3).

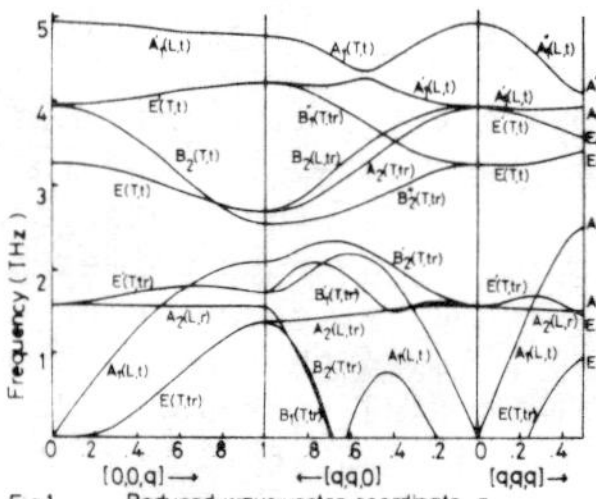

Fig.1 Reduced wave vector coordinate q
Phonon dispersion curves for $Ni(NH_3)_6Cl_2$ above T_c.
No reduction in cluster-cluster interaction.

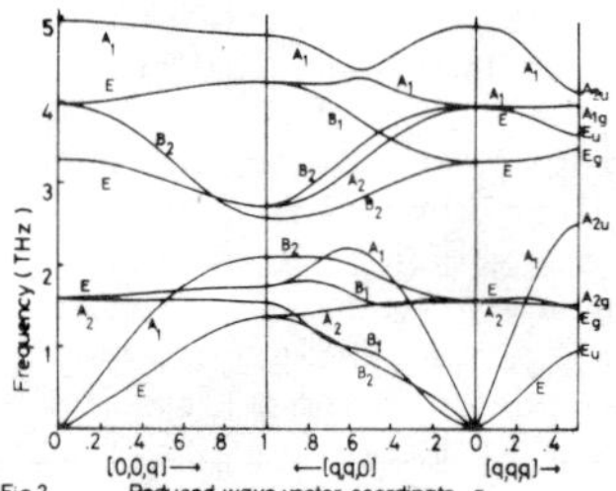

Fig.2 Reduced wave vector coordinate q
Phonon dispersion curves for $Ni(NH_3)_6Cl_2$ above T_c.
50% reduction in cluster-cluster interaction.

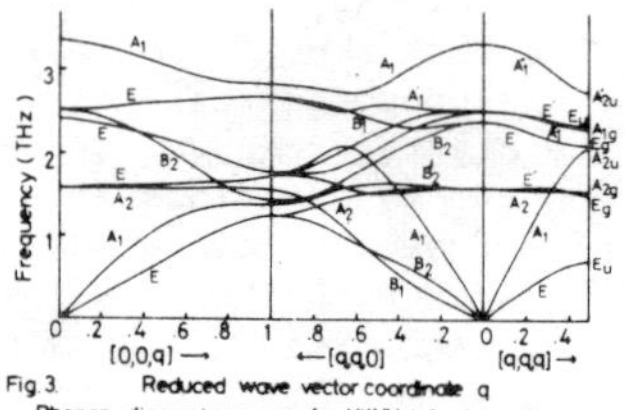

Fig.3 Reduced wave vector coordinate q
Phonon dispersion curves for $Ni(NH_3)_6I_2$ above T_c
50% reduction in cluster-cluster interaction

As the temperature is lowered, q_t increases and the charged triangles contribute to the multipole interaction. At the zone centre, this causes a splitting of the triplets by an amount which increases with q_t. Away from the zone centre, the mixed acoustic librational modes are very sensitive to the ratio q_t/q_r, which is perhaps not surprising in view of the sensitivity of this motion to the interaction between clusters. Along [1 1 0] the initial gradients of the transverse acoustic modes are reduced owing to coupling with the depressed librational doublet i.e. they have a tendency to become 'soft'. For the chloride lattice, when $q_t/q_r \sim 0.03$ the initial gradient of one of these modes becomes zero, whereas for the iodide lattice the modes are far less sensitive to this ratio and a value of $q_t/q_r \sim 0.2$ is required (figure 4).

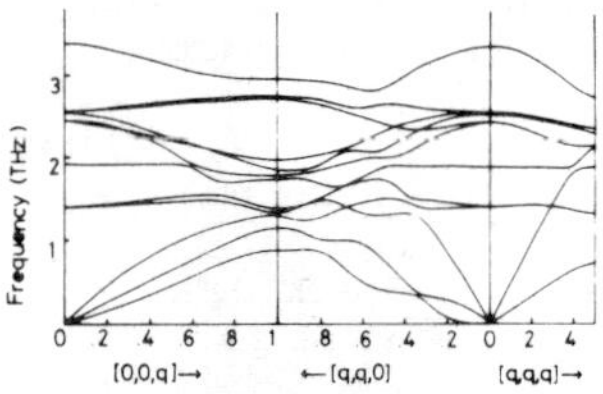

Fig.4 Reduced wave vector coordinate q
Phonon dispersion curves for $Ni(NH_3)_6I_2$ $q_t/q_r = 0.2$

If these 'soft' modes correspond to real instabilities, they will give rise to crystal structure changes within the lattice at characteristic temperatures. Furthermore, since an increase in q_t/q_r corresponds to a decrease in temperature, the phase transition for the iodide is expected to occur at a lower temperature than for the chloride.

Further work is needed to find the precise nature of the change in crystal structure and to confirm that the 'instabilities' arising in our calculations are relevant to the change in phase. If they are we should expect that the chloride crystal would suffer a larger distortion than the iodide lattice, a result which has been confirmed experimentally.

4. Conclusions. These preliminary results are consistent with the previously proposed mechanism explaining the phase transitions. Particularly encouraging is the sensitivity of the low-frequency modes to the magnitude of the cluster-cluster interaction and the result that these low-frequency modes tend to become unstable due to the H_3 triangle contributions in the multipole interactions. The model also correctly predicts that the iodide lattice is more stable than the chloride lattice, and that the critical phase transitions for the iodide crystal would occur at a lower temperature than for the chloride.

5. References

Bates A.R., Stevens K.W.H., 1969, J. Phys C: Solid St. Phys., 2, 1573-85
Bates A.R., 1970, J. Phys C: Solid St. Phys., 3, 1825-30.
Haussühl S., 1963, Phys. Stat. Sol., 3, 1072-76.
Kellermann E.W. 1940, Phil. Trans. Roy. Soc., A, 238 513-48
O'Leary G.P., Wheeler R.G. 1970, Phys. Rev. B. 1, 4409-39.
Palma-Vitorelli M.B. et al, 1960, Physica, 26, 922-30
Sacconi L. et al, 1964, Inorg. Chem, 3, 1772-74.
Trapp C., Chin-I Shyr, 1971, J. Chem Phys. 54, 196-203.

STUDIES OF NH_3 GROUPS DYNAMICS IN HEXAMMINE-CADMIUM FLUOROBORATE $Cd/NH_3/_6/BF_4/_2$ BY NMR RELAXATION

N.Piślewski, J.Stankowski and L.Laryś
Div.of Solid State Radiospectroscopy, Polish Academy of Sciences, Poznań, Poland

Abstract. Relaxation times T_1 and $T_{1\varrho}$ are studied vs temperature in hexammino-cadmium fluoroborate. From measurements of $T_{1\varrho}$ was determined the activation energy of tumbling motion of the complex cation and from T_1 obtained energy activation of tumbling motion of the complex cation and hindering rotation of NH_3 groups. In the phase transition region T_1 are the sharp discontinuity.

1. Introduction. In $NiA_6/BF_4/_2$ a transition point, $T_c = 135^oK$, was determined by EPR [1], whereas the specyfic heat anomaly detected by Janik et al. [2] was found to lie at $T_c = 141^oK$. These phase transitions in hexammino compounds can be related theoretically [3] with a change in the dynamics of NH_3 groups. The presend paper is devoted to the investigation of NH_3 group dynamics in hexammine cadmium fluoroborate. It is moreover aimed at ascertaining wheter the nuclear relaxation method permits to determine of T_c in diamagnetic hexammine salt.

2. Experimental. The NH_3 group dynamics in $Cd/NH_3/_6/BF_4/_2$ was studied with a Bruker SXP-4-100 pulsed spectrometer. Proton resonance was observed at 90MHz. The relaxation times T_1 and $T_{1\varrho}$ were measured vs temperature from 100 to 440°K and from 200 to 400°K, respectively /Fig.1/. Measurements of $T_{1\varrho}$ were carried out in field $H_1 = 15.5$ Oe. The temperature of the sample was checked and stabilized with an accuracy better than $\mp 1^o$K.

3. Discussion of the results. The relaxation time T_1 vs reciprocal temperature grows monotonically from decomposition temperature to about 200°K. The activation energy, evaluated in this range from the linear dependence of $\ln T_1$ vs the reciprocal temperature, amounts to 6.63 kcal/mole. At 200°K, the relaxation time is maximal, equaling /7.7$\pm$ 0.1/s. A further decrease in temperature leads to a minimum of T_1 at 105°K with a value of $T_1 = 400$ ms. However, the decrease in relaxation time is not monotonic: in the neighbourhood of 147°K, a marked jump in T_1 is observed. This anomaly is evidence for a change in the dynamics conditions of the NH_3 group, and is related with a transition temperature T_c. On the assumption of independient motion of the three-proton groups, one can write for the pure polycrystalline substance the formula derived by Bloembergen, Purcell and Pound, and supplemented by O'Reilly and Tsang [4], describing the dependence of T_1 on the correlation time. For an isolated NH_3 group, the correlation rate due to dipole interaction alone is given by the following expression:

$$\frac{1}{T_1} = \frac{9}{20}\,\frac{\gamma^4\hbar^2}{r^6}\left[\frac{\tau_c}{1+\omega^2\tau_c^2} + \frac{4\tau_c}{1+4\omega^2\tau_c^2}\right], \qquad /1/$$

where τ_c is the correlation time related with hindered rotation or oscilations about the equilibrium position of the NH_3 group, the remaining notation being conventional. Calculations for $r_{H-H} = 1.63$ Å /the distance between protons in the free NH_3 molecule/ and a Larmor frequency amounting to 90 MHz in the present case, lead with regard to the condition $\omega_0\tau_c = 0.616$ to a minimal value of $T_1 = 39$ ms. The same calculation carried out with Bildanov's formula [5], leads to $T_{1min.} = 77$ ms for the free molecule. Both these values are much smaller than the time T_1 determined experimentally. Assuming that the value of T_1 can be thought to be determined by one NH_3 group, the others contributing by way of spin diffusion, the formula for the temperature-dependence of T_1 is:

$$\frac{1}{T_1} = \frac{N}{n}\left(\frac{1}{T_1}\right)_{NH_3}, \qquad /2/$$

where N is the number of all protons in $Cd/NH_3/_6^{2+}$, and n=3 denotes the NH_3 group in its deep energy minimum. The value for T_{1min} obtained from Eq-/2/ amounts to 462 or 234 ms according to wheter T_1 of NH_3 group is calculated on the Bildanov or O'Reilly-Tsang theory. Complete agreement with experiment is achieved on Bildanov's theory if a smaller distance, 1.59Å, is assumed between the protons of NH_3 group.

The activation energy for the NH_3 group between 145 and 120°K amounts to 1.98 kcal/mole. This is in agreement with the value derived by Aiello and Palma-Vittorelli [6] from dielectric studies at temperatures below the phase transition point for $Ni/NH_3/_6\,Cl_2$.

The shortening of T_1 with groving temperatures from about 250°K to the decay temperature /440°K/ without exhibiting a minimum can point to isotropic motion in the $[Cd/NH_3/_6]^{2+}$ ion. The lack of a minimum of T_1 in the temperature range under consideration precludes the determination of the interproton distance or calculation of the correlation times. The two quantities, however, can be calculated by having recurse to the dependence of relaxation time $T_{1\varrho}$ on the reciprocal temperature exhibiting a marked minimum of about 350 μs. Assuming that due to rapid rotation of NH_3 groups all distances between interacting protons of

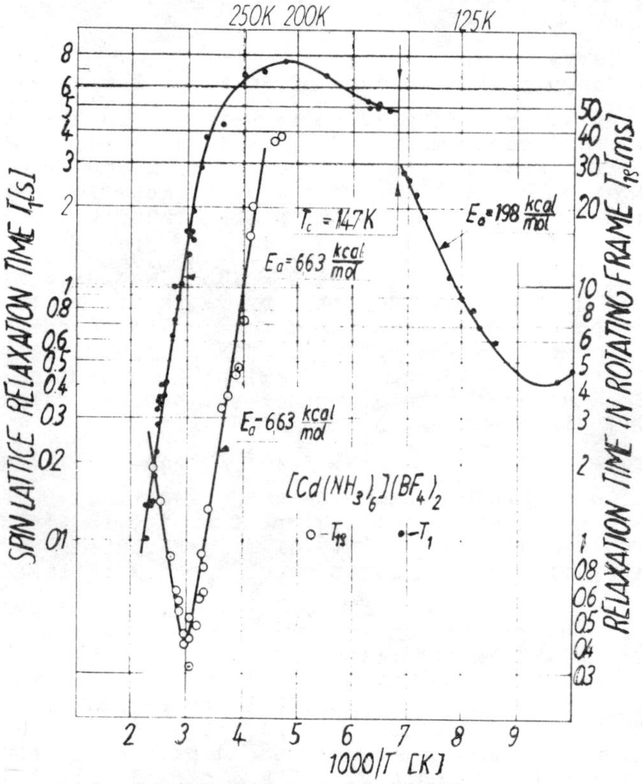

Fig.1.Experimental results for T_1 and $T_{1\rho}$ vs reciprocal temperature for Ni/NH$_3$/$_6$/BF$_4$/$_2$

neighbouring NH_3 groups are determined by the distances between the centres of the planes defined by the rotating protons,one can calculated correlation time of tumbling and the distance between central ion and the centre of the planes of hydrogens.Taking into consideration only interactions between NH_3 groups of the same complex ion and with regard to its symmetry,formula for $T_{1\rho\,min.}$ is:

$$\frac{1}{T_{1\rho\,min}} = \frac{3}{2}\,\frac{33}{64}\,\frac{\gamma^4\hbar^2}{a^6}\,\frac{1}{4\omega_1} \qquad , \qquad /3/$$

where a is the distance between the central ion and the centre of the plane of hydrogens.Insertion in /3/ of the experimental value $T_{1\,min.}$ = 350 μs and ω_1= = $65\cdot10^3$ rad$\cdot$s^{-1} yields a distance of a=2.68 Å. Since the preceding relaxation mechanism concerns an ion of the type $[MeA_6]^{2+}$, the value derived for the distance a is well adapted to a comparison with the value 2.5 Å assumed by Bates and Stevens for $[NiA_6]^{2+}$ in their theory.

The activation energy,evaluated from $T_{1\rho}$ measurements amounts to 6.63 kcal/ mole,in agreement with the result high temperature measurements of T_1. This is evidence that relaxation is defined by the same mechanism in both cases, namely by motion of the complex $[MeA_6]^{2+}$ as a whole. Calculated from $T_{1\rho}$ measurements the correlation times of tumbling of the complex ion range from 10^{-6} to 10^{-3} s and those for NH_3 below the phase transition lie between $5\cdot10^{-9}$ and $2\cdot10^{-10}$ s.

4. References.

1. J.Stankowski at al Phys.Stat.Sol./a/16 K 167 /1973/
2. J.Janik at al. Phys.Stat.Sol./a/16 K 165 /1973/
3. A.R.Bates and K.W.H.Stevens, J.Phys.C./Solid State Phys./ 2 1573 /1969/
5. M.M.Bildanov, Diss.Kazań 1968
6. G.Aiello and M.B.Palma-Vittorelli, Phys.Rev.Lett. 21, 137 /1968/
4. O.E.O'Reilly and T.Tsang, Phys.Rev., 157 417 /1967/

^{14}N NQR AND RELAXATION IN CD_3CN

Tzalmona A.*, and Kaplan A.**.
*Department of Physics, University of Nottingham, University Park, Nottingham NG7 2RD, ENGLAND.
**Department of Physics, The Hebrew University of Jerusalem, ISRAEL.

Pulsed NQR study of ^{14}N in CD_3CN has been performed from 77^oK up to the melting point (230^oK). The NQR frquencies at 77^oK are f = 2812.8 Kc/sec and 2804.2 Kc/sec. Spin-lattice relaxation time measurements reveal a minimum in T_1 due to the motion of the methyl group with activation energy of 2 Kcal/mole. The appearance of a phase transition has been found at 220^oK.

OBSERVATION OF ANOMALOUS SPIN-LATTICE RELAXATION FOR PROTONS IN SOLID $(NH_4)_2PbCl_6$

J.E. Tuohi, E.E. Ylinen, and L.K.E. Niemelä
Wihuri Physical Laboratory, University of Turku, SF-20500 Turku 50, Finland

Abstract. Nuclear spin-lattice relaxation times T_1 for protons in polycrystalline ammonium hexachloroplumbate, $(NH_4)_2PbCl_6$, were measured by the NMR pulse method at the Larmor frequencies 9 MHz and 22 MHz. Three minimum values of the experimental T_1 at 22 MHz and two minimum values at 9 MHz were observed. A complex frequency dependence of T_1 was detected below 70 K. The height of the rotational barriers was considered.

1. Introduction. The study of the motion of ammonium groups in some solid ammonium salts by NMR techniques at low temperatures has recently been actively pursued. The temperature dependence of the spin-lattice relaxation time T_1 for protons or deuterons has been found to be inexplicable by the aid of the classical reorientation model (Niemelä et al., 1970, Ylinen et al., 1974, and Tuohi et al., 1974). In order to get more information about ionic motions in solids at low temperatures the spin-lattice relaxation times of protons in the solid ammonium hexachloroplumbate(IV), $(NH_4)_2PbCl_6$, were measured in the present investigation at different Larmor frequencies. No previous works concerning the ionic motions in this salt or its physical properties at temperatures below liquid nitrogen temperature have come to the writers' attention. From a study of powder photographs of $(NH_4)_2PbCl_6$ it has been shown that this compound has the same type of atomic arrangement as a cubic $(NH_4)_2PtCl_6$ (Wychoff et al., 1926).

2. Experimental. Ammonium hexachloroplumbate was synthesized by passing chlorine gas into a suspension of lead chloride in hydrochloric acid. The lemon-yellow crystals were obtained on adding ammonium chloride into the product (Friedrich, 1893). The salt was dried in vacuum for 70 h and then sealed in a glass ampoule under He-atmosphere. Proton spin-lattice relaxation times T_1 in the polycrystalline sample were measured at the Larmor frequencies 9 MHz and 22 MHz by the NMR pulse method using Bruker SXP4-60 spectrometer with the pulse programmer constructed in our Laboratory. A saturation with three pulses $-t-90^{\circ}$ pulse trains were used to measure T_1. The signals were recorded by photographing. The temperature of the sample was controlled below the room temperature down to 4.2 K, either, by employing cold nitrogen and helium gases or by using liquid helium. The temperature was measured as described by Tuohi et al. (1974).

3. Results and discussion. The values of proton spin-lattice relaxation time are plotted in Figure 1. in the semilogarithmic scale versus temperature between 4.2 K and 100 K.

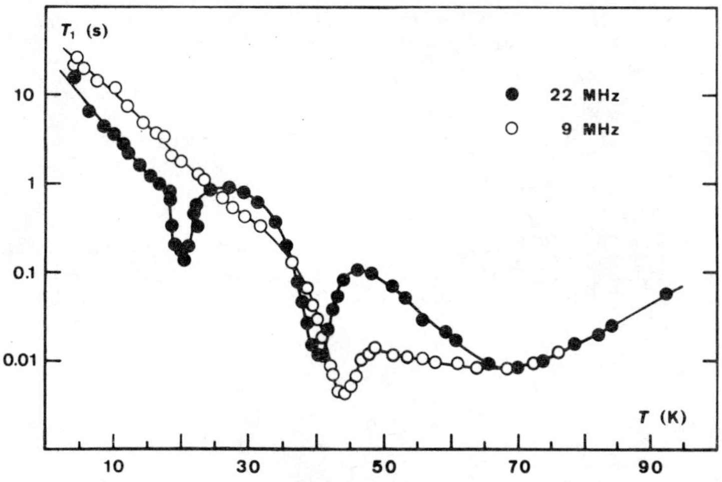

Figure 1. Proton spin-lattice relaxation time T_1 as a function of temperature in $(NH_4)_2PbCl_6$ at Larmor frequencies 9 MHz and 22 MHz.

Three minimum values of the experimental T_1 at 22 MHz can be observed: 8.5 ms at 69 K, 12 ms at 40 K and 150 ms at 20 K. The last minimum disappeared at 9 MHz and only 8.5 ms at 69 K and 4.2 ms at 44 K can be seen. Thus with the increasing Larmor frequency the second minimum anomalously shifts to lower temperatures and at the same time T_1 becomes longer. At the last minimum T_1 has also been found to behave in the same way at 22 MHz and 42 MHz (Tuohi et al., 1975).

The recovery of the magnetization of the nuclear spin system after the saturation was examined at various temperatures and it was proved to be non-exponential below 20 K. The behaviour of the T_1 minima and the non-exponentiality of the recovery of the magnetization are concluded to be due to the quantum-mechanical tunneling effects. Below 20 K at both frequencies the shape of the free induction signal was ascertained to differ from that found at upper temperatures and the signal seems to consist of two components. Both components gave approximately same values for T_1.

The shape of the T_1 minimum at 69 K proved to be frequency dependent at the Larmor frequencies used and is anomalously much broader at 9 MHz than at 22 MHz. Further studies have indicated, however, that the value of this T_1 minimum will also increase with increasing Larmor frequency (Tuohi et al., 1975).

When the proton relaxation is mainly controlled by proton-proton intra-ionic interaction modulated by isotropic NH_4^+ reorientation, with the correlation time τ_c, T_1 is given by the well-known expression (Abragam, 1960):

$$T_1^{-1} = \frac{9\gamma_H^4\hbar^2}{10r_{HH}^6}[(1+\omega_o^2\tau_c^2)^{-1} + 4(1+4\omega_o^2\tau_c^2)^{-1}]\tau_c, \quad (1)$$

where the parameters have their usual meanings. Eq. (1) gives the following minimum values of T_1 at 69 K : 1.75 ms at 9 MHz and 4.29 ms at 22 MHz. At 22 MHz the fitting of equation (1) at the T_1 minimum by using the experimental constant in front of the brackets gives very good agreement with experimental points, although the value of T_1 minimum differs from the theoretical one. The origin of this minimum is considered to be mainly reorientational motion, though not quite in the classical meaning, similar with recent results (Ylinen et al., 1974, and Tuohi et al., 1974). The correlation times for the reorientational motion of ammonium ions can be deduced from formula (1), and thus the numerical values of the apparent activation energies and pre-exponential factors can be obtained assuming that the motion is describable simply as a thermally activated Arrhenius process. By using the experimental minimum value and the measured data above 60 K for T_1, the values of 5.9±0.2 kJ/mol and $(3.4\pm0.4)\cdot10^{-13}$ s were calculated for these parameters. These values are in good agreement with those obtained for $(NH_4)_2GeF_6$ (Ylinen et al., 1974). The fitting of the form of equation (1) to the minima at 40 K at 22 MHz and at 44 K at 9 MHz separately and the use of the Arrhenius equation gives the apparent activation energy 11.3±0.8 kJ/mol, which is remarkably greater than that at 69 K.

4. Acknowledgements. The writers are obliged to Professor V. Hovi for the facilities of this Laboratory. Warm thanks are due to Mr. J. Mäkelä, B.A., for preparing the sample. We wish to express our gratitude to the National Research Council for Sciences (Valtion luonnontieteellinen toimikunta), Finland, for financial support in 1974.

5. References

Abragam, A., 1960, Principles of Nuclear Magnetism, Oxford University Press, London.
Friedrich, H., 1893, Ber. 26, 1434.
Niemelä, L., and Tuohi, J., 1970, Ann. Univ. Turkuensis A I, No. 137.
Tuohi, J.E., and Ylinen, E.E., 1975, to be publ.
Tuohi, J.E., Ylinen, E.E., and Niemelä, L.K.E., 1974, to be publ. in Chem. Phys. Letters.
Ylinen, E.E., Tuohi, J.E., and Niemelä, L.K.E., 1974, Chem. Phys. Letters 24, 447.
Wyckoff, R.W., and Dennis, L.M., 1926, Am. J. Sci. 12, 503.

MOLECULAR MOTION IN SOLID SILANE

G. Janssens[*], P. Van Hecke[**], L. Van Gerven
Laboratorium voor Vaste Stof-Fysika en Magnetisme, University of Leuven, 3030 Leuven (Belgium)

Abstract. Molecular motion in polycrystalline solid SiH_4 is studied between 90 K and 4.2 MHz. Measurements reflect translational diffusion between the melting point and the phase transition temperature, freezing of the molecules and reorientational motion below the transition temperature, and the appearance of spin isomerism effects below 25 K.

Very little is known about silane in the condensed phase (Klein et al., 1969; Jones, 1973). In an attempt to explain the motional behaviour of silane in the solid phase, we measured the spin-lattice relaxation time T_1 as a function of temperature, at 24 MHz. We also calculated second moments from the F.I.D.

The temperature dependence of T_1 (fig. 1) is characterized by discontinuities at the melting point (88.5 K) and at the solid-solid phase transition (64.5 K), by a minimum at 47 K, and a knee around 23 K.

The minimum in T_1 at 47 K is characteristic of a reorientational process described by (intramolecular contribution only):

$$T_1^{-1} = \frac{A}{\omega}\left(\frac{\omega\tau_c}{1+\omega^2\tau_c^2} + \frac{4\omega\tau_c}{1+4\omega^2\tau_c^2}\right) \qquad (1)$$

τ_c is the correlation time of the reorientation motion, and A is a constant. Expression (1) is maximum for $\omega\tau_c = 0.62$, from which we determine τ_c.

Using the activation energy E_a, determined from the slope of the T_1 vs. T curve, we calculate τ_c°, the preexponential factor.

For temperatures below the minimum at 47 K, E_a = 1.1 kcal/mol and $\tau_c^\circ = 1.3 \times 10^{-14}$ s; above the minimum we found E_a = 1.9 kcal/mol and $\tau_c^\circ = 8 \times 10^{-17}$ s.

The low temperature value (below 30 K) of the second moment calculated from the Gaussian F.I.D. is found to be $2.7 \pm 0.2\ G^2$. The theoretical expression for a rigid four-spin-1/2 system gives an intramolecular second moment $S_1 = 5.46\ G^2$ (SiH_4 is a tetrahedral molecule with Si-H bond length value of 1.477 Å), and an intermolecular moment $S_2 = 0.88\ G^2$. The latter value was found assuming that the low temperature lattice structure of SiH_4 is f.c.c., the same structure as CH_4 and CD_4, but not confirmed for SiH_4 (Klein et al., 1969), and a density of 0,80 g/cm^3 at low temperatures (Stewart, 1962). This disagreement with the experimental value is due to the effects of spin isomerism. For a spectrum dominated by a central meta component, the linewidth is known to be definitely narrower than for the four-proton-1/2 system (Tomita, 1953).

The para-modification has spin I = 0 and does not contribute, while the meta-modification (I = 2) has zero intramolecular dipolar interaction, which yields an unshifted line with intensity 5 in the absorption spectrum. The only intramolecular contribution to the spectrum and the second moment is due to the ortho-modification (I = 1), which gives 6 absorption lines around the Larmorfrequency, with total intensity 3.

For a polycrystalline sample, the intramolecular second moment is then calculated to be 1.75 G^2. Assuming a f.c.c. lattice, the intermolecular contribution to the second moment is 0.88 G^2 the total value is then 2.63 G^2 which is in good agreement with the measured rigid lattice value of 2.7 G^2.

Around 30 K this value decreases to a new constant value of 1.7 G^2, up to the phase transi-

---

[*] Research fellow of the Belgian "Interuniversitair Instituut voor Kernwetenschappen".

[**] "Aangesteld navorser" of the Belgian "NationaalFonds voor Wetenschappelijk Onderzoek".

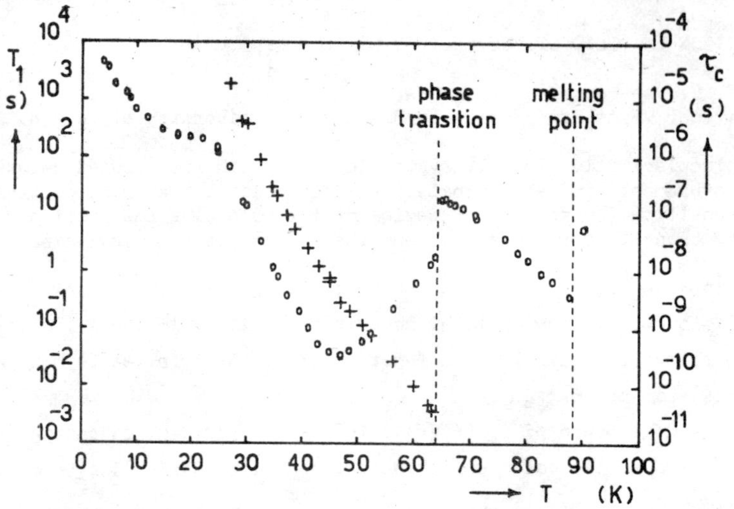

Fig. 1 Proton spin-lattice relaxation time T_1 (°) and rotational correlation time τ_c (+) vs. temperature, in solid SiH_4 at 24 MHz.

tion at 64.5 K, which is due to motional narrowing by reorientation of the molecules. This effect shows up when the correlation frequency becomes larger than the line width ($\delta = 1.18\ M_2^{1/2}$ for a Gaussian line), in our case when $\tau_c = 2.6 \times 10^{-6}$s, in agreement with the value from the τ_c vs. T curve at 31 K.

Below 25 K an anomaly in T_1 appears, also reported by Jones (1973), and is taken to be along with the linewidth (second moment) data as a good evidence for the existence of nuclear spin symmetry states.

At the phase transition (64.5 K) a discontinuity in the spin-lattice relaxation time and a jump in the second moment value occur. Above this temperature, T_1 is a decreasing function of the temperature: at 24 MHz, however, a minimum is not reached. The lineshape is Lorentzian, characterized by a transversal spin-lattice relaxation time T_2 which is also a decreasing function of temperature, with the same slope as the T_1 vs. T curve. This slow decay of the F.I.D. is due to intermolecular dipolar interactions modulated by translational diffusion of the molecules. For slow translational motions ($\omega\tau_c >> 1$) correlation times are proportional to T_1 (Resing et al., 1963), so that the value of the activation energy for molecular self-diffusion can be calculated from the slope of the T_1 vs. T curve. This gives 2.5 kcal/mol.

This value is to be compared to the ones predicted by other approaches (Niemela et al., 1970): 2.83 kcal/mol as calculated from the melting temperature 88.5 K, and 2.63 kcal/mol from the latent heat of melting (159.5 cal/mol, Eucken, 1939).

At the melting point T_1 values in the liquid phase agree with the measurements of Bozek et al. (1968).

References.

BOZEK M., HENNEL J.W., JASINKI A., KOZINSKI S., (1968), Acta Phys. Polon. 33, 337.
EUCKEN A., (1939), Ztschr. Elektrochem. 45, 2, 128.
JONES E.P., (1973), Phys. Lett. 43A, 1.
KLEIN M.L., MORRISON J.A., WEIR R.D., (1969), Disc. Farad. Soc. 48, 93.
NIEMELA L., NIEMELA M., (1970), Ann. Acad. Sc. Fenn. A341.
RESING H.A., TORREY H.C., (1963), Phys. Rev. 131, 3, 1102.
STEWART J.W., (1962), J. Chem. Phys. 36, 2, 400.
TOMITA K., (1953), Phys. Rev. 89, 429.

NUCLEAR MAGNETIC RELAXATION IN SOME TETRAMETHYLAMMONIUM SALTS

S.Jurga°, J.Depireux+ and Z.Pająk°
° *Instytut Fizyki, Uniwersytet im.A.Mickiewicza, POZNAŃ Poland*
+ *Institut de Physique, Université de Liège, LIEGE, Belgium*

ABSTRACT. Spin-lattice relaxation times have been measured in polycrystalline tetramethylammonium (TMA) nitrate, sulphate, perchlorate and fluoborate over a wide temperature range. The results have been interpreted in terms of various molecular motions occurring in those compounds.
A new phase transition for nitrate was found near 273 K; proton and fluorine relaxation time measurements on the fluoborate showed the existence of some cross-relaxation due to proton-fluorine interactions.

1. Introduction. The molecular motions in solid tetramethylammonium nitrate (TMA NO_3), sulphate ($(TMA)_2SO_4$), and perchlorate (TMA ClO_4) were studied by means of the proton resonance line shape and the second moments by Mahajan and Nageswara-Rao (1974). Stammler et al. (1966) carried out X-Ray diffraction and Differential Thermal Analysis which showed the perchlorate and tetrafluoborate salts to be isomorphous. In this note, we report a study of the proton spin-lattice relaxation times T_1 and $T_{1\rho}$ in TMA NO_3 and in $(TMA)_2SO_4$, T_1 in TMA ClO_4 as well as proton and fluorine T_1's in TMA BF_4 over a wide range of temperature. Our aim was to obtain much more reliable values of activation energies for the processes governing the relaxation and to look for some possible phase transitions in the nitrate and sulphate salts.

2. Experimental results and discussion. The nuclear magnetic relaxation times were measured using a BRUKER NMR pulse spectrometer (B-KR 322s) operated at 60 and 62 MHz. T_1 was measured with either $90-\tau-90$ or $180-\tau-90$ pulse sequences and $T_{1\rho}$ by spin locking technique. The samples were prepared in the Department of Chemistry of the A.Mickiewicz University in Poznań.

a. Tetramethylammonium nitrate - The results for TMA NO_3 are shown in Fig.1 as semi-log plots of T_1 and $T_{1\rho}$ vs. $1/T$. The T_1 curve shows a discontinuity around 274 K when the solid is cooled and around 285 K upon heating. Our thermogrammes (kindly recorded by Dr.P.Jérôme, University of Liège) similarly show peaks at 274.5 K and 287.5 K upon cooling and heating, indicating a first-order phase transition. Above these points, the spin-lattice relaxation time T_1 exhibits a minimum of 21.5 ms at 286 K and increases again with increasing temperature, a feature which we ascribe to the TMA ion rotation as a whole: this assumption implies that only the intermolecular term contributes to the modulated second moment, M_2(mod), which is easily deduced from the relation $M_2(mod) = M_2(\text{rigid lattice}) - M_2(\text{inter})$. Taking for the latter two quantities the experimental values of 27.5 G and 0.85 G given by Mahajan and Nageswara-Rao (1974), we calculate, at 60 MHz, T_1min to amount to 20.8 ms in excellent agreement with the observed value. The lack of crystal structure for TMA NO_3 does not allow an accurate calculation of M_2(rl) and of M_2(inter), but the values adopted from the literature seem reasonable (Dufourcq et Lemanceau (1970) and other references cited therein). Assuming that the motion responsible for relaxation is thermally activated, i.e., $\tau_c = \tau_o \exp(E_a/RT)$, values of E_a=5.4 kcal/mol and of $\tau_o = 1.2 \cdot 10^{-13}$s may be calculated for the isotropic rotation of the TMA ion as a whole. Below the phase transition, down to 220 K, the plot of log T_1 vs $1/T$ is non-linear; below 220 K, it becomes linear with an activation energy of 6.5 kcal/mol. We assume that in this phase there exist two different motions, viz. the rotation of the CH_3 groups around their C_3 axes and the tumbling of the whole TMA ion. In order to resolve these motions, $T_{1\rho}$ measurements were performed between 274 and 160 K and are shown in Fig.1. An H_1 field of 30 G was used. The $T_{1\rho}$ data give two minima with values of 0.50 and 0.34 ms at temperatures around 245 and 200 K. We associate these features of $T_{1\rho}$ with the overall tumbling of the TMA ion and with CH_3 group rotation. A modification by Albert et al. (1972) of Woessner's (1962) general expression for the spin-lattice relaxation of a proton pair undergoing simultaneously both anisotropic reorientation about an axis and a random tumbling of this axis leads to a relationship that allows a clear distinction between both types of motions. Assuming the TMA ion in the nitrate salt to have the same structure as in simple TMA halides, we may take from Dufourcq et Lemanceau (1970) the inter-proton distance,

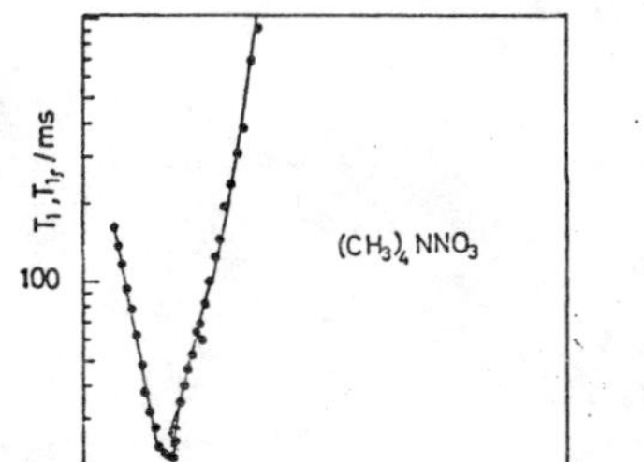

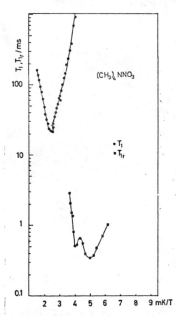

Fig.1 Relaxation times vs. inverse temperature

r=1.78 Å and the distance between protons in adjacent methyl groups, r°=3.04 Å, (the same values will be used throughout) which lead to T_1min(tumbl)=0.49 and T_1min(CH_3)=0.28 ms. The agreement with the observed values is good and the 20% discrepancy for CH_3 rotation may be due to large errors in measurements bound to the considerable narrowing of the FID. The activation parameters deduced from this study are the followings: E_a(tumbl)=10.2 kcal/mol; τ_o(tumbl)=0.6 10^{-15}s; E_a(CH_3)=6.5 kcal/mol and τ_o(CH_3)=0.5 10^{-13}s, which are all in good agreement with the same set of values estimated from previous line shape measurements.

b.Tetramethylammonium sulphate - T_1 in function of temperature reaches at 385 K a minimum of 31.5 ms which may be assigned to the CH_3 rotation. The estimated value for this motion amounts to 32.8 ms in good agreement with the experimental value. The temperature measurements of $T_{1\rho}$ at H_1=30 G exhibit two minima of 0.47 and 0.34 ms at temperatures of 347 and 160 K which may be attributed, following the model of Albert et al. (1972) to tumbling motion of the TMA ion and to rotation of CH_3 about its C_3 axis with estimated values of 0.49 and 0.28 ms respectively. The corresponding activation parameters are: E_a(tumbl)=14.1 kcal/mol; τ_o(tumbl)=0.9 10^{-15}s; E_a(CH_3)=5.4 kcal/mol and τ_o(CH_3)=1.4 10^{-12}s.

c.Tetramethylammonium perchlorate - The T_1 temperature curve is V-shaped with a minimum of 22 ms at 239 K. The behaviour of T_1 for this salt is quite analogous to that of the nitrate above its phase transition; $T_{1\rho}$ measurements give the same results, which implies that only isotropic motion of the TMA ion as a whole takes place. Data collected at temperatures below the phase transition mentioned by Stammler et al. (1966) at 613 K give E_a=6.7 kcal/mol and τ_o=0.8 10^{-13}s.

d.Tetramethylammonium tetrafluoborate - The results of ^1H and of ^{19}F relaxation times measurements against temperature are shown in Fig.2. The proton T_1 reaches minima of 21 and 275 ms at 256 and 104 K. The slopes of the curve on both sides of the high temperature minimum give activation energies of 4.9 and 4.3 kcal/mol; the pre-exponential factor amounts to about 2 10^{-13}s. The T_1 behaviour up to 160 K is nearly the same as that of the isomorphous TMA ClO_4: this is why we assign it to the isotropic rotation of the TMA ion as a whole. The fluorine T_1 shows some inflection around 247 K and reaches a minimum of 108 ms at 202 K. Upon lowering further the temperature, it passes through a maximum at 167 K after what it decreases. The ^{19}F T_1 minimum may be ascribed to the isotropic rotation of BF_4. The fluorine dipolar interaction gives, for a motion of this kind, with r(F-F)=2.34 Å a T_1min of 112 ms at 62 MHz (r cited by Huettner et al. 1968). It seems that the ^{11}B-^{19}F **intraionic** coupling doesnot play any rôle here contrary to what happens in inorganic tetrafluoborates (Huettner, 1968). The activation parameters for this motion are: E_a=3.1 kcal/mol and τ_o=8 10^{-13}s. We assume that from 333 K to 247 K, fluorine relaxation is dominated by the TMA rotation, slower than that of BF_4 and which modulates the interionic H-F dipolar interaction. In the region of 230-150 K, ^{19}F T_1 is non-exponential; the same occurs for proton between 172 and 130 K. Due to the same cross-relaxation which causes a minimum around 104 K in the proton T_1 curve and by the decrease in the fluorine T_1 from 167 K on, the correlation frequencies of TMA and of BF_4 become comparable with the difference between the proton and fluorine Larmor frequencies.

A more detailed analysis of relaxation processes in TMA BF_4 is now being completed and will soon be published.

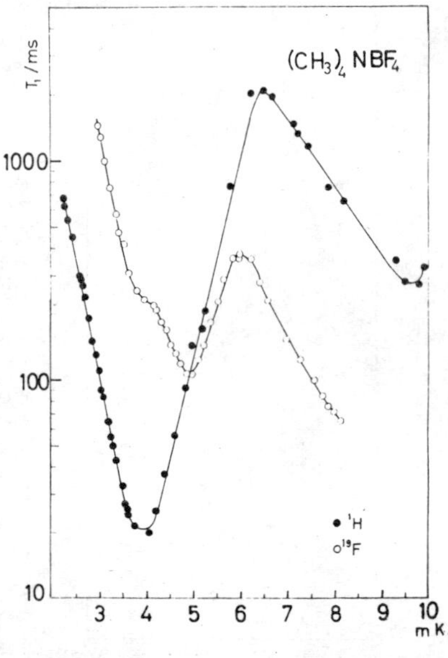

Fig.2 Relaxation times vs. inverse temperature

3. *References*.

Albert S., Gutowsky H.S. and Ripmeester J.A. 1972, *J.Chem.Phys.*, 56, 3672

Dufourcq J. et Lemanceau B., 1970, *J.Chim.Phys.* 67, 9.

Huettner D.J., Ragle J.L., Sherk L., Stengle T.R. and Yeh H.J.C., 1968, *J.Chem.Phys.*, 48, 1739.

Mahajan M. and Nageswara-Rao B.D., 1974, *J.Phys.C: Solid State*, 7, 995.

Stammler M., Bruenner R., Schmidt W. and Orcutt D., 1966, *Adv.X-Ray Anal.*, 9, 170.

Woessner D.E., 1962, *J.Chem.Phys.*, 36, 1.

4. *Acknowledgements*.

We gratefully acknowledge the help of Dr.M.Rinné; we thank Dr.A.Jankowski for having prepared the samples.

BROAD LINE NMR AND MOLECULAR DYNAMICS OF TETRAETHYLAMMONIUM BROMIDE

R.Goc, Z.Pająk and B.Szafrańska
Institute of Physics, A.Mickiewicz University, Poznań, Poland.

Abstract. PMR spectra of polycrystalline tetraethylammonium bromide revealed between 77°K and 473°K four linewidth transitions, two of which are acompanied by C_p anomalies, detected thermogravimetrically. The second moment values are discussed in terms of the most probable models of sterically hindered molecular reorientation suggesting the existence of correlated motions of the ethyl groups.

The structure and symmetry of the $N/Et/_4^+$ /TEA/$^+$ ion suggests the presence of molecular motions more highly complicated than in the case of $N/Me/_4^+$ ions /Polak et al.1973/. The relatively large dimensions of the Br^- ion result in weak intermolecular interactions largely simplifying the theoretical calculations.

NMR spectra of TEABr were recorded with a USSR made RYa2301 spectrometer /40MHz/. DTA was performed with a TA2000 Mettler instrument of sensitivity 60μV.s/mcal. The results of second moment measurements of NMR line /accuracy ±10%/ are shown in Figs 1 and 2. In all, about 700 spectra were recorded for 3 variously dried samples. The points in the graphs correspond to mean values from12 spectra on the average. The shape of M_2 in Fig.1 leads to the conclusion that the line narrowing observed troughout a wide range of temperatures occurs in four, clearly distinguishable steps. The value M_2=25,8G^2 at 77°K corresponds to a rigid structure of the $N/Et/_4^+$ ion with immobile methyl groups.The first low-temperature linewidth transition is caused by the emergence of sufficiently rapid rotation of $-CH_3$ groups. The activation energy, determined from the BPP formula, amounts to E=2,5kcal/mole,whereas the preexponential factor derived from the Arrhenius formula equals τ_0=4,5 10^{-11}s. The decrease in second moment in Region 2 /Fig.2/ is attributable to the emergence of vibrations of entire ethyl groups in conjunction with rapidly rotating $-CH_3$ groups. In fact, the plateau value M_2=13,6G^2 upwards of Region 2 corresponds to vibration motions of ethyl groups with amplitudes of about ±20°. The plateau of M_2=5,2G^2 following on Region 3 corresponds to full rotation of ethyl groups about the N-C bonds. Region 3 of the variations in linewidth is due to an increase, with rising temperature, in the frequency of this full rotation, which appears suddenly at 415°K presenting from the onset a rather high frequency. At the same temperature, moreover, an anomaly of the specific heat observed involving a change in enthalpy by ΔH=13cal/mole. This suggests that full rotation of the ethyl groups emerges as the result of a phase transition.For steric reasons, however, full rotation has to be ruled out. Therefore it is plausible that we have here a highly interesting case of mutual correlation of entire ethyl groups. For methyl groups such correlation has already been reported/ Chezeau et al. 1971/.

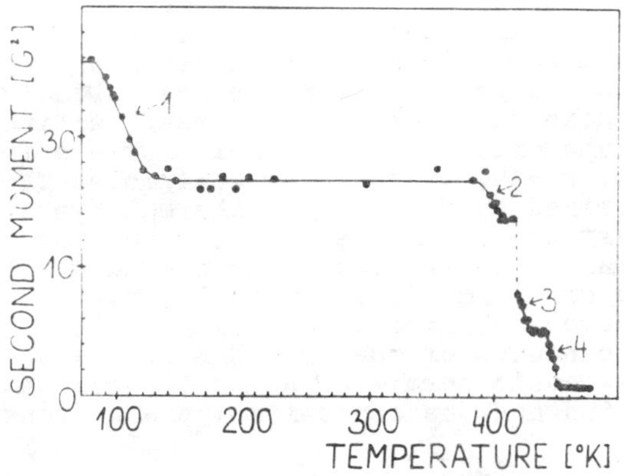

Fig.1.Second moment of TEABr vs.temperature.

The 4-th linewidth transition /M_2=0,6G^2/ corresponds to the appearance of isotropic rotation of the TEA^+ ion about its centre of mass. The activational /in the first stage of line narrowing/ process of increase in rotation frequency undergoes a perturbation due to a phase transition at 448°K, upwards of which rotation is so rapid that a further increase in its frequency is unable to affect the second moment value.

The DTA curve exhibits a complex anomaly of the specific heat in the vicinity of 448°K with a total change in enthalpy by ΔH=4 kcal/mole. Since isotropic rotation of molecules is highly specific for the cubic phase, it is plausible that, at 448°K, a transition occurs from tetragonal structure to the cubic one. At 468°K one more specific anomaly with ΔH=300cal/mole is observed, albeit

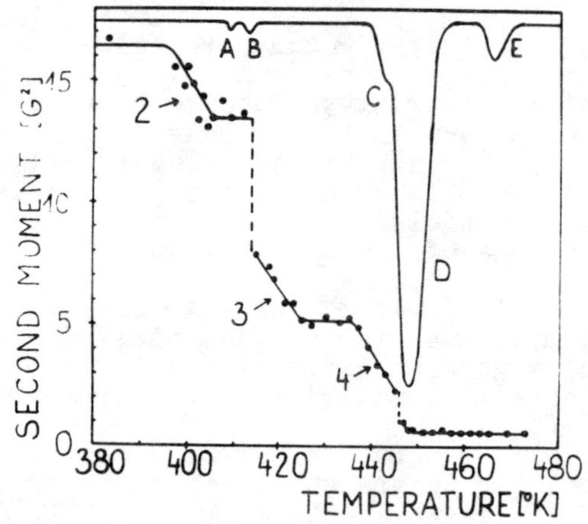

Fig.2. Second moment and DTA for TEABr plotted in extended temperature scale.

with no perceptible influence on the second moment. Nor it is easy to account for the specific heat anomaly at 408°K. The latter may be related with the emergence of vibrations of ethyl groups, since an attempt to apply the BPP formula to calculate the activation energy of this motion failed. This may suggest that we are not dealing here with a purely activational process. Nor are the further changes in linewidth accessible to explanation on the BPP formula. However, since the various changes in linewidth are in fact activational processes /at least within some range of temperatures/, activation energies were calculated from the approximate Waugh-Fiedin formula for the successive Regions of linewidth as: $E_2=15$; $E_3=16$; $E_4=17$kcal/mole. The linewidth transition in Region 4 is characterized by very strong thermal hysteresis, observed in thermal analysis too. At cooling, the specific heat anomalies C and D occur seperately, the anomaly D exhibiting a stronger hysteresis /it appears at 400°K at a cooling rate of 2°K/min/. The enthalpies, calculated from the DTA curve of the cooling process, amount to $\Delta H_c=1$ and $\Delta H_D=2$kcal/mole, respectively. Studies of further compounds of the type $N/Et/_4X$, with X=J, F and Cl are under way, and are expected to permit a fuller interpretation of the internal dynamics of the TEA^+ ion and the concemitant phase transitions.

Acknowledgements.
The authors are indebted to Mr.K. Vogel of Mettler Instruments AG, Zürich for performing the DTA. They also wish to thank R.Konieczka and J.Radomski for their part in carrying out the NMR measurements.

References

Chezeau,J.M.,Dufourcq,J.,Strange,J.H. 1971, Molec. Phys.,20,305.
Polak,M.,Sheinblatt,M. 1973, J. Mag. Res.,12,261.

SECTION K

Polymers

ELECTRON SPIN RESONANCE OF COMPLEXES CONTAINING SUBSTITUTED PYRIDINES AND $CuCl_2$ or CuNi $(CN)_4$

D. Inan and R. Morehouse*
Fizik Enstitusü, Hacettepe Ünivertesi, Ankara, Turkey

Abstract. Previous EPR studies of $Cu(NH_3)_2$ $Ni(CN)_4$ has assumed that the observed axis of symmetry was along the ammonia axis. Our studies of the inorganic polymers $Cu(py)_2$ $Ni(CN)_4$ and $Cu(py)_2$ Cl_2, where py is a substituted pyridine, show that the axis of Symmetry lies along one of the cyanide or chloride bonds.

1. Introduction. In his original preparation of nickel compounds, most of which turned out to be clathrates, Hofmann (K.A. Hofman and F. Hochtlen, 1903) produced a compound with the formula $Ni(py)_2$ $Ni(CN)_4$ which we refer to as Hofmann's Nickel pyridine complex. Analogous compounds were the first nickel has been replaced by other transition metals has been studied previously (Akynüz, et al. 1974). Of these compounds the Hofmann's copper pyridine complex was of particular interest since it showed a crystalline field of elongated tetragonal symmetry and high delocalization of the unpaired electron. Other atudies by Miyoshi et al. (1968) on Hofmann's copper-benzine type clathrate, $Cu(NH_3)$ $Ni(CN)_4$, C_6H_6, had found a compressed tetragonal crystalline field at the copper site and they had assumed that the axis of symmetry was along the ammonia axis. However infrared spectra (Akyüz et al 1973) showed in equivalent cyanide bonds. As discussed by Akyüz (1974) the probable axis of symmetry is along the cyanide bonds.

2. Results and Discussion. In the current work powder samples of $Cu(py)_2$ $Ni(CN)_4$ and $Cu(py)_2$ were py is a symmetrically substituted pyridine: -picoline 3,5 lutidine, 2,4,6 collidine, 4-tert, butyl pyridine, 3,5 dichloropyridine and 3,5 dibiomo pyridine. Not all of these would form satisfactory samples with the Hofmann pyridin complex sort but they did form the one dimensional chains of copper-pyridine-chloride. Electron Spin Resonance Spectra were measured at x-band with a standard Varian spectrometer. Many of these spectra showed a slight shoulder on the portion of the spectra associated with g perpendicular but only those spectra with clear splittings are listed with 3 g-factors. Measurements at higher frequencies will probably show that all the samples have three g-factors. A list of the g-factors is included below. Results for $Cu(NH_3)_2$ Cl_2 and $Cu(NH_3)_2$ Cu $(CN)_4$ have been included for comparison.

The infrared spectra of all these samples indicated a 2-fold axis at the copper site, for example a clear splitting of 40 wave numbers in the cyanide stretching frequency at 2100 wave numbers. This along with some x-ray cell sizes (Akyüz, 1974) indicate that the pyridine is much closer to the copper than to other transition metals and yet the symmetry is that of an elongated tetragon. It was noticed that a number of the spectra had a small extra bump at a position that corresponded to 45° from the axis of symmetry and in one case 3 lines of about 10 gauss splitting could be resolved there. This splitting was assigned to the nitrogen in the pyridine ring for an orientation of magnetic field perpendicular to the ring.

The extended tetragonal crystalline field in a sample where pyridine is particularly close to the copper, the fact that in general there three g-factors, and coupling of the electron to the pyridine nitrogen suggested to us that the axis of symmetry had to be along the cyanide or chloride bond. This is supported by the infrared cyanide splitting and the similarity between Hofmann pyridine type complexes and the copper pyridine chloride complexes where x-ray studies (Danitz, 1957) showed that the copper chloride bonds were the inequivalent and yet these compounds tend to show the same axial symmetry as the Hofmann pyridine type complexes.

* Currently at Physics Department, Pahlavi University, Shiraz, Iran.

Thus although single crystal x-ray results are not available for these compounds it seems certain that the symmetry axis is along the cyanide or chloride bond. It is unusual to have a compound with chemically identical bonds Vi7. Cu-CN that are of different lengths but not unheard of for copper complexes of water and ammonia (see Wells, I962)

It would thus seem that the Jahn-Teller effect invocod by Miyoshi and others for copper ammonia compounds and considered in the present case plays no direct role in the axial symmetry. That rather, there is sometimes a coincidental symmetry produced by similar bonds between copper pyridine and copper cyanide (or chloride) and the chemical identical copper cyanide (or chloride) bonds are of different strengths for other reasons.

	g_I	g_2	g_3
$Cu(pic)_2Cl_2$	2.2I	2.06	2.03
$Cu(t\text{-}butyl\ py)_2$	2.I9	2.II	2.05
$Cu(NH_3)_2\ Cl_2$	2.22	2.09	2.03
$Cu(col)_2\ Ni(CN)_4$	2.25	2.08	2.05
$Cu(NH_3)_2\ Cu(CN)_4$	2.I5	2.I2	2.07

	g_I	$g_{2,3}$
$Cu(lut)_2\ Cl_2$	2.I9	2.07
$Cu(col)_2Cl_2$	2.I4	2.07
$Cu(3,5\text{-}Cl_2py)_2$	2.20	2.08
$Cu\ (py)_2\ Ni(CN)_4$	2.2I	2.07
$Cu\ (pic)_2\ Ni(CN)_4$	2.27	2.07
$Cu\ (lut)_2\ Ni(CN)_4$	2.25	2.07
$Cu\ (3,5\ Cl_2py)_2\ Ni(CN)_4$	2.22	2.07
$Cu\ (NH_3)_2\ Ni(CN)_4$	2.24	2.08

References

Akyus, S. et al., I974, The Structure of Hofman Pyridine Complexes, submitted to the Dalton Transactions of the Chemical Society of London.
Akyuz, S. et al., I973, J. Mol. Structure I7, I05.

Danitz, J.D. I957, Acta Cryst. IO, 307.
K.A. Hofmann and F. Hochtlen, I903, Ber. der Deutschen Chem.
Miyoshi, T., et al. I968, Inorg. Chim. Acta, 2, 329.
Wells, A.F., I962 Structural Inorganic Chemistry

^{19}F NUCLEAR MAGNETIC RESONANCE OF VISCOUS POLYMER SOLUTION BY HIGH SPEED SAMPLE ROTATION

M. Rabii
Tehran Polytechnic, Department of Electrical Engineering, Tehran, IRAN
H. Benoit
Université de Paris VI, Tour 23, 75230 Paris, FRANCE

Abstract. Line narrowing by high speed sample rotation at magic angle is used for the investigation of ^{19}F N.M.R. in several viscous solutions of fluorine copolymers. Study of the line width variation with respect to the rotation frequency of the specimen permits an evaluation of the correlation times for segmental and interchain motion in these copolymers.

1. Introduction. The NMR spectra of macromolecule solutions is not usually well resolved for the chemical shift investigations. The rapid segmental motion of the polymer chain remove considerably the intragroup dipolar interaction, but intermolecular motions, even in solution, are relatively slow because of the large molecular weight of the polymer. Due to this intermolecular dipolar interaction broadening, the polymer NMR spectra are barely resolved. This slowly modulated interaction can however be removed if the specimen is rapidly rotated about an axis inclined at the magic angle $\theta \simeq 54°44'$ with respect to the applied magnetic field(Andrew et al.1959, 1971). The line narrowing will be effective when the rotation frequency of the specimen is comparable to $1/\tau_1$, where τ_1 represents the correlation time of slow interchain motion(Schneider et al.1970, Rabii 1972). The residual line width is principally determined by the short correlation time of the segmental motion(Waugh et al.1969, Rabii et al.1973).

2. Line Narrowing in Fluorine Copolymers. The samples studied here at room temperature are the vinylidene fluoride-hexafluoropropylene copolymers $-CH_2-CF_2-CF_2-CF(CF_3)-$ (usually called Viton) of various monomer ratio. The NMR spectra of copolymer viscous solution in ethyle acetate is not sufficiently resolved for chemical shift measurements(fig. 1a). Line narrowing by sample rotation at magic angle reveals five chemically shifted lines(fig. 1b). We obtain the same result with polymer swollen in benzene(fig. 2). The line intensities are directly related to the monomer ratio in the copolymers. By comparison with the chemical shift values usually measured in ^{19}F high-resolution N.M.R., it results that nearly all monomer bonds are in head to tail configuration.

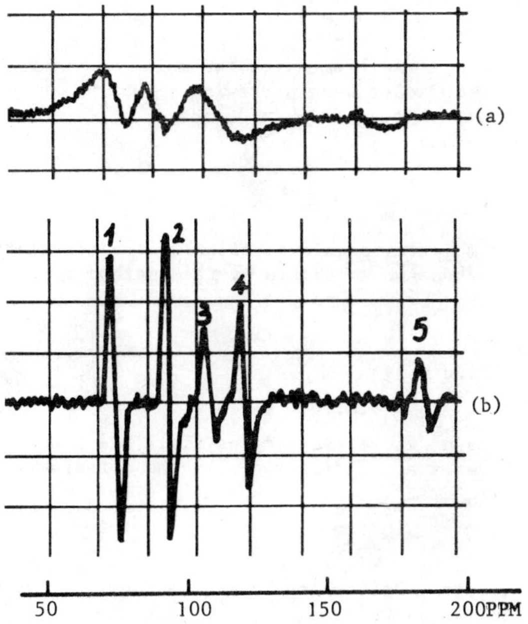

Fig. 1. ^{19}F NMR derivative spectra of a viscous solution,in ethyle acetate,of Viton A35$(-CH_2-CF_2^{2,3}-)_{77\%}$ $(-CF_2^4-CF^5\ CF_3^1-)_{23\%}$.

a)Static sample, b)sample rotating about the magic axis at 2200 Hz. Chemical shifts are relative to $CFCl_3$. The superscripts are related to line numbers.

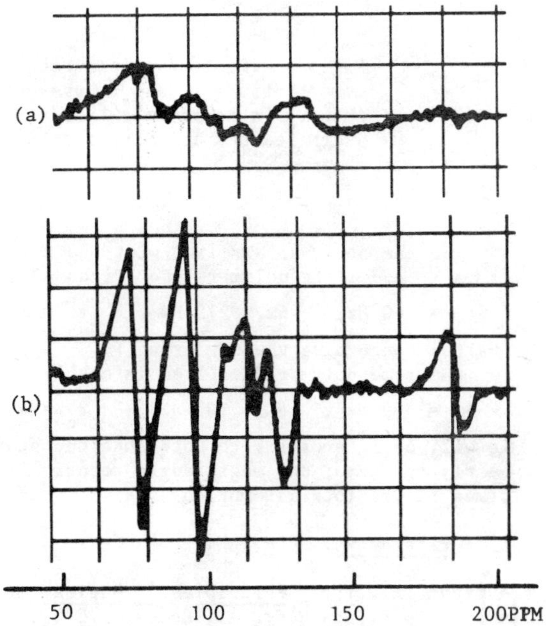

Fig.2 . ^{19}F NMR derivative spectra of a Viton A35 swollen in benzene.
a)static sample, b)rotated about the magic axis at 2200Hz.

3. Correlation Times of Segmental and Interchain Motion. For these samples, at room temperature, it will be assumed that the fluctuation of the local dipolar field arises from: i)the segmental motion characterized by a spectral density function $J_c(\omega)$ with a short correlation time τ_c , ii)the reorientational motion of the macromolecule, $J_1(\omega)$, with a much longer correlation time τ_1. The line width $\delta\omega_s$ of the static sample is then expressed as(Abragam, 1961):

$$(\delta\omega_s)^2 = (\delta\omega_c)^2 \int_{-\delta\omega_s}^{+\delta\omega_s} J_c(\omega)\, d\omega + (\delta\omega_1)^2 \int_{-\delta\omega_s}^{+\delta\omega_s} J_1(\omega)\, d\omega \qquad \text{(I)}$$

where $(\delta\omega_c)^2$ and $(\delta\omega_1)^2$ are respectively the intra and intermolecular contribution to rigid lattice second moment.

In a sample spinning at ω_r , we may consider the components near $\pm\omega_r$ of $J(\omega)$ as secular terms and the line width $\delta\omega_r$ is found to be :

$$(\delta\omega_r)^2 = (\delta\omega_c)^2 \int_{\omega_r-\delta\omega_r}^{\omega_r+\delta\omega_r} J_c(\omega)\, d\omega + (\delta\omega_1)^2 \int_{\omega_r-\delta\omega_r}^{\omega_r+\delta\omega_r} J_1(\omega)\, d\omega \qquad \text{(II)}$$

One must also consider the components near $\pm 2\omega_r$ in $J(\omega)$.

This result is illustrated in fig.3 . The line narrows when the rotation frequency ω_r is comparable to $1/\tau_1$; for higher frequencies the line width reaches a limiting value $\delta\omega_1$. By a BPP relaxation model of $J(\omega)$, this residual line width is found from (II) to be :

$$\delta\omega_1 = 2/\Pi\ (\delta\omega_c)^2\ \tau_c \qquad \text{(III)}$$

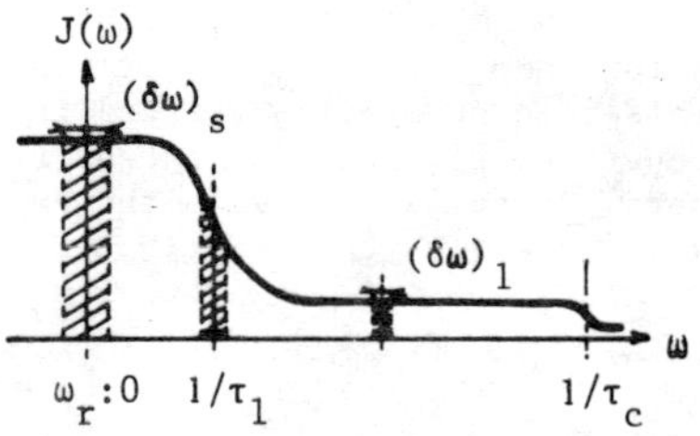

Fig.3 . Effect of sample rotation on the line width $\delta\omega$, with two molecular motions τ_c and τ_1 present in the sample. The line narrowing occurs when ω_r is comparable to $1/\tau_1$. Starting from $(\delta\omega)_s$ at $\omega_r = 0$, the line width reaches a limiting value $(\delta\omega)_1$ for $1/\tau_1 < \omega_r < 1/\tau_c$.

So the study of NMR line width as a function of ω_r permits an evaluation of the correlation times τ_c and τ_1 in polymers. For Viton A35 swollen in benzene we obtain by this method :

$\delta\omega_s = 540$ Hz, $\delta\omega_1 = 275$ Hz , $\tau_c \simeq 1{,}1.10^{-5}$ s, $\tau_1 \simeq 8.10^{-3}$ s

(values concerning the CF_3 group).

For the same polymer dissolved in ethyle acetate we find :

$\delta\omega_s = 500$ Hz , $\delta\omega_1 = 100$ Hz , $\tau_c \simeq 4.10^{-6}$ s , $\tau_1 \simeq 5{,}8.10^{-3}$ s

It will be interesting to note that the static spectra of these two samples(fig.1 and 2)are nearly the same; the magic angle rotation reveals however a net difference in the correlation times of the internal motion.

4. References.

Abragam, A., 1961, Principles of Nuclear Magnetism, Oxford University Press, London.
Andrew, A.R.,Bradbury, A., Eades, R.G., 1959, Nature,(183)1802.
Andrew, A.R., Jasinsky,A., Solid State Phys., 1971, (4)391.
Rabii, M., 1972, Thesis, Université de Paris VI, N° C.N.R.S. A07874.
Rabii, M., Benoit, H., 1973, Chem. Phys. Letters,(21/3)466.
Schneider, B., Doskocilova, D., 1970, Chem. Phys. Letters,(6/4)381.
Waugh, J.S., Haeberlen, U., 1969, Phys. Rev.,(185)420.

FORMATION AND TRANSFORMATIONS OF POLY(GLYCOL METHACRYLATE) POLYMER RADICALS PREPARED BY VIBRATION GRINDING

J. Pilař and K. Ulbert
Institute of Macromolecular Chemistry, Czechoslovak Academy of Sciences, 162 06 Prague 616, Czechoslovakia

Abstract. It was found that poly(glycolmethacrylate) free radicals generated by vibration grinding at liquid nitrogen temperatures in vacuo react with oxygen at -150°C. An anomaly in the radical concentration on the temperature decay curve reaching maximum at -60°C was observed which is attributed to polymeric tetroxide partially decomposing to nonparamagnetic products.

1. Introduction. The formation of polymer radicals after the mechanical degradation of polymers by vibration grinding was first observed by Bresler and Butyagin and co-workers (Bresler et al., 1959, Butyagin et al., 1959). Later, properties and transformations of polymer radicals thus formed (Bresler et al., 1963 and 1965; Butyagin et al., 1968) were studied in more detail. In this paper we would like to present some new results obtained in the study of the interaction of polymer radicals obtained by the vibration grinding of PGMA with oxygen at low temperatures (from -150 to -60°C).

2. Experimental. Poly(ethylene glycol methacrylate) (PGMA) crosslinked to 40% with ethyleneglycol dimethacrylate and prepared (Čoupek et al., 1972) in the form of porous globules under the trade name Spheron 300 was evacuated and ground at liquid nitrogen temperature. EPR spectra were recorded with a JES-3B JEOLCO Tokyo spectrometer. The radical concentrations were determined by double integration; the absolute concentrations were determined by comparing with a standard solution of diphenylpicrylhydrazyl. In the case of superposition of the spectra the concentrations of the individual components were determined by numerical simulation and integration of spectra on a WANG 614 computer.

3. Results. The EPR spectrum of polymer radicals formed by the vibration grinding of PGMA in vacuo recorded at -150°C is given in Fig. b. The radical concentrations in samples varied between 1 to $3\times10^{18}cm^{-3}$, depending on the conditions of grinding. This spectrum is similar to the known spectrum "5+4 lines" of poly (methyl methacrylate) and to that of X-ray irradiated PGMA (Szöcs and Ulbert, 1967). The EPR spectrum of PGMA radicals obtained by vibration grinding has somewhat different widths and intensity ratios of the individual components of the hyperfine structure. With increasing temperature of the sample the shape of the spectrum gradually approaches the spectrum "5+4 lines" (Fig. a), while the radical concentration in the sample decreases at the same time (on heating to -60°C to approximately 60% of the original value).

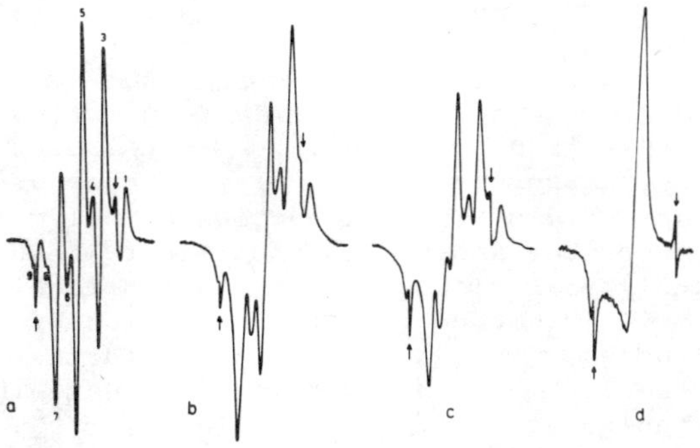

If oxygen is introduced into the sample at -150°C, the total radical concentration falls off steeply to c. 25% of the original value. The observed EPR spectrum (Fig. c) is a superposition of the spectrum of the polymer radicals R$^\cdot$ (Fig. b), which represent 20% of the original radical concentration, and of that of the polymer peroxy radicals ROO$^\cdot$ (Fig. d), which represent 5% of the original radical concentration in the sample. An increase in the sample temperature (in steps by 20°C up to -60°C) results in a fast decrease of the concentration of the radicals R$^\cdot$ to zero and to an increase in the concentration of the ROO$^\cdot$ radicals, which at -60°C attains a maximum value of about 40% of the original radical concentration in the sample, thus considerably exceeding the total radical concentration measured in the sample immediately after the introduction of oxygen. At temperatures -60°C and higher the concentration of the ROO$^\cdot$ radicals rapidly decreases with both time and temperature. At lower constant temperatures the total radical concentration does not vary within the limits of experimental error (c. 15%). In order to be sure that the steep decrease in radical concentration after the introduction of oxygen is not due to the heating of the sample by warm (room temperature) oxygen, warm nitrogen was introduced into the sample. The original concentration of the R$^\bullet$ radicals remained unchanged.

4. Discussion. An explanation of the facts observed here can be based on conclusions offered by Ingold and Thomas (Ingold, 1969; Thomas and Ingold, 1968) who assume, in the low-temperature range, the possibility of formation of nonparamagnetic tetroxides from low-molecular tertiary peroxy radicals

$$ROO^\bullet + ROO^\bullet \rightleftharpoons ROO\text{-}OOR\ .$$

The EPR method has revealed (Bartlett and Guaraldi, 1967) that with increasing temperature the equilibrium of this reaction is shifted in favour of peroxy radicals; concentration changes of these peroxy radicals in the temperature range from -150 to-70°C are quite reversible. It also follows from the above conclusions by Ingold and Thomas that after supply of certain activation energy the low-molecular tetroxides can decompose into nonparamagnetic final products, such as peroxides of the ROOR type, oxygen, etc.

We believe that under suitable conditions polymer tetroxides may also be formed and behave similarly. To explain the behaviour of our samples it is sufficient to assume that after oxygen has been introduced into the sample at -150°C, and polymer peroxy radicals has been formed, polymer tetroxide is formed rapidly in such a way that some 75% of polymer radicals R$^\bullet$ are transformed into nonparamagnetic polymer tetroxide. With increasing temperature of the sample in the range between -150°C and -60°C the remaining polymer radicals react with oxygen while giving rise to polymer peroxy radicals, and possibly another polymer tetroxide. At the same time the polymer tetroxide is decomposed again to yield some polymer peroxy radicals, while its greatest part decomposes directly into nonparamagnetic products. At temperatures -60°C and higher, the prevailing reactions are those which transform polymer peroxy radicals into final nonparamagnetic products. In contrast with low-molecular peroxy radicals the concentration changes of polymer radicals with temperature are not completely reversible, which may be due to their lower mobility and to the decomposition of polymer tetroxide into nonparamagnetic products already at very low temperatures.

The objective of our further experiments was to investigate the influence of the sample temperature at which oxygen is introduced, and its partial pressure on the interaction of polymer radicals with oxygen. The conditions of formation of polymer tetroxide are impaired with increasing temperature. The reason should be sought above all in that with increasing temperature of the evacuated sample the concentration of the polymer radicals R$^\bullet$ decreases as described above, probably owing to their recombination on the surface of microscopic polymer particles. At a higher temperature the portion of polymer radicals transformed into nonparamagnetic products by decomposition of the polymer tetroxide also decreases: If the sample is heated in vacuo to -60°C, the concentration of the polymer radicals R$^\bullet$ drops to 60% of its original value. After introduction of oxygen at this temperature some 60% of these polymer radicals are transformed into polymer peroxy radicals. If air is introduced into the sample at -150°C instead of oxygen, only some 50% of polymer radicals R$^\bullet$ are transformed into the nonparamagnetic tetroxide form, which is obviously due to the lower partial concentration of oxygen.

Experiments were also carried out in order to reveal the role played by the antioxidant in the reaction of polymer radicals with oxygen under conditions described above. The preliminary results show that a pronounced interaction of polymer radicals with the antioxidant takes place at temperatures about -60°C.

The mechanism of interaction of polymer radicals with oxygen at low temperatures outlined here calls for further experimental proofs. The investigation of the problems involved is under way. The results obtained up to now will be published in full extent later.

5. References

Bartlett, P.D., Guaraldi, G., 1967, J.Amer.Chem.Soc., 89, 4799.
Bresler, S.E., Zhurkov, S.N., Kazbekov, E.N., Saminskii, E.M., Tomashevskii, E.E., 1959, Zh.Tekh.Fiz., 29, 358.
Bresler, S.E., Kazbekov, E.N., Fomitchev, V.N., Szöcs, F., Smejtek, P., 1963, Fiz.Tverd.Tela, 5, 675.
Bresler, S.E., Kazbekov, E.N., Fomitchev, V.N., 1965, Kinetika i Kataliz,6, 820.
Butyagin, P.Yu., Berlin, A.A., Kalmanson, A.E., Bljumenfeld, A.A., 1959, Vysokomol.Soed., 1, 865.
Butyagin, P.Yu., Kolbanev, I.V., Dubinskaya, A.M., Kisluk, M.U., 1968, Vysokomol.Soed., 10A, 2265.

SECTION L

Electron Spin–Lattice Relaxation

SPIN-LATTICE RELAXATION IN MgO, Ni^{2+}

By P. LOPEZ and J. PESCIA
Laboratoire de Magnétisme et d'Electronique Quantique, Université Paul Sabatier, 31077 TOULOUSE Cédex, France.

Abstract: Spin-lattice relaxation of $\Delta m = 1$ and $\Delta m = 2$ transitions was investigated in the 4.2 - 160 K temperature range. Furthermore, experiments at 77 K were carried out for several crystal orientations. The obtained results fit well with the phonon shell model, for $T > 40$ K.

1. Introduction. The spin-lattice relaxation rate of MgO, Ni^{2+} was measured at helium temperatures by JONES and al. (JONES and al. 1967). SMITH and al. (SMITH and al. 1969) gave an estimation of it at 123 and 130 K using the broadening of the E.S.R. line. All these experiments were devoted to $\Delta m = 1$ transition. They seem to indicate marked differences from what is usually found in RAMAN process. We present here results for $\Delta m = 1$ and $\Delta m = 2$ transitions, obtained from the modulation method, fitted for the determination of short times (PESCIA 1965). Temperature dependence and orientation dependence have been investigated within the 4.2 - 160 K temperature range.

2. Experimental. The crystal is held in a suprasil cryostat along the axis of a cylindrical X - band cavity, operating on the TE 011 mode. The microwave field is amplitude modulated at a circular frequency Ω. A pick up coil coaxial with the magnetic field is wounded around the sample. The signal $S(\Omega)$ induced in the coil is proportional to the time rate of change of magnetization. The modulating frequency is varyied over the range $0 < \Omega < 2/T_1$, keeping all other parameters constant. The value of T_1 is easily obtained from the variations of S with Ω (PESCIA 1965; LOPEZ 1974).

3. Results. We have investigated a monocrystal with nickel-ion concentration $f \sim 2 \times 10^{-3}$. The overall impurity concentration was below to 500 ppm. The values of T_1, obtained at various temperatures for both transitions are presented in fig.1.
T_1 has been also measured at 77 K for various crystal orientations.

θ^0	0	45	90	135	180
$\Delta m = 1$ $10^{-7}.T_1^{-1}.s^{-1}$	2.46	0.83	2.52	0.82	2.72
$\Delta m = 2$ $10^{-7}.T_1^{-1}.s^{-1}$	0.60	0.55	0.63	0.57	——

Table 1: Spin-lattice relaxation time of MgO, Ni^{2+} at 77 K versus orientation. θ is the angle between magnetic field and the axis (100) of the crystal.

At least, we have found at 4.2 K and $\theta = 0$: - $T_1 \sim 125 \times 10^{-6}$ sec. $(\Delta m = 1)$
- $T_1 \sim 26 \times 10^{-6}$ sec. $(\Delta m = 2)$

4. Discussion. (i) T_1^{-1} varies as T^n $(2.76 < n < 3.15)$ within the 40 - 160 K temperature range for both transitions. The dependence is markedly different from the usual T^7 law for $T \ll \theta_D$ ($\theta_D \sim 980$ K here). A shell model of phonons seems to be in good agreement with our results. (One ion is a rigid system with respect to DEBYE model. On the contrary, for a phonon shell model, the outer electronic shell can move relatively to the ion core). We have computed the relaxation rates, assuming a phonon shell model. Good agreement is found with experimental results (see fig. 1). This confirm the model to be adequate here. Another confirmation is given by the neutron scattering experiments (PECKAM 1969).

(ii) At 77 K, the $\Delta m = 2$ relaxation rate is found to be isotropic. The $\Delta m = 1$ relaxation rate is very anisotropic with : $T_1^{-1} - T_{10}^{-1} \propto \sin^2\theta.\cos^2\theta$. Both results are well explained by the calculation of transition probabilities induced by spin-phonon interaction (KUMAR and al. 1968).
For the double spin transition : $T_1^{-1} \sim W_{|0> \to |+1>} + 2W_{|-1> \to |+1>}$. A calculation outlined with phonon integrals and spin-phonon coupling coefficients shows T_1 to be independent on θ. For the single spin transition, the relaxation rate can be expressed as :

$$T_1^{-1} = (6/\pi)\ \{10.7 \times 10^5 - 11.78 \times 10^5 \cos^2\theta \sin^2\theta\}\ \text{sec}^{-1}$$

Our results are in fair agreement with this value since we obtain :

$$\frac{T_{1\,max}}{T_{1\,min}}\ \text{calculated} \sim 2.1 \qquad \frac{T_{1\,max}}{T_{1\,min}}\ \text{experimental} \sim 2.7$$

Let us consider T_1-values measured at 4.2 K for both transitions. Their ratio is equal to 4.8. The same ratio can be calculated, assuming the DEBYE model and zero concentration (LEWIS and al. 1967; LOPEZ 1974). It is found to be equal to 5. The DEBYE model is adequate here.

In conclusion, our relaxation measurements seem to show that the lattice of Ni^{2+}, MgO is well described by the DEBYE model at helium temperatures. The phonon shell model would be more adequate above 40 K.

The authors are grateful to Professor J. JOFFRIN for the loan of samples and Professor J. PH. GRIVET for helpful discussions.

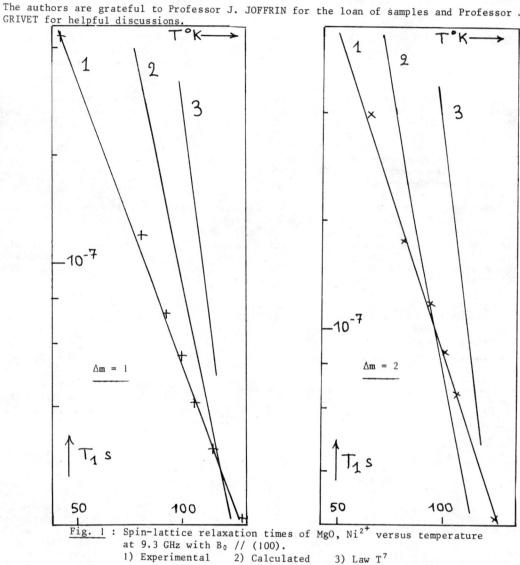

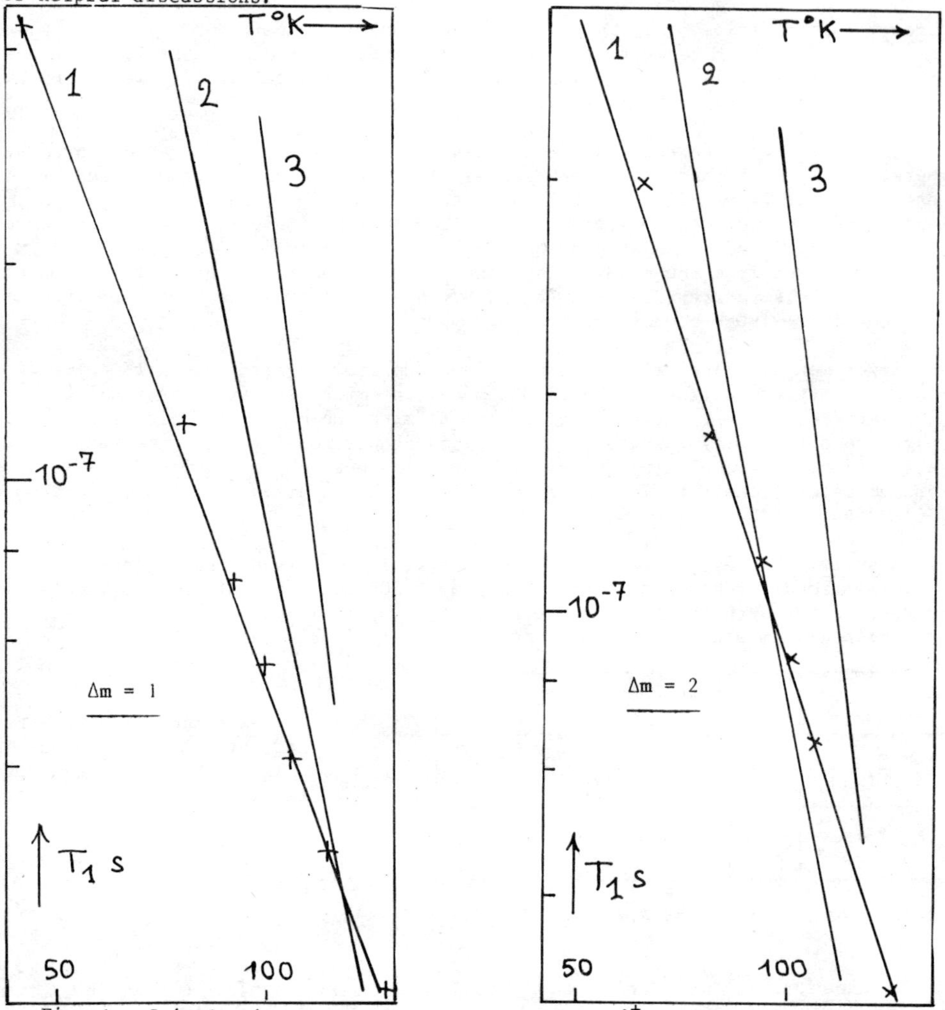

Fig. 1 : Spin-lattice relaxation times of MgO, Ni^{2+} versus temperature at 9.3 GHz with B_0 // (100).
1) Experimental 2) Calculated 3) Law T^7

5. References.

JONES, J.B. and LEWIS, M.F., 1967, Sol. Stat. Com., 5, 595.

KUMAR, S., RAY, T. and RAY, D.K., 1968, Phys. Rev., 176, 489.

LEWIS, M.F. and STONEHAM, A.M., 1967, Phys. Rev., 164, 271.

LOPEZ, P., 1974, Thesis, Université Paul Sabatier, Toulouse, France.

PECKHAM, G., 1967, Proc. Phys. Soc., (London), 90, 657.

PESCIA,J., 1965, Ann. Phys., 10, 389.

SMITH, S-R.P., DRAVNIEKS, F. and WERTZ, J.E., 1969, Phys. Rev., 178, 471.

MEASUREMENT OF SPIN-RELAXATION RATES IN INHOMOGENEOUSLY BROADENED LINES INCLUDING CROSS-RELAXATION

J.-P. Korb and J. Maruani
C.M.O.A. du C.N.R.S., 23, rue du Maroc, 75019-Paris, FRANCE.

Abstract. The spin-packet model has proven useful in measuring spin-relaxation rates for inhomogeneously broadened lines. Cross-relaxation between spin packets can be included using a model of coupled thermodynamic reservoirs. It is shown that the measured spin-lattice relaxation rate decreases when one includes cross-relaxation between all packets.

1. Introduction. Measurement of spin-spin T_2 and spin-lattice T_1 relaxation times for magnetic resonance lines is often rendered difficult by the presence of inhomogeneous broadening adding to the ever-present homogeneous broadening. The naive method, which consists of treating the first broadening as being composed of a continuous distribution of independent spin packets (Portis,1953),was developed by various authors in the case of continuous saturation (Castner,1959; Zhidkov,1967;Maruani,1972).However,this model cannot be retained at high microwave powers and when both broadenings are of the same order,because then cross-relaxation mechanisms (Bloembergen et al.,1959) generate spectral diffusion of energy, which cannot be neglected in interpretations of measurements.The previous treatments,both within (Buishvili et al.,1968) and without (Clough and Scott,1968) the spin-packet model,did not aim at determining a method for extracting T_1 and T_2 when there are no independent spin packets.We have used a model of coupled thermodynamic reservoirs (Boucher and Nechtschein,1970) to study the effects of cross-relaxation on the values of T_1 and T_2 obtained by the naive method.

2. Model. The model of coupled thermodynamic reservoirs is essentially based on the possibility of dividing a physical system into subsystems which reach internal thermal equilibrium in times much shorter than those necessary to reach equilibrium between the subsystems.One can then separate the total Hamiltonian $\mathcal{H}$ into a principal part $\mathcal{H}_0$ equal to a sum of terms $\mathcal{H}_K$ to which,under certain conditions (Boucher and Nechtschein,1970), one can associate thermodynamic reservoirs, and a perturbation part $\mathcal{H}_1$ equal to a sum of couplings $\mathcal{H}_i(K,L,\ldots,Q)$ between the reservoirs.The reservoirs are characterized by an internal energy $U_K=\mathrm{Tr}_{\mathcal{H}_0}\{\rho(t)\mathcal{H}_K\}$, a self-entropy $S_K=-k\,\mathrm{Tr}_{\mathcal{H}_0}\{\rho(t)\mathrm{Ln}\rho_K(t)\}$ and a temperature parameter $\beta_K\equiv 1/kT_K=\partial S_K/k\partial U_K$,where $\rho(t)$ is the density matrix of the whole system.It can be shown (Boucher and Nechtschein,1970) that the temperature parameters obey the kinetic equations :

$$\dot{\beta}_K(t)=-\sum_{L\neq K} W_{KL}^{\mathcal{H}_1}[\beta_K(t)-\beta_L(t)] \quad (1), \qquad W_{KL}^{\mathcal{H}_1}=\frac{-1}{\hbar^2\,\mathrm{Tr}_{\mathcal{H}_0}\{K^2\}}\int_{-\infty}^{+\infty}\mathrm{Tr}_{\mathcal{H}_0}\{e^{i\mathcal{H}_0\tau/\hbar}\times$$

$$\times[K,Y]e^{-i\mathcal{H}_0\tau/\hbar}[Y^{\dagger},L]\}\,d\tau \quad (2), \qquad K\equiv\mathcal{H}_K, \quad Y+Y^{\dagger}=\mathcal{H}_1 .$$

Consider, now, a set of N identical paramagnets ($S=\frac{1}{2}$) subjected to a static magnetic field H_0 along z and to static local fields ΔH_α which represent other magnetic interactions.A spin packet α is the subset of N_α spins $S_{\alpha i}$ with the same Larmor frequency $\omega_\alpha=\omega_0+\Delta\omega_\alpha$.The application of a microwave field $H_1 e^{i\omega t}$ induces transitions which perturb the thermal equilibrium of the spin system with the surroundings, while thermally-induced local field fluctuations tend to restore equilibrium.For symmetry reasons, and also because cross-relaxation is the more efficient as spin packets are closer in frequency, we may assign a thermodynamic reservoir to each of the n spin packets composing the line. More specifically, we write :

$$\mathcal{H}=\mathcal{H}_0+\mathcal{H}_1 \; (3), \quad \mathcal{H}_0=\sum_{\alpha=1}^{n}\mathcal{H}_{Z\alpha}+\mathcal{H}_D+\mathcal{H}_P+\mathcal{H}_L \; (4), \quad \mathcal{H}_1=\mathcal{H}_{CR}+V+R \; (5),$$

where $\mathcal{H}_{Z\alpha}$ is the Zeeman term for the packet α, $\mathcal{H}_D$ is the part of the dipolar interactions which commutes with $\sum\mathcal{H}_{Z\alpha}$, $\mathcal{H}_P$ and $\mathcal{H}_L$ are the specific Hamiltonians of the photon ($\beta_P=0$) and phonon ($\beta_L=C$) reservoirs (Buishvili et al.,1968), $\mathcal{H}_{CR}$ is the part of the dipolar interactions which does not commute with $\sum\mathcal{H}_{Z\alpha}$ but conserves total spin projection, V is the spin-microwave field interaction, and R writes as a sum of products of spin and phonon operators which depends on the processes involved. From this decomposition and the commutation relations between the different terms of $\mathcal{H}$, one can build a network of n+3 reservoirs and the couplings (Fig. 1).

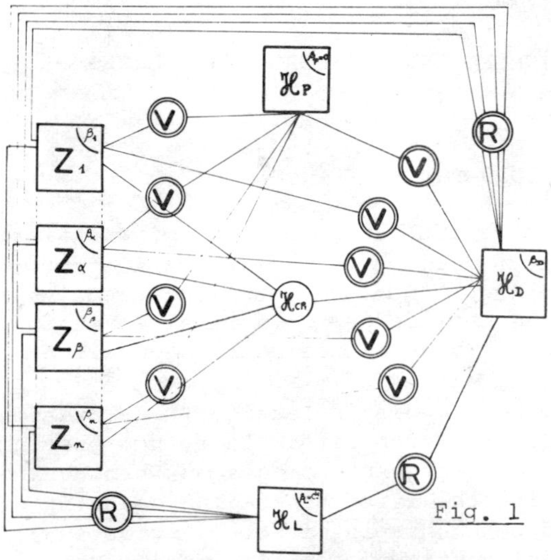

Fig. 1

3. Treatment. With the help of the previous network, and using the expressions of the corresponding terms in the total Hamiltonian $\mathcal{H}$, we can write n+1 kinetic equations of type (1), where the $W_{KL}^{\mathcal{H}}$'s can be expressed using relation (2). In continuous saturation, the β_K's do not depend on time: $\dot{\beta}_K=0$. Besides, we can make the following simplifying assumptions : 1) all spin packets have the same absorption form, assumed to be Lorentzian, with the same width, T_2^{-1} ; 2) the spin-lattice relaxation probability for all Zeeman reservoirs has the same value, T_1^{-1}; 3) the dipole-lattice relaxation time, T_1', is equal to $T_1/2$ (Provotorov, 1962); 4) the relaxation rate $(T_1'')^{-1}=W^R_{Z_\alpha\mathcal{H}_D}$ is negligible with respect to T_1^{-1} (Boucher, 1972). If, then, we introduce the following parameters : $c=T_1/T_2$, $h=\omega_0/\Delta\omega_G$, $a=T_2^{-1}/\Delta\omega_G$, $r=(\omega-\omega_0)/T_2^{-1}$ and $s=(\gamma H_1)\times(T_1T_2)^{\frac{1}{2}}$, where $\Delta\omega_G$ is the inhomogeneous broadening of the line, the coefficients of the system of equations of type (1) we have obtained reduce to combinations of these parameters, and the absorption Ust(c,h,a,r,s) writes as $as/\sqrt{\pi}$ times a sum of products of the $p_n(\alpha)=N_\alpha/N$ and the solutions, $x_n(\alpha)=\beta_{Z_\alpha}/\beta_L$, of these equations; the choice of a binomial distribution for $p_n(\alpha)$, with n increasing for achieving convergence, allows simulation of a Gaussian envelope.

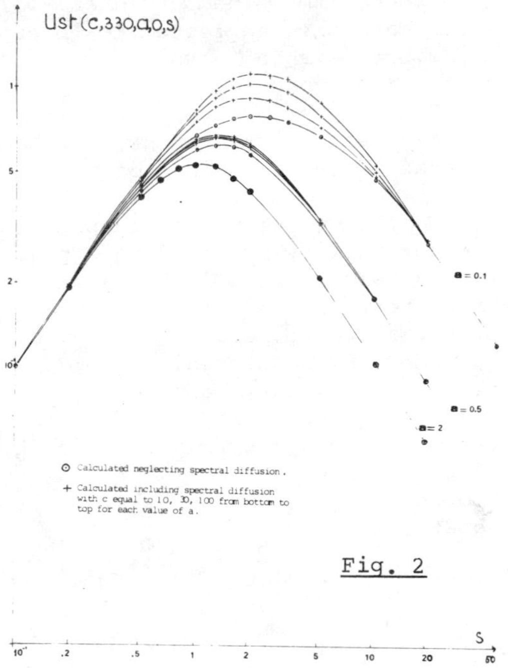

Fig. 2

4. Results. Preliminary computer calculations show two main results. First, Ust(c,h,a,r,s) tends to (s/t)u(ar,at) for all values of c with large h in the limit of $s\rightarrow 0$, u representing the Voigt profile. As a result, the previously described method for extracting values of T_2 and $\Delta\omega_G$ from the unsaturated line shape (Maruani, 1972) remains unchanged when spectral diffusion is included. Second, the maximum of the saturation curve for the center of the resonance lies above that relative to the Voigt profile, the difference increasing with c and with decreasing a (Fig. 2). As a result, the value of T_1T_2 obtained neglecting spectral diffusion (Maruani, 1972) is smaller than the corrected value. For $a\sim 0.1$ and $c\sim 30$, we find $\Delta T_1/T_1\sim 50\%$. Our computer program will help in producing corrections to the diagrams used to extract spin-relaxation rates from experimental data.

5. References.

Bloembergen,N., Shapiro,S., Pershan,P.S., and Artman,J.O., 1959,Phys. Rev.,114,445.
Boucher,J.P., and Nechtschein,M.,1970, J. Physique,31, 783; Boucher,J.P., 1972, Thesis, Grenoble.
Buishvili,L.L., Zviadadze, M.D., and Khutsishvili,G.R., 1968, J.E.T.P., 27, 469.
Castner, T. G., 1959, Phys. Rev., 115, 1506.
Clough, S., and Scott, C.A., 1968, J. Phys., C1, 919.
Maruani, J., 1972, J. Magn. Res., 7, 207.
Portis, A.M., 1953, Phys. Rev., 91, 1071.
Provotorov, B.N., 1962, J.E.T.P., 14, 1126.
Zhidkov, O.P., 1967, Sov. Phys.-Solid State, 9, 1095.

CROSS-RELAXATION MEASUREMENTS AND THEIR INTERPRETATION

C. A. Bates*, A. Gavaix†, P. Steggles*, A. Vasson† and A.-M. Vasson†
*Department of Physics, University of Nottingham, University Park, Nottingham NG7 2RD, ENGLAND.
†Laboratoire de Radioélectricité et Théorie du Solide, Université de Clermont-Ferrand, UER des Sciences Exactes et Naturelles, B.P. 45, 63170-Aubière, FRANCE.

Abstract. V^{3+} and V^{4+} ions cause resonant reductions in the measured spin-lattice relaxation times T_{1m} of Cr^{3+} ions in Al_2O_3. A quadrupole-electric-field interaction is postulated to explain the Cr^{3+}-V^{4+} cross-relaxation. From measurements of the temperature dependence of T_{1m}, the zero field splitting of V^{3+}:Al_2O_3 is also deduced.

While it is often very difficult to detect fast-relaxing magnetic impurities by EPR directly, their presence and properties may often be determined by spin-lattice relaxation time measurements (Raoult et al., 1971).

Several ruby crystals have been studied which also contain some vanadium in the V^{3+} and V^{4+} states. Preliminary results were reported by one of us earlier for V^{3+} (Raoult and Duclaux, 1966; Duclaux, 1969) but results for V^{4+}, stabilised by the addition of magnesium in the original powder are reported for the first time. The energy levels of the three ions are shown in Figure 1 and the isofrequency resonance curves at 9.3 GHz are shown in Figure 2. The isofrequency curves for V^{3+} are

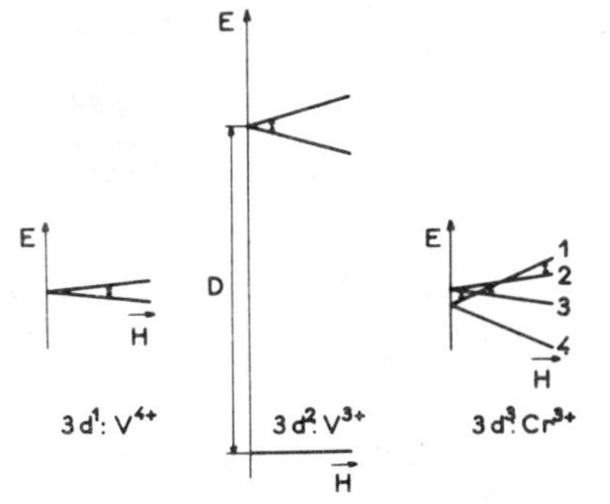

Figure 1

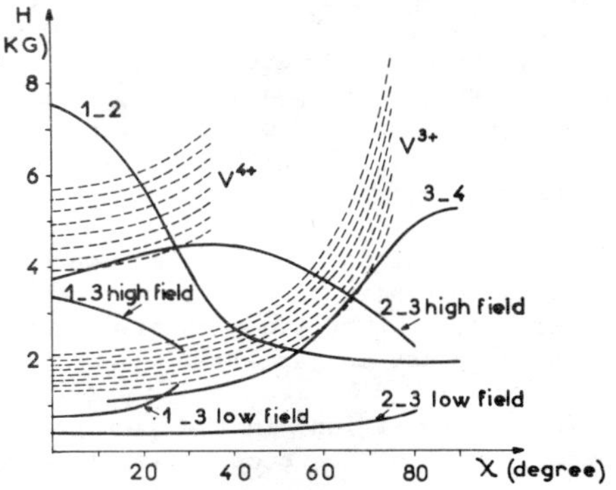

Figure 2

deduced from EPR measurements and those for V^{4+} are obtained from the APR results of Abou-Ghantous et al. (1972). Some of the Cr^{3+} isofrequency lines intersect at least some of the V^{3+} and V^{4+} hyperfine lines and it is these regions which concern us.

Spin-lattice relaxation measurements T_{1m} of the Cr^{3+} ion have been carried out at liquid helium temperature at 9.3 GHz at various angles χ of H to the crystal c-axis. Apart from a χ-independent reduction in T_{1m}, there are also significant dips in T_{1m} whenever the isofrequency curves intersect. Figure 3 readily shows the eight hyperfine lines of V^{3+} and Figure 4 shows one of the V^{4+} hyperfine lines twice (see Figure 2) for two different samples. In all cases, the shape of the dip (T_{1m} versus χ) is determined by the shape of the vanadium resonances and the angle between the isofrequency curves when they intersect.

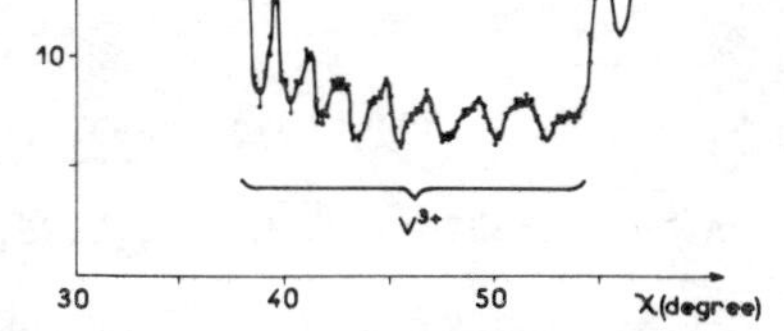

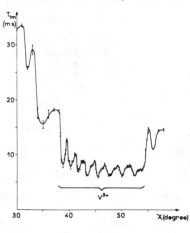

Figure 3

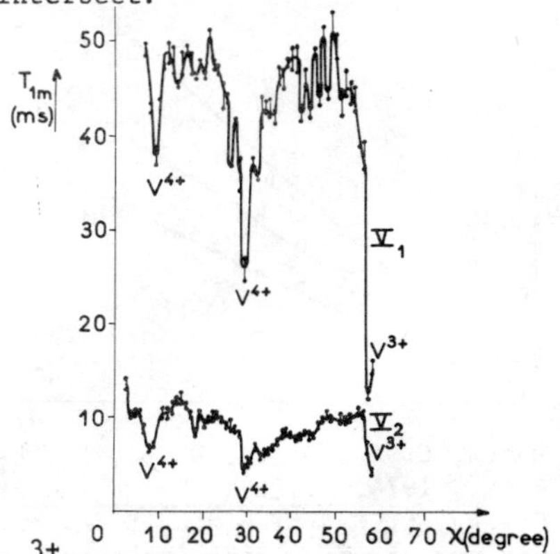

Figure 4

V^{3+} and V^{4+} have relaxation times much shorter than those for Cr^{3+} and thus act as fast-relaxing impurities. They are able to stimulate transitions within Cr^{3+} whenever the energy level separations are exactly equal. It has been customary to assume that this takes place via a magnetic dipole-dipole interaction. This interaction could explain the observed Cr^{3+}-V^{3+} cross-relaxation in Figure 3. However, for V^{4+}, $g_\perp$ is reported to be ~ 0.2 (Dranov et al., 1971) and thus any magnetic dipole-dipole effects will be small. Furthermore, the V^{4+} concentration will be much less than that of V^{3+}. Thus a dipole-dipole mechanism for Cr^{3+}-V^{4+} cross-relaxation would appear to be very much less efficient than for V^{3+}-Cr^{3+} in contradiction to experiment.

We believe that a quadrupole-electric-field interaction explains the cross-relaxation for V^{4+}-Cr^{3+}. Such an interaction was postulated earlier (Bates et al., 1974) for cross-relaxation between Cr^{3+} and Fe^{2+} in Al_2O_3. The basic idea is as follows: neglecting hyperfine effects, transitions within the V^{4+} ground doublet may be induced by acoustic-type perturbations such as those obtained in APR or in thermally-detected electric-field-induced EPR (EF-EPR) (Moore et al., 1973). The V^{4+}-lattice relaxation time is fast and thus the possibility exists of the radiation field from V^{4+} stimulating emission from Cr^{3+}. Any transition in Cr^{3+} is accompanied by a reorientation of its electric quadrupole moment (as well as its magnetic dipole moment) and thus it generates an electric field at the V^{4+} cluster. This in turn causes a transition within the V^{4+} doublet directly, in exactly the same way as a transition is induced in an EF-EPR experiment. If the probability of the transition Cr^{3+}-V^{4+}-lattice is greater than that for the direct transition Cr^{3+}-lattice, and that the Cr^{3+} and V^{4+} splittings are exactly equal a sharp reduction in T_{1m} occurs. This is always observed.

The Hamiltonian for the quadrupole-electric field interaction to explain cross-relaxation can be written in the form:

$$\mathcal{H}_\varepsilon \propto [E_- \sum_n O_m^{(2)} F(m,\underline{R}) + c.c.] \tag{1}$$

where $O_m^{(2)}$ are the normalised tensor operators defined by Smith and Thornley (1966) for Cr^{3+}, $E_\pm = E\theta \pm E\varepsilon$, where $E\theta$, $E\varepsilon$ are the orbital operators for the V^{4+} ion transforming like the (θ,ε) components of the C_{3V} group and $F(m,\underline{R})$ is a constant. The cross-relaxation time T_{1X} determined by $\mathcal{H}_\varepsilon$ is a function of the V^{4+} concentration. The calculation of T_{1X} from Cr^{3+}-Fe^{2+} is in good agreement with experiment, but, although the calculations for T_{1X} fro Cr^{3+}-V^{4+} are not complete, agreement is expected.

Transitions in $V^{3+}:Al_2O_3$ can also be induced by APR. We may then suppose that the quadrupole-electric-field mechanism could also act for the Cr^{3+}-V^{3+} system as well as the magnetic dipole-dipole interaction. The two mechanisms cannot be distinguished in our experiments and we thus suppose that they act in parallel.

If the relevant transitions in the fast relaxing-ion occur in an excited doublet, T_{1X} is temperature dependent whatever the cross-relaxation mechanism (Vasson et al., 1974). For a given sample,

$$1/T_{1X} \propto f(T) \tag{2}$$

where

$$f(T) = \frac{2\cosh(\sigma/T)}{2\cosh(\sigma/T) + \exp[(D+\varepsilon)/T]} \tag{3}$$

$2\sigma = 9.3$ GHz and $\varepsilon = (g_\perp \beta H \sin\chi)^2\{(g_\parallel H\beta\cos\chi)^2 - D^2\}/2D^3$.

If $1/T_{1X}$ is plotted against $f(T)$, a straight line should result provided D is known. Values of T_{1X} at one of the dips have been deduced from our measurements of T_{1m} as a function of T. A linear regression programme on a Wang 600 has been used to test both equation (2) and from it deduce a value of D for $V^{3+}:Al_2O_3$. Figure 5 shows the results for one Cr^{3+} transition. We deduce $D = 8.45 \pm 0.35\ cm^{-1}$ in good agreement with other measurements of this quantity (Abou-Ghantous et al., 1975).

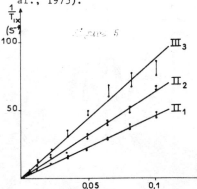

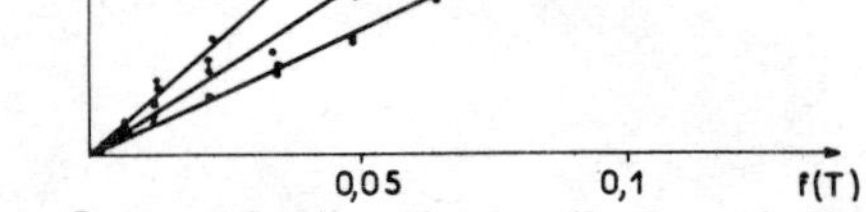

In conclusion, we have found pronounced dips in T_{1m} for Cr^{3+} in crystals containing V^{3+} and V^{4+} whenever the isofrequency curves cross. We postulate that cross-relaxation for Cr^{3+}-V^{4+} is by a quadrupole-electric-field interaction and that for cross-relaxation in Cr^{3+}-V^{3+}, both quadrupole-electric-field and dipole-dipole mechanisms probably occur. We have deduced a value for the zero field splitting of $V^{3+}:Al_2O_3$ in a region of the EM spectra for which direct measurements are very difficult.

References

Abou-Ghantous, M., Disdier, F. and Locatelli, 1972, Phys. Letts., 39A, 53-4.

Abou-Ghantous, M., Bates, C. A. and Goodfellow, L. C., 1975, to be published.

Bates, C. A., Moore, W. S., Al-Sharbati, T. M., Steggles, P., Gavaix, A., Vasson, A. and Vasson, A.-M., 1974, J. Phys. C: Solid State Phys., 7, L83-7.

Dranov, L. N., Kichigin, D. A. and Chernina, E. A., 1971, Soviet Phys. Solid State, 12, 2220.

Duclaux, A. M., 1968, Ann. Phys., (Paris), 3, 89-144.

Moore, W. S., Bates, C. A. and Al-Sharbati, T. M., 1973, J. Phys. C: Solid State Phys., 6, L209-14.

Raoult, G., Gavaix, A., Vasson, A. and Vasson, A.-M., 1971, Phys. Rev., 4, 3849.

Raoult, G. and Duclaux, A.-M., 1966, C.R. Acad. Sc., 263, 391.

Smith, B. and Thornley, J. H., 1966, Proc. Phys. Soc., 779-81.

Vasson, A.-M., Vasson, A., Gavaix, A., Yalcin, S., Steggles, P., Moore, W.S. and Bates, C.A., 1974, J. Phys. (Paris) Letts, 35, L73-6.

ON THE LOW FREQUENCY LONGITUDINAL PARAMAGNETIC SUSCEPTIBILITY IN THE PRESENCE OF THE HIGH FREQUENCY TRANSVERSE MAGNETIC FIELD

I.G.Shaposhnikov and M.Kacimi*, M.El Qacemi*
Department of Theoretical Physics, University of Perm, Perm 614005, USSR
and *Laboratory of Electronics, Institute of Nuclear Research, University of Algiers, Algiers 1017, Algeria

Abstract. The increasing of the low frequency non-resonant longitudinal paramagnetic susceptibility due to the presence of the strong high frequency transverse magnetic field near the resonance condition (Atsarkin effect, Atsarkin, 1973) is considered theoretically. The scheme of an experiment aimed at testing the predictions done is given.

1. Let us consider a sample without the natural anisptropy possessing the normal electron or nuclear paramagnetism (C and b being Curie and Van Vleck constants respectively) placed into the applied magnetic field

$$\vec{H} = \{H_2 \cos \Omega t,\ H_2 \sin \Omega t,\ H_0 + H_1 \cos \omega t\},\quad H_0 \gg H_{1,2}. \tag{1}$$

Our consideration will be limited to the case where, firstly, the spin system may be treated as isolated from the lattice and, secondly, the relaxation of the transverse components $M_{x,y}$ of the macroscopic magnetic moment $\{M_x, M_y, M_z\}$ in the constant field $\{0, 0, H_0\}$ is the most rapid process in the spin system. In this case, in the framework of a rather general approach to the magnetic dynamics problems (see Shaposhnikov and Kadyrov, 1970) the equations

$$\dot{M}_z = -(\varkappa_1 + \tilde{W}) M_z + (\varkappa_1 \varphi + \tilde{W}\psi)\beta, \tag{2}$$

$$\dot{\beta} = q(\varkappa_1 \varphi + \tilde{W}\psi) M_z - q(\varkappa_1 \varphi^2 + \tilde{W}\psi^2)\beta \tag{3}$$

are valid for time values larger as compared with the transverse relaxation time; in these equations β is the interaction subsystem inverse temperature, $\tilde{W} \equiv \omega_2^2 \varkappa_2 (\varkappa_2^2 + \psi^2)^{-1}$, $\varphi \equiv 1 + \omega_1 \cos \omega t$, $\psi \equiv \Delta + \omega_1 \cos \omega t$, $\Delta \equiv 1 - \Omega$, $q \equiv \omega_i^{-2}$, $\omega_{1,2,i} \equiv -\gamma H_{1,2,i}$, $H_i \equiv (b/C)^{1/2}$ is the inner field constant, $\varkappa_{1,2}^{-1} \equiv \tau_{1,2}$ are the Bloch relaxation times and all the quantities are dimensionless, the following quantities with dimension: lattice inverse temperature β_0, $CH_0\beta_0$, $\omega_0 \equiv -\gamma H_0$, and ω_0^{-1} taken as units of β, M_z, the frequencies Ω, ω and the quantities $\omega_{1,2,i}$, $\varkappa_{1,2}$, and time respectively; the quantity $(\tilde{W})_{\omega_1=0} \equiv W = \omega_2^2 g$, $g \equiv \varkappa_2(\varkappa_2^2 + \Delta^2)^{-1}$ is important for what follows, g characterizing the transverse resonance line shape in the linear response approximation.

2. Using the standard quantum mechanical formalism, one can immediately see that there are two causes of the time variation of M_z : the non-secular part of the spin-spin interactions, and the spin-spin interactions (all of them) on condition that the transverse applied field is present. According to (2) and (3), two quantities characterize these causes: $\varkappa_1$ and $\tilde{W}$ respectively.

3. In the absence of the transverse field (2) and (3) give the following expression of the longitudinal susceptibility:

$$\chi = \chi_\infty (1+q)^{-1} (1 + i\tau_{1a}\omega)^{-1} \tag{4}$$

with

$$\chi_\infty \equiv (1+q)^{-1}[q M_z(0) + \beta(0)],\quad \tau_{1a} \equiv (1+q)^{-1}\varkappa_1^{-1}; \tag{5}$$

here and in what follows χ is dimensionless, $\chi_0 \equiv C\beta_0$ taken as a unit.

4. To study the role played by the transverse field, let us begin with the case where the non-secular interactions may be neglected. The equations (2) and (3) with $\varkappa_1 = 0$ give

$$\dot{M}_z = -\tilde{W}(M_z - \psi\beta), \tag{6}$$

$$\dot{\beta} = q\tilde{W}\psi(M_z - \psi\beta) \tag{7}$$

which is Provotorov type equations. The susceptibility corresponding to (6) and (7) is

$$\chi = \chi_\infty (1+p)^{-1} (1 + i\tilde{\tau}_{1a}\omega)^{-1} \tag{8}$$

with

$$\chi_\infty \equiv (1+p)^{-1}[p\Delta^{-1}M_z(0)+\beta(0)]\,, \quad \tilde{\tau}_{1a} \equiv (1+p)^{-1}W^{-1} \tag{9}$$

where $p \equiv q\Delta^2 = (\Delta/\omega_i)^2$.

5. If $\varkappa_1 \neq 0$, the result is rather complicated. Let us limit our discussion to the situation where the following approximations:

$$q \gg 1\,, \quad |\Delta| \ll 1\,, \quad p \sim 1\,, \quad \tau_{1a} \gg \tilde{\tau}_{1a} \tag{10}$$

are acceptable, the latter corresponding to the case where the non-secular interactions, though not negligible, are small enough (obviously the possibility of $|\Delta| \ll 1$ must be guaranteed by the proper signs of γ and Ω). In this situation one gets (8) once again for frequencies $\omega \sim \tau_{1a}^{-1}$ and

$$\chi = \chi_\infty (1+i\tau_{1a}\omega)[1+i(1+p)\tau_{1a}\omega]^{-1} \tag{11}$$

with χ_∞ given by (9) if frequencies belong to the range $\omega \sim \tilde{\tau}_{1a}^{-1}$; in both cases it is the time interval $\tau_{1a} \gg t \gg \tilde{\tau}_{1a}$ in which the solution of the equations (2) and (3) is considered as steady.

6. Now the Atsarkin effect may be discussed. Let the variable part of $\vec{H}$ be switched on at $t=0$, the sample being in equilibrium; so $M_z(0) = \beta(0) = 1$. In the absence of the transverse field for frequencies $\omega \sim \tau_{1a}^{-1}$ (4) gives $|\chi_{H_2=0}| \sim 1$ for $q \sim 1$ and $|\chi_{H_2=0}| \ll 1$ for $q \gg 1$ or $q \ll 1$; that is, $|\chi_{H_2=0}|$ doesn't exceed 1. If the transverse field is present and some parameters have proper values, the susceptibility increases. The most simple case is where $\varkappa_1 = 0$, $|\Delta| \ll 1$, $p \sim 1$; then for frequencies $\omega \sim \tau_{1a}^{-1}$ it follows from (8) $|\chi| \sim |\omega_i|^{-1} \gg 1$. This is just what was predicted by Atsarkin (Atsarkin, 1973) and then observed experimentally (Mefed and Atsarkin, 1973). In the situation (10) the same takes place if $\varkappa_1 \neq 0$, as well. But in this case (11) gives $|\chi| \sim |\omega_i|^{-1}$ for frequencies $\omega \sim \tilde{\tau}_{1a}^{-1}$, too. So the general equations (2) and (3) lead to the conclusion that there are two different frequency ranges in which the Atsarkin effect is expected to take place, instead of the only one considered previously.

7. Here is the scheme of an experiment aimed at testing the theoretical predictions done. There are cases where the electron non-resonant absorption lines in the field $\{0, 0, H_0 + H_1 \cos \omega t\}$ can be described by the theoretical formula (4) which makes it possible to get experimentally H_i and τ_1 for some fixed value of H_0. One can obtain τ_2 for the same value of H_0 by means of the usual resonant absorption measurement. Now the formula (11) can be tested experimentally provided the situation (10) is guaranteed by the proper values of all the parameters concerned. The experiment of this kind is in progress.

References

Atsarkin, V.A., 1973, Zh.Eksp.i Teor.Fiz., 64, 1087.
Mefed, A.E. and Atsarkin, V.A., 1973, Zh.Eksp.i Teor.Fiz., Letters, 18, 683.
Shaposhnikov, I.G. and Kadyrov, D.I., 1970, C.R.Acad.Sc.Paris, Série B, 271, 611.

THE DIRECT SPIN-LATTICE RELAXATION PROCESS IN MANGANESE SULPHATE TETRAHYDRATE

C.L.M. Pouw and A.J. van Duyneveldt
Kamerlingh Onnes Laboratorium der Rijksuniversiteit, Leiden, The Netherlands

Abstract. A.c. susceptibility measurements on manganese sulphate tetrahydrate in magnetic fields up to 50 kOe show a direct relaxation process. This direct process is compared to the theoretical predictions based on the Kronig-Van Vleck and Waller-Al'tshuler interaction mechanism. The best agreement is achieved by using the Kronig-Van Vleck interaction mechanism, suggesting the direct process relaxation time to be given as: $\tau^{-1} = 0.9\times10^{-11}H^3\coth(g\beta H/2kT)\ s^{-1}$.

1. Introduction. At low temperatures and strong external magnetic fields, the spins of manganese compounds relax to the equilibrium state by the direct, one phonon process. In manganese fluosilicate and manganese Tutton salt, the observed relaxation times agree with the predictions based on the Kronig-Van Vleck (KvV) interaction mechanism between the lattice vibrations and the magnetic ions (Gorter et al.,1972). The Waller-Al'tshuler (WA) interaction mechanism, involving the magnetic dipolar interaction, is supposed to be confirmed by ESR measurements on magnetically diluted manganese sulphate tetrahydrate (Turoff et al., 1967). In this paper, relaxation times are presented for manganese sulphate tetrahydrate derived from differential susceptibility measurements in magnetic fields up to 50 kOe, in order to examine the possible occurrence of different relaxation mechanisms in equivalent manganese compounds.

2. Experimental data. Manganese sulphate tetrahydrate belongs to a group of isomorphous monoclinic crystals, but apart from Mn and Fe (II), other ions of the iron group occur in metastable form. The structure has been shown to be composed of isolated $M''_2(SO_4)_2\cdot8H_2O$ ring molecules, there are four metal ions in the unit cell (Baur, 1962/64). The present experiments were performed on powdered samples with an average crystal diameter of 0.1 mm. Measurements on other hydrated paramagnetic salts showed that in such powders the phonon-bottleneck effects are sufficiently reduced (Soeteman et al., 1974). Two concentrated samples were used (100% P and 100% NP), while the only diluted sample we obtained contained 5% zinc sulphate (sample 95% P). For clarity, only some typical results are given in the figure. At liquid hydrogen temperatures, the field dependence of the relaxation times on sample 100% P (symbol ●, T = 14.1 K) is in agreement with earlier results (Bijl, 1950, symbol □). An asymmetric broadening in the Argand diagrams (χ'' versus χ' plots from which the relaxation time τ is obtained) occurs at weak magnetic fields; above 14 kOe these deviations from a simple relaxation vanish. At liquid helium temperatures, a double relaxation is observed below 15 kOe. Re-examining this sample showed the double relaxation to disappear in the course of time, a result that strongly indicates the effect of physical impurities on the observed relaxation times in weak fields. Some average times are given for T = 4.2 K (symbol ▼). The deviations in the Argand diagrams are observed also on a newly prepared specimen (100% NP). At strong fields the relaxation times of both samples coincide, see for example the results at 14.1 K (symbol ◐). The magnetically diluted sample, 95% P, does not show strong deviations in the Argand diagrams. The measurements yield the same result for τ in strong fields as in the nondiluted samples; in weak fields the times are slightly shorter (○: T = 14.1 K, ▽: T = 4.2 K, △: T = 2.0 K).

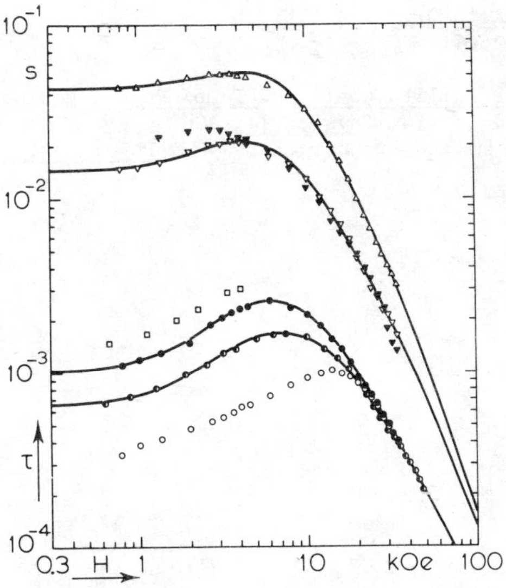

Constants determining the direct relaxation process in $MnSO_4\cdot4H_2O$ samples.

samples	T (K)	A_{KvV} ($kOe^{-3}s^{-1}$)	$A_{WA\ I}$ ($kOe^{-3}s^{-1}$)	$A_{WA\ II}$ ($kOe^{-3}s^{-1}$)
100% P	2.00	.008	.009	.011
	4.23	.008	.011	.016
	14.10	.0095	.017	.029
	15.96	.0096	.017	.030
100% NP	14.03	.0088	.015	.025
	14.95	.0088	.015	.025
	16.00	.0088	.015	.026
95% P	2.04	.0065	.0075	.0086
	4.25	.0074	.011	.015
	14.06	.008		
	15.92	.008		

3. Discussion. A conclusion to be drawn from the above described measurements is that at strong magnetic fields, the observed relaxation times approach the general behaviour of the direct relaxation process. A more detailed analysis is necessary in order to distinguish between the various interaction mechanisms. On basis of the theory of Kronig and Van Vleck (interaction involving the electrical crystal field) the direct process relaxation time for an isolated manganese ion can be written as (Blume et al., 1962):

$$\tau^{-1}_{direct} = A_{KvV}\, H^3 \coth\,(g\beta H/2kT)\ . \qquad (1)$$

If the contact between lattice and spins is to occur via the magnetic dipolar interaction (Waller-Al'tshuler), the following expression is derived (Al'tshuler, 1956):

$$\tau^{-1}_{direct} = A_{WAI}\; H^3\exp(g\beta H/kT)/(\exp(g\beta H/kT)-1) + A_{WAII}\; H^3\exp(2g\beta H/kT)/(\exp(2g\beta H/kT)-1)\ , \qquad (2)$$

the first term representing a single spin flip, the second one arises if two spins undergo a simultaneous transition. If $g\beta H \ll kT$, the terms in (1) and (2) can be simplified to $\tau^{-1} \propto TH^{\alpha}$, with $\alpha = 2$ in all cases. If this condition is not fulfilled, α increases and for $g\beta H \gg kT$, $\alpha = 3$. In the KvV theory, $\alpha = 2.2$ if $g\beta H/kT = 1.35$, while eq. (2) gives the same result for α if $g\beta H/kT = 0.8$ and 0.4, respectively. This implies that, for instance at 2 K, in the WA theory α amounts to 2.2 above 12 or 6 kOe resp., while in the KvV theory this value occurs above 20 kOe. At 2 K, our measurements between 20 and 35 kOe can be given as $\tau^{-1} \propto TH^{\alpha}$, with $\alpha = 2.2$, a value that already supports the description with eq. (1). In performing a more complete analysis, the relaxation in weak fields has to be considered also. It is shown that this relaxation is due to the occurrence of fast relaxation centres. In the description of the observed field dependences of τ, the detailed relaxation behaviour of the fast relaxation centres need not be known. The magnetic system is considered to consist of two systems, one being the bulk of manganese ions (relaxation time τ_{direct}, internal field b/C), the other related to the fast relaxation centres (relaxation time τ_2, internal field b'/C'). Using rather general properties, this 'two spin system' model leads to:

$$\tau^{-1}_{observed} = \tau_2^{-1}\; C'(b'/C'+H^2)/C(b/C+H^2) + \tau^{-1}_{direct}\ . \qquad (3)$$

Our results are fitted to eq. (3) by means of a computer programme in which for τ^{-1}_{direct} alternatively the terms of eqs. (1) and (2) are chosen. The resulting values of A are given in the table. Some computer fits, using the expression for the KvV direct process, are shown by the drawn lines in the figure. If one of the WA terms is used, the experimental points deviate more from the computer calculations. These larger deviations are related to the fact that the observed relaxation times at strong magnetic fields do not tend towards $\tau^{-1} \propto H^{\alpha}$ with $\alpha = 3$. Both theories predict constants A that are independent of temperature. This independency is best approached by the constants A_{KvV}, leading to $A_{KvV} = 0.009$ kOe^{-3}s^{-1}. This value coincides with the results in other manganese compounds (Gorter et al., 1972) and also with the theoretical estimate for Mn^{2+} in a cubic environment (Blume et al., 1962). A theoretical estimate for A_{WAI} and A_{WAII} can be obtained using the crystallographic data on $MgSO_4.4H_2O$, which is isomorphous to the manganese sulphate (Baur, 1962/64). The numerical agreement is less convincing than for the KvV theory. Considering all aspects, the conclusion to be put forward is that, in contrast to the ESR results (Turoff et al., 1967), our present measurements suggest the KvV interaction mechanism to be the cause of the direct relaxation process observed in manganese sulphate tetrahydrate.

4. References

Al'tshuler, S.A., 1956, Izv. Akad. Nauk USSR 20, 1207.
Baur, W.H., 1962, Acta Cryst. 15, 815 and 1964, Acta Cryst. 17, 863.
Blume, M. and Orbach, R., 1962, Phys. Rev. 127, 1587.
Bijl, D., 1950, Physica 16, 269.
Gorter, C.J. and Van Duyneveldt, A.J., 1972, Proc. XIIIth Conf. Low Temp. Phys., Boulder, 621.
Soeteman, J., Bevaart, L. and Van Duyneveldt, A.J., 1974, Physica 74, 126.
Turoff, R.D., Coulter, R., Irish, J., Sundquist, M. and Buchner, E., 1967, Phys. Rev. 164, 406.

SPIN-LATTICE RELAXATION OF PHOTOEXCITED TRIPLET STATE MOLECULES

By P. LOPEZ, D. BOURDEL, P. BOUJOL, J. PESCIA
Laboratoire de Magnétisme et d'Electronique Quantique, Université Paul Sabatier, 31077 TOULOUSE Cédex, France.
And J.-PH. GRIVET
Département de Physique, Université d'Orléans, 45045 ORLEANS Cédex, France.

Abstract: Spin-lattice relaxation of phosphorescent aromatic molecules was investigated in rigid ethanol solutions at 77 K. Values of the relaxation time T_1 are similar for the $\Delta m = 1$ and the $\Delta m = 2$ transitions. T_1 is increased by deuteration and shows an approximately linear concentration dependence.

1. Introduction. The phosphorescent state of organic molecules is paramagnetic, with a total spin of one. It has been thoroughly studied by electron spin resonance (GUERON 1972). However, a complete understanding of the spin-lattice relaxation mechanisms is still lacking. Many results have been obtained at helium temperatures, where T_1 is longer than 10^{-3} s (KONZELMANN and al. 1973, WOLFE 1971, ANTHEUNIS 1974). Spin-lattice relaxation (SLR) times were also measured at 77 K by the progressive saturation technique (CHACHATY and al. 1971, KITE 1974). Results ranged from 10^{-7} to 10^{-8} s. Such short relaxation times can be conveniently measured with our modulation technique (PESCIA 1965, GOURDON and al. 1973). We present here results for phosphorescent aromatic molecules in glasses at 77 K.

2. Experimental. The microwave field is amplitude modulated at a circular frequency Ω. The sample is held in a quartz Dewar, along the axis of a cylindrical TE 011 cavity. A coil, coaxial with the Zeeman field, is wound around the sample. The signal $S(\Omega)$ induced in this coil is proportional to the time rate of change of the sample magnetization. Slits in the cavity wall allow the light of a high pressure mercury arc to reach the sample. Solutions in ethanol were outgassed and sealed in quartz tubes under argon.

3. Determination of the relaxation times. In order to find T_1, we vary the modulating frequency Ω over the range $0 < \Omega < 2/T_1$, keeping all other parameters constant. At the same time, we record the amplitude $S(\Omega)$ versus Ω, from which the value of T_1 is readily derived (PESCIA 1965, GOURDON and al. 1973, LOPEZ and al. 1973, 1974), assuming a reasonable model for the resonance phenomenon. The $\Delta m = 2$ line is weakly anisotropic. At resonance, the three levels are equidistant. The resonance is therefore well described as homogeneous, in spite of the presence of an unresolved hyperfine structure. Furthermore, the spin system can be described by a spin temperature.

On the other hand, the $\Delta m = 1$ transitions are quite anisotropic: only stationary peaks can be observed in a glass. The measurements reported here concern the low-field absorption shaped singularity of χ''. In our experiments, Cross-relaxation occurs and this singularity can be described again as homogeneous. Thus, values of T_1 are always found on the basis of an homogeneous broadening of the line. At last, the modulation method is not affected by the forbidden aspect of transitions.

4. Results. We have found that the linewidths are independant of concentration; they amount to 6-7 gauss for the deuterated compounds and to 12-13 gauss for the hydrogenated isomers (FWHM of $\Delta m = 2$ line). Our T_1 values are given in the table. The uncertainty is of the order of 15%. Measurements were performed at X - band.

Species and fine structure parameters	Concentration mole/liter	T_1 ($\Delta m = 2$) 10^{-8} second	T_1 ($\Delta m = 1$) 10^{-8} second
Deuteronaphthalene, $C_{10}D_8$	4.7×10^{-2}	5.9	4.45
$D = 0.099\ cm^{-1}$	10^{-2}	9.15	9.4
$E = -0.0137\ cm^{-1}$	4.7×10^{-3}	10.8	10.5
	10^{-3}	13.4	11.5
	5×10^{-4}	15.1	----
Naphthalene, $C_{10}H_8$	4.7×10^{-2}	4.2	----
Deuterophenanthrene, $C_{14}D_{10}$	5×10^{-2}	8.4	----
$D = 0.100\ cm^{-1}$	2×10^{-2}	11.7	10.6
$E = -0.046\ cm^{-1}$	10^{-2}	12.3	11.2
	5×10^{-3}	13.5	11.7
	2.5×10^{-3}	14.7	12.8
	10^{-3}	20	16
	5×10^{-4}	23.4	----
Phenanthrene, $C_{14}H_{10}$	2 10^{-2}	6.15	----

Table 1: Spin-lattice relaxation times of:
- naphtalene,
- deuteronaphtalene,
- phenanthrene,
- deuterophenanthrene

in frozen ethanol solutions at 77 K.

5. Discussion. (i) For all samples investigated here, we find that SLR is about as effective for the $\Delta m = 1$ transition as for the $\Delta m = 2$ line. The transition probabilities are thus approximately equal for the three pairs of levels.

(ii) T_1 values are increased by a factor 1.4 upon deuteration. If thermal modulation of the hyperfine interaction were the dominant SLR mechanism, deuteration would increase T_1 by a factor $(\gamma_H/\gamma_D)^2 \times 3/8 \simeq 15$. This mechanism is therefore inadequate in the present case. We note that WOLFE (WOLFE 1971) found that deuteration increased SLR rates for naphthalene in a durene single crystal below 4.2 K.

(iii) By extrapolation to zero concentration, we obtain relaxation times for the $\Delta m = 2$ transition of isolated molecules :

$$T_{10} \text{ (deuteronaphthalene) } = 1.9 \times 10^{-7} \text{ s}$$
$$T_{10} \text{ (deuterophenanthrene) } = 2.7 \times 10^{-7} \text{ s}$$

The transition probability is approximately proportional to $D^2 + 3E^2$. This seems to be consistent with the model of thermal modulation of the fine structure (MARUANI 1970).

(iv) Let c be the concentration and let us write $W = 1/T_1$, $W_0 = 1/T_{10}$. We then define: $\Delta W = W - W_0$. From the values given in the table, it is readily seen that :

$$\Delta W \sim c^n \quad \text{with} \quad 1 < n < 1.3 \quad \text{for} \quad c < 2 \times 10^{-3} \text{ M/l.}$$

The intermolecular mechanisms of relaxation should lead to $1 < n < 2$ in good agreement with our experimental results. For $c > 2 \times 10^{-3}$M/l we find $\Delta W \sim c$. But, the concentration in molecules is no longer proportional to the concentration in triplets and the problem becomes very complex.

6. References.

ANTHEUNIS, D., 1974, Thesis, LEIDEN.

CHACHATY, C. and MARUANI, J., 1971, C. R. Ac. Sci., 273 B, 1119.

GOURDON, J.C., VIGOUROUX, B. and PESCIA, J., 1973, Phys. Lett., 45 A, 69

GRIVET, J.PH., 1970, Mol. Phys., 19, 389.

GUERON, M., 1972, Electron Spin Resonance of Triplet States, Edited by LAMOLA, A. and DEKKER, M..

KONZELMAN, U. and SCHWOERER, M., 1973, Chem. Phys. Lett., 18, 143.

LOPEZ, P., GRIVET, J.PH. and PESCIA, J., 1973, C. R. Ac. Sci., 277 B, 703.

LOPEZ, P., 1974, Thesis, Université Paul Sabatier, TOULOUSE, France.

MARUANI, J., 1970, Chem. Phys. Lett., 7, 29.

PESCIA, J., 1965, Ann. Phys., 10, 389.

WOLFE, J.P., 1971, Chem. Phys. Lett., 10, 212.

KITE, T.M., 1974, Ph. D. Thesis, LARAMIE, USA.

FIELD-ORIENTATION DEPENDENCE OF SPIN-LATTICE RELAXATION RATES FOR TRIPLET-STATE MOLECULES

X. Gille and J. Maruani

C.M.O.A. du C.N.R.S., 23, rue du Maroc, 75019-Paris, FRANCE.

Abstract. The anisotropy of spin-lattice relaxation rates for triplet-state molecules has been the subject of various investigations. We have used a semi-classical treatment to investigate the first-order effects of modulations of the fine-structure Hamiltonian, and obtained a fit of the continuous-saturation results for pyrene in crystals.

1. Introduction. In recent years there has been increased interest in the spin-lattice relaxation rates, T_1^{-1}, of triplet-state molecules dissolved in solids (Schwoerer et al., 1972 ; Denison, 1973). One of the most striking results obtained in crystalline matrices, whether by pulse saturation below 4.2°K (Wolfe, 1971; Konzelmann and Schwoerer, 1973) or by continuous saturation above 77°K (Denison and Fischer, 1968, 1969 ; Kite, 1974), is the 1 : 10 anisotropy of the T_1's, contrasting with the 1 : 2 anisotropies measured in an alcohol glass (Chachaty and Maruani, 1971). We have extended a previous approximate, semi-classical calculation of the T_1's (Maruani, 1970) and tried to account for the field-orientation dependences observed for photo-excited pyrene in room-temperature crystals.

2. Model. The main assumptions in our previous treatment were the following. 1) The Hamiltonian responsible for spin-lattice relaxation may be represented by a sum of stochastic, scalar spin-operators, and first-order, time-dependent perturbation theory may be used (Abragam, 1961). 2) All these scalar operators have the same time dependence ; actually, identical results would be obtained under the less stringent condition that all inter- and auto-correlation functions be equal. 3) The effective correlation time, τ_c, is much smaller than the free-electron angular frequency inverse, ω_0^{-1}, so that the product $4\omega_0^2\tau_c^2$ be negligible. 4) The lattice thermal energy, kT_L, is much larger than $\hbar\omega_0$, so that $W^L_{ji}=W^L_m$. 5) There is a spin temperature ; in such a case, the expressions of the T_1's as functions of the W^L's are the same for all the lines and for both pulse and continuous saturation. 6) The perturbation of the Zeeman states by the fine-structure term, $\bar{S}.\bar{\bar{D}}.\bar{S}$, may be neglected in calculating the W^L's. 7) The set of stochastic spin-operators may be written $\bar{S}.\bar{\bar{D}}(t).\bar{S}$ (Wolfe, 1971), with only the first-order terms in the expansion of the components of $\bar{\bar{D}}(t)$ included (Denison and Fischer, 1968, 1969).

In the present treatment, only the assumptions 1 and 7 are retained. Assumption 2 is replaced by the more likely assumption (2') that only the auto-correlation functions are equal, all inter-correlation functions being zero (Abragam, 1961) ; in this case, different results are obtained by performing the treatment in terms of D' and E' or in terms of X', Y' and Z': the anisotropies obtained by neglecting all cross-terms coming from assumption 2 will be different (Gille, 1974). Assumptions 3 and 4 are not explicitly made, although the latter would hold whenever the semi-classical treatment would be appropriate, and the former also will hold for the system of interest here (Kite, 1974). Assumption 5 is replaced by the more realistic assumption (5') that cross-relaxation coupling of the $\Delta m=1$ transitions is negligible ; the expressions of the T_1's are, consequently, different for pulse and continuous saturation (Gille, 1974). Finally, assumption 6 is replaced by the approximation (6') of first-order perturbation of the Zeeman states by $\bar{S}.\bar{\bar{D}}.\bar{S}$.

3. Treatment. Using the previous assumptions, we find the following expressions for calculating the T_1's of the different lines.

$$T_1^{-1}{}_{,ij}(i>j\neq k) = 2\,\frac{W^L_{ik}W^L_{jk} + W^L_{ik}(W^L_{kj}+W^L_{ji}) + W^L_{jk}(W^L_{ki}+W^L_{ij}) + (W^L_{ij}+W^L_{ji})(W^L_{ki}+W^L_{kj})}{(W^L_{ik}+2W^L_{ki})+(W^L_{jk}+2W^L_{kj})},$$

with $W^L_{ij}=W^L_{i,j}\exp(-\hbar\omega_{ij}/2kT_L)$. This expression includes terms neglected in (Maruani, 1972), and is invariant under permutation of i and j. Under assumption 4, it reduces to :

$$T_1^{-1}{}_{,ij} = 2\left[W^L_{i,j} + W^L_{j,k}W^L_{i,k}/(W^L_{j,k}+W^L_{i,k})\right], \qquad (1)$$

simpler than in (Maruani, 1972). The $W^L_{i,j}$ are given by the expressions :

$$W^{L}_{1,2} = (U_{10} + u)\, 2\tau_c/(1 + \omega_{21}^2 \tau_c^2), \quad W^{L}_{2,3} = (U_{10} - u)\, 2\tau_c/(1 + \omega_{32}^2 \tau_c^2), \quad W^{L}_{1,3} = (U_{1-1})\, 2\tau_c/(1 + \omega_{31}^2 \tau_c^2). \qquad (2)$$

Here $\omega_{ij} = \omega_i - \omega_j$; $\omega_1 = -\omega_0 + \tilde{\omega}$, $\omega_2 = -2\tilde{\omega}$, $\omega_3 = \omega_0 + \tilde{\omega}$; $\omega_0 = g\beta H_0$ in $\hbar$ units and $\tilde{\omega} = -\sum_i X_i^2 l_i^2/2$, the X_i's being the principal values of $\bar{\bar{D}}$ in $\hbar$ units and the l_i's being the direction cosines of $\bar{H}_0$ on the principal axes of $\bar{\bar{D}}$; finally :

$$U_{10} = \sum_{\substack{i=1 \\ k\neq j\neq i}}^{3} \left[X_i'^2 l_i^2 (1-l_i^2) + \alpha_i'^2 (X_j - X_k)^2 (l_j^2 + l_k^2 - 4 l_j^2 l_k^2)\right]/2,$$

$$U_{1-1} = \sum_{\substack{i=1 \\ k\neq j\neq i}}^{3} \left[X_i'^2 (1-l_i^2)^2 + 4\alpha_i'^2 (X_j - X_k)^2 (1-l_j^2)(1-l_k^2)\right]/4, \qquad (3)$$

$$u = \sum_{\substack{i=1 \\ k\neq j\neq i}}^{3} (X_i/4\omega_0)\Big\{3X_i'^2 l_i^2 (1-l_i^2)^2 + \alpha_i'^2 (X_j - X_k)^2 \left[(1-l_i^2)(1+3l_i^2) - 4 l_j^2 l_k^2 (1-3l_i^2)\right] +$$

$$+ \sum_j \left[X_j'^2 l_j^2 (1 - l_j^2 - 11 l_i^2 + 9 l_i^2 l_j^2) - \alpha_j'^2 (X_k - X_i)^2 \left(3 l_i^2 (1-l_i^2) + l_k^2 (1+4l_i^2)(1-3l_i^2)\right)\right]\Big\} ;$$

the X_i''s are the modulation amplitudes of the X_i's in $\hbar$ units and the α_i''s are the modulation amplitudes of the Eulerian angles α_i in radians.

4. Results. We have performed a set of computer calculations in order to obtain a fit of the experimental results from (Denison and Fischer, 1968, 1969), summarized in the empirical, dashed lines of the figure. Our results should be linear in τ_c up to $\tau_c \sim 10^{-11}$ sec, for $\omega_0 \sim$ 9.25 GHz. Using the expressions (1) and (2), neglecting u, one sees from expressions (3) that one cannot account for the non-zero T_1^{-1}'s along the principal axes if one neglects the modulation of orientation of principal axes. One can then estimate α'_x, α'_y, α'_z from the measured values of T_1 along these axes. Now, the calculations show that modulations of the angles alone do not allow fitting the negligible angular dependence in the molecular plane ; the "hole" in the calculated curves can be "filled" only if modulations of the principal values are included. These affect only slightly the shape of the anisotropy in the other two symmetry planes and the values of the α'_i's. The best overall fit is obtained for the set of values given in the figure, and is shown by the solid lines. The agreement is acceptable, in view of the large experimental errors.

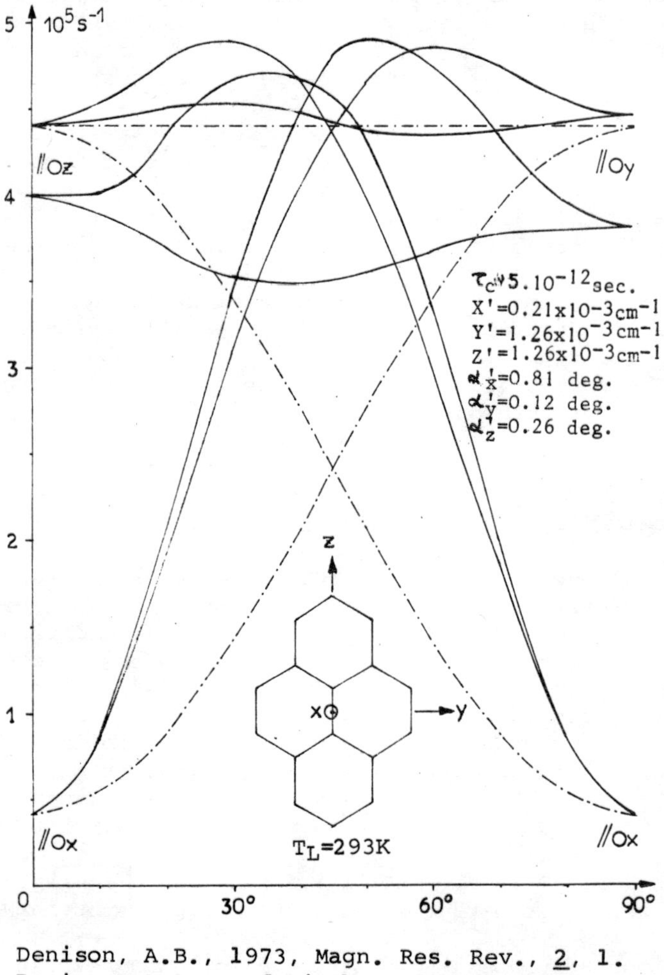

5. References.

Abragam, A., 1961, The Principles of Nuclear Magnetism, Oxford University Press, London, chap. VIII.
Chachaty, C., and Maruani, J., 1971, C.R. Acad. Sci. (Paris), B273, 1119.
Denison, A.B., 1973, Magn. Res. Rev., 2, 1.
Denison, A.B., and Fischer, P.H.H., 1968, Proc. XV. Coll. AMPERE (Grenoble), North-Holland, Amsterdam, p. 455 ; 1969, Mol. Phys., 17, 297.
Gille, X., 1974, Master's thesis, Paris.
Kite, T.M., 1974, Thesis, Laramie (Wyoming).
Konzelmann, U., and Schwoerer, M., 1973, Chem. Phys. Lett., 18, 143.
Maruani, J., 1970, Chem. Phys. Lett., 7, 29 ; 1972, J. Magn. Res., 7, 207.
Schwoerer, M., Konzelmann, U., and Kilpper, D., 1972, Chem. Phys. Lett., 13, 272.
Wolfe, J.P., 1971, Chem. Phys. Lett., 10, 212.

PULSED ENDOR IN DILUTED COPPER TUTTON SALTS

H. Hoogstraate, W.Th. Wenckebach and N.J. Poulis
Kamerlingh Onnes Laboratorium der Rijksuniversiteit Leiden, Leiden, Nederland

Abstract. The method of pulsed ENDOR is used to investigate the thermodynamical model that describes dynamical polarization and ENDOR in diluted Cu-Tutton salts.

In our previous experiments ENDOR was applied in diluted Copper Tutton salts without any modulation. This was possible because the signals were very strong. When the continuous r.f. field, that saturated the protons, was put on, the ENDOR signal appeared and when it was turned off, the equilibrium value of the E.S.R. signal was restored. It is also possible to apply a 180° r.f. pulse instead of a continuous r.f. field. The advantage of this pulse method is that much more information can be obtained about the dynamic behaviour of the ENDOR mechanism and about the contacts between the different systems. In particular it is possible in this way to make very fast measurements. Moreover, a situation exists after the pulse, that can be described theoretically in a relatively simple way.

The measurements were all carried out in Copper Tutton salts*. The formula is $MN_2(SO_4)_2.6H_2O$ where 2% - 0.5% of M consists of Cu^{2+} and the rest Zn^{2+}, while N is Rb^+ or Cs^+.

When placed in a static magnetic field H_o of ∿3 kOe, the E.S.R. spectrum of a Cu^{2+}, which has an effective spin ½, consists of four absorption lines due to the hyperfine interaction with the Cu nucleus that has spin $^3/2$. Because two non equivalent sites exist in the Tutton salts, the spectrum has 8 resolved lines. However, we always chose the direction of H_o such that the spectra of the non equivalent sites coincided so that, finally the spectrum consisted of only four absorption lines.

In order to explain the ENDOR and dynamic polarization a thermodynamical model is used (fig. I). In this model, applied to the Copper Tutton salts each of the four absorption lines corresponds with an energy reservoir with a temperature of its own[(2)]. We call these the four electron Zeeman systems (EZS). These correspond each with one absorption line and with one direction of the Cu nuclear spin in the magnetic field H_o. The dipolar energy of the electrons is represented by the electron dipole system (EDS) and the proton Zeeman system (PZS). These two systems have also a temperature of their own. Several contacts exist between the systems

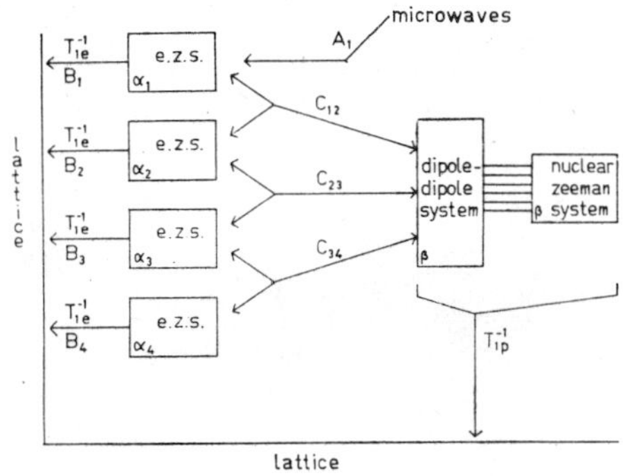

Fig. I Model representing the spin systems of a diluted Cu-Tutton salt

and between the systems and the lattice. The electron spin relaxation tends to equalize the temperatures of the four EZS and the lattice. Moreover a direct relaxation path exists between the EDS and the lattice, whereas the proton relaxation directly to the lattice is so slow that it can be neglected. A very strong contact between the PZS and the EDS has been shown to exist[(3)].

In the cross relaxation process, two electrons of which the quantumnumbers of the accompanying nuclear spin are different, make a flip flop. The small difference in energy of these two electrons in the magnetic field H_o is transferred to the EDS. When one of the transitions of the electrons is irradiated with saturating microwave power, the temperature of the corresponding system increases. Due to the cross relaxation process also the temperature of the EDS changes and may either increase or decrease. Because a good contact exists between the EDS and the PZS, the temperature of the last changes also, which leads to a positive or negative dynamic polarization of the protons. In our experiments the ESR signal level is observed in this situation and with a 180° r.f. pulse the proton magnetization is inverted. After this pulse the ESR signal changes as a function of the time. This time behaviour can be analysed within the measuring accuracy as the sum of three exponential functions. The ENDOR signal builds up with a time constant τ_1 and successively returns to the previous value with the time constants τ_2 and τ_3. The intensity of the observed ESR line is proportional to the inverse spin-temperature so that we measure the temperature of this

* For other work in this field see references 1 and 4.

EZS as a function of the time.

The following remarks can be made about each of the time constants:

τ_3 depends very little on the saturating microwave power. It is related strongly to the electron spin relaxation time T_{1e} because it has about the same value and because τ_3 changes like T_{1e} at different temperatures in the region of 1.2 to 4.2 Kelvin.

τ_2 is inversely proportional to the microwave power over a great region. In the limit of very small power τ_2 reaches τ_3 and in the limit of great power τ_2 reaches τ_1.

τ_1 is determined by the cross relaxation process. There are four arguments to acknowledge this:

τ_1 depends little on both the temperature and the microwave power.

The cross relaxation time depends entirely on the hyperfine splitting of the E.S.R. spectrum and may thus be changed by choosing different orientations of the sample in the magnetic field. The value of τ_1 varies in a similar way as the cross relaxation time with this orientation. Depending on the orientation of the sample we found values for τ_1 of 20 msec to 150 msec for H_o along the k_2 axis and the k_3 axis respectively, which agrees very well with calculated values of the cross relaxation time and with other measurements(2).

It is impossible that τ_1 would be related to the contact between the PZS and the EDS for two reasons. It is very improbable that this contact can be described by an exponential function and the time in which the PZS and the EDS reach an equilibrium should not depend strongly on the orientation of the sample in the magnetic field H_o.

Apart from the fact that these measurements and their interpretation agree very well with the theory and that the cross relaxation time can be measured directly with the aid of pulsed ENDOR, we conclude that the equilibrium between the PZS and the EDS is established in a time shorter than 20 msec.

References

(1) Atsarkin, V.A. and Rodak, M.I., 1972, Uspekhi Fiz. Nauk. 107, 3; 1972, Sov. Phys. Uspekhi 15, 251.
(2) Hoogstraate, H., van Houten, J., Schreurs, L.A.H., Wenckebach, W.Th. and Poulis, N.J. 1973, Physica 65, 347.
(3) Wenckebach, W.Th., van den Heuvel, G.M., Hoogstraate, H., Swanenburg, T.J.B. and Poulis, N.J., 1969, Phys. Rev. Lett. 22, 581.
(4) Wenckebach, W.Th., Swanenburg, T.J.B. and Poulis, N.J., to be published in Physics Reports.

ON A STRONG ESR SATURATION UNDER THE CONDITION OF PHONON BOTTLENECK

L.L.Buishvili, N.P.Giorgadze, M.D.Zviadadze, A.I.Ugulava
Institute of Physics, Academy of Sciences of the Georgian SSR,Tbilissi,USSR

Abstract. The effects of phonon heating and of phonon avalanche furmation are considered in the paper theoretically under conditions of strong and super strong ESR saturation.

1. The theory of phonon bottleneck (PB) with account for dipole-dipole reservoir (DDR) in the case of Provotorov (Intermediate saturation) has been recently developed (Borghini, 1968; Buishvili, Giorgadze, 1971; Altshuler et al,1972). In this work we consider the effects of phonon heating and phonon avalanche formation in alternating fields of the order of the local field (the case of Redfield) and higher ones (super strong saturation).

2. The Hamiltonian of the spin-phonon system is written in the form (Buishvili et al. 1971)

$$\mathcal{H}=\mathcal{H}_z+\mathcal{H}_d+\mathcal{H}_{ph}+\mathcal{H}_{sh}+\mathcal{H}_{s,ph}, \quad \mathcal{H}_z=\omega_s S^z, \quad \mathcal{H}_{ph}=\sum_{\bar{k}j}\omega_{kj}a^+_{\bar{k}j}a_{\bar{k}j} \tag{1}$$

$\mathcal{H}_z$ is Zeeman energy of the spin system (SS), $\mathcal{H}_d$ is the secular part of the dipole-dipole interaction, $\mathcal{H}_{ph}$ is the phonon Hamiltonian, $\mathcal{H}_{sh}$ is the SS interaction energy with the circularly polarized alternating magnetic field with an amplitude ω_1/γ and frequency Ω, $\mathcal{H}_{s,ph}$ is the spin-phonon interaction corresponding to the one phonon relaxation mechanizm; $a^+_{\bar{k}j}$ and $a_{\bar{k}j}$ are the creation and annihilation operators of phonons of the j-th branch with the wave vector $\vec{k}$.

By means of the unitar transformation

$$\mathcal{U}=\exp\{i\Omega(S^z+\sum_{\bar{k}j}a^+_{\bar{k}j}a_{\bar{k}j})t\}$$

a transition to the new reference system which is called rotating frame (RF) is made. The Hamiltonian of the system in RF is explicitely independent of time and can be represented as

$$\mathcal{H}^*=\mathcal{H}^*_s+\sum_{\bar{k}j}\mathcal{H}^*_{\bar{k}j}+\mathcal{H}_{s,ph}, \quad \mathcal{H}^*_s=(\omega_s-\Omega)S^z+\omega_1 S^x+\mathcal{H}_d, \quad \mathcal{H}^*_{\bar{k}j}=(\omega_{\bar{k}j}-\Omega)a^+_{\bar{k}j}a_{\bar{k}j}. \tag{2}$$

3. Let us assume that the considered spin-phonon system represents a set of the subsystems which are in the internal equilibrium and weakly interacting between each other: SS with the inverse temperature β^*_s and the phonon packets of definite frequencies with the inverse temperatures $\beta^*_{\bar{k}j}$.

The further evolution of the system to the stationary state in RF on the account of the spin-phonon interaction can be described by the method of nonequilibrium statistical operator (Zubarev, 1971). The initial equation set is obtained for β^*_s and $\beta(\omega)=\frac{\omega-\Omega}{\omega}\beta^*(\omega)$ in the form:

$$\frac{d\beta^*_s}{dt}=\frac{1}{T_{sp}}\frac{1}{(\omega_s-\Omega)^2+\omega_1^2+\omega_d^2}\int_{-\infty}^{+\infty}d\omega\left(\frac{\omega}{\omega_s}\right)^2(\omega-\Omega)^2 f^*(\Omega-\omega)\frac{\beta_0}{\beta(\omega)}\left\{\frac{\omega}{\omega-\Omega}\beta(\omega)-\beta^*_s\right\},$$

$$\frac{d\beta(\omega)}{dt}=-\frac{C_s}{\bar{C}_p}\frac{1}{T_{sp}}\frac{\omega(\omega-\Omega)}{\omega_s^2}f^*(\Omega-\omega)\frac{\beta(\omega)}{\beta_0}\left\{\frac{\omega}{\omega-\Omega}\beta(\omega)-\beta^*_s\right\}-\frac{\beta(\omega)}{\beta_0}\frac{\beta(\omega)-\beta_0}{T_0}, \tag{3}$$

Here

$$\omega_d^2=\frac{Sp\,\mathcal{H}_d^2}{Sp\,S_z^2}, \quad f(\omega)=\frac{1}{2\pi}\int_{-\infty}^{+\infty}dt\, e^{i\omega t}\frac{Sp\{S^- e^{i\mathcal{H}^*_s t}S^+ e^{-i\mathcal{H}^*_s t}\}}{Sp\{S^+S^-\}},$$

and $C_s/\bar{C}_p$ is the ratio of SS heat capacity to the density of the phonon distribution at the frequency ω_s. $\beta(\omega)$ is the inverse phonon temperature in the laboratory frame system. The correlator $f^*(\omega)$ essentially differs from that in the theory of the intermediate ESR saturation. The last term in the right hand side of the second equation of set (3) describes phenomenologically the simpliest form of the phonon relaxation to the thermostat with the relaxation time T_0 independent of frequency. β_0 is the inverse temperature of the thermostat.

4. Let us consider different limiting cases.

a) Assuming that there is PB effect and the same temperature β_p^{-1} may be attributed to phonons, interacting with SS.

The stationary solution of the equation system obtained in this case has a form

$$\beta_s^{*st} = \frac{\omega_s(\omega-\Omega)}{(\omega_s-\Omega)^2+\frac{1}{2}\omega_1^2+2\omega_d^2}\beta_0, \quad \beta_p^{st} = \frac{(\omega_s-\Omega)^2+\frac{1}{2}\omega_1^2+2\omega_d^2}{(\omega_s-\Omega)^2+(\frac{1}{2}\omega_1^2+2\omega_d^2)(\sigma+1)}\beta_0, \qquad (4)$$

$\sigma = \frac{C_s T_0}{2\Delta C_p T_{sp}}$ is the PB coefficient. In the absence of PB ($\sigma \ll 1$) (4) turns into a well-known Redfield's result for the stationary value of the spin temperature in RF. Formula (4) contains Borghini's result (Borghini, 1968) as a particular case, which relates to the intermediate saturation ($\omega_1 \ll \omega_d$) under the condition of PB.

b) Let us now consider the case when the spectral diffusion in the phonon spectrum is absent under PB. Assume that the SS is excited by a rather short HF-pulse in the course of which it can be considered isolated from the lattice. Then at small times after the pulse is finished, the deviation $z(\omega)=[\beta_0-\beta(\omega)]/\beta(\omega)$ of the resonance phonon number from the equilibrium value in the frequency range

$$(\omega_s-\Omega)(\omega_s-\omega)+\omega_1^2+\omega_d^2>0, \quad (\omega_s-\Omega)(\Omega-\omega)>0$$

increases at first linearly and then exponentially with the time. In other words, a phonon avalanche arises in the given frequency range. This result is similar to that obtained for the case of the intermediate saturation (Borghini, 1968).

5. Super strong saturation ($\omega_1 \gg \omega_d$) is considered by means of the transition to the effective frame (EF) in which the axis is directed along the effective magnetic field $\vec{H}_e=\gamma^{-1}[\vec{k}(\omega_s-\Omega)+\vec{i}\omega_1]$ ($\vec{k}$ and $\vec{i}$ are unit vectors of the axes z and x in RF).

In the EF the spin system is consisting of two subsystems - Zeeman and DDR - with inverse temperatures β_s and β_d respectively. The analysis of the equation system for β_s , β_d and $\beta(\omega)$ shows that the SS interacts with two distinctly divided groups of phonons distributed near the frequencies $\Omega \pm \omega_e$, ($\omega_e=\sqrt{(\omega_s-\Omega)^2+\omega_1^2}$) the frequency width of these distributions being small in comparison with the ESR line width.

The investigation of the pulsed super strong ESR saturation shows that at $\omega_s-\Omega \lessgtr 0$ the avalanche covers the resonance phonons of $\Omega \pm \omega_e$ frequencies. The generation of the phonon avalanche in Provotorov's case is caused by DDR playing an active role at ESR saturation. The sharp variation of DDR temperatureleads to the fact that depending upon the sign of deviation of frequency from the resonance one, the effective spin-temperature for part of the resonance phonons aquires a negative value. Particularly, the avalanche develops in phonons lying at the line wing close to Ω . In strong alternation fields when $\omega_1 \gg \omega_d$ the Zeeman energy and the DDR defined in EF are not connected to each other thermodynamically. Now DDR role is reduced to the source of chaotization inside the spin-system, whereas DDR temperature in the absence of the rotational saturation at the frequency ω_e practically does not change. Nevertheless, the effective spin-temperature for the group of resonance phonons lying at the line wing close to Ω remains negative causing the generation of the avalanche.

References

Borghini,M., Phys.Lett., 20, 419 (1968)
Buishvili,L.L., Giorgadze N.P., Radiofizika, 15, 1493 (1971)
Al'tshuler S.A., Valishev R.M., Kochelaev B.N., Khasanov,A.Kh. Zh.Exper. Teor.Fiz., 62, 639 (1972).
Zubarev D.N. Nonequilibrium Statistical Thermodynamics, Nauka, 1971.

SECTION M

NMR Image Formation

'DIFFRACTION' AND MICROSCOPY IN SOLIDS AND LIQUIDS BY NMR.

P. Mansfield, P.K. Grannell and A.A. Maudsley.

Department of Physics, University of Nottingham, Nottingham NG7 2RD, England.

Abstract. A new approach to the study of structure in solids and liquids by pulsed NMR is described. Examples of diffraction by ordered and disordered systems are given together with examples of two-dimensional image formation, in which spin density micrographs are formed from their one-dimensional density projections. Factors affecting resolution are discussed.

1. Introduction. Recently we introduced a new method (Mansfield and Grannell, 1973) for the determination of spatial structures in solids, which relies on NMR diffraction effects, in which identification of the lattice sites in the frequency spectrum is obtained by applying a linear magnetic field gradient to the sample. The usual effect of a gradient, first shown by Gabillard (1952) is to produce a free induction decay (FID) which, through a Fourier transform, reflects the bulk shape of the material, assumed to be a continuous distribution of spins. This idea has also recently been reactivated and extended independently by Lauterbur (1973) and applied to the study of localized liquid regions in biological systems.

In this paper, we show that NMR diffraction may be applied to partially disordered systems, yielding the one dimensional spatial distribution function. The two dimensional spatial Fourier transform for non-regular structures has also been performed yielding a two-dimensional reconstructed image or micrograph.

2. Diffraction from Disordered Materials. We wish to consider the diffraction pattern and its spatial Fourier transform when we have partially disordered lattices, like for example, biological membranes, fibrous structures or mosaic crystals. Of course, x-ray studies in liquids and amorphous solids do in general reveal some structure, the origin of which was first explained by Zernike and Prins (1927). Their idea and subsequent developments in x-ray studies have allowed the so-called radial distribution function to be measured in liquids and amorphous solids.

As a model approximation to a real system we consider a random collection of spheres of resonant nuclei with constant spin density ρ_0 confined within a rectangular container such that one edge coincides with the linear field gradient which is applied along the z axis.

The results for such a system of spheres are shown in Fig (1). The experimental results were got by observing a complementary signal, (Babinet's principle) obtained by immersing a random collection of 0.2mm diameter glass spheres in water. The spheres were supported in a matrix of fine 470μm beads to ensure that they did not touch on average. The theoretical curves are based on a simplified 'random' model using the one dimensional distribution function of Zernike and Prins (1927).

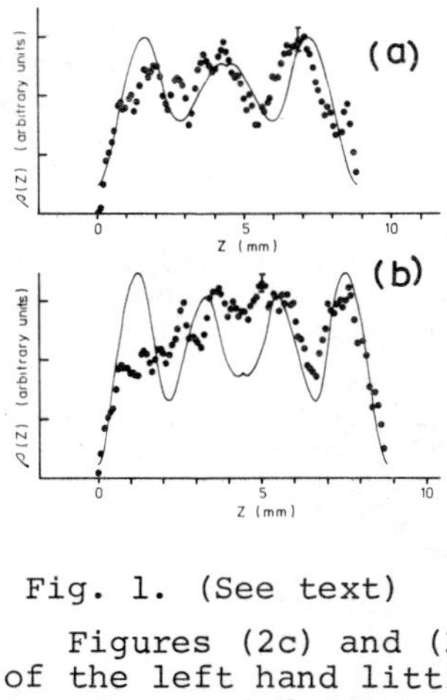

Fig. 1. (See text)

3. NMR Microscopy. The one dimensional density distribution may be useful as it stands if the general form of the object is known. Figure (2) illustrates this point. Figure (2a) is the proton spin distribution for a cylinder of water 13.4mm in diameter while Fig (2b) is the spin distribution for the same object but with a co-axial occlusion 9.8mm in diameter containing no spins. The axes of these distributions are transverse to the cylindrical axis in each case. The spatial resolution achieved here is about 0.5mm and is limited by the non-uniformity of the field gradient (Tanner, 1965).

Figures (2c) and (2d) are one dimensional spin density distributions of the tip of the left hand little finger of P.K.G. recorded in vivo. Strong proton signals are received from the skin, subcutaneous tissues, synovial sheath, blood vessels etc., while weaker signals come from the distal phalanx. The NMR coil was positioned over the region where the nail merges with the eponychium. The general similarity between Fig. (2b) and Figs, (2c) and (2d) is consistent with the finger tip being approximately cylindrical with the bone centrally placed. Also apparent in Figs.(2c) and (2d) is a small distortion caused by the nail.

Two dimensional spin density distributions, $\rho(x;y)$ are more useful and represent cross-sectional pictures or micrographs of the object. Micrographs corresponding to the annulus data and the finger data of Figs.(2b, c and d) are shown in Fig (3) and are constructed using ten grey levels. The matrix reconstruction method for

Fig. (3a) is described elsewhere (Mansfield and Grannell, to be published). The first algorithm of Gordon and Herman (1971) was used to produce Fig. (3b). The approximate outline of the finger and the expected position of the bone are sketched in.

4. Resolution Limits. In both solids and liquids the resolution is directly related to the uniformity of both the static magnetic field B_0 and the applied linear field gradient $\underset{\sim}{G}$. The intrinsic lifetime of the signal, T_2 or $T_{2\varepsilon}$, plus translational diffusion also play an important role.

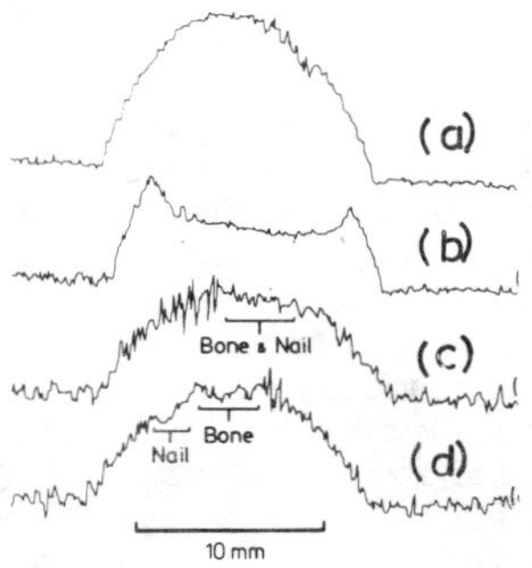

Fig. 2. (See text)

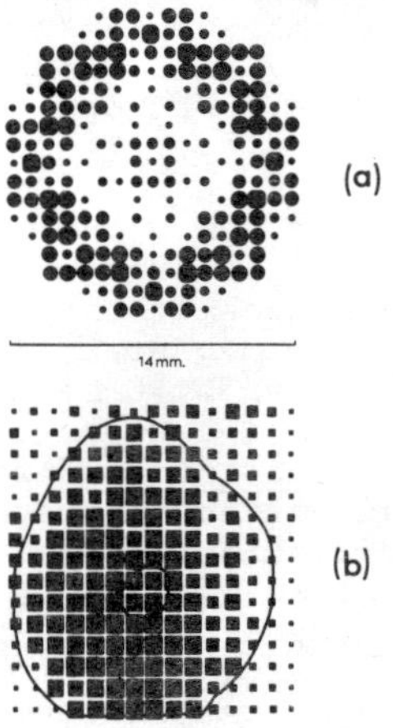

Fig. 3. (See text)

For high resolution, chemical shift variations and changes of the bulk electronic susceptibility will clearly have an important effect as well. The fact that spatial resolution depends on position in the field gradient implies that movements of both sample and gradient coil will be important.

Finally, the texture of the sample material itself and the presence of polycrystals will impose some limitations.

If we take diffusion as the limiting mechanism of resolution in liquids and consider protons in water in particular as the material most common in, for example, biological specimens, then for $D = 1.85\ 10^{-5}\ cm^2\ sec^{-1}$ and $G = 100\ G\ cm^{-1}$ we find a resolution limit of 6μm. Since the water must remain liquid, diffusion is not an experimental variable, so that resolution improvements must come from larger field gradients.

In dipolar broadened solids, the resolution limit is directly related to the degree to which this broadening can be reduced using multiple-pulse techniques. A conservative value for the multiple pulse decay time constant $T_{2\varepsilon}$ is 2.0 msec for protons, and in order to keep within this value, the total static field variation across the sample should not exceed about 1 gauss (Garroway, Mansfield and Stalker). For a sample 0.1 mm thick, this corresponds to a maximum field gradient $G_z = 100\ G\ cm^{-1}$. If $T_{2\varepsilon}$ is the only broadening source then the best spatial resolution obtainable for protons in a solid is ∿4.0 μm.

5. Conclusions. We have shown that NMR of disordered systems in the region of a boundary can be made to yield the spin density distribution analogous to the radial distribution function in X-ray studies. With a technologically reasonable field gradient of $100\ G\ cm^{-1}$, the best resolution obtainable in a liquid (water) is about 6 μm, the principal limit being translational self-diffusion. For solids, using multiple-pulse line narrowing techniques, the resolution limit is currently 4 μm. In this case, however, the limit is imposed by the intrinsic line width due to imperfections in the multiple-pulse sequence used.

Procedures for forming the two dimensional Fourier transform from its projections have been tried using point matrix reconstruction techniques. Since the specimens studied here are continuous functions of the spatial co-ordinates, these reconstruction methods are always inexact.

Two dimensional images or micrographs allow examination of internal and external structures and characteristics of specimens. Application to biological materials is stressed and an example using a physiological specimen *in vivo* is given.

6. Acknowledgements. We wish to thank A.N. Garroway and D.C. Stalker for their general assistance in performing these experiments and C.J. Doran for implementing the reconstruction algorithm of Gordon and Herman.

7. References

Gabillard, R., 1952, C.R. Acad. Sci. Paris 232, 1551.
Gordon, R. and Herman, G.T., 1971, Comm. A.P.M. 14, 759.
Lauterbur, P.C., 1973, Nature 242, 190.
Mansfield, P. and Grannell, P.K., 1973, J. Phys. C6, L 422.
Tanner, J.E., 1965, Rev. Sci. Instrum. 36, 1086.
Zernike, F. and Prins, J.A., 1927, Z. für Phys. 41, 184.
Garroway, A.N., Mansfield, P. and Stalker, D.C., Phys. Rev. (in press)

THE APPLICATION OF TIME DEPENDENT FIELD GRADIENTS TO NMR SPIN MAPPING

W. S. Hinshaw*
Department of Physics, University Park, Nottingham NG7 2RD, ENGLAND

Abstract. It has been possible to obtain the spatial dependence of parameters measured by nuclear magnetic resonance experiments by adding a small time dependent inhomogeneous component to the static magnetic field. The introduction of time dependence provides a means of solving many of the problems inherent in static field NMR imaging.

1. Introduction. Recent work points the way toward a potentially useful application of nuclear magnetic resonance to the production of images of the nuclear spin distribution in heterogeneous samples. (Lauterbur, 1973; Mansfield, 1973). This work has been based on the fact that if a uniform gradient is applied across the sample and if other broadening mechanisms are negligible. The resonance spectrum is inhomogeneously broadened and is a projection of the sample density onto a line in the direction of the gradient. By obtaining spectra for various gradient directions and combining the spectra in accordance with image reconstruction techniques (Smith et al. and references therein, 1973), an image of the spatial distribution of the resonant spins in the sample can be generated. Although simple in principle, this method of image production suffers from many practical difficulties.

Introducing time dependence to the field gradient during the measurement provides a means of overcoming many of the limitations inherent in the reconstruction method (Hinshaw, 1974). If three sets of gradient coils producing orthogonal gradients are used simultaneously, the result is a single gradient in some intermediate direction. However, if these sets of coils have time dependent currents with a different time dependence in each set, the resulting gradient is time dependent in its direction and magnitude. The NMR signal from a sample in such a field contains spatial information in its time dependence. The extraction of this information from the signal has been accomplished in various ways. The most simple is based upon the fact that if the currents are properly chosen there is only one point in the sample which experiences a time independent, zero, contribution from the gradient coils. The signal from all other regions has a time dependence and thus can be filtered out with a simple low pass filter. The actual time dependence of the inhomogeneous field is not critical. Noise, square wave, and sinusiodal currents, for example, have been used successfully. The "sensitive point" can be moved anywhere in the sample by adjusting the current ratios in the gradient coils. We have added three orthogonal gradient coils to the probe of a standard broad line pulse spectrometer and by sweeping the sensitive point in a regular pattern have produced images, or spin maps, of the sample. The modified spectrometer also has the capability of making many of the usual NMR measurements, such as T_1, flow, and molecular diffusion, at easily selected regions of the sample.

2. Comparison with Earlier Method. The sensitive point mehtod requires much less uniform and stable fields than the image reconstruction method. Image reconstruction alorithms assume that the field gradient is uniform, both in direction and magnitude, throughout the sample. Any deviation from uniformity rapidly degrades the image, not only by introducing distortion but by destroying the image itself. For example, if the surfaces of constant field produced by the gradient coils deviate slightly from flat planes, the resonance spectra no longer represent simple projections and the reconstruction methods fail. For similar reasons, H_0 must be uniform throughout the sample. Many samples of interest have a high paramagnetic ion content and thus distort the field through bulk susceptability effects. These induced field distortions change as the sample orientation is changed, making it impossible to reconstruct a clear image. The reconstruction method would seem to be limited in its resolution by the uniformity of the fields since the total spatial distortion of the gradient field must be less than the resolvable distance. By contrast, the sensitive point method does not depend upon the uniformity of the gradient. The time dependent inhomogeneous field can have any spatial variation as long as only one stationary region experiences a constant zero field. By a proper selection of irradiating pulses, even the value of H_0 at that sensitive region is not critical. Thus neither the H_0 nor the gradient field needs to be uniform.

A second comparison between the two methods which should be considered is the time taken to produce equivalent images. The information rate, or signal to noise ratio per unit time, is difficult to reliably establish. Experience in both techniques and direct comparisons are necessary. With the proper application of pulses, the sensitive point method can produce an almost continuous signal output and the system bandwidth can be reduced with a resulting increase in signal to noise. Thus the method is closely analogous to CW spectroscopy where the signal of interest is selected and scanned while the others are filtered out. A naive approach might indicate that selecting information form one point at a time would be slower than methods where information from all the sample is accepted. However, if bandwidth, lineshape, and other factors are considered, the advantage disappears (Ernst and Anderson, 1966; Ernst, 1966). It is tempting to use time dependent gradients or similar techniques to localize the sensitivity to a line, apply a static gradient along that line, and use Fourier transform techniques to determine the distribution of spins along that line. However, with the broad distributions usually encountered, FT techniques have no advantage over scanning methods. Space prohibits a detailed discussion of relative information rates which would have to include T_1 limitations, noise or clutter generated by the reconstruction algorithum, computation times, pulse sequences used, etc.

3. Results. Three spin maps produced by the scanning sensitive point method are shown. They are all copies of an XY plotter output without alteration. The pen of the plotter was scanned across

the page synchronously with the scanning of the sensitive point through the sample. They are not projections but are cross sections or "slices".

The upper map in figure 1 is a map of a multi-tube sample containing doped water with T_1's as indicated in milliseconds. The apparent height is proportional to the sample density. The two samples with the weaker signals were smaller tubes. The three traces across the top were taken using only the X and Z gradients and thus represent the projection in the Y direction of the slice shown. They illustrate the difficulty that can be encountered in trying to obtain the sample image from projections. This particular projection seems to contain very little information. The lower map in figure 1 is the same sample "photographed" in the same way except for the addition of a 180 degree rf pulse which was applied every 100 ms throughout the scan. This rf pulse destroyed the signal from the long T_1 sample but did not affect the signal from the short T_1 sample. The intermediate T_1 samples produced a signal that was modulated at 10 Hz by the 180 degree pulses. This pair of pictures illustrates one method of distinguishing between volumes of the sample on the basis of relaxation times.

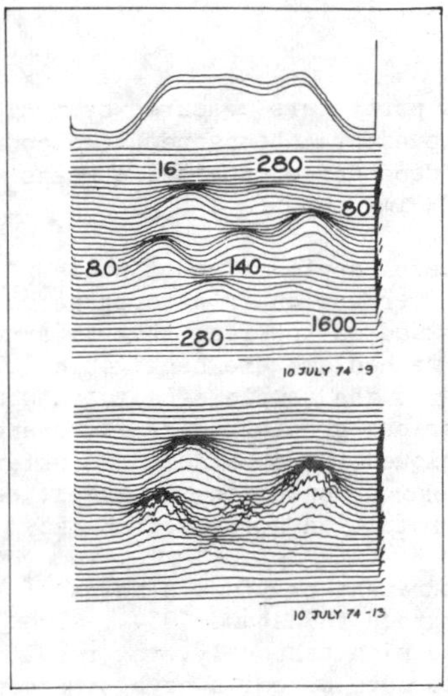

Figure 1

Figure 2 is a scan through water filled holes drilled in a PTFE rod. To produce this map, the pen of the plotter was made to rapidly oscillate with an amplitude proportional to the signal as the pen was scanned. Thus the water appears dark. This method of display is easier to interpret but does not have the dynamic range possible with the method used for figure 1. The gradient strength used for this picture was about 3 gauss/cm which, experience showed, produced a resolution of about 0.3 mm using Taylor's resolution criterion of classical optics. The resolution can not be determined directly from this figure since the method of display is not linear. The gradient strength was limited by the power available to drive the gradient coils. Each pen trace took about 30 seconds.

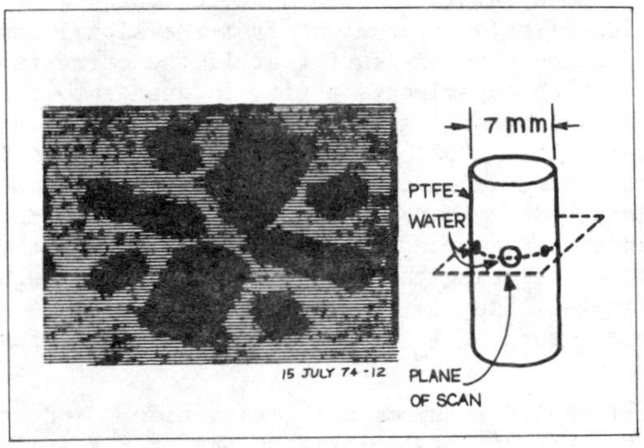

Figure 2

Figure 3 is the cross section of a 6 mm diameter glass tube filled with water and containing an ordinary brass screw. This picture may not be very clear, but it is a striking illustration of the insensitivity of the sensitive point method to rf, H_0, and gradient field inhomogenieties. The method of display is the same as the previous map but with considerably slower pen oscillations to improve the display linearity.

Many other spin maps have been produced from samples such as spring onions and chicken bones. When the system is set up, a sample can be dropped in and a few minutes later the image is complete. Maps of biological samples were not shown because they do not demonstrate the capabilities of the technique. It is difficult to tell from the image alone if there is any correspondence between it and the actual biological sample used.

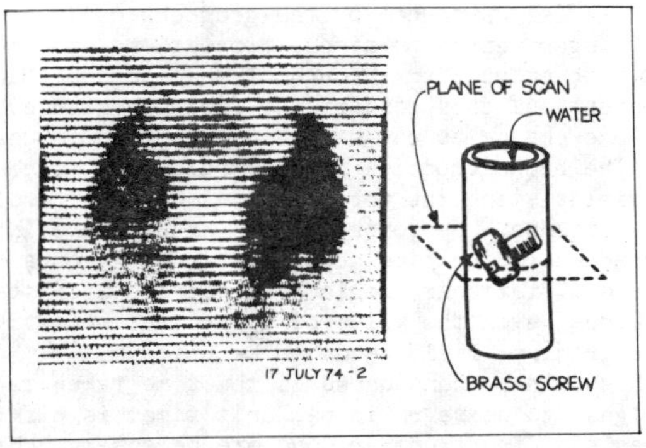

Figure 3

4. References

R. R. Ernst, "Sensitivity Enhancement in Magnetic Resonance"in "Advances in Magnetic Resonance". Vol. 2, Academic Press, New York, 1966.

R. R. Ernst, and W. A. Anderson, Rev. Sci. Instr. 37, 93 (1966).

W. S. Hinshaw, Phys. Lett. 48A, 87 (1974).

P. C. Lauterbur, Nature 242, 190 (1973).

P. Mansfield and P. K. Grannell, J. Phys. C 6, L422 (1973).

P. R. Smith, T. M. Peters, and R. H. T. Bates, J. Phys. A 6, 361 (1973).

*Present address: Department of Physics,University of Pittsburgh,Pittsburgh,Pennsylvania,15260,USA.

VELOCITY PROFILE MEASUREMENTS BY NMR

A.N. Garroway
Department of Physics, Nottingham University, Nottingham NG7 2RD, ENGLAND.

Abstract. We describe how the spatial velocity profile of a flowing fluid can be determined by NMR. An improved method is also presented for directly measuring the velocity distribution function. Both employ a linear magnetic field gradient. Experimental results for the velocity profile of water flow in circular and rectangular pipes are presented, as well as for the velocity distribution function in a circular pipe.

1. Introduction. Application of a magnetic field gradient produces a spatially dependent Larmor frequency and so labels the nuclear spins along the preferred direction of the gradient by a spatial variable. Detailed information about the (macroscopic) velocity of flowing fluids may then be obtained.

2. Spatial Profile of a Velocity Distribution: The 90-τ-90 Sequence. A static magnetic field H_0 and linear gradient $g = dH_0/dx$ determine the z direction and the fluid flows along x with a velocity $v(y,z)$. A 90° rf pulse is applied and, for simplicity, is assumed to produce a 90° rotation of all spins within the interval x = 0 to L and that outside the other spins are unaffected. A second 90° pulse is applied at the time τ, chosen greater than T_2* and much less than T_1. By this time, fresh spins, unaffected by the first pulse, have entered the active region and an NMR signal is observed following the second pulse. In the special case of laminar flow these fresh spins flow out to the point $x = v(y,z)\tau$. The free induction decay (FID) following the second pulse is of the form

$$F(t) = \tau\iint dz\ dy\ \rho(y,z)v(y,z)\cos\left((\Omega_0 + \gamma gz)t\right) , \quad (1)$$

where $\rho(y,z)$ is the spin density assumed independent of x, γ the gyromagnetic ratio, and Ω_0 is the resonance offset frequency. We arrange that $V_0\tau<L$, where V_0 is the maximum velocity. Provided that the gradient is sufficiently large, motion during the FID may be ignored; we also neglect effects of self-diffusion.

The cosine Fourier transform of the FID is then

$$G(\omega) \propto \tau\int dy\ \rho(y,z)v(y,z) \quad , \quad (2)$$

where $z = (\omega-\Omega_0)/(\gamma g)$. Hence the Fourier transform measures the integral over the plane $z = (\omega-\Omega_0)/(\gamma g)$ of the product of the spin density and velocity; we call $G(\omega)$ the velocity-density lineshape. With a knowledge of $\rho(y,z)$, the velocity profile $v(y,z)$ can then be determined from $G(\omega)$. We consider two examples.

For laminar flow between two parallel plates at z = 0 and z = 2h, the velocity profile is parabolic. Since the density is independent of y, the predicted velocity-density lineshape is directly proportional to the velocity profile:

$$G_p(\omega) \propto V_0\tau\left(1 - (\omega-\Omega_p)^2/(\Delta\omega_p)^2\right), \quad \text{for } 0<|\omega-\Omega_p|<\Delta\omega_p, \quad (3)$$

where $\Delta\omega_p = \gamma gh$, $\Omega_p = \Omega_0 + gh$, V_0 is the maximum velocity, and the subscript p denotes the parallel plate case.

As a second example, a circular pipe also has a parabolic velocity profile which depends on the radius R. The predicted lineshape is then

$$G_c(\omega) \propto \tfrac{2}{3}V_0\tau\ \left(1 - (\omega-\Omega_c)^2/(\Delta\omega_c)^2\right)^{3/2}, \quad \text{for } 0<|\omega-\Omega_c|<\Delta\omega_c \quad (4)$$

where c indicates the circular pipe case and $\Delta\omega_c$ and Ω_c are defined similarly.

For reference note that the Fourier transform of the signal following the first pulse is proportional to the number of spins in the xy plane and we call this the density lineshape. For parallel plates the density lineshape is constant for $0<|\omega-\Omega_p|<\Delta\omega_p$, whereas for a circular pipe it is semicircular, centred about Ω_c.

In the following experiments the proton NMR signal was observed. The active region was defined by using copper shields so that approximately 11 mm of the pipe were exposed to the rf irradiation.

Figure 1a shows the Fourier transform for a 90-τ-90 sequence applied to tap water flowing in a 2x8 mm rectangular pipe with the applied gradient perpendicular to the 2 mm side. The experimental density lineshape is shown and, as expected, is rectangular. Also shown is the density-velocity lineshape and, for illustrative purposes, we compare it to the theory (solid curve) for parallel plate flow; the agreement is reasonable.

In Fig. 1b are shown the results for water flow (average velocity 19 cm/sec) in an 8.2 mm id circular pipe. The observed density lineshape agrees well with the theoretical semicircle (dashed curve) and the density-velocity lineshape is reasonably well matched by the predictions of Eq.(4), shown as the solid curve. (The velocity-density lineshapes in Figs. 1a and 1b have been normalised to allow direct comparison with the density lineshape results.)

This method can be extended to flow studies in more complicated geometries. By varying the direction of the applied gradient, one finds the projections of $\rho(y,z)$ and $\rho(y,z)v(y,z)$, and then reconstructs the three dimensional image (Lauterbur, 1973; Mansfield and Grannell, 1973) corresponding to these projections. The velocity profile can then be separated from the velocity-density profile.

3. Velocity Distribution Function: The $90_0-(\tau-90_{90})_n$ Sequence. In this experiment a linear magnetic gradient is applied in the direction of the flow, so $g = dH_0/dx$. The spins are irradiated by the sequence $90_0-(\tau-90_{90})_n$, where the phase of the n pulses following the initial one has been shifted by 90° and we set $\tau > T_2^*$. The signal is sampled immediately prior to each rf pulse and, provided a spin's velocity is constant over the interval $n\tau$, the observed signal is

$$F(t) = \int_0^\infty dv\, \rho(v)\cos(\tfrac{1}{2}\gamma g v \tau t), \quad t = n\tau \quad , \tag{5}$$

where $\rho(v)$ is the velocity distribution function, defined as the number of spins with velocity v (parallel to g). The complications of relaxation, self-diffusion and sample leakage ($nv\tau > L$) are ignored. Hence the cosine Fourier transform of the signal gives directly the even part of $\rho(v)$. For example, for laminar flow in a circular pipe $\rho(v)$ is constant for $0 < v < 2\bar{v}$, where $\bar{v}$ is the average velocity. This sequence is less sensitive to the inhomogeneity of the rf field than a similar sequence, $90_0-(\tau-180_0)_n$, previously proposed (Packer, 1969; Hayward et al 1972).

The cosine Fourier transform of the response to the $90_0-(\tau-90_{90})_n$ sequence is shown in Fig. 2 for the case of water flow through two parallel circular pipes (3.5 mm id), with average velocities of $\bar{v}_1 = 1.2$ and $\bar{v}_2 = 4.0$ cm/sec in a gradient of 1.49 G/cm with $\tau = 1.273$ ms. The expected velocity distribution functions are shown and agreement is fair.

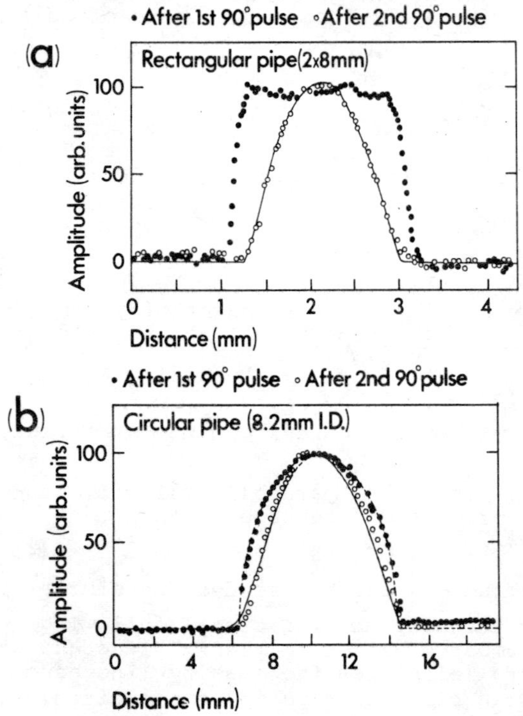

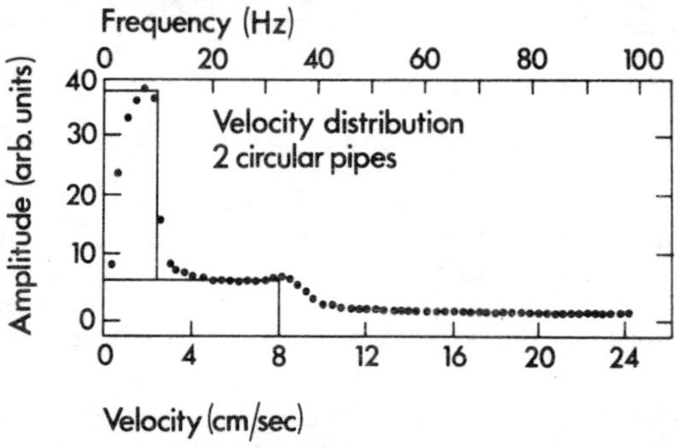

Fig. 1. (a) Lineshapes for flow in a rectangular pipe for the 90-τ-90 sequence. The solid curve represents the velocity profile for flow between parallel plates. (b) Lineshapes for flow in a circular pipe. The dashed curve represents the predicted density profile and the solid curve indicates the theoretical velocity-density lineshape.

Fig. 2. The Fourier transform of the echo envelope for water flow in two circular pipes with $\bar{v}_1 = 1.2$ and $\bar{v}_2 = 4.0$ cm/sec. The theoretical velocity distribution functions are shown as the solid curves.

We acknowledge the Science Research Council for providing a postdoctoral assistantship. W.S. Hinshaw and P. Mansfield have provided helpful comments.

4. References.

Hayward, R.J., Packer, K.J. and Tomlinson, D.J., 1972, Molec. Phys. 23, 1083-1102.
Lauterbur, P.C., 1973, Nature 242, 190-191.
Mansfield, P. and Grannell, P.K., 1973, J. Phys. C 6, L442-6.
Packer, K.J., 1969, Mole. Phys. 17, 355-368.

SECTION N

Pair Interactions

ESR STUDIES ON THE BONDING AND EXCHANGE INTERACTION IN TRANSITION METAL COMPLEXES

B.Jeżowska-Trzebiatowska, J.Jezierska, A.Jezierski, A.Ożarowski, T.Cukierda and J.Baranowski
Institute of Chemistry, University, Wrocław, POLAND

Abstract. The ESR method was applied to systemmatic studies on the chemical bonding character in salicylaldimine type /Cu/II/,VO/II// and nitrosyl /Cr/I/,Mn/II// complexes with sulfur chelates. Exchange interaction in VO/II/,Cu/II/ and Fe/III/ complexes with oxygen bridge in powder and monocrystalline form was also examined by this method.

1. Introduction. Transition metal nitrosyl complexes are characterised by the presence of a strong bonding M-NO, in oxovanadium /IV/ complexes V-O bond considerably influences on the properties. However, as we have proved, the electron density in these complexes can be changed within certain limits, by means of the organic ligands. We have determined the spin Hamiltonian parameters for a couple of dozens of that type complexes / from ESR spectra of their liquid and frozen solutions/. From those data we have calculated the molecular orbital parameters and we have estimated the isotropic contact term. Even if from the standard methods calculating of these values /Neiman and Kivelson, 1961, Kuska and Rogers, 1968, McGarvey, 1967, etc.,/ the absolute values of these parameters can not be obtained, these methods are very helpful to define the trend of changes in a series of similar compounds.

2. Chemical bonding character. In oxovanadium/IV/ complexes with bidentate Schiff bases /the condensation products of primary aliphatic and aromatic amines with salicylaldehyde/ the bonding character proved to be dependent on the electrophilic character of substituents connected with the coordinating nitrogen. The β^2 parameter, the measure of the π bonding in a plane $(B_2 = \beta d_{xy} - \beta' \phi_{Lig})$, occured to be the good covalency indicator. Its changes are small, but out the limits of the measurement error. The β^2 parameters calculated both from isotropic and anisotropic spin Hamiltonian parameters growes with the electron withdrawing character of these substituents. / $\beta^2_{iso.}$ is 0.80 for aliphatic and 0.82 for aromatic substituents/. The lowest β^2 is observed when coordinate tetradentate Schiff bases /the condensation products of diamines and salicylaldehyde/. Identical direction of changes has been stated for the analogous Cu/II/ complexes. The β_1^2 values / $B_{2g} = \beta_1 d_{xy} - \beta_1' \phi_{Lig}$/ changing within the broader limits /to 0.92 for p-nitrophenyl substituent/ was found to be the better covalency measure than the α^2/ $B_{1g} = \alpha d_{x^2-y^2} - \alpha' \psi_{Lig}$/ / Jeżowska-Trzebiatowska and Jezierska, 1973/. Furthermore, the trend in the β_1^2 values is in good agreement with polarographic half-wave potentials. The dependences of the bonding character on those substituents, connected with coordinating nitrogen, were also confirmed for the numerous Schiff base complexes with derivatives of substituted salicylaldehyde. No such explicity dependence of the bonding on basicity of dithiocarbamate parent secondary amine in a series of dialkyldithiocarbamate nitrosyl complexes, $Cr/NO//dtc/_2X$, where X is a monodentate ligand. It appears that the bonding properties of the $S_2CNR_2^-$ groups depend significantly on the natмre of the substituents on the nitrogen atom. The ability to delocalize the electron in the chelate ring is essentially influenced by C-N-C angle in this ligand. Increase of that angle, to begin with pyrrolidine-N-carbodithioate, where C-N-C angle is considerably different from normal trigonal angle, then morpholine-N-carbodithioate, piperidine-N-carbodithioate etc., just to the diisopropyldithiocarbamate / of large volume alkyl groups/ effects reduction the β^2 from 0.74 to 0.70. The last value is observed also for analogous complexes with alkyl xanthate or thioxanthate ligands. In these cases electron delocalization in the sulfur chelate ring is the greatest and Cr-S distance must be shorter. The capacity to the expansion of the d-orbitals is characteristic for sulfur ligands, the rings formed by dithiochelates offer additional pathways of delocalization through π-back bonding. The ESR measurements can be find these dependences quantitatively, calculating β^2 parameters. Such a course of changes was observed for the nitrosyl manganese complexes with the same ligands. Although β^2 for these complexes is almost constant, distinct changes of isotropic contact term are observed. For interpretation Cr-nitrosyl and Mn-nitrosyl ESR spectra energy levels ordering $e/xz,yz/ < b_2/xy/ < e\ /\pi^*NO/ < b_1\ /x^2-y^2/ < a_1/z^2/$ was used.
Besides of the interactions in the xy plane very essential was found to be the influence of a ligand in trans position to V-O or M-NO, mainly for the chromium nitrosyl complexes, where β^2 is diminished by sulfur ligands /thiourea derivatives, mercaptans/ coordinating in that position, about 0.08 in relation to H_2O and 0.13 in relation to nitrogen bases /aliphatic amines, diazoles/. It should be noted, that for the discussed VO/II/,Cr/I/ and Mn/II/ complexes

the isotropiccontact term is a good measure of the bonding covalency character and remains nearly a linear dependence between this term and the β^2 coefficients.

3. Dimeric complexes. The metal-metal interactions were found in a number of VO/II/ and Cu/II/ complexes with tridentate Schiff bases, of the C=N & O & O-C \\ C-O & M & O & M & N=C  type, where M is Cu/II/ or VO/II/. The exchange integrals calculated from the magnetic susceptibility measurements are of the order 100-300 cm^{-1}, both for the copper and vanadyl complexes. The ESR spectra could be described by the spin Hamiltonian with S=1. The calculated D parameters for copper complexes are about 0.2 cm^{-1}. The smallest D parameter /0.19 cm^{-1}/ was found for 2-/acetylpropylidenatoamino/phenolato Cu/II/. The substitution of acetyl group by larger groups /benzoyl, naphthoyl etc./ results in the increase of the D parameter to 0.23 cm^{-1}, as for 2-/salicylidenatoamino/phenolato Cu/II/ and similar compounds with substituted salicylaldehyde. Contrary to the copper complexes, which powder ESR spectra present the series of $H_{i\,1,2}$ lines /i =x,y,z/ /these ESR transitions were analysed using Wasserman's, Snyder's and Yager's /1964/ method/ the similar oxovanadium/IV/ complexes present the ESR spectra composed from a central line and the "forbidden" transition in half field region with a clear hyperfine structure. The D parameter estimated for the latter complexes amounts about 0.03 cm^{-1}. Hence it should be assumed, that in VO/II/ complexes, contrary to the analogous dimeric copper complexes where exchange is going mainly via the bridge, the direct overlapping of the d_{xy} orbitals is also essential.

To determine the spin Hamiltonian parameters of dimers of a spin > 1 /e.g. iron/III/ dimers with oxygen bridge/ the angular dependences of ESR spectra of monocrystals of the chosen compounds were examined. In that way has been examined /Fesalen/$_2$O·$CHCl_3$ and /FeMequin$_2$/$_2$ O·$CHCl_3$, where salen is bis-/salicylideneethylenediamine/ and Mequin is 2-methyl-8-hydroxyquinoline. In calculations of the spin Hamiltonian parameters the angular dependence of the resonance field corresponding to the triplet state /S=1/ was used. The resonance lines were attributed to the respective spin states on the basis of the temperature dependence analysis of the line intensities. The parameters were computed on the JEC-6 type computer solving the system of the equations with a given resonance fields in two orientations because of the D and E. For the obtained D and E values the angular dependence of the ESR spectrum was simulated. The monocrystal of the first complex was rotated around the x axis bonding the iron ions and y axis, the bisector of the Fe -O- Fe angle is the z axis /Jeżowska-Trzebiatowska et al.,1973 /. The monocrystal of the second complex was rotated around the direction θ =53° and φ =76.7° /around the crystallographic a axis, approximately/ in the coordinate system, where the y axis is the bisector of the Fe -O- Fe angle, and x axis is bonding the iron/III/ central ions. The D =1.87 cm^{-1}, E =0.13 cm^{-1} and D =3.90 cm^{-1}, E =0.12 cm^{-1} for these complexes, respectively.

This work was supported by the Polish Academy of Sciences.

References

Mc Garvey, B.R., 1967, J.Phys.Chem., 71, 51
Jeżowska-Trzebiatowska, B. and Jezierska, J. , 1973, J.Mol.Structure, 19, 627
Jeżowska-Trzebiatowska, B. , Kozłowski, H., Cukierda, T., and Ożarowski, A., 1973, J.Mol.Structure, 19, 663
Kuska, H.A. and Rogers, M.T. , Electron Spin Resonance of First Row Transition Metal Complex Ions, Radicals Ions, Interscience Publishers, 1968,
Neiman, R. and Kivelson, D., 1961, J.Chem.Phys., 35, 156
Wasserman E., Snyder, L.C., and Yager, N.A., J.Chem.Phys., 1964, 41, 1763

RESONANCE MAGNETIQUE DU CHROMITE DE TERRE RARE $GdCrO_3$ A 9,3 ET A 33 GHz ENTRE 2,5 ET 300 K

A. Marchand et J.P. Bongiraud
Laboratoire de Magnétisme, C.N.R.S., B.P. 166, 38042-Grenoble-Cedex, France.

Abstract. Les chromites de terres rares peuvent être décomposés en un réseau antiferromagnétique d'ions chrome et en un autre réseau antiferromignétique d'ions terres rares. Les interactions terres rares-chrome sont faibles. Le champ d'échange est de l'ordre de $1,9.10^6$ Oe, le champ de Dzialoshinskii de l'ordre de 35.10^3 Oe.

1. Introduction. Bertaut et Forrant (1956) ont montré que le composé $GdCrO_3$ cristallisait dans la structure pérovskite distordue rhomboédrique de groupe d'espace D_{2h}^{16} - Pbnm avec 4 molécules par maille. Aléonard et al (1966), Bertaut et al (1965) ont montré qu'il convenait de distinguer deux températures de Néel Θ_{N1} = 170 K et Θ_{N2} = 4 K auxquelles s'ordonnent respectivement les ions Cr^{3+} et les ions Gd^{3+}. Entre ces deux températures il existe, superposé à l'antiferromagnétisme du réseau du chrome, un faible ferromagnétisme dû à Cr^{3+}, lequel polarise à son tour la terre rare par suite d'un couplage négatif Cr-T. Le faible ferromagnétisme dû à Cr^{3+} est dirigé suivant l'axe c, l'axe a étant l'axe d'antiferromagnétisme.

2. Réalisation des mesures. Les mesures ont été faites principalement sur $GdCrO_3$ et accessoirement sur $YCrO_3$ qui ne présente qu'une température de Néel Θ_{N1} = 140 K. Les échantillons sont soit des poudres, soit des sphères monocristallines de diamètre 1,8 et 0,5 mm. Le champ directeur varie entre 0 et 85 kOe.

3. Résultats.

3.1. Résonance au dessus de Θ_{N1} = 140 K. Nous observons pour tous les échantillons de $GdCrO_3$ une raie de résonance de facteur g = 2,00 ± 0,01. Cette raie est due à la contribution des ions Cr^{3+} et Gd^{3+} dont les facteurs g sont voisins de 2,00. Pour déterminer l'importance relative de chacune de ces contributions nous considérerons $YCrO_3$ déjà étudié par Sanina et al. Y^{3+} ayant un moment magnétique nul, seul Cr^{3+} intervient dans la raie de résonance de ce composé. L'intensité de la raie $YCrO_3$ est environ cinquante fois plus petite que celle de $GdCrO_3$. La contribution de Cr^{3+} à l'intensité de la raie de $GdCrO_3$ serait donc de l'ordre de 2 %., celle de Gd^{3+} de l'ordre de 98 %.

3.2. Résonance entre Θ_{N1} = 170 K et Θ_{N2} = 4 K. Ce composé étant fortement anisotrope, les champs de résonance dépendront beaucoup de la forme des échantillons et des fréquences utilisées. Dans tous les cas, au fur et à mesure que la température décroît d'une valeur supérieure à Θ_{N1} à une valeur inférieure, la raie de résonance paramagnétique se déplace vers les champs faibles sans changement notable de son intensité lors du passage par la température de Néel Θ_{N1}.

A 9,3 GHz nous présentons figure 1, les valeurs du champ de résonance en fonction de la température, le champ directeur étant appliqué suivant l'axe a, b ou c du monocristal. Ces résultats s'interprètent d'une manière identique à celle que nous avions proposée lors de la résonance d'une poudre de $GdCrO_3$ (Bongiraud et Marchand (1973)). La raie de résonance observée dans $GdCrO_3$ quelques dizaines de degrés au-dessous de Θ_{N1} est uniquement le fait de Gd^{3+}.

A 33 GHz, les résultats que nous avons publiés précédemment (Bongiraud et Marchand (1973)) nous ont permis de montrer que les interactions terre rare-chrome étaient faibles. Elles varient de 5500 à 2500 Oe quand la température varie de 77 à 143 K. Aléonard et al ont montré que l'angle β dont s'éloigne l'aimantation spontanée des Cr^{3+} à partir de la direction d'antiferromagnétisme est à toute température de 9.10^{-3} radian. D'après la théorie d'Hermann (1963)

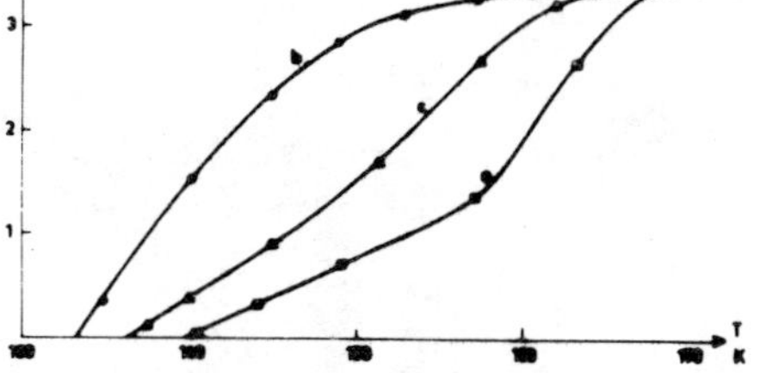

Figure 1

$$tg\, 2\beta \sim \frac{A_{ac} + D}{E}$$

où A_{ac} est l'une des composantes du tenseur d'anisotropie, D le champ de Dzialoshinskii, E le champ d'échange. Judin et Shermann (1966) ont déterminé E par des méthodes statiques dans le cas de $YCrO_3$ et obtiennent $1{,}6.10^6$ Oe. Co Comme le champ d'échange est sensiblement proportionnel à la température de Néel, dans le cas de $GdCrO_3$ il serait de l'ordre de $1{,}9.10^6$ Oe. En négligeant A_{ac} devant D, D est alors de l'ordre de 35.10^3 Oe.

3.3. Résonance au voisinage de Θ_{N2} = 4 K. A 9,3 GHz, l'échantillon est une poudre. Nous présentons figure 2 le champ de résonance de la raie observée en fonction de la température. La raie au-dessous de 4 K est attribuée à la résonance antiferromagnétique du gadolinium.

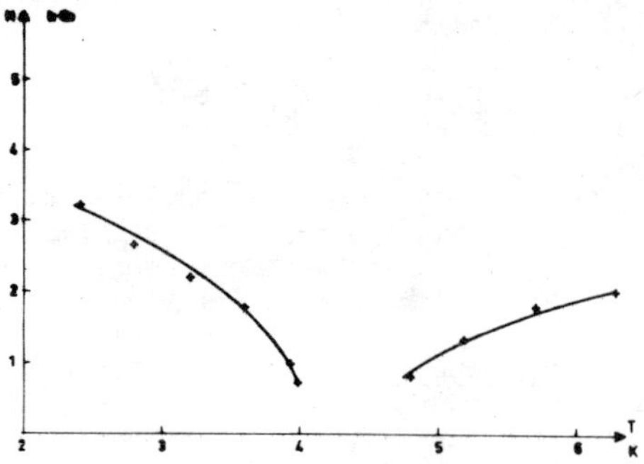

Figure 2

4. Références

Aléonard, R., Mareschal, J., Pauthenet, R., Rebouillat, J.P., Zarubica, V., 1966, C.R. Acad. Sc. Paris, 262, p. 799 et 866.

Bertaut, E.F., Forrat, F., 1956, J. Phys. Rad., 17, P. 129.

Bertaut, E.F., Mareschal, J., Pauthenet, R., Rebouillat, J.P., 1965, International Conference on Magnetism and Magnetic Materials, San Fransisco.

Bongiraud, J.P., Marchand, A., 1973, C.R. Acad. Sc. Paris, 277B, p. 399.

Hermann, G.F., 1963, J. Phys. Chem. Solids, 24, p. 597

Judin, V.M., Shermann, A.B., 1966, Solid State Communications, 4, p. 661.

Sanina, V.A., Golovenchits, E.I., Fomina, T.A., Gurevich, G.A., 1971, J. de Physique, 32 C1, p. 1149

g-SHIFT CALCULATION FOR EXCHANGE-COUPLED PAIRS OF Co^{2+} IONS IN MgF_2

C. A. Bates, M. C. G. Passeggi*, K. W. H. Stevens and P. H. Wood
Department of Physics, University of Nottingham, University Park, Nottingham NG7 2RD, ENGLAND.
*Centro Atomico Bariloche, CNEA, San Carlos de Bariloche, Rio Negro, ARGENTINA.

Abstract. An orbitally-dependent exchange Hamiltonian isotropic in the spin, is used to represent the Co^{2+}-Co^{2+} interaction in MgF_2. The changes in the g-values of one of the Co^{2+} ions in a pair are calculated and are found to be in reasonable agreement with the existing experimental data.

Spectroscopic Properties of Co^{2+} Ions in MgF_2. MgF_2 is of the rutile-type crystal structure which is shown in Figure 1. The Mg^{2+} ions form two inter-penetrating tetragonal sub-lattices each characterised by an environment rotated by 90° about the c(z) axis with respect to each other. Each Co^{2+} ion, substituting for Mg^{2+}, is surrounded by a tetragonally distorted octahedron of fluorine ions. Further rhombic distortions give a site symmetry of D2h. After spin-orbit coupling is included, the ground states consist of six Kramers doublets each belonging to the representation E of the D2h double group.

Form of the Exchange Interaction. Using the techniques of second quantisation similar to those outlined in Stevens (1974a) the dominant term in the exchange interaction between the Co^{2+} ions, each with a $^4F(^4T_1)$ ground state, can be written as (Passeggi and Stevens, 1973):

$$\mathcal{H}_{exch} = \sum_{\substack{\ell,m,\\ \ell',m'}} (\tilde{J}^{\ell\ell'}_{mm'} + J^{\ell\ell'}_{mm'}\underline{S}_A\cdot\underline{S}_B)O^{\ell}_{m}(L'_A)O^{\ell'}_{m'}(L'_B) \qquad (0 \le \ell,\ell' \le 2) \tag{1}$$

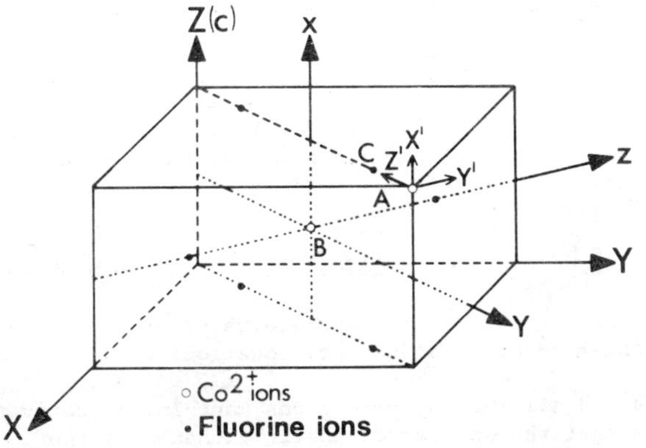

Fig.I. Rutile Structure.

Calculations of the parameters $\tilde{J}^{\ell\ell'}_{mm'}$ and $J^{\ell\ell'}_{mm'}$ up to and including second order of perturbation theory have been made for nearest neighbour ions. For a pair of next-nearest neighbour (nnn) ions (see Figure 1), the symmetry group contains only two elements, the identity operator and a reflection in a vertical plane containing the pair (the XBY plane of Figure 1). $\mathcal{H}_{exch}$ is again given by equation (1); in this case J^{00}_{00}, $J^{11}_{11}, J^{1\,1}_{1-1}, J^{11}_{10}, J^{11}_{01}, J^{1\,1}_{0-1}$ and also the $\tilde{J}$ parameters having the same combinations of ℓ,ℓ',m,m' dominate (Stevens, 1974b, Cowley et al., 1973).

g-Shift Due to Exchange Coupling. It is assumed that the local symmetry of each Co^{2+} ion is not affected by its neighbour. That is, the additional crystal field on one of the Co^{2+} from the other ion in the pair is neglected. The ground state and first excited state for each ion can thus be specified simply. The change in the g-value of a single ion which is a member of a pair is known as the g-shift. It can be calculated by perturbation theory and we find

$$\Delta g_{xx} \equiv \Delta g_{x'x'A} = \Delta g_{xxB} = -\frac{1.838}{\Delta}(J^{11}_{11} - J^{1\,1}_{1-1}) \qquad \Delta g_{yy} \equiv \Delta g_{y'y'A} = \Delta g_{yyB} = -\frac{0.318}{\Delta}(J^{11}_{11} - J^{1\,1}_{1-1})$$

$$\Delta g_{zz} \equiv \Delta g_{z'z'A} = \Delta g_{zzB} = -\frac{1.638}{\Delta}(J^{11}_{11} - J^{1\,1}_{1-1}) \tag{2}$$

with $\Delta = 152\ cm^{-1}$ while A and B label the ions in a pair. Thus, $\Delta g_{xx}/(\Delta g_{yy} + \Delta g_{zz}) = 0.904$ and is independent of the $J^{1\,1}_{1\pm1}$ parameters. This is in good agreement with the experimental value for this ratio 0.744 obtained from Belorizky et al., 1969.

Calculation of the Exchange Parameters J^{11}_{11} and $J^{1\,1}_{1-1}$. For transition metal ions the dominant contributions to these parameters are likely to come from an antiferromagnetic term of the form:

$$\mathcal{H}_{exch} = \sum_{\substack{mm'\\ pp'}} \Big(\frac{2}{\Delta_{a\to b}}\langle b_{m'}|\tilde{h}|a_p\rangle\langle a_m|\tilde{h}|b_{p'}\rangle \times \sum_{\tau\tau'} a^*_{m\tau}a_{p\tau'}b^*_{m'\tau'}b_{p'\tau}\Big) \tag{3}$$

(orbital components) (spin components)

where $\Delta_{a\to b}$ represents the transfer energy for an electron between the sites a and b. $\tilde{h}$ is the one-particle operator of the many-electron Hamiltonian dealing with essentially one-particle excitations, and (a^*,b^*) and (a,b) representing creation and annihilation operators respectively, of electrons in particular states. This second quantised expression is proportional to the angular momentum and spin operators in equation 1 (Passeggi and Stevens, 1973).

The one-electron eigenstates are expanded in terms of the atomic basis states for ions A, B and C such that

$$\Phi_A = U_A - \lambda_{AC}U_C \qquad \Phi_B = U_B - \lambda_{BC}U_C \qquad \Phi_C = U_C + \gamma_{AC}U_A + \gamma_{BC}U_B$$

in a 'three-centre four-electron' picture of the problem. The constants λ and γ are admixture coefficients. The ion C is shown in Figure 1 and it is assumed such an arrangement gives the maximum contribution.

$$\lambda_{AC} = S_{AC} + \gamma_{AC} \qquad \lambda_{BC} = S_{BC} + \gamma_{BC} \text{ , where S is the overlap.}$$

Using these orbitals, equation 3 reduces to

$$H_{exch} = \sum_{\substack{mm' \\ pp'}} \left(\frac{2}{\Delta_{a\to b}} . \lambda_{b_{m'}c_{m'}} \lambda_{a_p c_p} \lambda_{a_m c_m} \lambda_{b_{p'}c_{p'}} . (\Delta\varepsilon_c)^2 \times \sum_{\tau\tau'} a^*_{m\tau} a_{p\tau'} b^*_{m'\tau'} b_{p'\tau}\right)$$

where $\Delta\varepsilon_c$ is of the order of the one-electron ionization energy of the anion found by retaining one-centre terms only.

For the rutile structure, some angular dependence of the λ-admixture parameter is required. These angular factors can be found from a knowledge of the Euler angles at each site such that the symmetry axes are parallel. In this second quantised formulation of the problem, the angular momentum dependence of the parameters J^{11}_{11} and $J^{1\,1}_{1-1}$ must be introduced into the problem and a total summation made over all components. We find

$$J^{11}_{11} = - J^{1\,1}_{1-1} = \frac{(\Delta\varepsilon_c)^2}{100\Delta_{a\to b}} 2\sqrt{6}\, \lambda^{\sigma}_{AC}\lambda^{\pi}_{AC} (2\sqrt{2} \sin\theta \sin 3\theta (\lambda^{\pi}_{BC})^2 + \sqrt{6} \cos\theta \cos 3\theta\, \lambda^{\pi}_{BC}\lambda^{\sigma}_{BC} + \sqrt{6} \cos^2\theta (\lambda^{\sigma}_{BC})^2) \times 8.06 \times 10^3 \text{ cm}^{-1}$$

with $\theta = 49.8^{\circ}$ (Bavr, 1956), $\Delta_{a\to b} = 5.35$ eV (Messick et al., 1972) and $\Delta\varepsilon_c \simeq 15$ eV and the labels σ and π denote the components. Using the above method and the experimental information available, it is possible to fit the g-shifts within the experimental range of the admixture coefficients. From Gladney (1966), the ranges of these coefficients are

$$\lambda^{\sigma}_{AC} = 0.205 \leftrightarrow 0.189 \text{ , } \lambda^{\pi}_{AC} = 0.109 \leftrightarrow 0.75 \text{ , } \lambda^{\sigma}_{BC} = 0.235 \leftrightarrow 0.187 \text{ , } \lambda^{\pi}_{BC} = 0.109 \leftrightarrow 0.000 \text{ .}$$

A computer fitting procedure of λ^{σ}_{AC}, λ^{σ}_{BC}, λ^{π}_{AC} and λ^{π}_{BC} is made to the average value of J^{11}_{11} (= 14.97 cm^{-1}) found from the substitution of the experimental g-shift values into equations 2.

Despite the uncertainty of some of the parameters and all the approximations made in the derivation of the g-shift, it seems reasonable to assume that the antiferromagnetic exchange is the dominant contribution and an estimation of its magnitude can be made within the limits of existing experimental information.

References

Baur, V. W. H., 1956, Acta Cryst., 9, 515-520.
Belorizky, E., NG., S. C. and Phillips, T. G., 1969, Phys. Rev., 181, 467-477.
Cowley, R. A., Buyers, W. J. L., Martel, P. and Stevenson, R. W. H., 1973, J. Phys. C: Solid State Phys., 6, 2997-3019.
Messick, L., Walker, W. C. and Glosser, R., 1972, Phys. Rev. B, 6, 3941-3949.
Passeggi, M. C. G. and Stevens, K. W. H., 1973, J. Phys. C: Solid State Phys., 6, 98-108.
Stevens, K. W. H., 1974a, Proc. XVIIIth Ampere Congress (Nottingham), 1974 (Eds. E. R. Andrew, P. S. Allen and C. A. Bates, pp.61-4.).
Stevens, K. W. H., 1974b, Phys. Letters, 47A, 401-403.

MOLECULAR FIELD INTERACTIONS AND g-VALUE SHIFTS FOR TRANSITION METAL IONS IN PARAMAGNETIC HOST LATTICES

Michael R. St. John and Rollie J. Myers
Department of Chemistry, University of California and Inorganic Materials Research Division, Lawrence Berkeley Laboratory, Berkeley, California 94720 U.S.A.

Abstract: A simplified 1st order theory of the g-value shift for the EPR of ions in paramagnetic hosts is developed and applied to Cu^{2+} and Co^{2+} doped into the α-$NiSO_4 \cdot 6H_2O$ lattice. The g-value shifts are interpreted in terms of isotropic exchange. This is compared with a molecular field model.

Introduction. The EPR of paramagnetic impurities in paramagnetic host lattices is usually broadened away by the host-impurity spin interaction. The time dependence of the interaction may be eliminated when it is possible to populate a single state of the host. This may be accomplished by low temperatures and high fields and is most easily done for ions with a singlet ground state. When the above conditions are met, sharp EPR spectra can be observed, and the spin interaction only results in shifts of the impurity ion spectral parameters from those that would be obtained if no interaction was present. For ions which possess only small orbital moments the EPR parameters without exchange should be well-approximated by those determined from an isostructural diamagnetic host. For these ions, the difference between diamagnetic host and paramagnetic host may be related to the exchange interaction.

Theory. Hutchings, et al. derived an impurity-ion spin Hamiltonian by an extension of the conventional method. Starting with a pair Hamiltonian all matrix elements except those involving impurity spin operators are evaluated, and then a sum is taken over an appropriate number of neighbors. Keeping only first-order interaction terms, which are the dominant ones for our host, the impurity ion Hamiltonian becomes

$$\mathcal{H}(A) = \beta \underset{\sim}{H} \cdot \underset{\sim}{g}(A) \cdot \underset{\sim}{S}(A) + (-2J)\left(\sum_{i}^{NN} \langle G_i | \underset{\sim}{S}(B_i) | G_i \rangle\right) \cdot \underset{\sim}{S}(A) \qquad (1)$$

where A denotes the impurity ion and B the host ions. The exchange has been assumed isotropic, the sum is over nearest neighbors, and $|G_i\rangle$ is the ground state wavefunctions of the host ions. Eqn. 1 has the advantage that it is easily amendable to more complicated spin interactions but has the disadvantage of the need for precise wavefunctions.

An alternative approach is to use the molecular field theory and substitute an average spin value for the sum of the host spins (Moriya and Obata). With this done, the average spin may be replaced by expressions for the magnetization. This results in a Hamiltonian given by

$$\mathcal{H}(A) = \beta \underset{\sim}{H} \cdot \underset{\sim}{g}(A) \cdot \underset{\sim}{S}(A) + \frac{(-2J)Z\, \underset{\sim}{m}(B) \cdot \underset{\sim}{S}(A)}{g(B)\beta} \qquad (2)$$

where Z equals the number of equivalent nearest neighbors and $\underset{\sim}{m}(B)$ is the magnetization per ion. This magnetization may be obtained, under some circumstances, from the macroscopic magnetization or from the energy level scheme.

In both cases, the coefficient preceding the impurity spin in the exchange term can be shown to be linearly dependent on the external field. This makes it appear as an additional Zeeman term or equivalently a g-value shift. Thus, with certain knowledge of the host, the spectra of the impurity may be interpreted and the exchange interaction evaluated by either comparison with an appropriate diamagnetic host or by experimental fit when enough independent data is available.

Experimental. α-$NiSO_4 \cdot 6H_2O$ is a paramagnetic lattice which meets the requirements of the Introduction. It has a singlet ground state separated from an excited doublet by 4.74 cm^{-1} (Fisher and Hornung). The α-$NiSO_4 \cdot 6H_2O$ was doped with approximately 1% Cu^{2+}, and spectra of the Cu^{2+} were easily resolved at 1.3°K, but were entirely broadened away at 4.2°K. The qualitative features of the Cu^{2+} spectra were the same as those in the isostructural diamagnetic $ZnSeO_4 \cdot 6H_2O$ lattice (Jindo and Myers). The most obvious difference, outside of line broadening, is that the g-values are 0.2 to 0.6 units larger in the α-$NiSO_4 \cdot 6H_2O$. This increase and the failure of the spectra to conform to a convential spin Hamiltonian indicated that the Ni-Cu interactions needed to be taken into account.

The Cu^{2+} spectra in α-$NiSO_4 \cdot 6H_2O$ can be characterized by the four constants, $g_{\parallel}$, $g_{\perp}$, ϕ, and (-2J) where ϕ is the angle between the tetragonal axis of the Cu^{2+} and the c crystal axis. Using equations (1) and (2), four independent measurements would suffice to specify the four constants. Only three could be readily obtained, and ϕ was then taken as an adjustable parameter. The expectation is that the $g_{\parallel}$, $g_{\perp}$, and ϕ in the Ni lattice should be close to those in the Zn lattice. With this in mind, ϕ was varied until the g-values most closely corresponded to the diamagnetic lattice. Table I gives the best g-values and the resulting exchange interaction.

Page Two - "Molecular Field Interactions and g-Value Shifts ... in Paramagnetic Host Lattices"

Table I

Lattice	ϕ	$g_{\parallel}$	$g_{\perp}$	$-2J(cm^{-1})$
$ZnSeO_4 \cdot 6H_2O$	43.3°	2.4295	2.0965	---
$NiSO_4 \cdot 6H_2O$ (equation 1)	45°	2.40	2.12	.146±.007
$NiSO_4 \cdot 6H_2O$ (equation 2)	45°	2.44	2.05	.152±.008

In the case of Co^{2+}, the orbital contributions to the g-values are very large. This yields g-shifts even without exchange interaction. In addition, the orbital moment can yield both anisotropic and antisymmetric exchange. Other ions investigated have been Mn^{2+} and V^{2+}.

References

Fisher, R. A. and Hornung, E.W., J. Chem. Phys., 48, 4284 (1968).

Hutchings, M. T., Windsor, C. G., and Wolf, W. P., Phys. Rev., 148, 444 (1966).

Jindo, Akira and Myers, R. J., J. Mag. Res., 6, 633 (1972).

Moriya, Turo and Obata, Yukio, J. Phys. Soc. Japan, 13, 1333 (1958).

ENDOR OF GaP:Mn

P. van Engelen
Philips Research Laboratories, Eindhoven, The Netherlands

Abstract. ENDOR measurements were made on the cubic Mn^{2+} centre in GaP. Ga transitions were found, which we attribute to Ga on next nearest neighbour sites along ⟨110⟩ directions. The hyperfine- and quadrupole tensors are nearly uniaxial. For a [110] pair the unique axes are rotated from [110] in $(1\bar{1}0)$ by 32.5° and $\sim 42^\circ$ respectively.

1. Introduction. Supertransferred hyperfine interactions (i.e. interactions with further than nearest neighbours) are of current interest (Taylor et al, 1973; Simanek and Sroubek, 1972). ENDOR measurements of the cation-cation interaction were made in the systems $LaAlO_3:Fe^{3+}$ and Cr^{3+} (Taylor et al, 1973) and in $CdTe:Cr^+$ (Ludwig and Lorenz, 1963). The first authors made a detailed analysis of the hyperfine interaction between the magnetic ion and ^{27}Al in terms of a transfer path involving one oxygen ligand. Their calculations which assume that the spin densities on ^{27}Al are only caused indirectly, by overlap with spin polarized oxygen orbitals, lead to satisfactory agreement for the isotropic hyperfine interaction A_s, but disagree seriously for A_p. Ludwig and Lorenz (1963) found that in $CdTe:Cr^+$ none of the principal axes of the hyperfine interaction tensor of Cd coincide with the Cr-Cd axis. In fact the unique axis of the nearly uniaxial tensor lies approximately along the [111] direction for a [110] pair.
In this paper we will give preliminary results for the hyperfine- and quadrupole interaction parameters of ^{69}Ga and ^{71}Ga (both I=3/2) in the system $GaP:Mn^{2+}$. Mn substitutes for Ga and is tetrahedrally surrounded by P ligands. Charge compensation of the cubic centre is apparently remote. The EPR of this Mn centre is described by Title and Plaskett (1969). We could not find ENDOR lines attributable to P. This is analogous to what happens in $CdTe:Cr^+$ for which Te lines were not detected (Ludwig and Lorenz, 1963).

2. Experimental. We used an X band ENDOR spectrometer at 4.2 K. The EPR spectrum consists of six inhomogeneously broadened (25 Oe) hyperfine lines. When one of these EPR transitions is partially saturated an ENDOR spectrum is seen, consisting of more than two hundred lines between 0 and 50 MHz. The linewidth is typically 100 kHz. We used solution grown and L.E.C. (i.e. pulled) crystals both of which were additionally doped with S, which increases the solubility of Mn (Vink and van Gorkom). The ENDOR spectrum of solution grown crystals is further complicated by crystal twinning.
We concentrated on the high frequency side of the spectrum where the lines are generally reasonably well resolved. The identification procedure consisted of a) measuring the angular dependence of the lines when H rotates in a (110) plane, b) measuring the shift of the lines when successively setting the magnetic field on different Mn hyperfine lines, and c) looking for equivalent lines for which the ratio of the transition frequencies corresponds to the ratio of the magnetic moments of ^{69}Ga and ^{71}Ga.

3. Results. The pattern of the observed angular diagram of the ENDOR lines is characteristic of interactions between Mn on (0,0,0) and Ga on twelve equivalent {n,n,m} sites with n≠0 and m≠n (Hale and Mieher, 1969). Because the isotropic hyperfine interaction is relatively large, it is most probable that the interaction is between Mn and Ga on nearest neighbour cation sites along ⟨110⟩ directions. The angular diagram is complicated because the principal axes of the hyperfine- and quadrupole interaction tensors do not coincide. The spin Hamiltonian for the nth Ga nucleus may be written

$$\mathcal{H}^n = -g_N\beta_N \bar{H}\cdot\bar{I}^n + \bar{I}^n\cdot\bar{\bar{A}}\cdot\bar{S} + Q'\left[I_X^{n\,2} - \tfrac{1}{3}I(I+1) + \tfrac{\eta}{3}(I_Y^{n\,2} - I_Z^{n\,2})\right]$$

in which the first term is the nuclear Zeeman interaction with the external field, the second term is the hyperfine interaction with the magnetic ion and the third term represents the quadrupole interaction. The lettering of the principal axes of both tensors was chosen in accordance with the convention introduced by Hale and Mieher (1969).
The results are:

— The isotropic transferred hyperfine field per spin 1, $H_s = A_s/(g_N\beta_N)$ = +11.49 kOe. The anisotropic components along the principal axes are $H_{x'}$ = +1.34, $H_{y'}$=-0.54 and $H_{z'}$=-0.80 kOe. The z'axis is along $[\bar{1}10]$ and the angle between x' and $[\bar{1}10]$ is found to be 32.5° for a Mn-Ga pair along [110]. Apparently the same situation occurs in $CdTe:Cr^+$ (Ludwig and Lorenz, 1963) where this angle is $37^\circ \pm 5^\circ$.

Dipolar interaction between Mn electron spins and the Ga nuclear spin contributes to the anisotropic part of the hyperfine interaction. Assuming point dipoles, the dipolar interaction tensor is calculated and subtracted from the measured tensor. The remaining tensor, which describes the transfer and overlap part of the hyperfine interaction, is orthorhombic; $H_{x''}$=+1.79, $H_{y''}$=-0.67 and $H_{z''}$=-1.12 kOe. The principal x" axis makes an angle of 22° with [110] and z"=z' for a Mn-Ga pair along [110].

The principal axes of the quadrupole interaction tensor are rotated relative to those of the hyperfine interaction tensor by approximately 10° about the z'=Z axis, i.e. the angle between the X axis and [110] is ~42° for a Mn-Ga pair along [110]. The parameters for ^{71}Ga are: $|Q'/h|$ =1.55 MHz, $\eta \sim 0.15$.

A full account of this work will be published elsewhere.

4. References.

Hale E.B. and Mieher R.L., 1969, Phys. Rev. 184, 739

Ludwig G.W. and Lorenz M.R., 1963, Phys. Rev. 131, 601

Simanek E. and Sroubek Z., 1972, in Electron Paramagnetic Resonance, ed. S. Geschwind, Plenum Press, New York-London

Taylor D.R., Owen J. and Wanklyn B.M., 1973, J.Phys.C: Solid St.Phys. 6, 2592

Title R.S. and Plaskett T.S., 1969, Appl. Phys. Lett. 14, 76

Vink A.T. and van Gorkom G.G.P., private communication.

ANALYSIS OF ESR LINEWIDTHS IN PdEr and PdDy

W. Zingg, J. Buttet and M. Hardiman
Département de Physique de la Matière Condensée, Université de Genève, Geneva, Switzerland.

Abstract. Linewidths measurements by ESR on PdEr and PdDy dilute single crystals are analysed in terms of Redfield-Hirst's theory as a function of temperature and magnetic field orientation. In PdEr the results fit well with an isotropic exchange parameter $|J_{sf}| = 6.1 \pm .4$ meV ($6.4 \pm .7$ meV in PdDy).

1. Introduction. Electron spin resonance (ESR) investigations on dilute PdEr (R.A.B. Devine et al.(1972)) and PdDy (R.A. Devine et al.(1972)) single crystals have been reported. In both systems the electronic ground state of the Rare Earth impurity is a quartet Γ_8 resulting from the cubic crystalline electric field (CEF) of the host. The fundamental property of a Γ_8 multiplet is its anisotropic behaviour in a magnetic field. The observed resonances were fitted with the hamiltonian

$$\mathcal{H}_o = W\left[x\frac{O_4}{F_4} + (1-|x|)\frac{O_6}{F_6}\right] + g_J'\mu_B\vec{H}.\vec{J} \qquad (1)$$

where x, W are the two cubic CEF parameters, O_4 and O_6 are fourth and sixth order angular momentum operators and F_4 and F_6 associated constant factors; in the Zeman term, $\vec{J}$ is the total angular momentum of the ion and g_J' is an effective Landé g-factor which takes into account an isotropic exchange interaction J_{sf} between the Rare Earth and conduction electrons. The line shift induced by the exchange interaction has been discussed previously. This work is specifically devoted to the study of the linewidths as a function of temperature and orientation of magnetic field. According to Hirst (1969) there are exponential contributions to the linewidths of an ESR multi-level system (in our case CEF levels) in addition to the usual linear Korringa terms. Such deviations have recently been observed in AuEr (D. Davidov et al.(1973)). In PdEr and PdDy, these effects were not measurable in the limited range of temperature where the resonance lines could be observed. Nevertheless, we applied Redfield-Hirst's relaxation theory (C.P. Slichter (1963); L.L. Hirst (1969)). Thus in PdEr the linewidths of transition 2→3 (see below) in different orientations are fitted with an isotropic exchange parameter $|J_{sf}|=6.1\pm.4$meV. Also the linewidth of the excited Γ_6 doublet is well understood with this analysis. The situation for PdDy is more delicate. Only one anisotropic resonance has been observed and its behaviour can be reproduced with two sets of CEF parameters x and W.

2. Relaxation formulae. The system which we consider consists of diluted magnetic ions with CEF levels (typical overall splitting of the order of 100°K) and conduction electrons. The relaxation mechanism between crystalline states is due to the exchange interaction

$$\mathcal{H}' = -(g_J-1)\,J_{sf}\,\vec{J}.\vec{s} \qquad (2)$$

where g_J is the Landé factor of the free ion, J_{sf} the coupling constant and $\vec{s}$ the conduction electron spin. In Redfield's theory (C.P. Slichter (1963)), one calculates the evolution of the angular momentum $\vec{J}$ by second order time dependent perturbation theory. Calling ρ the density matrix the relaxation equation is written

$$\frac{d\langle J^r\rangle}{dt} = \sum_{\alpha,\alpha',\beta,\beta'} R_{\alpha\alpha'\beta\beta'}\,\rho_{\beta\beta'}\,\langle\alpha'|J^r|\alpha\rangle \qquad (3)$$

where α,α', etc. are eigenstates of the uncoupled system of impurities + conduction electrons. The Redfield coefficients $R_{\alpha\alpha'\beta\beta'}$ involve the correlation integrals of the time dependent perturbation $\mathcal{H}'$. One can write them as products of angular momentum matrix elements J^q_{ab} (of component q of $\vec{J}$ between crystalline states $|a\rangle$ and $|b\rangle$) and the conduction electron spectral density (Hirst 1969). With the two assumptions that 1° there is no resonance degeneracy, i.e. no other transition except a→b possesses an energy difference equal to the microwave energy $\hbar\omega$, 2° only transverse relaxation contributes to the linewidth, the relaxation equation (3) takes the simple Bloch form. One can then identify the relaxation rate $1/T_2$. Using $g\mu_B\Delta H_{\frac{1}{2}}=\hbar/2T_2$, the linewidth is expressed as

$$\Delta H_{\frac{1}{2}} = \frac{\pi}{4\mu_B g}\left[(g_J-1)J_{sf}\eta(E_F)\right]^2\frac{K(\alpha)}{(1-\alpha)^2}\sum_q\Big\{(J^q_{aa}-J^q_{bb})^2kT + J^q_{ab}J^q_{ba}\hbar\omega\coth\left(\frac{\hbar\omega}{2kT}\right) +$$

$$+\sum_{c\neq a,b}J^q_{ca}J^q_{ac}\frac{\Delta ca}{e^{\Delta ca/kT}-1} + \sum_{c\neq a,b}J^q_{cb}J^q_{bc}\frac{\Delta cb}{e^{\Delta cb/kT}-1}\Big\} \qquad (4)$$

$\eta(E_F)$ is the conduction electron density of states (1.6 st/eV.at.spin for Pd), $K(\alpha)$ and $(1-\alpha)^{-1}$ correct for electron-electron interactions (for Pd resp. 0.17 and 7), Δ_{ik} are energy splittings between levels i and k. The term $\coth(\hbar\omega/2kT)$ yields practically a contribution linear in temperature if $kT>\hbar\omega$. If $\Delta>0$ the exponentials become important only at values kT of order Δ. However, if $\Delta<0$ (i.e. $|a\rangle$ and $|b\rangle$ are not ground states) a finite width remains at T=0.

3. Experimental results and analysis. All measurements were done on a superheterodyne reflection type spectrometer at 9.5 GHz. For linewidth calculations the necessary matrix elements in formula (4) were computed after diagonalization of the hamiltonian (1).

A. PdEr. Using x=0.47 and W=-0.16°K, we calculated the linewidths of the ground state quartet Γ_8 Both experimental and theoretical results are given in Figure 1. The levels of the quartet are denoted by 1,2,3 and 4 with increasing energy. The linewidths of transition 2→3 are fitted with one single parameter $|J_{sf}|$=6.1meV in the three orientations [001], [111] and [110]. It should be noted that a residual width extrapolated from experiment is taken into account for each orientation. These residual widths increase with concentration and also depend on sample preparation. The thermal broadening does not depend appreciably on concentration in the investigated range (500 to 1500 ppm). The given value for J_{sf} also yields a linewidth for the excited doublet Γ_6 (isotropic) in good agreement with the measured one. An important contribution at T=0 (56G) comes from the saturation of the exponential terms with negative value of Δ ($\simeq$-16°K). Only a very small residual width (5G) was added to formula (4). The exchange parameter $|J_{sf}|$=6.1meV deduced from linewidths appears considerably smaller than that found from resonance field shift (J_{sf}(q=0)$\simeq$-28meV). Qualitatively, we attribute this descrepancy to different averages of J_{sf} in each case (R.Orbach et al. in Proceed. of Conf. on EPR of Ions in Metals, Haute-Nendaz, Switzerland (1974)).

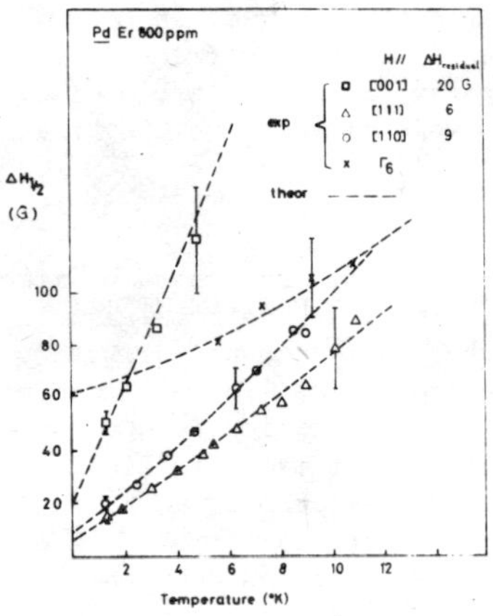

Fig.1. Experimental (symbols) and theoretical (---) linewidths (half height halfwidth) in PdEr. A unique value of exchange $|J_{sf}|$ = 6.1meV was used for all calculated curves. At T = 0 the Γ_6 linewidth consists of 56 G due to exponential terms (see text) and only 5 G empirical residual width.

B. PdDy. A strongly anisotropic resonance line has been observed in several PdDy single crystals at very low temperatures together with some weak and broad lines at low field (R.A. Devine et al. (1972)). As for PdEr the anisotropic resonance is attributed to a Γ_8 quartet as the ground state. We explored the possible values of parameters x and W for which a transition fits the experimental angular dependence. Assuming isotropic exchange, we find two solutions x_1=-0.91,W_1=-0.09°K and x_2=-0.559,W_2=-0.24°K. In both cases the transition 2→3 (levels labeled 1,2,3 and 4 with increasing energy) gives the correct behaviour. Both solutions are equivalent except for the exchange interaction deduced from resonance shifts and for transitions 1→2 and 3→4 near the [001]-axis. These last transitions were not observed. The linewidth of transition 2→3 was calculated with formula (4) with the same result for either set of parameters. As shown for PdDy 500 ppm in Figure 2 the fit between experiment and theory with a constant value of the exchange parameter $|J_{sf}|$=6.4meV is reasonable. Residual widths of the order of 40G have been taken into account. The analysis shows that up to 10°K deviations from linear behaviour are small. As for PdEr the exchange value deduced from linewidths is much less than the values obtained from resonance shifts, $J_{sf}\simeq$-28meV (with x_1,W_1) and $J_{sf}\simeq$-19meV (with x_2,W_2).

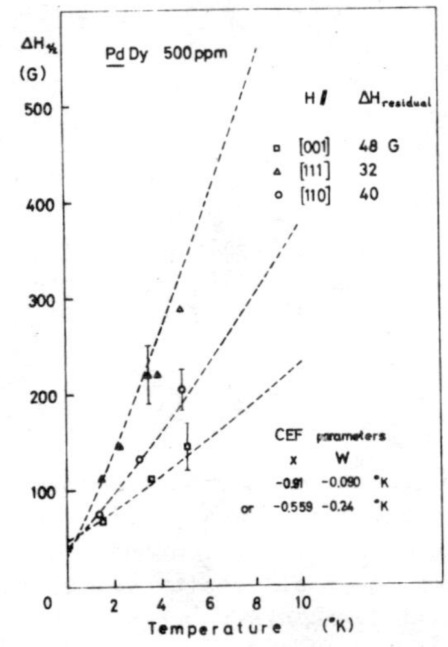

Fig.2. Experimental (symbols) and theoretical (---) linewidths of transition 2 → 3 in PdDy. Calculations were done with a constant value of exchange $|J_{sf}|$=6.4meV. Empirical residual widths are indicated.

4. Conclusion. With the analysis of linewidths presented in this communication, the experimental results for PdEr and PdDy can be interpreted with an isotropic exchange parameter. Deviations from the Korringa dependence for transitions in the ground state Γ_8 would take place only above the limit of temperature up to which the lines were observable. The calculated linewidth of the excited Γ_6 in PdEr gives support to the application of Redfield-Hirst's relaxation theory. A study of residual linewidth still has to be done and calls for careful measurements as a function of concentration.

References.

Davidov, D., C. Rettori, A. Dixon, K. Baberschke, E.P. Chock and R. Orbach, Phys.Rev., B8, 3563 (1973).

Devine, R.A.B., W. Zingg and J.-Moret, Solid State Commun., 11, 233 (1972).

Devine, R.A., J.-M. Moret, J. Ortelli, D. Shaltiel, W. Zingg and M. Peter, Solid State Commun.10., 575 (1972).

Hirst, L.L., Phys.Rev. 181, 597 (1969).

Proceed. of Conf. on EPR of Ions in Metals, Haute-Nendaz, Switzerland, Ed.G.Cohen and B.Giovannini, Univ. of Geneva (1974).

Slichter, C.P., Principles of Magnetic Resonance, p.142 ff, Harper and Row, New York (1963).

Zingg, W., H. Bill, J. Buttet and M. Peter, Phys.Rev.Lett. 32, 1221 (1974).

FAR WING LINESHAPE IN THE PRESENCE OF EXCHANGE

C. Cusumano* and G.J. Troup†
*Istituto di Fisica dell' Universita, via Archirafi 36, 90123 Palermo, Sicily.
†Physics Department, Monash University, Clayton, Victoria 3168, Australia.

Abstract. An EPR study of the far-wing lineshape of solid solutions of DPPH in polystyrene is reported, at concentrations where the dominant non-exchange interaction is the hyperfine one. The wing shape is exponential. This is explained as high-temperature Heisenberg ferromagnet relaxation. Implications for non-amorphous systems are indicated.

For an exchange-narrowed EPR line, it is of interest to know how far into the wings the lineshape is Lorentzian; what function the lineshape then becomes; and whether this function makes the moments converge. For solid solutions of DPPH in polystyrene, the exchange interaction can be strongly varied by varying the concentration. The spin dynamics of this system had already been partially studied (Boscaino et al, 1973), and hence it was chosen for this study.

The EPR measurements were made with a conventional Pound-stabilized spectrometer at ∿3 cm wavelength. The necessary sensitivity was obtained by dividing the absorption line into several regions: then each region was recorded with a different magnetic field modulation amplitude, chosen to obtain maximum sensitivity without distortion. Overlap of each region with the adjoining neighbours allowed iterative calibration of each record. Small samples were used for the central portion of the line, and larger ones for the wings.

The sample studied were those with an exchange field H_{ex} lying between 0 and 100 gauss, otherwise good data were not obtainable outside the Lorentzian region. Many samples were studied and all showed the general features to be reported; those that turned out to be inhomogeneously diluted were rejected. For brevity, the results for only 3 samples are reported. These were among the most homogeneous, and are well representative of other samples whose concentration lies close to one of them. Sample No. 1 had 2.0×10^{20} spins/cm^3; No. 2 had 2.5×10^{20} spins/cm^3; and No. 3 had 3.2×10^{20} spins/cm^3.

The derivatives $\frac{d\chi''}{dH}$ of the absorption lines with respect to magnetic field H are plotted in figure 1, on a log-linear scale, against the separation $\Delta H = (H-H_0)$ from line centre at H_0.

Figure 1

Derivatives $\frac{d\chi''}{dH}$ of lineshapes plotted against deviation from line centre.

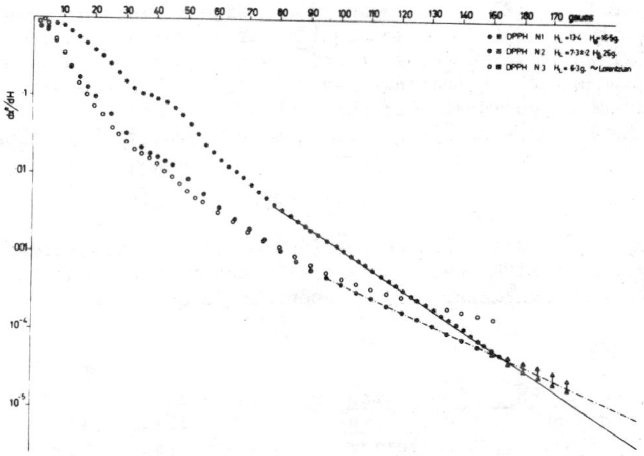

All lines are Lorentzian near the centre: for samples 1 and 2 the lines become of exponential form, $I(\Delta H)\propto\exp(-H/H_*)$, in the wings, while for sample 3 the Lorentzian behaviour is maintained over the entire range. The small structure apparent in the lines is attributable to hyperfine coupling (Lord and Blinder 1961). We characterise the Lorentzian portion of the lines by H_L, the inverse of the Lorentzian relaxation times. H_L, H_* and the second and fourth moments for samples 1 and 2, assuming the exponential behaviour continues to infinity, are reported in table 1. Only $H_L \doteqdot 6.3$ gauss can be deduced for sample No. 3.

Table 1

Sample	H_L (gauss)	H_* (gauss)	M_2 (gauss^2)	M_4 (gauss^4)
No. 1	13.4	18	340±20	11×10^5
No. 2	7.5	25	340±20	16×10^5

It is clear that H_* depends on the spin concentration: in fact, we shall see that H_* is directly proportional to H_{ex}. At these concentrations, the major contribution to the second moment is the hyperfine interaction; the calculated dipolar contribution is of the same order as our uncertainty in determining M_2. (Boscaino et al, 1973).

The theory of exchange-narrowed lines states that the Fourier transform $\phi(t)$ of the lineshape $I(\omega)$ is given by

$$\phi(t) = \exp - \omega_p^2 \int_0^t d\tau(t-\tau)\phi_{\Delta\omega}(\tau) \quad (1)$$

where $\omega_p^2 \equiv M_2$, and $\phi_{\Delta\omega}(t)$ is a correlation function (e.g. Anderson, 1954). $\phi_{\Delta\omega}(t)$ is readily recovered from $\phi(t)$: (Gulley et al, 1970; Cusumano and Troup, 1973a). Assuming a wholly exponential line, $\phi(t) = (1+\omega_*^2t^2)^{-1}$, and

$$\phi_{\Delta\omega}(t) = (2\omega_*^2/\omega_p^2)(1-\omega_*^2t^2)(1+\omega_*^2t^2)^{-2}. \quad (2)$$

The Fourier transform $\phi_{\Delta\omega}(\omega)$ of this is

$$\phi_{\Delta\omega}(\omega) = (|\omega|/\omega_*)\exp(-|\omega|/\omega_*). \quad (3)$$

Blume and Hubbard (1970) from their auto-correlation function $F(t)$ for the ith spin of a high temperature Heisenberg ferromagnet obtain the asymptotic spectrum $(\omega\to\infty)$

$$F(\omega) = [\omega/(\omega_+\sqrt{2})]\exp\{-(\pi\omega)/(2\omega_+)\} \quad (4)$$

where $\omega_+ = [(\frac{1}{3})S(S+1)]^{\frac{1}{2}}\hat{J}$, and $\hat{J}^2 = \sum_{i,j} J_{ij}^2$, the familiar exchange parameter. If we make the identification $\omega_* = 2\omega_+/\pi$, (4) becomes identical with (3), apart from a constant factor, and it is clear that ω_* and H_* depend directly on the exchange. This identification can be justified as follows. The correlation function $\phi_{\Delta\omega}(t)$ in our case, where the hyperfine interaction dominates, will be the two-spin one. Because our system is amorphous and disordered, and because of the isotropy of the Heisenberg exchange Hamiltonian in spin-space, the spin-correlations then satisfy (Soos et al, 1973)

$$\langle S_i^\alpha(o)S_j^\beta(t)\rangle = \delta_{\alpha\beta}\langle S_i^\alpha(o)S_j^\beta(t)\rangle$$

for $\alpha,\beta = x,y,z$, and hence our $\phi_{\Delta\omega}(t) \equiv F(t)$ of Blume and Hubbard. (Our previous explanation (Cusumano and Troup 1973b) in terms of the dipolar interaction was incorrect). Since the calculation of $\langle S_i^z(o)S_i^z(t)\rangle$ for small t is almost independent of the symmetry, and of the distance over which the exchange takes place (Blume and Hubbard, 1970), we would expect that for nonamorphous exchange-narrowed systems in which the hyperfine interaction dominates, the far-wing lineshape would also be exponential.

From the $F(t) = \mathrm{sech}^2\omega_+t$ of Blume and Hubbard, $\phi(t)$ is found to be $[\mathrm{sech}\,\omega_+t]^{(\omega_p/\omega_+)^2}$ from (1). As $t\to\infty$, this has the form

$$I(t\to\infty)\propto\exp\{-(\omega_p/\omega_+)^2t\}$$

which yields a Lorentzian centre portion for $I(\omega)$ of half-width $\omega_L = \omega_p^2/\omega_+ = \omega_p^2/[(\pi/2)\omega_*]$, in rough agreement with the experimentally determined quantities. In view of the necessary theoretical approximations, the agreement could not be expected to be exact.

References

Anderson, P.W., J. Phys. Soc. Japan, 9, 1954, 316.
Blume, M., and Hubbard, G.D., Phys. Rev. B1, 1970, 3815.
Boscaino, R., Brai, M., Ciccarello, I., and Strandberg, M.W.P., Phys. Rev. B7, 1973, 50.
Cusumano, C., and Troup, G.J., Physics Letters 44A, 1973(a), 441.
Cusumano, C., and Troup, G.J., Physics Letters 45A, 1973(b), 283.
Gulley, J.E., Hone, D., Scalapino, D.J., and Silberngel, B.G., Phys. Rev. B3, 1970, 1020.
Lord, N.W., and Blinder, S.M., J. Chem. Phys. 34, 1961, 1693.
Soos, Z.G., Huang, T.Z., Valentine, J.S., and Hughes, R.C., Phys. Rev. B8, 1973, 993.

SPIN COUPLING BETWEEN TWO IDENTICAL COMPLEX IONS

G.Amoretti and V.Varacca
Gruppo Nazionale di Struttura della Materia del C.N.R., Istituto di Fisica dell'Università, Via M.D'Azeglio 85, Parma, Italy.

Abstract. A theoretical interpretation of the ESR spectra of spin coupled ion pairs is given in terms of a Heitler-London model. The importance of the exchange interaction, via the bridging ligands, and the relative conditions on the wave functions of the complex ions are pointed out.

1. Introduction. In this work we want to explain the different behaviour of similar dimeric complex ions, in dependence on the nature of the ligands. In particular, we want to justify why in complexes such as $Cu(aebg)(C_2H_4N_4)\,SO_4 \cdot H_2O$ and $Cu(aebg)(NCS)\,(SCN)$, where the bridging ligands are N atoms, the ESR spectrum is typical of an effective spin S'=1 (Buluggiu et al., 1971), while complexes such as $Cu(NH_3)_2C_8H_4O_4$, where the bridging ligands are O atoms, show a spectrum of a spin $S=\frac{1}{2}$ (Dascola et al., 1972). Moreover, in the former complexes, the g and fine structure tensors have different principal axes. In particular, the g tensor has the same symmetry (and the same values) as in an isolated complex, while the fine structure principal axis lays almost along the Cu-Cu joining line.

2. Heitler-London approach (Judd, 1959). As starting point, we assume as "unperturbed" wave functions of the single complex ion those for a hole in a $3d^9$ complex under D_{2h} symmetry, but already mixed by the spin-orbit coupling. moreover, to take into account the geometrical arrangement of the dimeric complex (see Fig. 1), we build for each octahedron wave functions "adapted" to the $\overline{C_{2h}}$ symmetry of the whole dimer (we use the double group $\overline{C_{2h}}$, because the spin is no more a good quantum number for spin-orbit mixed functions). With an almost obvious notation:

$$\Phi^{a,b}_{0\uparrow} = A\,\varphi^{a,b}_{x^2-y^2\uparrow} + B\,\varphi^{a,b}_{z^2\uparrow} + C\,\varphi^{a,b}_{yz\uparrow}$$
$$\Phi^{a,b}_{0\downarrow} = A\,\varphi^{a,b}_{x^2-y^2\downarrow} + B\,\varphi^{a,b}_{z^2\downarrow} + C\,\varphi^{a,b}_{yz\downarrow} \qquad (1)$$

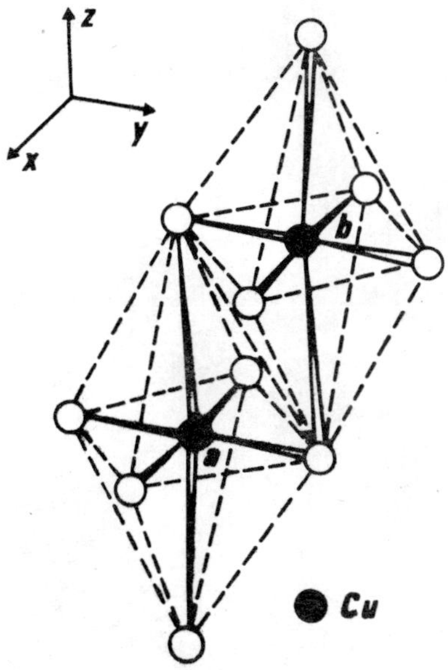

Fig. 1, Dimeric complexes typical structure.

From (1) we may now build two particle wave functions transforming as the irreducible representations of $\overline{C_{2h}}$. Applying the Zeeman interaction operator, we obtain rather complicated expressions for the elements of the g tensor. It turns out that the g tensor becomes diagonal in the reference frame of the single octahedron only if C=0, that is no φ_{yz} mixing in the functions (1). With this condition on the wave functions, we calculate the matrix elements for the electrostatic and dipole-dipole magnetic interactions. First of all, it turns out that it is possible to use a TT_z quantization scheme for an effective spin S'=1 only if both B≠0 and there is a not negligible s-character at the bridging ligands orbitals. In fact, only these conditions ensure a remarkable splitting between the "singlet" and the "triplet". Moreover, the same conditions are important to give both a fine structure splitting among the triplet levels and a rotation of this tensor as said in the Introduction. Therefore, we justify the different behaviour of the compounds under study in terms of different ligand

orbitals. Namely, the N ligands must have a remarkable s-character at their orbitals, while the O ligands a negligible amount only. In this second case, the dipole-dipole interaction, owing to its opposite sign, may cancel the feeble residual exchange interaction.
A first comparison with the experimental ESR data greatly supports the validity of our calculations. For a complete and detailed treatment of this subject, impossible here owing to the little space at disposal, one can refer to a paper of ours (Amoretti and Varacca, 1974).

3. References.

Amoretti, G. and Varacca, V., 1974, The Journal of Chemical Physics, (in press).

Buluggiu, E. et al., 1971, Physica Status Solidi (b), 45, p.217.

Dascola, G. et al., 1972, Physica Status Solidi (b), 52,p.K99.

Judd, B.R., 1959, Proceedings of the Royal Society, 250, p.110.

THE EPR SPECTRA CAUSED BY DISLOCATIONS IN SILICON

H. Alexander, B. Nordhofen and E. Weber
Abteilung für Metallphysik im II. Physikalischen Inst. Univ. Köln, W.-Germany

Abstract. The research on the EPR spectra of plastically deformed Silicon has been carried on. At higher temperatures appear new lines additionally to the lines of the Si-K1 and Si-K2 centers. Four of these lines can be attributed to a new center of the dangling bond type Si-K3 with the g_1-axis in the $\langle 111 \rangle$ - directions.

1. Introduction. The arrangement of atoms in the core of a dislocation is fundamentally different from that in undisturbed regions of the crystal. This becomes especially evident in crystals with directed and localized bonds such as Silicon. Considering the local crystal structure around the atom in the very core of the dislocation one finds not only strains but in general also a change of the number of nearest - neighbour (NN) atoms. A dislocation line s seperates two regions of the glide plane G (in the diamond structure of type $\{111\}$) relatively displaced one against the other by the Burgers vector $\vec{b}$ (type $a/2 \langle 110 \rangle$, in G) of the dislocation. By that division of the glide plane into two seperated parts a row of atoms in the neighbouring (unchanged) (111) plane loses their NN-atoms in the (111) direction (fig.1a). These atoms with three instead of four covalently bonded NN-atoms are called dangling bond atoms in the following. Their distance d along the dislocation line corresponds to the distance d' between two bonds linking the (111) planes and depends therefore on the orientation of s in G. It turns out that $d = b\sqrt{3}/(2\sin\alpha)$, α being the angle between s and $\vec{b}$. As there are only discrete values of d' in the crystal structure the dislocation line on an atomic scale may have only specified orientations, it must hold: $\tan\alpha = \sqrt{3}/(2n + 1)$, $(n=0,1,2,3,\ldots)$. Any dislocation must be composed of segments of these basic dislocation types. It is to be noticed that only one type - the screw dislocation ($\alpha = 0$) - does not form dangling bond atoms. The diamond lattice is composed of pairs of (111)planes. If the displacement is made not between two pairs of planes but between the two planes of one pair there originate three dangling bond atoms instead of one. The so defined dislocations of type II (glide set dislocations (Hirth and Lothe 1968)) are able to dissociate into two parallel partial dislocations with Burgers vectors b_1 and b_2 ($b_1 + b_2 = b$) bordering a stacking fault between them (fig.1b). The partial dislocations contain one and two dangling bond atoms respectively (per lattice plane along the dislocation). Recently splitting of most of the dislocations in plastically deformed Silicon was demonstrated by weak beam electron microscopy (Ray et al. 1971). From this observation follows that most dislocations are of type II in spite of the larger density of broken bonds. Dangling bonds imply unpaired electrons and should act as paramagnetic centers in an otherwise diamagnetic matrix. Introduced in a sufficient density dislocations therefore should be an object of EPR. If bond rearrangements in the dislocation core produce pair bonds not occurring in the undisturbed diamond structure these may have acceptor or donor character giving rise as well to uncompensated spins in a certain temperature range. In 1965 Alexander et al. found an EPR signal caused by plastic deformation. Later on this signal proved to be the stronger the lower the deformation temperature was (Wöhler et al. 1970).

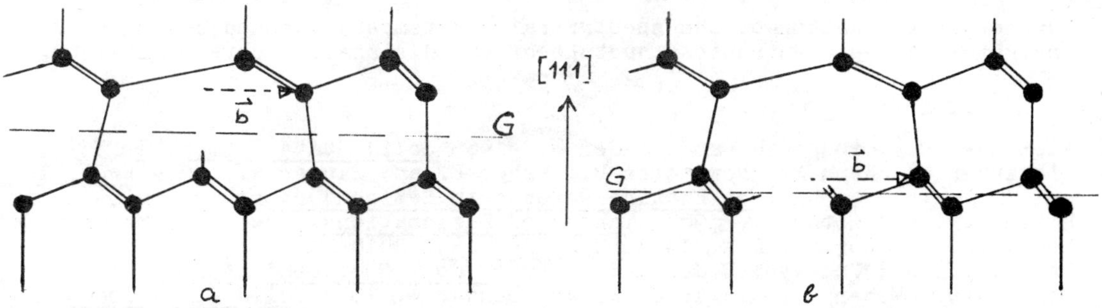

Fig.1. Perfect dislocations of type I (a) and type II (b) in the diamond structure (projected on the $(1\bar{1}0)$ plane). $\vec{b} = a/2[01\bar{1}]$.

2. Experiments. The spectra to be reported were equally found in a rather pure Silicon as well as in lightly p-doped crystals. The crystals were compressed along a [213] axis in an atmosphere of forming gas (92% N_2, 8% H_2) at 650°C. The orientation of the compression axis was selected so that most of the dislocations (80%) originated during the deformation got the same Burgers vector ("single glide"). Working with deformed crystals the difficulty of inhomogeneous

orientation arises and limits the accuracy of results. The dislocation density as revealed by electron microscopy was of the order of magnitude 10^9 cm/cm^3. The spectra were recorded as differentiated absorption curves by a Varian E-6 spectrometer in the X-band.

3. Results. The most prominent resonance lines to be seen between 6K and 70K originate from at least two different centers Si-K1 and Si-K2 (Schmidt et al.1974) the g tensors of which could be determined. The center Si-K1 is found in three and with minor certainty in a fourth equivalent orientations. Its g values (g_1=2.0017, g_2=2.0092, g_3=2.0075 $\pm 3\cdot 10^{-4}$) fall into the class "negative charge state of a broken bond or of parallel pair bonds" following the classification of Lee and Corbett (1973). The orientation is of an unusual type: the g_1 axes point in $\langle 123 \rangle$ directions, two of them in the conjugate glide system. The (peak to peak) width of the lines belonging to Si-K1 is 1.3 Oe. The intensity of the Si-K1 lines decreases with increasing temperature until at 70K they are screened by other lines now to be described. At higher temperatures beginning at 60K new lines emerge which are somewhat sharper ($\sim$ 0.8 Oe) and cover a region of the g scale more extended to higher values (Alexander et al. 1974). At present four of these lines can be followed through the spectra of four deformed crystals with different axes of rotation. They turn out to belong to a further center (Si-K3) with its axis of extremal g ($\sim g_0$) parallel to one of the four $\langle 111 \rangle$: g_1=2.0025 $\langle 111 \rangle$, g_2=2.0129 $\langle \bar{1}10 \rangle$, g_3=2.0125 $\pm 3\cdot 10^{-4}$. The g tensor belongs to the same class as Si-K1. The spectrum of the center Si-K2 is more complicated and shows a zero field splitting. It may be described by a Spin Hamiltonian:

$$\mathcal{H} = \mu_B \underline{S}\,\underline{\underline{g}}\,\underline{H} + \underline{S}\,\underline{\underline{D}}\,\underline{S}$$

The g tensor (g_1=2.0004, g_2=2.0094, g_3=2.0058 $\pm 5\cdot 10^{-4}$) is far from rotational symmetry and does not fit well into the classification of Lee and Corbett (1973). Choosing the largest g value as $g_{\parallel}$, as the authors suggest, the center falls into the class of the "bent pair bond".As we observe a fine structure of the spectrum S must be $\geq$ 1. But the spectrum consists of 9 strong and several weaker pairs of lines with common D-axes instead of one pair. The distances of all pairs δH_i are -with small deviations - multiples of one basic distance δH_0. When H is parallel $\vec{b}$= [011] holds: $\delta H_i = i\cdot 24.7$ Oe (i=1,2,3....). The series of even i ends with i=12, the odd numbers reach i=19, but some of these lines are weaker than others. Schmidt et al. (1974) discussed several possible interpretations of this zero field splitting effect. One may think of a true crystal field multiplet arising from giant moments S formed by groups of neighbouring spins coupled along limited dislocation segments. The odd-i lines then must be attributed to segments whose S is an integer, the even-i lines to S half numbered. In this model holds: $|D_{11}| = 23.4\cdot 10^{-4}$cm^{-1}, $|D_{22}| = 11.4\cdot 10^{-4}$cm^{-1}, $|D_{33}| = 12.0\cdot 10^{-4}$cm^{-1}. The strong lines correspond to maximum groups of 16 spins (dangling bonds). On the other hand one may ascribe the pairs of lines to a center with S=1 situated on different sites in a crystal field caused by the dislocation ($D^i_{11} = i\cdot 23.4\cdot 10^{-4}$cm^{-1}). It cannot be excluded that there is a central line common to all pairs which is hidden by the Si-K1 lines. If this is the case S is 3/2 and $D^i_{11} = i\cdot 11.7\cdot 10^{-4}$ cm^{-1}. The connection of Si-K2 with the dislocations is indicated most directly by the fact that the D_{11}-axis coincides with the Burgers vector of the dislocations. The lines belonging to the dislocations of the second stressed glide system can be found too. With increasing temperature D_{11} decreases. This can be seen till 220K where $D_{11} \sim 0.6\, D_{11}$(30K). The peak to peak width of the Si-K2 lines amounts to 10 Oe.

Further investigation of the spectra raising temperature and doping should give insight into the electronic structure of the dislocation core of Silicon.

4. References.

Alexander H., Labusch R. and Sander W., 1965 Solid State Commun. 3, 357
Alexander H., Kenn M., Nordhofen B., Weber E. and Sander W., 1974 Proc. Int. Conf. on Lattice Defects in Semiconductors in Freiburg, in press
Hirth,J.P. and Lothe J., 1968, Theory of Dislocations, New York: Mc Graw Hill
Lee Y.-H. and Corbett J.W., 1973 Phys. Rev. B8, 2810
Ray J.L.F. and Cockayne D.J.H., 1971 Proc. Roy. Soc. Lond. A325, 543
Schmidt U., Weber E., Alexander H. and Sander W., 1974 Solid State Commun. 14, 735
Wöhler F.D., Alexander H. and Sander W., 1970 Phys. Chem. Solids 31, 1381

EXCHANGE INTERACTIONS AND MIGRATION OF EXCITATIONS IN Cr(III) AND Cu(II) PAIRS

O.F.Gataullin, M.M.Zaripov, L.V.Mosina, Yu.V.Ryzhmanov, Yu.V.Yablokov
Kazan Physico-Technical Institute of the Academy of Sciences, USSR.

Abstract. The study of exchange interactions between limited number of magnetic atoms enable to study the mechanism of exchange and the nature of the paths for exchange, which provide unpaired electron delocalization. This work is concerned with aforementioned questions and illustrates the great possibilities of EPR in their detailed analysis.

The study of exchange-coupled pairs of Cr(III) atoms in guanidine aluminium sulfate hexahydrate (GASH) single crystals gives an example of isotropic exchange interactions via system of σ-bonds. The GASH unit cell has three formula units with dimensions a=11.738 Å and c=8,951 Å, space group C^2_{3v} -P31m (Schein et al.,1967) (Fig.1).

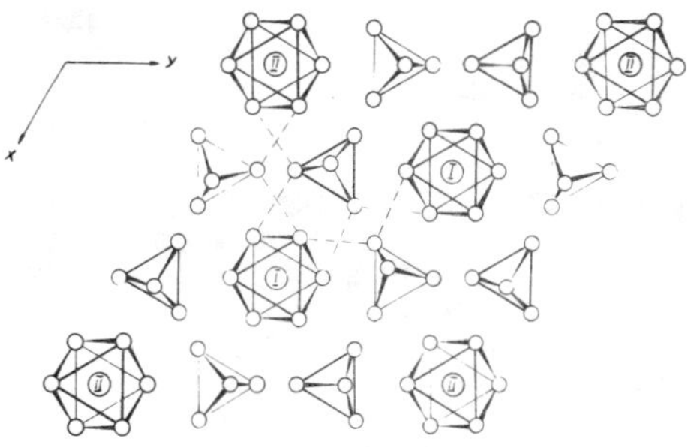

Figure 1

The peculiarities of the behaviour of some satellite lines in the EPR spectrum: a)60-angular periodicity about the C_3 axis, $\vec{H} \perp C_3$; b)the temperature dependent shift of satellite lines, comparable with the change in the fine structure parameter D; c)the symmetry of the spectrum about its centre etc., and the crystallographical data prove that additional lines in the spectrum are due to the exchange-coupled Cr(III) pairs with the constant of the isotropic interaction J smaller than D and Zeeman energy.

There are two equally probable types of Cr(III) pairs with radius-vector r_{12} =6.77 Å in the GASH unit cell: I-I exchange and I-II exchange (Fig.1). The angle between the pair radius-vector and the C_3 crystal axis is very close to $\pi/2$ in both cases. The 16-order secular equation was solved on the computer and the spin hamiltonian parameters were fitted to the $C_3 \parallel H$ orientation. The best fit parameters are given in Table 1.

Table 1

Cr(III) in GASH	T=293°K				T=100°K			
lattice	D_1,cm^{-1}	D_2,cm^{-1}	A,cm^{-1}	J,cm^{-1}	D_1,cm^{-1}	D_2 cm^{-1}	A,cm^{-1}	J,cm^{-1}
I-I exchange	-0.0547	-0.0547	+0.0053	+0.0129	-0.0790	-0.0790	+0.0055	+0.0148
I-II exchange	-0.0536	-0.0684	+0.0053	+0.0119	-0.0781	-0.1019	+0.0055	+0.0139

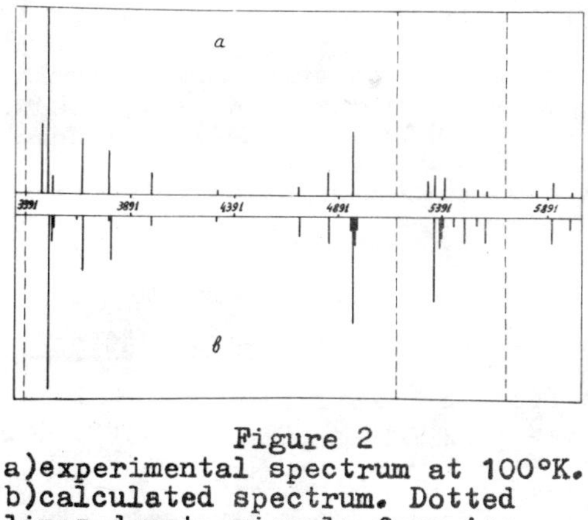

Figure 2
a)experimental spectrum at 100°K. b)calculated spectrum. Dotted lines denote signals from isolated Cr(III) atoms.

The spectrum calculated with given parameters for I-I and I-II exchange at 100°K (transition probabilities taken into account) and the experimental EPR spectrum are shown on Fig.2.

Here the superexchange realized via elements of intermediate structures by the system of hydrogen bonds is observed. In the GASH crystal the O-H...O hydrogen bond with 2.54-2.64 Å length (Zheludev, 1973) connects the oxygen atoms of sulfate ion and crystallization water forming the Cr(III) nearest environment. The experiments on substituting hydrogen for deiterium unambiguously prove that. The exchange in Cr-Cr pair reduced (J_{I-I} =0.0122cm^{-1} and J_{I-II}=0.0112cm^{-1} at T=293°K), because the O-D...O length increased (Zheludev, 1973). The parameter of magnetic dipole-dipole interaction A was not changed. Some satellite lines in good approximation are described by only magnetic dipole-dipole interaction with A=0.0023±0.0003cm^{-1}.

These are the results of the investigation of the exchange interactions between atoms with ground state magnetism. Interesting effects are discovered when the possibility of the interactions between centres with excited magnetic states is realized in crystal.

We shall consider several compounds whose lattice contains fragments with two Cu(II) atoms: Cu(II) carboxilates with amines $(RCH_2COO)_2$ Cu·L, where R=CH_3; CH_3CH_2 and L= p-NH_2-C_6H_4-Cl; μ-NH_2-C_6H_4-CH_3. The structure of such compounds is given on Fig.3. They are coordination polymers crystallographically, but the

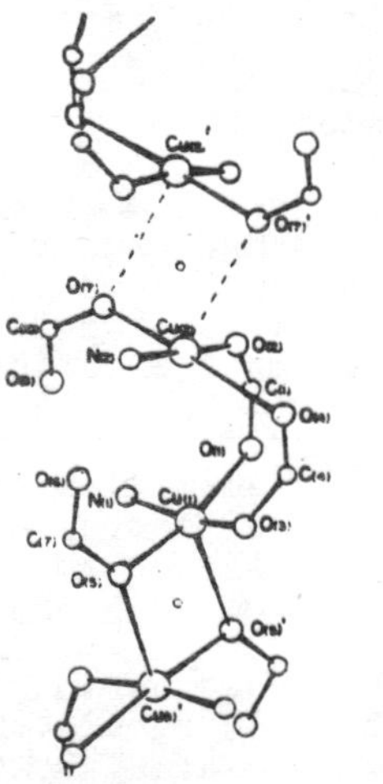

Figure 3.

conditions of the coupling of each Cu(II) atom with its neighbors are different and magnetically these compounds are coupled dimers (Yawney & Doedens, 1970; Simonov et al., 1973)

Exchange interactions $J\hat{S}_1 \cdot \hat{S}_2$ between Cu(II) atoms in the dimeric fragment give rise to ground singlet state and excited triplet state (antiferromagnetic exchange). Because of the translational symmetry and coordination between the dimeric fragments the excitations migrate through the crystal lattice, being essentially triplet excitons (Gavrilov et al., 1971; Mosina & Yablokov, 1974).

The migration and collisions of excitons result in intermolecular exchange interactions which give rise to some peculiarities in the EPR spectrum. These interactions are enhanced as the temperature increases, because the concentration of magnetic centres raises. Therefore the resolved triplet fine structure is observed only at sufficiently low temperatures, and then the spectrum is transformed into a single line. The transition region and spectrum resolution at given temperature depend on the singlet-triplet interval J and on the efficiency of the intermolecular exchange J', i.e. depend on the properties of the compound.

Analysing the temperature dependence of the fine structure signals intensity $I \sim 1/T(3+\exp J/kT)$ and the exchange signal intensity I_{ex} (when polycrystals are studied; the change in the spectrum splittings is measured when single crystals are studied) $I_{ex} \sim \nu I = \nu_0 \exp(-\Delta E_a/kT) \cdot I$, J and the activation energy for exchange ΔE_a are determined from the dependence of ln(IT) and $\ln\nu$ on T^{-1} (Table 2) The EPR spectra parameters determined according to (Yablokov et al., 1973) are given in Table 2 also.

Table 2.

Nº	Compound	$g_\perp$	$g_\parallel$	D, cm^{-1}	E, cm^{-1}	J, cm^{-1}	ΔE_a, cm^{-1}	J', cm^{-1} (77°K)
I	$(C_2H_5COO)_2Cu \cdot p\text{-}NH_2\text{-}C_6H_4\text{-}Cl$	2.061	2.289	0.124	<0.001	93	146	0.6
II	$(C_2H_5COO)_2Cu \cdot \mu\text{-}NH_2\text{-}C_6H_4\text{-}CH_3$	2.071	2.283	0.124	<0.001	70	160	0.5
III	$(C_3H_7COO)_2Cu \cdot \mu\text{-}NH_2\text{-}C_6H_4\text{-}CH_3$	2.071	2.290	0.124	0.022	82	110	1

Note the following: a)the excitations are Frenkel-type excitons because only two neighboring spins (within dimeric fragment) are strongly exchange coupled; b)the values of D and g-tensor components are practically unchanged. J values are close also (Table 2). It might have been supposed that ΔE_a is entirely determined by J. But ΔE_a is larger than J. The possible processes of exciton-lattice defect collisions, exciton-exciton annihilation or the diffusional motion with some activation energy decrease the effective concentration of triplet excitons, increasing ΔE_a; c)the increase of the distortions of the dimeric fragment in compound III ($E=0.022 cm^{-1}$) sufficiently decreases ΔE_a. It means that the increase of the state mixing when the claster symmetry is lowered leads to the improvement of the conditions for exciton migration and exchange interactions between them; d)the results obtained show that $J \gg J'$. Therefore the found effects are manifested only in the EPR spectra. EPR enables the reliable determination of the considered exciton characteristics and their detailed analysis.

References

Gavrilov, V.V., Yablokov, Yu.V., Milkova, L.N., Ablov, A.V., phys.stat.sol.(b), 45, 603, 1971

Mosina, L.V., Yablokov, Yu.V., phys.stat.sol.(b), 62, K51, 1974

Schein, B.J.B., Ligafeltev, E.C., Stewart, M.J., J.Chem.Phys., 47, 5183, 1967

Simonov, Yu.A., Ablov, A.V., Suntsova, S.P., Milkova, L.N., Simonov, M.A., Doklady AN SSSR, 211, 611, 1973

Yablokov, Yu.V., Gaponenko, V.A., Mosina, L.V., Kogan, V.A., Zhuchenko, T.A., Zh.Strukt.Khimii, 14, 216, 1973

Yawney, D.B.W., Doedens, R.J., J.Am.Chem.Soc., 92, 6350, 1970

Zheludev, M.S., Osnovy segnetoelektrichestva, Atomizdat, 1973

SECTION O

Theory of NMR Relaxation and Lineshapes

RELAXATION THEORY AND THE STOCHASTIC LIOUVILLE EQUATION

Alexander J. Vega and Daniel Fiat

Department of Structural Chemistry, Weizmann Institute of Science, Rehovot, Israel.

A new formalism of the theory of magnetic relaxation is given for systems where the random molecular motions are describable as a Markoff process. The master equation for the density matrix is derived from the stochastic Liouville equation. This method gives some new insight into the mechanism of relaxation and has several advantages over the conventional theories. The reason for this is that the correlation between the states of the lattice and the spins is taken into account. It is also possible to obtain the expression for T_1 under slow motion, while the conditions for its validity can be formulated.

A "semi-classical" treatment can not explain the approach to thermal equilibrium, as it could not in the conventional theory. However, through a suitable modification of the stochastic Liouville equation this problem can be solved. The modified stochastic Liouville equation is generally justified by using a quantum mechanical description of the lattice, but for the case of diffusional motion a much simpler picture can be retained. The modified stochastic Liouville equation also simplifies the treatment of relaxation in the presence of a time dependent external field.

FUNDAMENTAL TEST OF NUCLEAR MAGNETIC RELAXATION THEORY

Robin L. Armstrong and Kenneth E. Kisman
Department of Physics, University of Toronto, Canada, M5S 1A7

Abstract. Longitudinal nuclear spin relaxation rates are reported as a function of gas density in molecular hydrogen at 296 and 385 K and in mixtures of hydrogen with the noble gases helium and argon at 298 K. The measurements provide a significant test of nuclear magnetic relaxation theory.

The Bloom-Oppenheim theory (cf Bloom and Oppenheim, 1967) relates nuclear spin relaxation rates in molecular hydrogen to the spectral densities of certain time correlation functions which are evaluated in terms of the intermolecular forces. Molecular hydrogen has two modifications, ortho-hydrogen, which has total nuclear spin I = 1 and odd molecular rotational states J = 1, 3, 5 ... , and para-hydrogen, which has I = 0 and J = 0, 2, 4 ... . Only the protons of ortho-hydrogen contribute to the nuclear magnetic resonance signal. At temperatures below 100 K only the J = 1 rotational state of ortho-hydrogen has a significant population. In this situation intermolecular collisions cause transitions only between states of different M_J within the J = 1 manifold and the theory yields an expression for the longitudinal nuclear spin relaxation rate with a single adjustable parameter. This parameter may be selected as the asymptotic value of the slope S of the relaxation time versus density curve in the extreme narrowing regime. Convincing experimental confirmation of this result was first provided by Hardy (1966).

For molecular hydrogen at temperatures above 100 K but below 500 K only the two rotational states J = 1 and J = 3 of the ortho-modification contain appreciable numbers of molecules. Intermolecular interactions can now cause transitions between these two rotational states as well as within a single J-manifold. According to the Bloom-Oppenheim theory the longitudinal nuclear spin relaxation rate is given by

$$T_1^{-1} = P_{J_1}(T_1^{-1})_{J_1} + P_{J_3}(T_1^{-1})_{J_3}$$

where P_J is the fractional population in state J and $(T_1^{-1})_J$ is the relaxation rate for a collection of molecules which are all initially in state J. The quantities $(T_1^{-1})_J$ are given by

$$(T_1^{-1})_J = \frac{2}{3}\gamma^2 H_{sr}^2 J(J+1)\sum_\alpha \frac{C_{J_i}^{1\alpha}\Lambda_{1\alpha}}{\Lambda_{1\alpha}^2 + (\omega_I - \omega_J)^2} + \frac{6}{5}\gamma^2 H_d^2 \frac{J(J+1)}{(2J-1)(2J+3)}$$

$$\sum_\alpha C_J^{2\alpha}\left[\frac{\Lambda_{2\alpha}}{\Lambda_{2\alpha}^2 + (\omega_I - \omega_J)^2} + \frac{4\Lambda_{2\alpha}}{\Lambda_{2\alpha}^2 + 4(\omega_I - \omega_J)^2}\right]$$

In this expression H_{sr} = 26.752 G is the strength of the spin-rotation magnetic field, H_d = 33.862 G is the strength of the dipolar magnetic field, γ is the proton magnetogyric ratio, ω_I is the nuclear Larmor frequency and ω_J is the rotational Larmor frequency. The rate constants $\Lambda_{1\alpha}$, $\Lambda_{2\alpha}$ and the coefficients $C_{J_i}^{1\alpha}$, $C_{J_i}^{2\alpha}$ are calculated for an anisotropic intermolecular potential V_A which is a sum of two body terms of the form

$$V_A = b^{(1)}(R_{12})\, P_2(\cos\theta_1) + b^{(2)}(R_{12}) \sum_{q=-2}^{+2} a_q Y_{2q}(\Omega_1)\, Y_{2q}^*(\Omega_2)$$

where the $b^{(\alpha)}(R_{12})$ are functions of the separation of the centers of mass of the two particles, $P_2(\cos\theta_1)$ is the second order Legendre polynomial of the angle θ_1 which the molecular axis of molecule 1 makes with the separation vector $\underline{R}_{12}$, $Y_{2q}(\Omega_\beta)$ are the second order spherical harmonics with Ω_β the orientation of the symmetry axis of molecule β with respect to $\underline{R}_{12}$ and the a_q are constants. Finally, the expression for T_1^{-1} may be expressed in terms of two adjustable parameters. These may be chosen as the asymptotic value of the slope S of the relaxation time versus density curve in the extreme narrowing regime and as the ratio R which is a measure of the contribution to relaxation of collisions involving transitions between the J = 1 and J = 3 states and of collisions involving only transitions within the J = 1 manifold. A non-zero value of R implies that inelastic collisions play an important role in the relaxation process. For the limiting cases, corresponding to negligible importance and dominant importance of inelastic collisions, R takes the values 0 and ∞, respectively. The raison d'être of the present research was to test this theory experimentally for pure molecular hydrogen gas and for mixtures of hydrogen gas with several noble gases.

The relaxation rate measurements were carried out using a phase coherent pulsed nuclear magnetic resonance spectrometer incorporating a high speed signal averager and designed specifically for gas phase investigations. Particular attention was directed to the design of the probe and the network coupling the probe to the transmitter and to the receiver (Kisman and Armstrong, 1974). An operating frequency of 49 MHz was selected. Each data point was obtained from measurements of magnetization $M(\tau)$ versus pulse spacing τ using a π-τ-$\pi/2$ pulse sequence. Plots of [M(∞) - M(τ) /M(∞)] versus τ were all linear to within experimental error indicating exponential decays of the nuclear magnetization. Considerable effort was expended to eliminate systematic errors caused by instabilities and non-linearities in the spectrometer and great care was taken to avoid spurious signals or coil ringing from contributing to the measured nuclear signal.

Fig. 1 shows the data obtained at two temperatures for pure hydrogen. Theoretical curves were fit to the data using the statistical χ^2 method (Deming 1943) to select the best values of S and R. For the data taken at 296 K the minimum value of χ^2 occurs for S = 106 μs/amagat and R = 2. The solid line in the figure represents the theoretical prediction for these values of the two parameters. It provides an excellent description of the data. The value of S deduced from the analysis is in good agreement with the average value of 107 μs/amagat deduced from six independent studies carried out in the extreme narrowing regime and reported in the literature. Although the theoretical curves are not particularly sensitive to the value of the parameter R, curves for R < 1 and R > 10 do provide fits which, even to the eye, are notably poorer than the one shown. In particular, curves corresponding to the limiting cases of R = 0 and ∞ do not provide reasonable representations of the data.

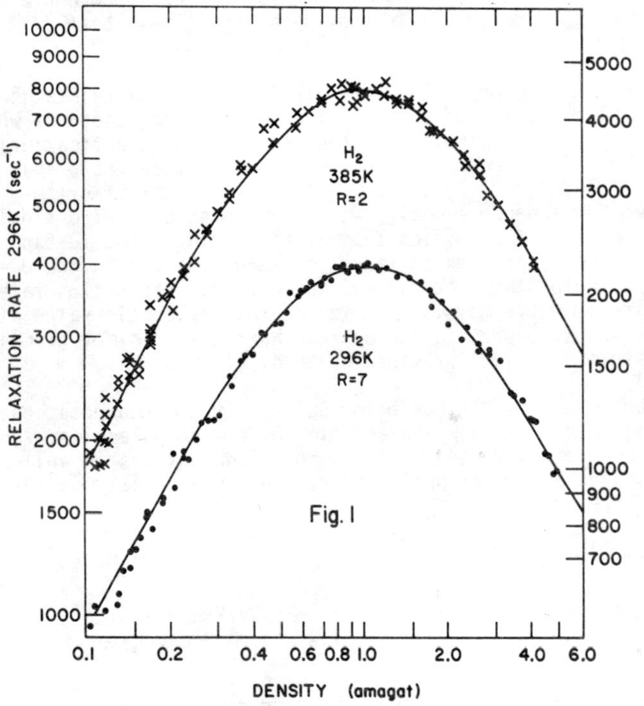

Fig. I

For the data taken at 385 K the corresponding values for the parameters are S = 98 μs/amagat and R = 7. The solid line in the figure represents the theoretical prediction for this case. It provides an excellent description of the data. It is interesting to note that R is expected to increase with temperature, reflecting the increased importance of inelastic collisions for the relaxation process.

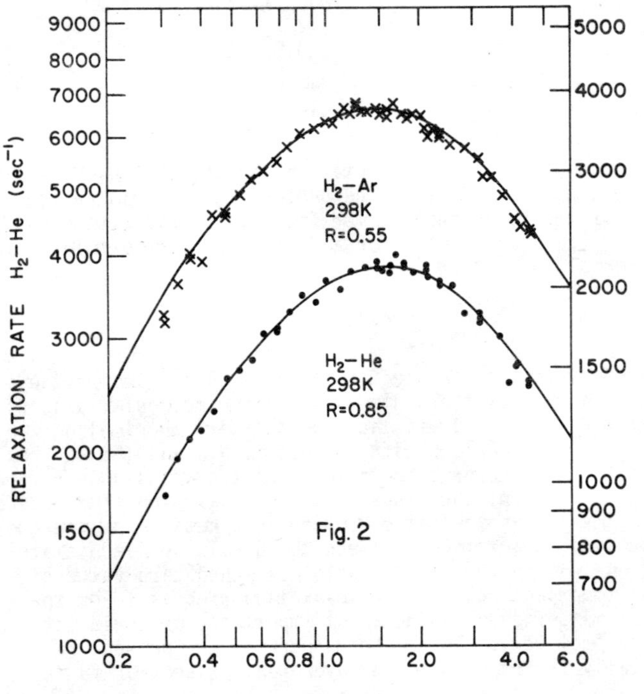

Fig. 2

Fig. 2 shows the data obtained in mixtures of molecular hydrogen with helium and with argon. For each case the mixture contains 25% hydrogen molecules. Again theoretical curves were fit to the data to obtain values of S and R. The parameters so obtained are S = 72 μs/amagat, R = 0.85 for the hydrogen-helium mixture and S = 75 μs/amagat, R = 0.55 for the hydrogen-argon mixture. The corresponding theoretical curves are shown in the figure; they provide good descriptions of the respective data. For orthohydrogen, noble gas molecular interactions, it is expected that R will approach 0; therefore, in a mixture containing 75% rare gas molecules, it is not surprising that R is less than for pure hydrogen at the same temperature.

The experimental results and analysis summarized in this contribution demonstrate that the Bloom-Oppenheim theory is able to account for longitudinal nuclear spin relaxation measurements in molecular hydrogen, both in the pure gas and in mixtures with the noble gases.

References.

Bloom, M. and Oppenheim, I. 1967. Intermolecular Forces, Interscience Publishers, New York, 549.
Deming, W.E. 1943. The Statistical Adjustments of Data, John Wiley & Sons, New York.
Hardy, W.N. 1966. Can. J. Phys. 44, 265.
Kisman, K.E. and Armstrong, R.L. 1974. Rev. Sci. Instru. to be published.

ON THE THEORY OF NUCLEAR MAGNETIC RELAXATION BY TRANSLATIONAL DIFFUSION

G. Held and F. Noack
Physikalisches Institut der Universität Stuttgart, Wiederholdstr.13, Germany

Abstract. The Harmon-Muller theory of nuclear spin relaxation by translational diffusion has been eveluated for arbitrary values of Larmor frequency and jump time; the results will be compared with Torrey's original treatment.

In his classical paper on nuclear spin relaxation by translational diffusion, Torrey (T)[1] calculated the spectral density $J^{(1)}$ of randomly walking spins and thus obtained the relaxation rates as a function of the Larmor frequency ω, the mean jump time τ and the step width parameter $\sigma^2 = \langle r^2 \rangle / d^2$, where $\langle r^2 \rangle$ represents the mean squared jump distance and d the closest approach between resonant nuclei. The Torrey theory makes the assumption, that there exists a homogeneous spin density, i.e. one uses a simple radial pair distribution function $g(r)=1$ for spin distances $r \geq d$. More recently Harmon and Muller (HM)[2] proposed a refinement by utilizing the well-known radial pair distribution function for hard spheres tc first order in the molecular number density n', namely[3] $g(r)=4/3\pi d^3 n'(1-3r/(4d)+r^3/(16d^3))$ if $d<r<2d$ or $g(r)=1$ if $r>2d$. But until now the related spectral density has only been eveluated in the 'low-frequency' limit, i.e. for $\omega\tau \ll 1$. Since the HM theory is frequently applied in the present literature we have solved the improved model for arbitrary values of ω, τ and σ^2. The results will be discussed with respect to Torrey's original findings.

1. General dependence of $J^{(1)}$ on the Larmor frequency ω. Whereas the spectral density of the Torrey theory can be written as[1)]

$$J_T^{(1)} = \frac{8\pi n}{15 d^3} \frac{\tau}{3} \frac{\sigma^2}{\omega^2\tau^2} \frac{1}{u} \left\{ t(u-1) + e^{-2t} \left\langle \left[2u + t(u+1) \right] \cos 2s + s(u-1) \sin 2s \right\rangle \right\} ; \; \binom{s}{t} = \left[\frac{3}{\sigma^2} \frac{\omega^2\tau^2}{4+\omega^2\tau^2} \left(\sqrt{1 + \left(\frac{\omega\tau}{2} \right)^2} \mp 1 \right) \right] ; \; u = s^2 + t^2 \quad (1a)$$

(n=concentration of the spins), we instead found for the HM model, after lengthy calculations[4)], a relation of the form

$$J_{HM}^{(1)} = J_T^{(1)} + \Omega \cdot F \; ; \; \Omega = \frac{4}{3}\pi d^3 n' \quad (1b)$$

$F(\omega, \tau, \sigma^2)$ is a complicated function of ω, τ and σ^2. The numerical analysis of equ.(1b) shows, that the second term which describes the deviation from Torrey's formula on account of the inhomogeneous distribution of the spin-spin distances, essentially has no influence on the frequency dependence but predominantly shifts the absolut value of $J_{HM}^{(1)}$ as compared with $J_T^{(1)}$. This is illustrated in Fig.1 for $\sigma^2=1$, $\Omega=2$, $\tau_M \equiv \frac{\tau}{2}\left[1 + \frac{6}{5\sigma^2} f_{T\,(HM)}\right] = 10^{-7}$s, $f_T=2$, $f_{HM}=1.433$[4,5)]. More details can be seen by considering some limiting cases.

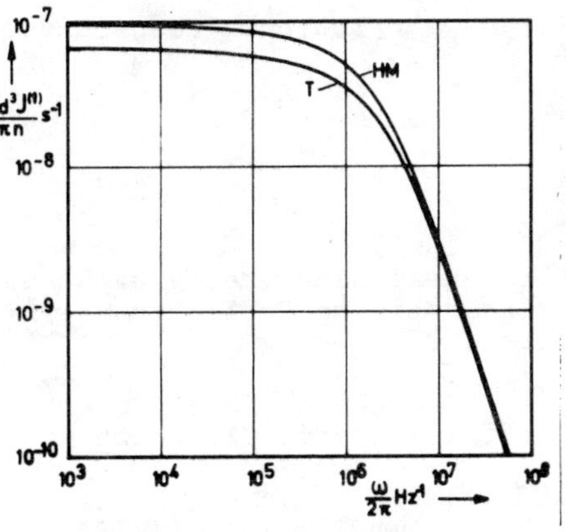

Fig.1: Dependence of $J_T^{(1)}$ and $J_{HM}^{(1)}$ from the Larmor frequency ω.

2. Large and small jump distances. In the limit, that σ^2 is large, i.e. for $\sigma^2 \gg 1$, one can establish the relations

$$J_T^{(1)}\Big|_{\sigma^2\gg 1}=\frac{8\pi n}{15d^3}\frac{\tau}{3}\left[1+\frac{\omega^2\tau^2}{4}\right]^{-1} \quad \text{(2a) and} \quad J_{HM}^{(1)}\Big|_{\sigma^2\gg 1}=\frac{8\pi n}{15d^3}\frac{\tau}{3}\left[1+\frac{\Omega}{32}(1+6\ln 2)\right]\left[1+\frac{\omega^2\tau^2}{4}\right]^{-1} \quad \text{(2b)}$$

which no longer depend on σ^2. The F-term of the general equ.(1b) is easily veryfied.

In the limit of small jump distances, i.e. for $\sigma^2 \ll 1$, equ.(1a) leads to the spectral density[1)]

$$J_T^{(1)}{}_{\sigma^2\ll 1}=\frac{8\pi n}{15d^3}\frac{\tau}{4}\frac{1}{\omega\tau}\frac{1}{x^3}\left\{2x^2-1+e^{-2x}\left\langle(2x^2+4x+1)\cos 2x+(2x^2-1)\sin 2x\right\rangle\right\};\ x=\left[\frac{2}{3\sigma^2}\omega\tau\right] \quad (3)$$

which has also been derived more directly by Pfeifer[6)] and Abragam[7)] with the aid of Fick's diffusion law instead of the random walk formalism. Equ.(1b) cannot be reduced to a comparable short form, however numerically shows a similar behavior like equ.(1b) as is expected from section 1.

3. Low and high frequency limits. Of particular interest for most practical applications are the proporties of equs.(1a) and (1b) for $\omega\tau \ll 1$ and $\omega\tau \gg 1$. In the low frequency limit, series expansions to the leading term in ω yield a rather similar behavior, namely[6,2)]

$$J_T^{(1)}{}_{\omega\tau\ll 1}=\frac{8\pi n}{15d^3}\frac{\tau}{3}\left\{1+\frac{12}{5\sigma^2}-\frac{1}{\sigma^2}\sqrt{\frac{6\omega\tau}{\sigma^2}}\right\} \quad (4a)$$

$$J_{HM}^{(1)}{}_{\omega\tau\ll 1}=\frac{8\pi n}{15d^3}\frac{\tau}{3}\left\{1+\frac{12}{5\sigma^2}+\Omega\left[\frac{1}{32}(1+6\ln 2)+\frac{1}{8\sigma^2}\right]-\frac{1}{\sigma^2}\sqrt{\frac{6\omega\tau}{\sigma^2}}\right\} \quad (4b)$$

Formally these equations can be written as $J^{(1)}=A-B\omega^{1/2}$ (A, B appropriate constants). In the last years, this very special form of the HM spectral density, known as the $\omega^{1/2}$-law, was used for the interpretation of relaxation measurements in glycerol[8)], liquid crystals[9)] and water[10)]; however it has never been emphasized that the special $\omega^{1/2}$-dependence is not at all typical for the HM model and that the simpler Torrey concept works as well. Moreover the validity of the expansions is seen to depend on the parameter σ^2, since for large values the exact results are given by equs.(2a) and (2b), involving expansions $J^{(1)}=A-B\omega^2$. Both viewpoints are in accordance with the fact, that a $\omega^{1/2}$-law also follows from Pfeifer's treatment[6)], which by principle implies small values of the step width parameter.

The high frequency limits ($\omega\tau \gg 1$) are easily seen from equs.(2) and (3), giving for small or large values of σ^2 either $\omega^{-3/2}$- or ω^{-2}- relations, respectively.

4. References.

1) Torrey, H.C.: Phys.Rev.92, 962 (1953)
2) Harmon, J.F., Muller, B.H.: Phys.Rev.182, 400 (1969)
3) Hirschfelder, J.O., Curtiss, C.F., Bird, R.B.: Molecular Theory of Gases and Liquids. New York, Wiley Interscience Publishers (1964)
4) Held, G.: Dissertation. Universität Stuttgart (1974)
5) Krüger, G.J.: Z.Naturfschg.24a, 560 (1969)
6) Pfeifer, H.: Ann.Phys.8, 1 (1961)
7) Abragam, A.: The Principles of Nuclear Magnetism. Oxford, Clarendon Press (1962)
8) Harmon, J.F., Burnett, L.J.: Z.Phys.253, 183 (1972)
9) Vilfan, M., Blinc, R., Doane, J.W.: Solid State Comm.11, 1073 (1972)
10) Parker, D.S., Harmon, J.F.: Chem.Phys.Lett.25. 505 (1974)

ANALYTICAL SOLUTION OF SOME PROBLEMS OF EXTERNAL RELAXATION AND SPIN DIFFUSION

I. V. Aleksandrov

Institute of Chemical Physics, Academy of Sciences of the USSR, Moscow, USSR.

A Schroedinger type equation with an imaginary potential is obtained for the two-level model of the external relaxation phenomena with diffusional modulated perturbation. The relaxation time and shift are expressed in terms of the complex 'scattering lengths', and the corresponding quantities are calculated for some spherical-symmetric interactions. A similar expression is found to be valid for the nuclear spin-lattice relaxation time in the spin-diffusion theory, with the only difference that the potential in the diffusional equation is real. Some analytical expressions valid both for high and low diffusional barriers are obtained.

A STUDY OF THE ULTRASLOW MOTION CORRELATION FUNCTIONS VIA DIPOLAR SPIN–LATTICE RELAXATION

S. Žumer

Department of Physics, University of Ljubljana, 61000 Ljubljana, Jadranska 19, Yugoslavia.

Abstract. In the case of ultraslow motions in solids a relation between T_{1D} and the autocorrelation function of the dipolar Hamiltonian is derived. This relation can be used either for correlated or uncorrelated ultraslow motions. The result is applied to a crystal of ice, which can be approximately described by an Ising model.

1. Introduction. Spin–lattice relaxation is a powerful method for studies of ionic and molecular motions in solids. With a choice of a proper frequency fast or slow motions can be explored. It is with the help of the dipolar spin lattice relaxation that we shall study ultraslow motions, that are the motions whose correlation times are longer than spin–spin relaxation time of the system.

As such motions couple strongly the dipolar energy reservoir to the lattice, perturbation approximation can not be used. Slichter and Ailion (1964) have derived expressions for T_{1D} due to several types of uncorrelated ultraslow motions, using sudden quantum approximation. In the following we shall derive a general relation between T_{1D} and the autocorrelation function of the dipolar coupling, which can be applied either to correlated or uncorrelated motions.

The usefulness of the relation is demonstrated on the dipolar spin–lattice relaxation in hexagonal ice.

2. Theory. Let us take a system of N identical nuclei with spin I coupled only by the magnetic dipolar interactions. The secular part of the coupling is the important part of the Hamiltonian which corresponds to the dipolar energy reservoir

$$\mathcal{H}' = \frac{1}{2} \sum_{i \neq j} F_{ij}^{(0)} C_{ij}^{(0)} \quad , \tag{1}$$

where standard notations (Abragam, 1960) have been used. The dipolar spin order can be well described by a density matrix, which has the following form in the high temperature approximation (Goldman, 1970):

$$\rho(t) = 1 - \mathcal{H}'_D(t)\beta(t) \ . \tag{2}$$

Ultraslow motions are in fact fast but infrequent ionic jumps or molecular reorientations. Therefore we can use sudden quantum approximation, which assumes that imediately after each jump the density matrix of the system is still unchanged. As the correlation time τ is longer than T_2, the internal equilibrium, which corresponds to a slightly different spin–temperature is established before the next jump of the same ion or molecule.

In a time interval short compared to T_2 the change of the dipolar energy due to jumps, which occure meanwhile, can be writen as

$$\Delta E_D(t) = -\langle \mathrm{Tr}\{\rho(t)[\mathcal{H}'_D(t) - \mathcal{H}'_D(t+\Delta t)]\}\rangle \ . \tag{3}$$

Using the expression (2) and introducing the autocorrelation function

$$G(t') = \langle \mathrm{Tr}\{\mathcal{H}'_D(0)\mathcal{H}'_D(t)\}\rangle \tag{4}$$

one gets

$$\lim_{\Delta t \to 0} \Delta E_D(t)/\Delta t = -\frac{\partial G(t')}{\partial t'}\Big|_{t'=0} \beta(t) \ . \tag{5}$$

Equalizing the expression (5) whith the time derivative of the dipolar energy, we get the kinetic equation for the spin–temperature of the dipolar energy reservoir

$$\frac{\partial \beta(t)}{\partial t} = \frac{1}{G(0)} \left(\frac{\partial G(t')}{\partial t'}\right)_{t'=0} \beta(t) \ . \tag{6}$$

The coefficient in this equation is just the dipolar spin–lattice relaxation rate

$$T_{1D}^{-1} = -\frac{1}{G(0)} \left(\frac{\partial G(t')}{\partial t'}\right)_{t'=0} \ . \tag{7}$$

This relation between the transport coeficient T_{1D} and the autocorrelation function G(t) is quite general and can be used for any kind of ultraslow motions. It helps us to study the short time behavior of the autocorrelation function of the dipolar coupling via T_{1D} measurements. One must not forget that this can be done only for $T_2 \ll \tau$, while for faster motions where $T_2 \simeq \tau$, the same information can be obtained from the free precession signal, for example with the Lowe et al. (1973) zero time resolution technique.

3. Application. Hexagonal ice crystal (for the structure see Peterson and Levy, 1957), can be approximately described by a dynamic Ising model in the disordered phase where each quasi spin represents a hydrogen bond between two oxygens. The model is not sensitive to the nature of the proton motions. We cannot discriminate between water molecule reorientations and intrabond proton jumps. Both can change the position of the proton in a bond which we shall describe by a two valued variable $p_i(t)$. According to Bernal and Fowler (1933) we assume that only two protons are near to each oxygen, which means that only half of $p_i(t)$ are independent. One can proof (Blinc and Žekš, 1972) that such a case can be nevertheless well described by an Ising model with properly selected coupling constants.

Expressing the time dependence of $F^{(0)}_{ij}(t)$ with variables $p_i(t)$

$$F^{(0)}_{ij}(t) = A_{0ij} + A_{1ij}\,p_i(t) + A_{1ji}\,p_j(t) + A_{2ij}\,p_i(t)\,p_j(t) \tag{8}$$

one gets for the autocorrelation function (4)

$$G(t) = \mathrm{Tr}\{C^2_{ij}\}\sum_{i\neq j}[A^2_{0ij} + A^2_{1ij}g_{ii}(t) + A^2_{1ji}g_{jj}(t) + 2A_{1ij}A_{1ji}g_{ij}(t) + 2A_{0ij}A_{2ij}g_{ij}(0) + A^2_{2ij}g^2_{ii}(t)] \tag{9}$$

where $g_{ij}(t) = \langle p_i(0)\,p_j(t)\rangle$. It is useful to introduce Fourier transformed variables $p_{\vec q}(t)$ whose autocorrelation functions have simple forms in (Suzuki and Kubo, 1966) the RPA approximation

$$\langle p_{\vec q}(0)\,p_{-\vec q}(t)\rangle = \langle p^2_{\vec q}\rangle \cdot e^{-t/\tau_{\vec q}}, \quad \text{where} \quad \tau_{\vec q} = \langle p^2_{\vec q}\rangle\tau \tag{10}$$

and τ would be the correlation time of an isolated hydrogen bond. Using this relation we get the following expression for the derivative of the G(t) at zero time

$$\frac{dG(t)}{dt}\Big|_{t=0} = \mathrm{Tr}\{C^2_{ij}\}\sum_{i\neq j}(A^2_{1ij} + A^2_{1ji} + 2A^2_{2ij})/\tau\,. \tag{11}$$

To obtain G(0) we simply assume that only protons bonded to the same oxygen are correlated. For such a proton pair $g_{ij}(t) = \pm 1/3$ while for the rest $g_{ij}(t) = 0$.

Evaluating the lattice summes and using the expression (7), we get $\tau = 0.4\ T_{1D}$ for the magnetic field which is orthogonal to the crystal c axes. Applying this relation to the preliminary T_{1D} measurements done by dr. G. Lahajnar in our laboratory we obtain $\tau = \tau_0 \exp(E_a/kT)$, where $\tau_0 = 2.3.10^{-17}$ s and $E_a = 0.6$ eV. The agreement of this two values with the two obtained from $T_{1\rho}$ measurements (G. Lahajnar et al. 1974) is very good. The value obtained for G(0) in our approximation gives for the second moment

$$M_2 = \frac{9}{16}\gamma^4\hbar^2\,G(0)/N\,\mathrm{Tr}\{C^2_{ij}\} = 30.5\ g^2\,, \tag{12}$$

which agrees with the data of Bruno and Pintar (1973).

4. Acknowledgements. The author wishes to thank to Prof. R. Blinc for helpful discussions.

5. References

Abragam A., 1960, Principles of Nuclear Magnetism, Oxford University Press, London.
Bernal J.D. and Fowler R.H., 1933, J. Chem. Phys. 1, 515.
Blinc R. and Žekš B., 1972, Adv. Phys. 21, 693.
Bruno G.F. and Pintar M.M., 1973, J. Chem. Phys. 58, 5344.
Goldman M., 1970, Spin temperature and Nuclear Magnetic Resonance in Solids, Clarendon, Oxford.
Lahajnar G., Zupančič I., Blinc R. and Gränicher, 1974, to be published.
Lowe I.J., Vollmers K.W. and Punkkinen M., 1973, Proceedings of the First Specialized "Collque AMPERE", Krakow.
Slichter C.P. and Ailion D.C., 1964, Phys. Rev. 135, A1099.
Suzuki M. and Kubo R., 1966, J. Phys. Soc. Japan 24, 51.

NMR AND ULTRA-SLOW MOTIONS IN CRYSTALS: A PERTURBATION APPROACH

D. Wolf and P. Jung
Max-Planck-Institut für Metallforschung, Institut für Physik, Stuttgart, Germany.

Abstract. A general perturbation formalism is developed to calculate the rotating-frame spin-lattice relaxation time $T_{1\rho}$ determined by the diffusion-induced time dependence of the dipolar spin-spin couplings. The theory includes strong collisions (so-called Slichter-Ailion theory).

I. Introduction. For the study of rather slow atomic motions in crystals spin-lattice relaxation has to be observed in a weak relaxation field which is usually accomplished by measuring the spin-lattice relaxation time $T_{1\rho}$ in the rotating frame (rf). The theoretical approaches to relate $T_{1\rho}$ to the temperature-dependent mean time τ between successive jumps of an atom (and thus to the self-diffusion coefficient) may be classified into two groups: (i) If the rotating-field amplitude H_1 is large compared to the local field $H_{L\rho}$ in the rf ($H_1 >> H_{L\rho}$) the dipolar Hamiltonian may be treated as a small perturbation on the Zeeman Hamiltonian in the rf [1]. (ii) In a weak relaxation field ($H_1 \lesssim H_{L\rho}$) the conventional perturbation method fails. According to Slichter and Ailion (SA) $T_{1\rho}$ may now be directly related to the energy change of the spin system due to jumping [2]. One of their basic assumptions is that $\tau >> T_m$, where T_m denotes the time constant characterizing the thermal mixing process of Zeeman and dipolar heat reservoirs (strong-collision region). For $H_1 >> H_{L\rho}$ T_m is very long and $\tau << T_m$ (weak-collision region). In the rest of this paper perturbation theory is applied to a modified rf Hamiltonian which allows for both weak and strong collisions.

II. Motion-Induced Spin-Lattice Relaxation in the Rotating Frame. At exact resonance the total spin Hamiltonian in the rf is given by [2]

$$H_\rho = H_{z\rho} + H_d^o(t) \tag{2.1}$$

with

$$H_{z\rho} = \gamma H_1 \sum_j I_{jz} \quad \text{and} \quad H_d^o = 1/2 \sum_i \sum_j A_{ij}^{(o)} F_{ij}^{(o)}(t) \tag{2.2}$$

(in units of $\hbar$). The operators $A_{ij}^{(o)}$ and the lattice functions $F_{ij}^{(o)}(t)$ are, in the nomenclature of Abragam [3],

$$A_{ij}^{(o)} = \gamma^2\hbar\, [I_{iz}I_{jz} - 1/4\ (I_i^+I_j^- + I_i^-I_j^+)]\ ,\quad F_{ij}^{(o)} = r_{ij}^{-3}(1-3\cos^2\theta_{ij}) \tag{2.3}$$

So far the Hamiltonian (2.1) has been used to calculate $T_{1\rho}$ in the weak collision (high-field) region [1]. To include strong collisions we modify (2.1) in the following way:

$$H_\rho = H_{o\rho} + H_{1\rho}(t) \tag{2.4}$$

with

$$H_{o\rho} = H_{z\rho} + f(\tau)\, H_d^{or\ell}\ ,\quad H_{1\rho}(t) = H_d^o(t) - f(\tau)\, H_d^{or\ell} \tag{2.5}$$

$H_d^{or\ell}$ denotes the secular part of the dipolar Hamiltonian in a rigid lattice. The function $f(\tau)$ is assumed to be zero for $\tau << T_m$ and one for $\tau >> T_m$. In the intermediate region $f(\tau)$ is assumed to vary steadily from zero to one. Since T_m is a constant for a given value of H_1 the parameter τ/T_m depends on temperature but not on time. As we see from (2.4) and (2.5) the choice of $f(\tau)$ allows to apply perturbation theory even in zero field H_1 for $\tau >> T_m$, since then $H_d^o(t) - H_d^{or\ell}$ is still a small perturbation on $H_d^{or\ell}$. On the other hand, for $\tau << T_m$ automatically the conventional perturbation treatment is recovered. From (2.4) $T_{1\rho}$ may be calculated using a method recently developed by Jacquinot and Goldman [4]:

$$1/T_{1\rho} = 1/[\mathrm{Tr}(H_{o\rho}^2)] \Big\langle \int_0^\infty [H_{o\rho}, \tilde{H}_{1\rho}(t)][H_{1\rho}(t'), H_{o\rho}]\, dt \Big\rangle_{t'} \tag{2.6}$$

where the brackets $\langle\ \rangle_{t'}$ indicate an average over the time t'. The interaction representation of $H_{1\rho}(t)$ has been introduced by

$$\tilde{H}_{1\rho}(t) = \exp(iH_{o\rho}t)H_{1\rho}(t)\exp(-iH_{o\rho}t) = 1/2 \sum_{ij}[F_{ij}^{(o)}(t) - f(\tau)F_{ij}^{(o)r\ell}]\tilde{A}_{ij}^{(o)}(t) \tag{2.7}$$

Here, (2.2) was used . Substituting (2.7) and (2.5) into (2.6) and using the equation of motion of $A_{ij}(t)$ we obtain after some calculation

$$1/T_{1\rho} = 1/[2\mathrm{Tr}(H_{o\rho}^2)] \sum_{ij}\{-[dG_{ij}(t)/dt]_{t=0}\mathrm{Tr}(A_{ij}^{(o)})^2 - \int_0^\infty d^2G_{ij}(t)/dt^2\, \mathrm{Tr}[\tilde{A}_{ij}^{(o)}(t)A_{ij}^{(o)}]dt\} \tag{2.8}$$

$\tilde{A}_{ij}^{(o)}(t)$ may be interpreted as a transformation of $A_{ij}^{(o)}(t)$ into the doubly rotating frame (rotating around $\vec{H}_{eff} = (H_0 - \omega/\gamma)\hat{z} + H_1\hat{x}$. Therefore we write

$$\tilde{A}_{ij}^{(\omega)}(t)=\sum_{p=-2}^{+2} a^{(p)} A_{ij}^{(p)\prime}(t)\exp[i\omega^{(p)}t] \ , \ A_{ij}^{(p)\prime}(t)=\exp[if(\tau)H_{d\rho}^{or\ell}t]A_{ij}^{(p)\prime}\exp[-if(\tau)H_{d\rho}^{or\ell}t] \quad (2.9)$$

(The prime indicates the representation in the doubly rotating frame). The coefficients $a^{(p)}$ have been calculated by Douglass and Jones [5] for arbitrary orientation of the doubly rf with respect to the rf. At exact resonance they are [5]

$$a^{(o)}=1/2 \ , \ a^{(\pm1)}=0 \ , \ a^{(\pm2)}=-1/2; \text{ also: } \omega^{(o)}=0; \ \omega^{(\pm1)}=\pm\omega_1; \ \omega^{(\pm2)}=\pm2\omega_1 \quad (2.10)$$

Inserting (2.9) into (2.8) and replacing $A_{ij}^{(o)}$ by $\sum_{-2}^{+2} a^{(p)}A_{ij}^{(p)}$ [see Eq.(2.9)] the correlation function $Tr(A_{ij}^{(p)}(t)\ A_{ij}^{(p)})$ over spin operators enters the second term on the right of (2.8). This function describes the process of establishing a common spin temperature of the spin system through thermal mixing between Zeeman and dipolar Hamiltonian. Therefore we make the following assumption:

$$Tr(A_{ij}^{(p)\prime}(t)\ A_{ij}^{(q)\prime}) = Tr(A_{ij}^{(p)}\ A_{ij}^{(q)})\ \exp(-t/Tm) \quad (2.11)$$

After evaluation of the traces involved we obtain from (2.8) to (2.11)

$$\frac{1}{T_{1\rho}}=\frac{\gamma^2\hbar^2 I(I+1)}{H_1^2+f(\tau)H_{L\rho}^2}\{-\frac{1}{4}(\frac{dG}{dt})_{t=0}-\frac{1}{16}\int_0^\infty\frac{d^2G(t)}{dt^2}e^{-\frac{t}{Tm}}dt-\frac{3}{32}\int_{-\infty}^{\infty}\frac{d^2G(t)}{dt^2}\exp[-(\frac{1}{Tm}-2i\omega_1)t]dt\} \quad (2.12)$$

where the correlation function G(t) was introduced by $G(t)=1/N\ \sum\sum G_{ij}(t)$ which was calculated earlier by Wolf for random-walk diffusion [6] and for point-defect mechanisms of self-diffusion [7] in crystals. (2.12) is valid for any source of the time dependence of the dipolar interactions (like, e.g., molecular reorientations or diffusion), since no specific assumptions concerning the form of G(t) have been made except $G(\infty)=0$ and $dG/dt|_\infty=0$.

III. Discussion. From (2.12) the limiting cases of weak and strong collisions may easily be derived.

1. Weak Collisions. For $\tau<<Tm$ the function $\exp(-t/Tm)$ remains practically unchanged during a time interval of several τ, where G(t) faces its major time dependence. With $f(\tau)=0$ (see Sect.II) and $\exp(-t/Tm)\approx1$ we thus obtain from (2.12)

$$(\frac{1}{T_{1\rho}}) = \frac{\gamma^2\hbar^2 I(I+1)}{H_1^2}\{-\frac{1}{4}(\frac{dG}{dt})_{t=0}-\frac{1}{16}\int_0^\infty\frac{d^2G}{dt^2}dt-\frac{3}{16}\int_0^\infty\frac{d^2G}{dt^2}e^{2i\omega_1 t}dt\} \quad (3.1)$$

Integrating the third term in (3.1) by parts and taking the real part we finally have

$$1/T_{1\rho} = 3/8\cdot\gamma^4\hbar^2 I(I+1)\ J^{(o)}(2\omega_1) \quad (3.2)$$

where $J^{(o)}(\omega)$ is the Fourier-transform of the correlation function G(t). (3.2) is identical with the result obtained by Look and Lowe [1] and Douglass and Jones[5]

2. Strong Collisions. For $\tau>>Tm$ the function d^2G/dt^2 in (2.12) may be regarded as constant while $\exp(-t/Tm)$ decays to zero. Assuming further that $\omega_1 Tm<<1$ we can write, with $f(\tau)=1$, instead of (2.12)

$$(\frac{1}{T_{1\rho}})_{SA} = \frac{\gamma^2\hbar^2 I(I+1)}{H_1^2+H_{L\rho}^2}\{-\frac{1}{4}(\frac{dG}{dt})\Big|_{t=0}-\frac{1}{4}(\frac{d^2G}{dt^2})\Big|_{t=0}\int_0^\infty e^{-\frac{t}{Tm}}dt\} \quad (3.3)$$

If the correlation time between changes of G(t) is denoted by τc we can approximate $(dG/dt)_{t=0}$ by $(dG/dt)_{t=0} = -[G(o)-G(\tau c)]/\tau c$. A similar approximation applied to (d^2G/dt^2) shows that the second term in (3.3) is smaller by a factor $Tm/\tau c<<1$ than the first and we obtain

$$(1/T_{1\rho})_{SA}=H_{L\rho}^2/(H_1^2+H_{L\rho}^2)\cdot[G(o)-G(\tau c)]/[G(o)\cdot\tau c]=H_{L\rho}^2/[H_I^2+H_{L\rho}^2]\cdot(1-p)/\tau c \quad (3.4)$$

where the local field has been introduced by $H_{L\rho}^2=1/(4N)\gamma^2\hbar^2 I(I+1)\ \sum\sum(F_{ij}^{(o)}(o))^2$. Since G(o) is proportional to the initial energy of the spin system before a jump, $(Ed)_i$, and $G(\tau_c)=1/N\ \sum\sum F_{ij}^{(o)}(o)F_{ij}^{(o)}(\tau_c)$ characterizes the dipolar energy after a jump, $(Ed)_f$, we see that for a random-walk diffusion mechanism ($\tau_c=\tau/2$, see, e.g.,[6]) (3.4) is identical with the SA-result [8]. For point-defect mechanisms of self-diffusion the correlation function calculated by Wolf [7] may be inserted into (3.4). Thus a result is obtained from (3.4) which is identical with Wolf's result in the SA-region for $\tau_d<Tm<\tau$, where τ_d denotes the mean time between consecutive jumps of a point defect.

Acknowledgement: The authors appreciate helpful discussions with Prof.A.Seeger.

References. [1] D.C.Look and I.J.Lowe, J.ChemPhys. 44, 2295 (1966)
[2] D.C.Ailion, in "Advances in Magnetic Resonance" 5, 177 (1972)
[3] A.Abragam, "The Principles of Nuclear Magnetism", Clarendon Press,Oxford 1972
[4] J.F.Jacquinot and M.Goldman, Phys.Rev. 8, 1944 (1973)
[5] D.C.Douglass and G.P.Jones, J.ChemPhys. 45, 956 (1966)
[6] D.Wolf, J.Magn.Res., Nov.1974; [7] D.Wolf, Phys.Rev.B, 1 Sept.1974(paper 1)
[8] D.Wolf, Phys.Rev.B, 1 Sept. 1974 (paper 2)

DETERMINATION OF RELAXATION PARAMETERS IN COUPLED NUCLEAR SPIN SYSTEMS BY COMPLETE SETS OF OVERHAUSER EXPERIMENTS

A. Höhener, G. Bodenhausen and R.R. Ernst
Laboratorium für physikalische Chemie, Eidg. Technische Hochschule, 8006 Zürich

Abstract. A direct, non-iterative method is described for the determination of relaxation parameters in coupled nuclear spin systems based on a complete set of Overhauser experiments. The relaxation parameters are calculated by means of a least squares procedure.

A procedure is proposed to determine the spin-lattice relaxation parameters in coupled nuclear spin systems. It is known that spin-lattice relaxation in a coupled spin system is determined by a set of coupled differential equations. The observable signal intensities are, in general, non-exponential functions of time. Conventional, non-selective inversion-recovery or saturation-recovery experiments are, therefore, not suited for the investigation of coupled spin systems. Tedious iterative simulations of recovery curves would be necessary to determine the relaxation parameters and often no unique solution can be found. It is however possible to determine relaxation parameters from the initial rates in selective spin-inversion experiments.

Another possibility is to derive the relaxation parameters from the steady state intensities in a selective or non-selective saturation experiment. A non-selective saturation experiment, known under the name "progressive saturation" in Fourier spectroscopy, does not produce sufficient information to deduce all relaxation parameters, in general. Selective saturation or Overhauser experiments, on the other hand, are informative enough to permit the computation of all parameters in a direct, non-iterative manner.

The proposed procedure is based on a complete set of Overhauser experiments. In each experiment, one particular line is saturated and all resulting line intensities are measured. For a system consisting of N_L non-overlapping lines, up to N_L^2 signal intensities can be measured. But in each of the N_L spectra, only (2^n-1) signal intensities are independent (for n spins 1/2). It is, therefore, at first necessary to adjust the measured signal intensities considering the various sum relations among them. In an arbitrary manner, (2^n-1) lines with independent intensities $\underset{\sim}{\Lambda}$ are selected and the remaining intensities expressed as linear combinations of them.

$$\Delta\underset{\sim}{L} = A\,\Delta\underset{\sim}{\Lambda} \qquad (1)$$

with $\Delta L_j = (L_j - L_{oj})/L_{oj}$ being the normalized deviations from thermal equilibrium. A least squares procedure delivers then the adjusted intensity deviations $\Delta\underset{\sim}{\Lambda}$.

$$\Delta\underset{\sim}{\Lambda} = (A^T C_L^{-1} A)^{-1} A^T C_L^{-1} \Delta\underset{\sim}{L} \quad . \qquad (2)$$

The intensity **deviations** can also be expressed by the population deviations $\Delta\underset{\sim}{P} = \underset{\sim}{P} - \underset{\sim}{P}_o$, $\Delta\underset{\sim}{\Lambda} = T\,\Delta\underset{\sim}{P}$, where T is a $(2^n-1) * 2^n$ matrix. An inversion of the following special form, which considers the auxilary condition $\sum_j \Delta P_j = 0$, delivers

$$\Delta\underset{\sim}{P} = \begin{bmatrix} T \\ 1 \ldots\ldots 1 \end{bmatrix}^{-1} \cdot \begin{pmatrix} \Delta\underset{\sim}{\Lambda} \\ 0 \end{pmatrix} \quad . \qquad (3)$$

The N_L population vectors of the N_L Overhauser experiments, distinguished by an upper index (l), form the basis for the computation of the relaxation parameters. The population vector $\underset{\sim}{P}^{(l)}$ is given by the master equation in dynamic equilibrium

$$d/dt(\underset{\sim}{P}^{(l)}) = 0 = W\ (\underset{\sim}{P}^{(l)} - \underset{\sim}{P}_o) + S^{(l)}\ \underset{\sim}{P}^{(l)} \quad . \tag{4}$$

W is the relaxation matrix of the spin system which contains the unknown relaxation parameters and $S^{(l)}$ is the saturation matrix which characterizes the particular saturation experiment. It is assumed that line l, which connects the energy levels n and m, is saturated, exclusively. The relaxation matrix W is linear in the various relaxation parameters R_j and can be written in the form

$$W = \sum_j W^{(j)}\ R_j \tag{5}$$

where $W^{(j)}$ are coefficient matrices given by the structure of the spin systems and by the particular relaxation mechanism. They can easily be determined. It is most important to notice that Eq.(4), together with Eq.(5), forms a linear, overdetermined system of equations for the desired relaxation parameters R_j

$$0 = \sum_j R_j \sum_k W_{ik}^{(j)}\ \Delta P_k^{(l)} + \sum_k S_{ik}^{(l)}\ P_k^{(l)} \quad . \tag{6}$$

$W^{(j)}$ and $S^{(l)}$ are singular matrices (caused by the invariance of the total population). It is therefore possible to eliminate one equation for each l. The equation with i=m will be selected. The remaining equations are of the form

$$0 = \sum_j M_{ij}^{(l)}\ R_j + V_i^{(l)} \qquad \text{for } i = 1,..,N_P, i \neq m \text{ and } l = 1,..,N_L \tag{7}$$

with

$$M_{ij}^{(l)} = \sum_k W_{ik}^{(l)}\ \Delta P_k^{(l)}$$

$$V_i^{(l)} = -\ s^{(l)}\ (P_n^{(l)} - P_m^{(l)})\ \delta_{i,n}$$

$$= -\ s^{(l)}\ ((\Delta P_n^{(l)} - \Delta P_m^{(l)}) + (P_{on} - P_{om}))\ \delta_{i,n} \tag{8}$$

where $s^{(l)}$ is the square of the effective rf-field acting on line l (including the corresponding transition matrix element). This system of equations can again be solved by the method of least squares with the result

$$\underset{\sim}{R} = (\ M^T C_{MV}^{-1}\ M\ )^{-1}\ M^T C_{MV}^{-1}\ \underset{\sim}{V} \quad . \tag{9}$$

From Eq.(7) it is clear that the experimental values appear in the inhomogenous part $\underset{\sim}{V}$ as well as in the coefficient matrix M. Both determine the covariance matrix C_{MV}.

This method has the great advantage that no iterations are involved. Each experiment delivers an unique solution. On the other hand, this procedure has the disadvantage that no limits for the expected relaxation parameters can be set in advance. An extension involving linear programming technique could easily eliminate this problem.

ON ANISOTROPY OF NUCLEAR RELAXATION TIME

G.R.Khutsishvili and B.D.Mikaberidze
Institute of Physics, Academy of Sciences of the Georgian SSR, Tbilissi, USSR

Abstract. The nuclear relaxation time is calculated taking into account the tensoric nature of the g factor of the magnetic ion.

The relaxation of the matrix nuclei in a nonmetallic diamagnetic crystal with a magnetic impurity occurs due to the (d-d) interaction of nuclear and impurity ion spins. The time direct relaxation of the nucleus T_{dir} was calculated in many works (see e.g. Khutsishvili, 1965). It was assumed that the g-factor of the ion is isotropic or the external field H is directed along a principal direction of the tensor $\hat{g}$. It was obtained

$$[T_{dir}(r)]^{-1} = \frac{2S(S+1)(g\beta\gamma_I)^2\tau}{5[1+(\omega_I\tau)^2]r^6}, \tag{1}$$

where β is the Bohr magneton, γ_I is the gyromagnetic ratio of the nucleus, $\omega_I = \gamma_I H$, S is the spin of the impurity ion (i.e. the electron spin), τ is the time of correlation of its longitudinal component, $\vec{r}$ is the vector connecting the nucleus with the magnetic ion.

In this paper it is assumed that the g-factor of the magnetic ion is a symmetric tensor and the case of an arbitrary orientation of the field is considered. A set of axes x, y, z is introduced along the principal directions of the tensor (the principal values are g_x, g_y, g_z). Direction cosines of the external field are l_x, l_y, l_z. The electron Zeeman Hamiltonian has the form

$$\beta H(g_x l_x S_x + g_y l_y S_y + g_z l_z S_z). \tag{2}$$

We introduce the quantity

$$g = (g_x^2 l_x^2 + g_y^2 l_y^2 + g_z^2 l_z^2)^{1/2}. \tag{3}$$

In the case of axial symmetry of the tensor $\hat{g}$ we have $g = (g_{\parallel}^2 \cos^2\vartheta + g_{\perp}^2 \sin^2\vartheta)^{1/2}$, where ϑ is the angle between the symmetry axis and the field.

The electron spin $\vec{S}$ is quantized along the axis ζ with the direction cosines $g_x/g\,l_x$, $g_y/g\,l_y$, $g_z/g\,l_z$ (Abragam and Bleaney, 1970). Further we are interested in nuclei located sufficiently far from the magnetic ion, hence $\hbar\gamma_I H$ considerably exceeds the nucleus-ion d-d interaction energy. Because of that the nuclear spin is quantized on the direction of the external field.

The energy of interaction d-d can be written in the form $\hbar\vec{I}\hat{A}\vec{S}$, where

$$A_{ik} = \frac{\beta\gamma_I}{r^3}(3n_i n_l g_{lk} - g_{ik}) \tag{4}$$

($i\,k\,l$ run the values $x\,y\,z$, summation is performed over the repeated index, $\vec{n} = \vec{r}/r$).

The Hamiltonian of the system: "an electron spin + nucleus" can be written as (the fine structure and the quadrupole interaction are neglected)

$$\mathcal{H} = g\beta H S_\zeta - \hbar\gamma_I \vec{H}\vec{I} + \hbar\vec{I}\hat{A}\vec{S}. \tag{5}$$

Since $g\beta H \gg A$ it is possible in the perturbation energy ($\hbar\vec{I}\hat{A}\vec{S}$) to limit oneself by taking into account the terms proportional to S_ζ. For the correlator of S_ζ an exponentional approximation is used. The calculations of the relaxation probability gives (averaging is carried out over the directions of the vector $\vec{n}$, see details of calculations in Mikaberidze and Khutsishvili, 1974)

$$[T_{dir}(r)]^{-1} = \frac{1}{15}S(S+1)\left(\frac{\beta\gamma_I}{r^3}\right)^2 \frac{\tau}{1+(\omega_I\tau)^2}\left[7g^{-2}(g_x^4 l_x^2 + g_y^4 l_y^2 + g_z^4 l_z^2) - g^2\right]. \tag{6}$$

In the axially symmetric case the square bracket in (6) gives

$$[7g^{-2}(g_{\parallel}^4\cos^2\vartheta + g_{\perp}^4\sin^2\vartheta) - g^2].$$

Further we limit ourselves by the case of a cubic crystal. Then the tensor of the nuclear spin diffusion coefficient is reduced to the scalar $\mathcal{D}$. Using (6) and the theory of spin diffusion (Khutsishvili, 1965), it is possible to obtain the expression for the relaxation time T_n, of the longitudinal component of the nuclear magnetic moment of the specimen.

Let us consider the case of diffusion limited ($\mathcal{D}L$) relaxation i.e. we consider that $\delta \ll \beta$ (δ is the radius of the diffusion barrier, β is the radius of the so called pseudopotential). In addition we consider that reorientations of electron spins are caused by spin-lattice relaxation, then $\tau = \tau_\ell$ (τ_ℓ is the electron spin-lattice relaxation time).

It is known that from $[T_{dir}(r)]^{-1} = Cr^{-6}$ it follows that $T_n^{-1} = 8,5\, N C^{1/4} \mathcal{D}^{3/4}$ (N is the number of impurity atoms per a volume unit). Thus

$$T_n^{-1} = 8,5\, N \mathcal{D}^{3/4} (\beta\gamma_I)^{1/2} \left\{ \frac{S(S+1)}{15} \frac{\tau_\ell}{1+(\omega_I\tau_\ell)^2} \left[7g^{-2} (g_x^4 \ell_x^2 + g_y^4 \ell_y^2 + g_z^4 \ell_z^2) - g^2 \right] \right\}^{1/4}. \quad (7)$$

If impurity spins occupy several non equivalent positions in the specimen, it is necessary to make averaging over these non equivalent ions in (7).

The relaxation of the nuclei F^{19} in $CaF_2:U^{3+}$ was studied at 80K and 300K, at the frequency 28.7MHz and for the field orientations [100] and [111] (Leppelmeier and Jeener,1968). The concentration of impurities corresponded to one uranium atom per $9 \cdot 10^3$ fluorine atoms. Under such conditions $\tau = \tau_\ell$ and, in addition, the case of the $\mathcal{D}L$ relaxation takes place.

The centre U^{3+} in CaF_2 is tetragonal, therefore the tensor $\hat{g}$ has axial symmetry. There are three types of centres with the axes along the three cubic axes. In the case $H \parallel [111]$ ϑ equals to the magic angle for all the three types of centers. And in the case $H \parallel [100]$ $\vartheta = 0$ ($g = g_\parallel$) for one third of the centres and $\vartheta = \pi/2$ ($g = g_\perp$) for the two thirds.

Measurements of the time behaviour of the specimen nuclear magnetic moment were made after strong saturation of fluorine nuclei in $CaF_2:U^{3+}$. At large t the recovery proceeds according to an exponential law. On the other hand, it is known (Blumberg,1960) that the magnetic moment at small t is proportional to $(Ct)^{1/2}$. Thus, it is possible to determine both T_n and C from the experimental data. However, note that the measured value of T_n^{-1} must be proportional to the mean (over the non-equivalent types of ions) value of $C^{1/4}$, while the magnetic moment at small t is proportional to the mean value of $C^{1/2}$.

If to assume that τ_ℓ and $\mathcal{D}$ are isotropic, there is a discrepancy between the theory and experiment. If there were data on the angular dependence of τ_ℓ, one might make a complete comparison of the theory and the experiment. It is possible to obtain from formula (6) the expression for the ratio of the mean values of $C^{1/2}$ for the field orientations [100] and [111]. Such an expression contains the values of τ_ℓ for ϑ equal to zero, $\pi/2$ and to the magic angle respectively. The same ratio can be obtained by measuring the magnetic moment at small t. Then, using formula (7) and the experimental data on the angular dependence of T_n and τ_ℓ, it is possible to determine the anisotropy of $\mathcal{D}$.

References

Abragam, A. and Bleaney B., 1970, Electron Paramagnetic Resonance of Transition Ions, Clarendon Press, Oxford.

Blumberg W.E., 1960, Phys.Rev. 119, 79.

Khutsishvili G.R., 1965, Uspekhi Fiz.Nauk 87, 211

Leppelmeier G.W. and Jeener J., 1968, Phys.Rev. 175, 498

Mikaberidze B.D. and Khutsishvili G.R., 1974, Fiz.Tver.Tela 16, 2376

CAN THE RELAXATION TIME T_1 BE SHORTER THAN T_2 ?

J.S. Blicharski
Institute of Physics, Jagiellonian University, Cracow
Institute of Nuclear Physics, Cracow, Poland.

Abstract. Nuclear magnetic relaxation times are calculated in the weak collision case in the presence of anisotropy of the chemical shift. It is shown that in the case of sufficiently high asymmetry of the chemical shift tensor the spin-lattice relaxation time T_1 can be shorter than spin-spin relaxation time T_2.

According to the theory of nuclear magnetic resonance (Abragam, 1961), spin-lattice relaxation time T_1 should be longer than spin-spin relaxation time T_2. The purpose of this paper is to show that in some cases the classical inequality $T_1 \geqslant T_2$ can be broken.

Consider a system of equivalent nuclei of spin I=1/2, magnetic moment $\mu = \gamma \hbar I$, which are singly located in diamagnetic molecules of a liquid. Moreover, we assume that electron screening of the resonant nuclei plays an important role whereas other nuclei of the molecules are non-magnetic or their magnetic moments are zero or small enough for dipolar interaction to be neglected. If the considered system is subjected to a sufficiently strong external magnetic field H then the anisotropy of the chemical shift may become a dominant interaction in the nuclear relaxation process. The anisotropy of the chemical shift causes a change of energy of Zeeman interaction between the nuclei and external magnetic field which may be expressed in the form

$$\Delta W = \mu \cdot \sigma \cdot H \equiv \sum_{\alpha,\beta} \mu_\alpha \sigma_{\alpha\beta} H_\beta ,$$

where $\sigma_{\alpha\beta}$ is the chemical shift tensor (screening tensor) which in general is the non-symmetric one (Buckingham et al. 1963, 1971). The asymmetry may be easily seen if we note that

$$\sigma_{\alpha\beta} = \frac{\partial^2 W}{\partial \mu_\alpha \partial H_\beta} \neq \frac{\partial^2 W}{\partial H_\alpha \partial \mu_\beta} = \sigma_{\beta\alpha} .$$

The non-symmetric tensor $\sigma_{\alpha\beta}$ may be separated into a symmetric part $\sigma^s_{\alpha\beta}$ and an antisymmetric part $\sigma^a_{\alpha\beta}$,

$$\sigma_{\alpha\beta} = \sigma^s_{\alpha\beta} + \sigma^a_{\alpha\beta} , \quad \text{where} \quad \sigma^s_{\alpha\beta} = \frac{\sigma_{\alpha\beta}+\sigma_{\beta\alpha}}{2} , \quad \sigma^a_{\alpha\beta} = \frac{\sigma_{\alpha\beta}-\sigma_{\beta\alpha}}{2} .$$

In the measurements of NMR spectra only symmetric part $\sigma^s_{\alpha\beta}$ plays an important role (Griffin et al., 1972), whereas the antisymmetric part is not observable there. However, the antisymmetric part of the screening tensor can produce an additional contribution to the relaxation rate which, conversely to the symmetric part contribution (McConnel, Holm, 1956), can lead to $T_1 < T_2$.

One should note that inequality $T_1 < T_2$ may appear also in a system of inequivalent spins in a liquid, where cross relaxation causes an increase in the effective value T_2 (Vold, Chan, 1972). However this effect depends on the distance between π pulses in Carr-Purcell method which is used for T_2 measurements and at a sufficiently high density of π pulses one gets $T_1 = T_2$. The relaxation time T_1 can be also shorter than T_2 in the case of anisotropic ferromagnetic substances. In this paper we consider the relaxation in isotropic diomagnetic liquid.

Using the method presented elsewhere (Blicharski, 1972), one can get the following expressions for the relaxation times T_1 and T_2 in the case of weak collision and in the presence of isotropic molecular motion:

$$\frac{1}{T_1} = \frac{1}{2}\omega_0^2(\delta\sigma)^2 \frac{\tau_c}{1+9\omega_0^2\tau_c^2} + \frac{2}{15}\omega_0^2(\Delta\sigma)^2 \frac{\tau_c}{1+\omega_0^2\tau_c^2} ,$$

$$\frac{1}{T_2} = \frac{1}{4}\omega_0^2(\delta\sigma)^2 \frac{\tau_c}{1+9\omega_0^2\tau_c^2} + \frac{1}{45}\omega_0^2(\Delta\sigma)^2\left[4\tau_c + \frac{3\tau_c}{1+\omega_0^2\tau_c^2}\right] ,$$

where

$$(\Delta\sigma)^2 = (\sigma_{zz}-\sigma_{xx})(\sigma_{zz}-\sigma_{yy}) + (\sigma_{xx}-\sigma_{yy})^2 + 3\left[(\sigma_{xy}^s)^2 + (\sigma_{xz}^s)^2 + (\sigma_{yz}^s)^2\right] ,$$

$$(\delta\sigma)^2 = (\sigma_{xy}-\sigma_{yx})^2 + (\sigma_{xz}-\sigma_{zx})^2 + (\sigma_{yz}-\sigma_{zy})^2 ,$$

τ_c is the correlation time, $\omega_0 = \gamma H$ is the Larmor frequency and $\Delta\sigma$ and $\delta\sigma$ are the anisotropy and asymmetry parameters of screening tensor respectively. In the case of extreme narrowing $\omega_0\tau_c \ll 1$, which takes place in liquids of low viscosity, one can get:

$$\frac{1}{T_1} = \frac{1}{2}\omega_0^2\tau_c\left[(\delta\sigma)^2 + \frac{4}{15}(\Delta\sigma)^2\right] ,$$

$$\frac{1}{T_2} = \frac{1}{2}\omega_0^2\tau_c\left[\frac{1}{2}(\delta\sigma)^2 + \frac{14}{45}(\Delta\sigma)^2\right] .$$

From the above expressions it follows that:

$$\frac{7}{6}T_1 \leqslant T_2 \leqslant 2T_1 .$$

This means that under appropriate conditions, if $(\delta\sigma)^2 > \frac{4}{45}(\Delta\sigma)^2$ one can get $T_1 < T_2$. It seems that such situations may appear for ^{19}F, ^{13}C, ^{31}P, and some heavy nuclei located in unsymmetrical and non-magnetic molecular environments. The asymmetry parameter should be different from zero for the following points group appropriate to the shielded nucleus (Buckingham, Malm, 1971): C_n, C_{nh} (n = 1, 2, 3, 4, 6) and S_2, S_4, S_6.

Investigation of the relaxation processes for the above systems may be a basis for discovery of the antisymmetric part of the screening tensor.

References

Abragam, A., 1961, Principles of Nuclear Magnetism, Clarendon Press, Oxford.
Blicharski, J.S., 1972, Z. Naturforsch., 27a, 1456.
Blicharski, J.S., 1972, Acta Phys. Polon., A41, 223.
Buckingham, A.D., Pople, J.A., 1963, Trans. Farad. Soc., 59, 2421.
Buckingham, A.D., Malm, S.M., 1971, Mol. Phys., 22, 1127.
Griffin, R.G., Ellet, J.D., Jr., Mehring, M., Bullit, J.G., Waugh, J.S., 1972, J. Chem. Phys., 57, 2147.
McConnel, H.M., Holm, C.H., 1956, J. Chem. Phys., 25, 1289.
Vold, R.L., Chan, S.O., 1972, J. Chem. Phys., 56, 28.

STATISTICAL TECHNIQUE FOR THE SYNTHESIS OF NMR LINESHAPES IN GLASSES

G. E. Peterson and C. R. Kurkjian
Bell Laboratories, Murray Hill, New Jersey 07974.

Abstract. A more systematic procedure for the synthesis of NMR spectra in amorphous materials has been developed. The randomness in the structure is represented by a joint probability density function in terms of either (1) the parameters in the resonance equation such as quadrupole coupling constants or (2) molecular parameters such as geometry or bonding orbitals. Once the density function is specified the lineshape may be calculated by mapping into a coordinate frame in which one of the axes is the resonance frequency.

We represent the collection of all geometries of atoms about the resonance center in the glass by Ω, and each individual geometry as ω_i. Here Ω is the sample space of our probability system. We define on Ω the k dimensional random vector $\vec{x}$ whose components are the assumed known random quantities. The vector $\vec{x}$ designates a mapping from the sample space Ω into Euclidean k space and assigns a particular point in k space to each sample point ω_i in Ω. The components of the random vector $\vec{x}$ can be described by a joint probability density function $p_{\vec{x}}(\vec{\alpha})$. As information is desired concerning the resonance frequency ν_r, the random vector $\vec{x}$ is transformed into a new random vector $\vec{y}$, one of whose components is ν_r. The joint probability density function, $p_{\vec{y}}(\vec{\beta})$ of the random vector $\vec{y}$ can be integrated as many times as is necessary to eliminate all random quantities except ν_r. Thus:

$$p_{\nu_r}(\gamma) = \int \cdots\cdots \int p_{\vec{y}}(\vec{\beta})\,d\beta_1 \cdots\cdots d\beta_{k-1}.$$

To determine the density function $p_{\vec{y}}(\vec{\beta})$ the standard techniques for reversible transformation of random vectors are employed. If random vectors $\vec{x}$ and $\vec{y}$ are related by a 1:1 transformation:

$$\vec{y} = \vec{f}(\vec{x});\ \vec{x} = \vec{g}(\vec{y})$$

the density functions are related by:

$$p_{\vec{y}}(\vec{\beta}) = p_{\vec{x}}[\vec{g}(\vec{\beta})]\cdot|J_{\vec{g}}(\vec{\beta})|$$

where $|J_{\vec{g}}(\vec{\beta})|$ is the absolute value of the Jacobian associated with the transformation g.

As an elementary illustration, consider ^{11}B, (I = 3/2) in a glass with $\eta = 0$. A second order quadrupole interaction is assumed and the quadrupole coupling constant, $\mathcal{Q} \equiv e^2qQ/h$ is random and described by a probability density function $p_{\mathcal{Q}}(\gamma)$. We easily obtain the following lineshape equations:

$$p_{\nu_r}(\gamma) = \int_{\alpha=0}^{\alpha=1} \frac{p_{\mathcal{Q}}(\beta)}{\left[\left(\frac{\nu_\ell-\gamma}{\nu_\ell}\right)\left(\frac{3}{64}\right)(1-\alpha^2)(9\alpha^2-1)\right]^{\frac{1}{2}}}\,d\alpha$$

where:

$$\beta = \left[\frac{\left(\frac{64}{3}\cdot\nu_\ell\right)\left(\nu_\ell-\gamma\right)}{(1-\alpha^2)(9\alpha^2-1)}\right]^{\frac{1}{2}} \qquad \nu_\ell \equiv \text{Larmor frequency}$$

These may readily be evaluated by means of a digital computer. A typical lineshape and its derivative spectrum for ^{11}B in glass is shown in Figure 1 (B) and (C). These were calculated from the above expression using the reasonably density function for $\mathcal{Q}$ shown in (A). Of course the same procedure may be applied to $\eta \neq 0$, or to randomness in other variables such as ν_ℓ.

In order to characterize the glass structure it is necessary to relate the observed lineshape, through its density function to the structural parameter whose randomness is actually responsible for the spectral broadness. For example, we might try to relate geometrical distortions, such as site to site bendings and stretchings to a particular density function for $\mathcal{Q}$ and/or η. This is best done using SCF methods as Townes and Dailey theory does not appear accurate enough to properly estimate $\mathcal{Q}$ and η for distorted molecules. There are a number of reasons for this. One which may be overlooked is that the electron cloud in a distorted molecule need not lie exactly along the geometric bond. In fact, it probably tends to curve away from it. Thus, the wave function for a distorted molecule is not simply a group of hybrids that point along the geometric bonds, but may be something rather different. A second difficulty is that the Townes and Dailey method cannot easily handle stretching distortions and these are likely to be of importance in glass.

Figure 2 shows a typical result of some SCF molecular wave function calculations in a Gaussian basis set appropriate to borate glasses. This work was undertaken in collaboration with L. C. Snyder of our laboratory. The particular molecule examined is BH_3 and it is a first approximation to 3 coordinated boron in glass. The electric field gradient is expressed in a.u.. Perhaps the most interesting observation is that the electric field gradient is quite insensitive to molecular bending. This suggests that ^{11}B NMR can tell us nothing about such distortions, if the model considered is reasonable.

The statistical technique may also be applied to electron spin resonance. For example, if $g_{\parallel}$ and $g_{\perp}$ are assumed to be random and described by a joint probability density function $p_{g_\parallel g_\perp}(\alpha,\beta)$ the lineshape as a function of g can also easily be obtained. Thus:

$$p_g(\gamma) = \iint \frac{\gamma p_{g_\parallel g_\perp}(\alpha,\beta)}{[(\alpha^2-\beta^2)(\gamma^2-\beta^2)]^{\frac{1}{2}}}\, d\alpha d\beta .$$

The lower limit on both integrations is γ while the upper limits are chosen so that outside this region the contributions from the density function are nil.

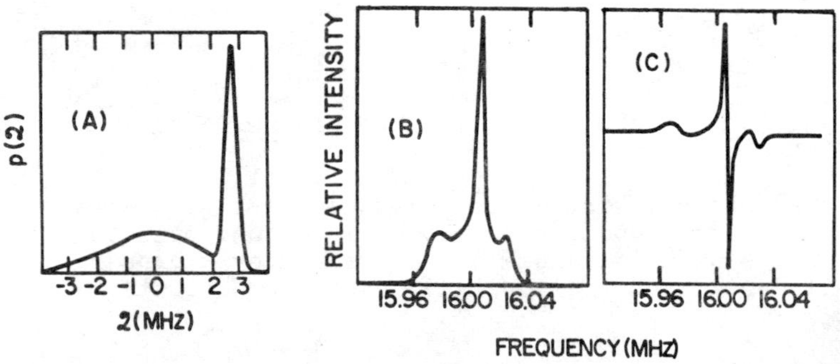

FIG. 1

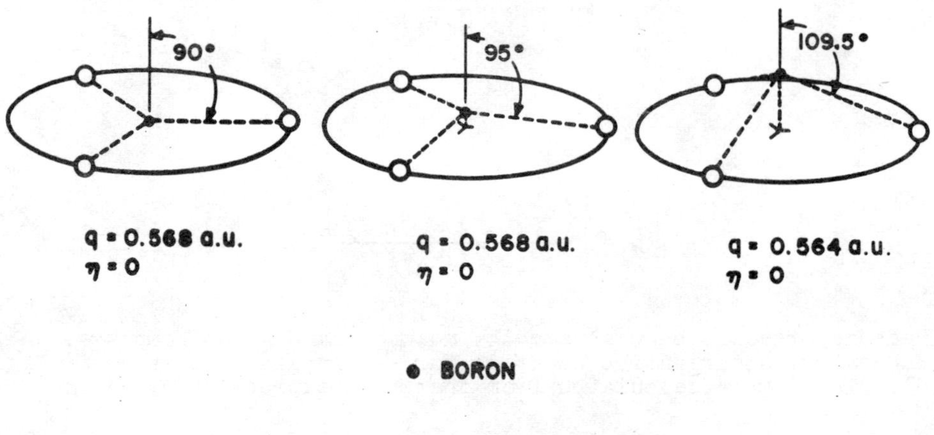

FIG. 2

CALCULATION OF ADIABATIC RAPID PASSAGE LINE SHAPES

Alexander J. Vega and Daniel Fiat

Department of Structural Chemistry, Weizmann Institute of Science
Rehovot, Israel.

If a magnetic resonance spectrometer operates with the field modulation technique, the signal is detected by the use of two phase detectors: the radio frequency detector and the modulation frequency detector. Thus, by the adjustment of the two reference phases four basic signal modes are detectable, i.e. 0° - 0° (dispersion - in phase), 0° - 90° (dispersion - out of phase), 90° - 0° (absorption - in phase), and 90° - 90° (absorption - out of phase). The line shape behavior of these signals is well known for the extreme cases of low modulation frequency (slow passage) and high modulation frequency (side bands). Recently it has been experimentally demonstrated that the signals with the modulation phase detector adjusted to 90°, show interesting features at intermediate values of the modulation frequency (rapid passage). In this paper a theoretical formalism for the description of the various detection modes is presented. The stochastic Liouville approach is employed to treat the effect of random molecular motions.

THE EFFECT OF LATTICE VIBRATIONS ON THE INTERSEGMENT CONTRIBUTION TO THE N.M.R SECOND MOMENT

M. Polak, U. Shmueli and M. Sheinblatt
Department of Chemistry, Tel-Aviv University, Tel-Aviv, Israel.

Abstract. Our previous approach of expressing the required motional average in Van Vleck's second moment formula, by means of the vibration tensors of the atoms or the molecules involved, has been expanded to include intersegment contributions coming from pairs of nuclei belonging to different rigid molecular fragments. In addition to the thermal tensors appearing in the vibrational corrections to the intermolecular contribution, in the present case tensors of translational vibrations are involved too.

The evaluation of the appropriate average in Van Vleck's second-moment expression (Van Vleck, 1948) over lattice vibrations for nuclei belonging to different rigid segments of the same molecule is closely related to the method proposed for the treatment of the vibrational corrections to the intermolecular contribution (Polak et al., 1974). In both cases we expand the relevant spatial factor in a power series of the instantaneous relative displacement of the two nuclei about the equilibrium position up to the second term. Taking the motional average of the expression derived leads to

$$\left\langle \frac{1-3\cos^2\Theta'}{r'^3} \right\rangle = \frac{1-3\cos^2\Theta}{r^3} + \frac{3}{r^4}\left[(5\cos^2\Theta-1)\hat{\underset{\sim}{r}}\cdot\langle\underset{\sim}{v}\rangle - 2\cos\Theta\,\underset{\sim}{H}\cdot\langle\underset{\sim}{v}\rangle\right] +$$

$$\frac{3}{2r^5}\left[(5\cos^2\Theta-1)\mathrm{Tr}(\underset{\sim}{V}) + 5(1-7\cos^2\Theta)\hat{\underset{\sim}{r}}\cdot\underset{\sim}{V}\cdot\hat{\underset{\sim}{r}} + 20\cos\Theta\,\underset{\sim}{H}\cdot\underset{\sim}{V}\cdot\hat{\underset{\sim}{r}} - 2\underset{\sim}{H}\cdot\underset{\sim}{V}\cdot\underset{\sim}{H}\right]$$

where r' and Θ' are instantaneous values of the internuclear distance and of the angle between $\underset{\sim}{r}'$ and the magnetic field direction (given by the unit vector $\underset{\sim}{H}$), Θ and r refer to the equilibrium configuation, and $\hat{\underset{\sim}{r}}$ is a unit vector along the equilibrium internuclear direction.

The vector $\langle\underset{\sim}{v}\rangle$ which expresses the relative atomic average displacement, becomes under the harmonic approximation

$$\langle\underset{\sim}{v}\rangle = \frac{1}{2}\left[(\underset{\sim}{L}^I - \mathrm{Tr}(\underset{\sim}{L}^I)\underset{\sim}{I})\cdot\underset{\sim}{\rho}_i^I - (\underset{\sim}{L}^{II} - \mathrm{Tr}(\underset{\sim}{L}^{II})\underset{\sim}{I})\cdot\underset{\sim}{\rho}_j^{II}\right]$$

where $\underset{\sim}{L}^I$ is the libration tensor of the I-th segment (or of the I-th molecule in the intermolecular case), $\underset{\sim}{I}$ is the unit dyadic, $\underset{\sim}{\rho}_i^I$ is the position vector of the i-th atom belonging to the I-th segment and referred to its center of libration, and $\mathrm{Tr}(\underset{\sim}{L})$ denotes the tensor's trace. The composite tnesor $\underset{\sim}{V}$ is given by

$$\underset{\sim}{V} = \underset{\sim}{U}_i + \underset{\sim}{U}_j - \underset{\sim}{U}_{ij} - \underset{\sim}{U}_{ji}$$

where the first two tensors are just the anisotropic vibration tensors of atoms i and j, which appear in the Debye-Waller factor, and the two others are the coupling tensors, whose components depend on the degree of correlation between the vibrations of the two atoms.

We suggest the following model for the relative vibrations of two segments. Since the segmented molecule translates as a rigid body, the coupling tensors, under the assumption of uncorrelated librations exhibited by the two segments, do not vanish completely (as they do in the case of two atoms belonging to different molecules which vibrate independently), but reduce to the translation tensors. Thus, the composite tensor $\underset{\sim}{V}$ is simplified and becomes calculable by means of ordinarily measurable thermal tensors:

$$\underset{\sim}{V}_{seg} = \underset{\sim}{U}_i - \underset{\sim}{T}^I + \underset{\sim}{U}_j - \underset{\sim}{T}^{II}$$

where $\underset{\sim}{T}^I$ and $\underset{\sim}{T}^{II}$ are the translation tensors of segments I and II, respectively. (For the intermolecular case $\underset{\sim}{V}_{uncor}$ equals just the sum of the atomic vibration tensors).

It can easily be shown that in the case of fully correlated vibrations $\underset{\sim}{V}$ reduces to the difference between the atomic vibration tensors. (Clearly, the same holds for the intermolecular case).

The vibration-affected second moment of a polycrystalline sample is obtained by integration of the square of $\langle 1-3\cos^2\Theta'/r'^3\rangle$ as described before (Polak et al., 1974). Adding the intramolecular correction (Shmueli et al., 1973) leads to the final expression

$$\bar{S} = K\left[\frac{4}{5}\sum\frac{1}{r^6} + \frac{12}{5}\sum_{intra}\frac{A}{r^6} - \frac{24}{5}\sum_{inter}\frac{B}{r^7} + \frac{24}{5}\sum_{inter}\frac{C}{r^8}\right]$$

where $A = \hat{\underset{\sim}{r}}\cdot\underset{\sim}{L}\cdot\hat{\underset{\sim}{r}}$, $B = \hat{\underset{\sim}{r}}\cdot\langle\underset{\sim}{v}\rangle$ and $C = 3\hat{\underset{\sim}{r}}\cdot\underset{\sim}{V}\cdot\hat{\underset{\sim}{r}} - \mathrm{Tr}(\underset{\sim}{V})$

Second moment calculations for p-terphenyl using the above expressions along with neutron diffraction data concerning libration tensors, translation tensors, atomic vibration tensors and hydrogen positions (Rietveld et al., 1970), as well as second-moment experimental measurements are now in progress in our laboratory.

References

Polak, M., Sheinblatt, M., Shmueli, U., 1974, J. of Magnetic Resonance, in press.
Rietveld, H.M., Maslen, E.N., Clews, C.J.B., 1970, Acta Crystallogr. B, 26, 693.
Shmueli, U., Polak, M., Sheinblatt, M., 1973, J. of Chem. Physics, 59, 4535.
Van Vleck, J.H., 1948, Phys. Review, 74, 1168.

SPIN ECHOES IN SYSTEMS CONTAINING SPIN-½ PAIRS

N. Boden, Y.K. Levine, D. Lightowlers and R.T. Squires.
Department of Physical Chemistry, The University, Leeds LS2 9JT.

Abstract. The proton spin echo responses to the $90^\circ - \tau - \beta_{90^\circ}$ and $90^\circ - \tau - \beta_{0^\circ}$ pulse sequences in powdered $CaSO_4.2H_2O$ are outlined and explained in terms of a model consisting of a planar arrangement of two spin-½ pairs. The behaviour is distinctly different from that predicted for solids containing single spin-½ nuclei.

When a resonant $90^\circ - \tau - 90^\circ_{90^\circ}$ rf pulse sequence is applied to a rigid dipolar coupled spin system, a spin echo occurs at time 2τ after the first pulse provided $\tau \ll T_2$. The formation and shape of this echo have been investigated by a number of workers [Powles and Mansfield, 1962; Powles and Strange, 1963; Mansfield, 1965] for the special case of a solid containing single identical spins. In most cases of practical importance the spins are not single but are arranged in groups. In these cases we expect distinctly different behaviour because of the differentiation between the intra- and inter- group interactions. The simplest case of such spin groupings is when the spins are arranged in equivalent pairs, e.g. H_2O in crystalline hydrates and $>CH_2$ in methylene chains.

The proton echo response to a resonant $90^\circ - \tau - \beta_{90^\circ}$ (XY) pulse sequence in powdered $CaSO_4.2H_2O$ exhibits a complex behaviour. The dependence of the maximum echo amplitude $E(\tau, \beta)$ on τ and β is illustrated in fig.1. The behaviour is due to the presence of two components, one of which is proportional to $\sin^2\beta$ and the other to $\sin^2\beta \cos^2\beta$ and with the former decaying much faster than the latter. In contrast, the $90^\circ - \tau - \beta_{0^\circ}$ (XX) sequence produces a single component echo of the form $\sin^2\beta \cos\beta$ as shown in fig.2. The τ dependences of $E(\tau, \beta)$ for the sequences $90^\circ - \tau - 90^\circ_{90^\circ}$ and $90^\circ - \tau - 54.7^\circ_{0^\circ}$ are plotted in fig.3; in both cases $E(\tau, \beta)$ has a Gaussian dependence on τ^2 over at least 95% of the decay. The decay constant for the $90^\circ - \tau - 90^\circ_{90^\circ}$ echo corresponds to $M_2(\text{inter}) = \frac{5}{6} M_2^{v.v.}(\text{inter})$, where $M_2^{v.v.}$ (inter) is the interpair second moment calculated by the van Vleck procedure.

This behaviour can be explained in terms of a planar arrangement of two spin-½ pairs on truncating the interpair dipolar Hamiltonian so that

$$[H_d^{o}(\text{intra}) + H_d^{o,t}(\text{inter}), H_Z] = 0 \text{ and } [H_d^{o}(\text{intra}), H_d^{o,t}(\text{inter})] = 0$$

It can be shown that provided the total dipolar Hamiltonian is used in the calculation of the echo responses, $E(\tau, \beta)$ is given by $\sin^2\beta + \frac{1}{3}\sin^2\beta\cos^2\beta$ for the XY sequence and $-\frac{2}{3}\sin^2\beta\cos\beta$ for the XX sequence in agreement with experiment. Importantly, the echo for the XX sequence and the $\sin^2\beta\cos^2\beta$ component of the XY sequence originate solely in the interpair dipolar interactions and contain no contributions from the intrapair interactions. The prediction of the echo behaviour requires the preservation of the spin-1 character of the eigenfunctions of $H_Z + H_d^{o}(\text{intra})$ in the presence of the interpair interactions. This also leads to the $\frac{5}{6}$ reduction in the van Vleck second moment of the system.

The incomplete refocussing of the interpair interactions by the XX and XY sequences causes the maximum echo amplitudes to decay with increasing pulse spacing according to

$$E(\tau, \beta)_{xy} \simeq \sin^2\beta \left\{1 - \tfrac{1}{2}(16/15)M_2(\text{inter})\,\tau^2 + \ldots\ldots\right\} + \frac{2}{7}\sin^2\beta\cos^2\beta\left\{1 - \tfrac{1}{2}(8/15)M_2(\text{inter})\,\tau^2 + \ldots\ldots\right\}$$

$$E(\tau, \beta)_{xx} \simeq -\frac{8\sqrt{3}}{19}\sin^2\beta\cos\beta\left\{1 - \tfrac{1}{2}\left(\frac{6}{13}\right)M_2(\text{inter})\,\tau^2 + \ldots\ldots\right\}$$

The appearance of the second order terms in τ in the above equations cause the decay. These terms arise from the truncation of the interpair dipolar Hamiltonian.

These experiments also provide an explanation for the proton echo behaviour in solid hydrogen reported by Metzger and Gaines (1966).

References

Mansfield, P., 1965, Phys. Rev., 137, A961.

Metzger, D.S., and Gaines, J.R., 1966, Phys. Rev., 147, 644.

Powles, J.G., and Mansfield. P., 1962, Phys. Letters, 2, 58.

Powles, J.G., and Strange, J.H., 1963, Proc. Phys. Soc. (London), 82, 6.

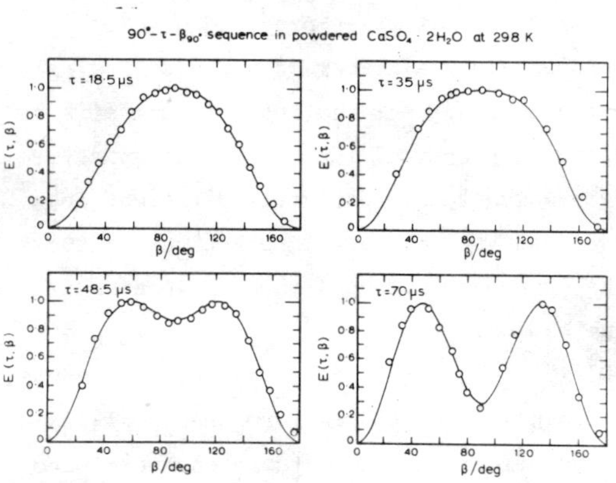

Fig. 1

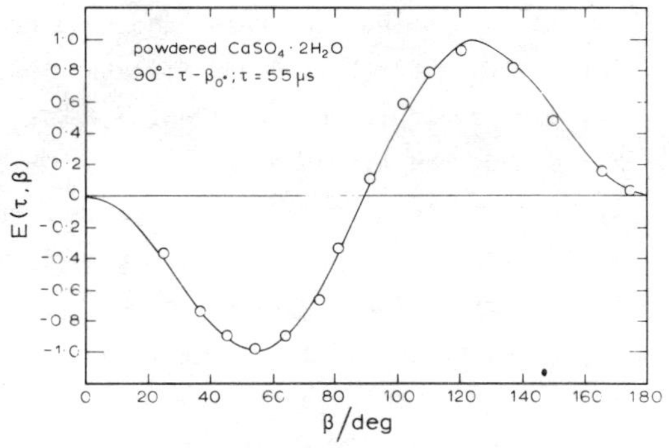

Fig. 2

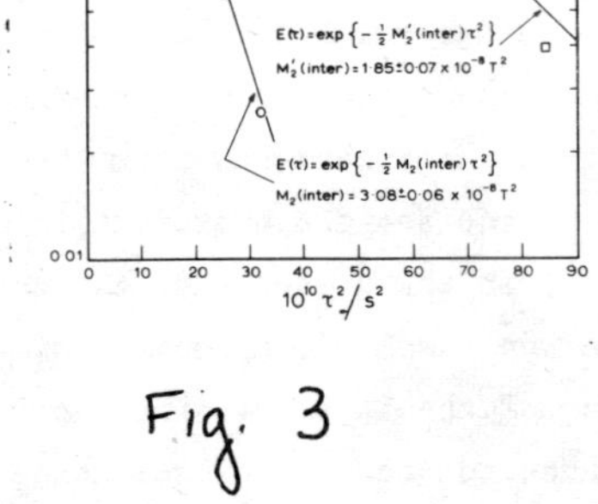

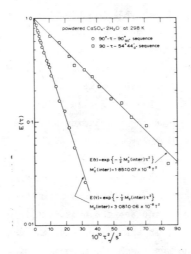

Fig. 3

FREE INDUCTION DECAY FOR A TRIANGULAR CONFIGURATION OF 1/2-SPIN NUCLEI IN A RIGID LATTICE

P. L. Indovina*, A. Rogani* and S. K. Ghosh+

* Laboratori di Fisica, Istituto Superiore di Sanità, Roma, Italy.

+ Istituto di Fisica, Università dell'Aquila, Italy.

Abstract. The FID for a rigid triangular configuration of 1/2-spin nuclei are calculated for varying interproton distance R and compared with the Fourier transforms of existing experimental lineshapes. Excellently good agreements are obtained between the theory and experiments for R = 1.825 Å rather than 1.79 Å used earlier for the lineshape calculations.

The proton NMR studies in many three-spin systems, such as methyl groups, show that the lineshapes at even lowest temperatures cannot be well reproduced from the simple consideration of dipolar interaction (Andrew and Bersohn 1950, hereafter referred to as AB); the experimental line is slightly narrow at the wings compared to the one calculated with the accepted interproton distance R. This failure to reproduce the lineshape has led to confirm the suggestion originally made from the observation of slightly reduced second moment (Gutowsky and Pake 1950, hereafter referred to as GP), that the quantum mechanical tunnelling is responsible for this discrepancy between the theory and experiment. In recent years, a renewed interest has developed to quantitatively understand the effect of tunnelling on the NMR lineshape (Apaydin and Clough 1968, hereafter referred to as AC; Mottley et al. 1971) and the second moment (Allen 1968; Clough 1971). The work af AC has essentially shown that the nuclear dipolar interaction is modified in the presence of tunnelling in a way to introduce an additional nuclear exchange interaction term, and the constant of this exchange term is related to the tunnelling frequency. Such an exchange interaction can qualitatively explain the observed reduction of second moment (Van Vleck 1948; Allen 1968; Clough 1971), but has not been more successful than the simple dipolar interaction in reproducing the lineshape, as can be evidenced comparing the results of AC with those of AB.

We report here the FID calculated for a rigid triangular configuration of 1/2-spin nuclei. This FID is then reduced appropriately for a polycrystalline system to compare with the existing experimental results (GP; Powels and Gutowsky 1953, hereafter referred to as PG). Our calculated FID is essentially the Fourier transform (FT) of the lineshape reported earlier by AB, as is expected from other theoretical considerations (Lowe and Norberg 1957). We have followed a more laborious course of directly calculating the FID from other considerations which have no immediate bearing on the present communication. For completeness, we give below a brief outline of our method which is essentially the same as used earlier by Das and one of us (1955, hereafter referred to as I) in a similar problem of two spins.

The Hamiltonian for our present problem is the same as that used by AB and is

$$H = -\gamma \hbar H_o \quad \Sigma_j^3 I_{jz} + (\hbar/2) \Sigma_{j<k} (\vec{I}_j \cdot \vec{I}_k - 3 I_{jz} I_{kz}) P_{jk} \tag{1}$$

where

$$P_{jk} = \gamma^2 \hbar (3 \cos^2 \theta_{jk} - 1) r_{jk}^{-3} \quad .$$

r_{jk} is the length of the radius vector joining the nuclei j and k, and θ_{jk} is the angle between magnetic field $H_o \hat{z}$ and $\vec{r}_{jk}$. We do not consider here the tunnelling which can be done, if one wishes, in a straighforward manner using the spin Hamiltonian given by AC.

As discussed in I, the FID is proportional to

$$< \mu_x (t) > = \mathrm{Tr} \left\{ \exp(-H/kT) R^+_{op}(t_w) T^+_{op}(t) \mu_x T_{op}(t) R_{op}(t_w) \right\} / \mathrm{Tr} \left\{ \exp(-H/kT) \right\} \tag{2}$$

where

$$\mu_x = \gamma \hbar \ \Sigma_j I_{jx}, \ R_{op}(t_w) = \exp(i\omega_o t_w I_z) \exp(i \gamma H_1 t_w I_y), \text{ and } T_{op}(t) = \exp(-iHt/\hbar).$$

In calculating the operator $R_{op}(t)$, the dipolar field has been considered negligibly small compared with the rf field H_1 of duration t_w

The trace Tr in Eq.(2) is most convenient to evaluate in a representation in which H is diagonal. Working in such a representation. we get for a 90 -pulse

$$<\mu_x(t)> = -(3\gamma \hbar^2 \omega_o/16kT) \left\{ (A+F)/F \cos\left[(3A+F)t/8\right] + (F-A)/F \cos\left[(3A-F)t/8\right] + \left[3(F^2-A^2)/2F^2\right] \cos(Ft/4) + (3A^2+F^2)/2F^2 \right\} \tag{3}$$

where $F^2 = A^2 + 2B^2 + 6C^2$, $A = P_{12} + P_{23} + P_{13}$, $B = P_{12} - 2P_{23} + P_{13}$, and $C = P_{12} - P_{13}$. Eq.(3) is exactly the same as the FT of the spectra given by AB.

To apply this result to a polycrystalline material, it is convenient to write Eq.(3) in the form:

$$<\mu_x(t)> = -(3\gamma \hbar^2 \omega_o/16\,kT) \left\{ (x+y)/y \cos\left[(3x+y)t/\hbar\right] + (y-x)/x \cos\left[(3x-y)\,t/\hbar\right] + \left[3(y^2-x^2)/2y^2\right] \cos(2yt/\hbar) + (3x^2+y^2)/2y^2 \right\} \tag{4}$$

where $x = \hbar A/8$ and $y = \hbar F/8$.

Following AB, we can write for an equilateral triangle:

$x = \mu\,\alpha(1 - 3\cos^2\psi)/2$, and $y = \mu\,\alpha\left[(27/4)\sin^4\psi - 3\sin^2\psi + 1\right]^{1/2}$, where ψ is the angle between H_o and the normal to the plane of the triangle and $\alpha = 3/2\ \mu R^{-3}$. The FID of these isolated triangular groups distributed isotropically is given by

$$F_3(t) = \int_0^{\pi} <\mu_x> \sin\psi \, d\psi/2 \quad . \qquad (5)$$

Approximating the interaction between the triangular groups by a Gaussian decay function, we get the FID for a polycrystalline material

$$F(t) = \exp(-<\Delta H^2> t^2/2) F_3(t) \qquad (6)$$

where $<\Delta H^2>$ is an adjustable parameter to get the best fits to the experimental data for each $F_3(t)$ obtained for different R. Two of these F(t) are shown together with the FT of GP data in Fig.1. For R = 1.825 A and $<\Delta H^2> = 1.92\,G^2$, we get the best fits with the GP data (taken at 90° K). Similar comparison has also been made with the PG data (taken at 77°K) which yield identical R for the best fit, although the fit is relatively poor. This may be due to the poor reproduction of their data points to get the lineshape used in our computation of the FID. It is interesting to note that our best-fit FID gives a second moment of 20.28 G^2 fairly in good agreement with 20.63 G^2 directly obtained from the lineshape used in our computations. These values should be compared with 18.7 G^2 reported by GP.

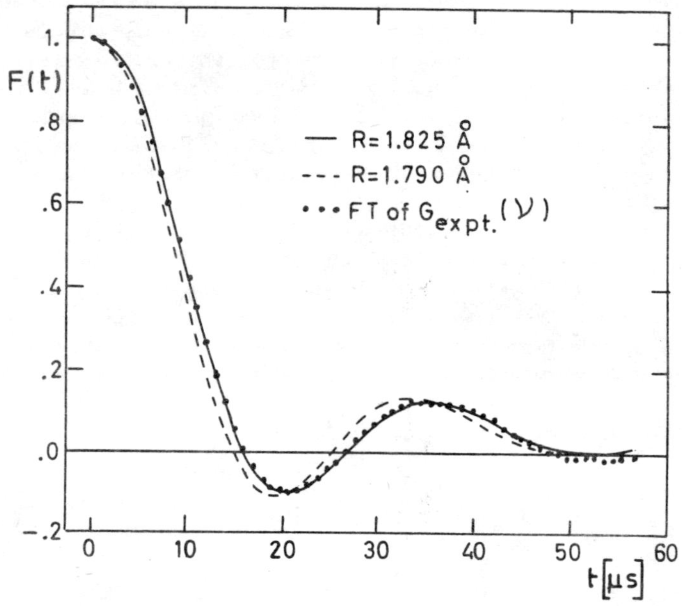

Fig.1. Comparison of the experimental and theoretical FID for two different R. Experimental FID is the FT of GP data for 1,1,1-trichloroethane obtained at 90°K.

The value of R that we obtain here from the NMR data in 1,1,1-trichloroethane is slightly higher (~2%) than the accepted value To our knowledge, we are not aware of any experiment giving R sufficiently accurate to compare with the present value. In conclusion we would like to point out that it would be more opportune to compare the FID with direct measurements, particularly for their higher accuracy. Such measurements of direct FID and the comparison with the theory may not only give precise proton positions, but may also yield precise tunnelling frequencies for properly chosen systems.

References

Allen, P.S., 1968, J.Chem.Phys., 48, 3031.
Andrew, E.R. and Bersohn, R., 1950, J.Chem.Phys., 18, 159.
Apaydin, F. and Clough, S., 1968, J.Phys.C (Proc.Phys.Soc.), Ser.2, 1, 932.
Clough, S., 1971, J.Phys.C: Solid St.Phys., 4, 1075.
Das, T.P. and Ghosh Roy, S.K., 1955, Indian J.Phys., 29, 272.
Gutowsky, H.S. and Pake, G.E., 1950, J.Chem.Phys., 18, 162.
Lowe, I.J. and Norberg, R.E., 1957, Phys.Rev., 107, 46.
Mottley, C., Cobb, T.B. and Johnson, C.S., 1971, J.Chem.Phys., 55, 5823.
Powels, J.G. and Gutowsky, H.S., 1953, J.Chem.Phys., 21, 1695.
Van Vleck, J.H., 1948, Phys.Rev., 74, 1168.

A GENERAL PROCEDURE FOR CALCULATIONS OF THEORETICAL SECOND MOMENTS AND RELAXATION TIMES IN DIPOLAR SOLIDS

R. O. I. Sjöblom*
Department of Physics, University of Nottingham, University Park, Nottingham NG7 2RD, ENGLAND.

Abstract. It is shown how the second moment may be written as products and sums of second order spherical harmonic functions. The second moment is completely determined by the direction of the magnetic field and the fifteen unique elements of the second moment tensor. Its dependence on the crystal structure and symmetry, and on assumptions about reorientational and vibrational motion is investigated quantitatively and an expression for the average over a powder is derived. Formulas are also given for the dependence of T_1 and $T_{1\rho}$ on the difference between two second moment tensors.

Introduction. In a solid the lineshape of the nuclear magnetic resonance signal from nuclei with spin one-half is often determined by the dipole-dipole interaction. Since the lineshape is difficult to calculate theoretically one usually uses the second moment which is given by the Van Vleck formula. Thus, by comparing experimental and calculated second moments it is possible to extract information about molecular reorientations, interatomic distances and librational motion (torsional oscillations). Similar and supplementary results may be obtained from relaxation studies. The often quite cumbersome calculations for the second moment can, if the structure is rigid, be greatly simplified if Van Vleck's formula is rewritten as a sum of products of functions,

$$M_2(\underline{H}) = \sum_{i=1}^{i=15} f_i g_i(\underline{H}) \quad ,$$

where f_i depends only on the positions of the atoms in the solid, and g_i depends only on the direction of the magnetic field (O'Reilly et al., 1962 and Gorskaya et al., 1968). Quite recently (Shmueli et al., 1973) formulas have been derived for the average over the vibrational motion of an otherwise rigid structure. Again recently (Soda et al., 1974) it has been shown how the relaxation time is related to the difference between the second moments for two motional situations in a polycrystalline sample. It is the purpose of this paper to demonstrate a general procedure for evaluations of second moments and relaxation times where in comparison with earlier formulas there are less restrictions on what may be done, and usually considerably less numerical labour is required.

The Second Moment Tensor. The second moment for a pair of interacting nuclei of the same kind can be calculated according to the Van Vleck formula:

$$M_2 = \tfrac{3}{4}\gamma^4\hbar^2 I(I+1)\langle\langle b\rangle^2\rangle \qquad \text{where } b = (3\cos^2\phi - 1)r^3 \quad . \tag{1}$$

The inner average is taken over all motions characterized by correlation times sufficiently short to cause the maximum possible reduction of the second moment. The outer average takes into account static disorder, or molecular motion which is too slow to affect the second moment. The formulas which will be introduced shortly refer to the orthonormal basis vectors, $\underline{i},\underline{j},\underline{k}$ where $\underline{i}$ // $\underline{a}$, $\underline{j}$ // $\underline{c^*} \times \underline{a}$ and $\underline{k}$ // $\underline{c^*}$ ($\underline{a},\underline{b},\underline{c}$ and $\underline{a^*},\underline{b^*},\underline{c^*}$ are the edges of the real and reciprocal crystallographic unit cells respectively). It is easy to show by algebraic manipulations that

$$b = \sum_{i=1}^{i=5} (10\pi/r^3)\phi_i(\underline{r})\phi_i(\underline{H})$$

where $\underline{r}$ is the internuclear vector, $\underline{H}$ is a normalized vector parallel to the magnetic field, and ϕ_i are the real second order spherical harmonics which are shown in Table 1. It is convenient to use the functions $\alpha_i = 5(C\pi)^{1/2}\,\phi_i(\underline{r})/r^3$ where C is equal to all the constants in (1), and $\beta_i = 2\pi^{1/2}\phi_i\underline{H}$. Formula (1) may now be written:

$$M_2 = C\langle\langle b\rangle^2\rangle = \langle\langle\sum_{i=1}^{i=5}\alpha_i\beta_i\rangle^2\rangle = \langle\sum_{i=1}^{i=5}\sum_{j=1}^{j=5}\overline{\alpha}_i\overline{\alpha}_j\beta_i\beta_j\rangle = \sum_{i=1}^{i=5}\sum_{j=1}^{j=5}\langle\overline{\alpha}_i\overline{\alpha}_j\rangle\beta_i\beta_j = \underline{\tilde{\beta}}\underline{\underline{\gamma}}\underline{\beta} \tag{2}$$

The symmetric matrix $\underline{\underline{\gamma}}$, which in general contains fifteen elements to be determined, will in the following be referred to as the second moment tensor.

The average over the different crystallites in a powder is:

$$\overline{M}_2 = \left(\int\sum_i\sum_j\beta_i\beta_j\gamma_{ij}\,d\tau\right)\Big/\int d\tau = \sum_i\sum_j\gamma_{ij}\left(\int\beta_i\beta_j\,d\tau\Big/\int d\tau\right) = \mathrm{Tr}\,\underline{\underline{\gamma}} \tag{3}$$

since $\int\beta_i\beta_j = 4\pi$ if $i = j$ and is equal to zero if $i \neq j$.

*Permanent Address: Institute of Chemistry, University of Uppsala, Uppsala, SWEDEN.

Symmetry Considerations. If symmetry is present it is possible to find a transformation of the functions $\beta_i\beta_j$ such that each new function transforms according to the symmetry species of the group. Since the second moment is totally symmetric with respect to all symmetry operations, all the coefficients of functions which do not belong to the totally symmetric species must vanish. This means that the coefficients of the untransformed functions must be related to each other in such a way that the total symmetry is preserved. The constraints on the elements of the second moment tensor for the possible second moment symmetries are listed in Table 2. The Laue symmetries 4/m and $\bar{3}$ are actually not symmetries of the second moment since it is possible to find a transformation of the basis functions $\beta_i\beta_j$, which corresponds to rotation of the three-dimensional basis vectors around the main symmetry axis, such that the constraints of 4/mmm and $\bar{3}$m become valid, respectively.

Table 1. The Spherical Harmonics Used

$\phi_1 = (5/16\pi)^{\frac{1}{2}}(3z^2 - 1)/r^2$
$\phi_2 = (15/16\pi)^{\frac{1}{2}}(x^2 + y^2)/r^2$
$\phi_3 = (15/4\pi)^{\frac{1}{2}}\ xy/r^2$
$\phi_4 = (15/4\pi)^{\frac{1}{2}}\ xz/r^2$
$\phi_5 = (15/4\pi)^{\frac{1}{2}}\ yz/r^2$

Table 2. The Constraints on the Second Moment Tensor for the Possible Second Moment Symmetries

Crystal System	Laue Symmetry		Nr. of Independent Parameters	The Non-Zero Elements of $\underline{\gamma}$ and Their Symmetry Constraints
Triclinic	1	S_2	15	All Fifteen
Monoclinic	2/m	C_{2h}	9	$\gamma_{11}\ \gamma_{12}\ \gamma_{14}\ \gamma_{22}\ \gamma_{24}\ \gamma_{33}\ \gamma_{35}\ \gamma_{44}\ \gamma_{55}$
Orthorhomic	mmm	D_{2h}	6	$\gamma_{11}\ \gamma_{12}\ \gamma_{22}\ \gamma_{33}\ \gamma_{44}\ \gamma_{55}$
Tetragonal	4/m	C_{4h}	5	$\gamma_{11}\ \gamma_{22}\ \gamma_{23}\ \gamma_{33}\ \gamma_{44} = \gamma_{55}$
	4/mmm	D_{4h}	4	$\gamma_{11}\ \gamma_{22}\ \gamma_{33}\ \gamma_{44} = \gamma_{55}$
Trigonal	$\bar{3}$	S_6	5	$\gamma_{11}\ \gamma_{22} = \gamma_{33}\ \gamma_{44} = \gamma_{55}\ \gamma_{24} = \gamma_{35}\ \gamma_{34} = \gamma_{25}$
	$\bar{3}$m	D_{3d}	4	$\gamma_{11}\ \gamma_{22} = \gamma_{33}\ \gamma_{44} = \gamma_{55}\ \gamma_{34} = \gamma_{25}$
Hexagonal	6/mmm	D_{6d}	3	$\gamma_{11}\ \gamma_{22} = \gamma_{33}\ \gamma_{44} = \gamma_{55}$
Cubic	m3m	T_d	2	$\gamma_{11} = \gamma_{22}\ \gamma_{33} = \gamma_{44} = \gamma_{55}$

Rigid Body Motion. If a molecule or a segment of a molecule is reorienting or librating as a rigid body, it can be shown that the internal second moment tensor $\underline{\underline{\gamma}}$ is reduced to $\underline{\underline{\gamma}}'$ by this motion according to the formula: $\underline{\gamma} = \tilde{\underline{\underline{\theta}}}\underline{\gamma}\underline{\underline{\theta}}$ (5)

Here θ depends only on the motion of the rigid body, and γ depends only on the internal structure and the internal motions of either kind. The rigid body approximation is often valid for the vibrations in a molecular solid, and it can be shown that θ in this case depends in a reasonably simple way on the librational tensor only. Formula (5) may also be suitable for calculations on clathrates and liquid crystals.

Relaxation Times. The second moment tensor can be used for calculations of relaxation times, or actually the relaxation constants since the correlation times usually have to be determined experimentally. One assumption here is that all configurations of the atom-atom vector have the same probability, and that the jumping rate is the same between any two of them. A derivation of the relaxation constants K, much along standard lines, yields:

$$K_m = \underline{\tilde{D}}_m^{(2)}\underline{\underline{\nu}}(\underline{\underline{\gamma}}_{rigid} - \underline{\underline{\gamma}}_{motion})\underline{\underline{\nu}}^*\underline{D}_m^{(2)*} \quad (6)$$

The unitary matrix $\underline{\underline{\nu}}$ transforms the real functions $\underline{\alpha}$ into the corresponding complex ones, and $D^{(2)}$ is the transformation matrix for the complex spherical harmonics, normalised similarly to β_i. The relaxation times T_1 and $T_{1\rho}$ are given by:

$$T_1^{-1} = (2/3)K_1\tau/(1 + \omega^2\tau^2) + (8/3)K_2\tau/(1 + 4\omega^2\tau^2) \quad (7)$$

$$T_{1\rho}^{-1} = K_0\tau/(1 + 4\omega_1^2\tau^2) + (5/3)K_1\tau/(1 + \omega^2\tau^2) + (2/3)K_2\tau(1 + 4\omega^2\tau^2) \quad (8)$$

The average over a powder is:

$$\overline{T}_1^{-1} = (2/3)\mathrm{Tr}(\underline{\underline{\gamma}}_{rigid} - \underline{\underline{\gamma}}_{motion})\cdot[\tau/(1 + \omega^2\tau^2) + 4\tau/(1 + 4\omega^2\tau^2)] \quad (9)$$

$$\overline{T}_{1\rho} = (2/3)\mathrm{Tr}(\underline{\underline{\gamma}}_{rigid} - \underline{\underline{\gamma}}_{motion})\cdot[(3/2)\tau/(1 + 4\omega_1^2\tau^2) + (5/2)\tau/(1 + \omega^2\tau^2) + \tau/(1 + 4\omega^2\tau^2)] \quad (10)$$

The difference ($\underline{\underline{\gamma}}_{rigid} - \underline{\underline{\gamma}}_{motion}$) may equally well refer to two motional cases.

Calculations. General computer programmes have been written which enable evaluations of the theoretical and experimental second moment tensors. Theoretical calculations have been carried out on some of the amino acids, a few methyl substituted ammonium and on oxalic acid dihydrate. In the latter case, where the structure contains flipping water molecules, experimental and theoretical second moment tensors have been used in least squares refinements of the relative atomic positions.

Acknowledgements. The author wishes to thank Professor E. R. Andrew, Dr. M. Punkkinen and Dr. J. Tegenfeldt for illuminating discussions and valuable suggestions and comments.

References. Gorskaya, N. V. and Fedin, E. I., 1968, Zhurnal Strukturnoi Khimii, 9, 560.
O'Reilly, D. E. and Tsang, T., 1962, Phys. Rev., 128, 2639.
Shmueli, U., Polak, M. and Sheinblatt, M., 1973, J. Chem. Phys., 59, 4535.
Soda, G. and Chihara, H., 1974, J. Phys. Soc. Japan, 36, 954.

SECTION P

Defect Centres

STUDY OF THE ELECTRIC DIPOLAR MOMENT OF THE V_1 CENTERS IN MgO

J-L. Ploix, A. Hervé and G. Rius - Section de Résonance Magnétique, Département de Recherche Fondamentale, Centre d'Etudes Nucléaires de Grenoble, BP 85 Centre de Tri, 38041 Grenoble-Cedex, France.

Abstract. We have tried to perform alignment experiments in an applied electric field in order to measure the dipolar moment of the V_1 center in MgO. No effect has been observed, and the dipolar moment of the defect is calculated to be less than 0.125 e Å. Simulated random electric field of mean value 100 kV/cm does not modify significantly this result.

1. Introduction. The V_1 center in alkaline-earth oxides consists of a hole trapped at an isolated cation vacancy. In its ground state, this defect undergoes a strong tetragonal distortion which localises the hole on a single oxygen ion adjacent to the vacancy. This defect was first identified in MgO by Wertz and al (5) who attributed the three line EPR spectrum at 77°K to the three possible <100> type orientations of the O^- vacancy axis. The validity of the "localized" model for the ground state has been verified by Halliburton et al (1). According to this model, the hole can occupy different positions relative to the trap and thus, the defect should behave as a dipole with several possible orientations.

Recently, one of us (2,3) has shown by ELDOR and uniaxial stress experiments that the defect is reorienting its axis between its equivalent distorted configurations at liquid helium temperatures. The tunneling matrix element was estimated to be 7.10^{-4} cm^{-1} (2). It turns out to be so small that the internal random electric fields and (or) strains in the crystal are strong enough to enforce a complete localisation of the spin on one of the six available oxygen sites adjacent to the vacancy. From uniaxial stress experiments, the "localisation temperature" was found to be very high, ranging from 10°K to 25°K in the samples investigated (2).

In order to verify if the localisation of the hole was due to the effect of internal electric fields, we have undertaken EPR experiments under static electric fields. In this work, we report the results of the experiments, from which we deduce an upper limit for the value of the electric dipole moment of the V_1 center.

2. Experimental. V_1 centers are produced in pure MgO crystals by a 3 MeV electron irradiation at room temperature. The samples are then annealed at 100°C for 3 minutes in order to reduce the concentration of V_M centers (holes trapped at vacancy impurity complexes, which give an EPR spectrum interfering with that of the V_1 centers (4)) to less than 1/10 of the V_1 concentration.

A 0,4 x 4 x 4 mm sample is fixed between two 50 μ copper sheets. Good electric isolation is achieved by a 50μ teflon sheet, screening the sample from one of the electrodes. The whole setting is immersed in liquid helium in a finger dewar. A varying voltage up to 3000 volts is applied to the electrodes, producing an electric field of 60 kV/cm in the sample. Measuring the intensity ratio of the corresponding EPR lines gives the population ratio of two types of sites.

3. Results. All the experiments were made at 4.2°K. The signal/noise ratio of the EPR lines was about 100. No change of the relative intensities of the lines was observed by applying the DC voltage across the sample.

4. Discussion. According to the model, the defect possesses C_{4v} symmetry about a cubic axis of the crystal, and the electric dipolar moment, if any, is directed along the vacancy O^- axis.

If the static electric field is applied in the z direction, the relative populations of the centers among the six possible orientations are :

$$\begin{cases} n_z = Z \exp (PE/kT) \\ n_{-z} = Z \exp (- PE/kT) \\ n_x = n_{-x} = n_y = n_{-y} = z \end{cases}$$

and the ratio r of the EPR lines intensities is :

$$r = \frac{n_z + n_{-z}}{n_x + n_{-x}} = \cos h \, (PE/kT)$$

From our measurements, we deduce that the upper value for the dipolar moment of the V_1 center is then :

$$P_{max} = 0.125 \text{ e Å}$$

If we assume that there exist in the crystal random electric fields (arising from charge-deficient impurity ions and lattice defects), the equations for the populations become :

$$n_z = \left\langle Z \exp \frac{P\,(E + E_{iz})}{kT} \right\rangle$$

$$n_{-z} = \left\langle Z \exp - \frac{P\,(E + E_{iz})}{kT} \right\rangle$$

$$n_x = \left\langle Z \exp \frac{P\,E_{ix}}{kT} \right\rangle \qquad \text{and so on.}$$

And

$$r = \frac{\left\langle Z \cosh \dfrac{P\,(E + E_{iz})}{kT} \right\rangle}{\left\langle Z \cosh \dfrac{P\,E_{ix}}{kT} \right\rangle}$$

Assuming a gaussian distribution for the internal field

$$p\,(E_i) = K \exp\,(- E_i^2/2\,E_m)$$

with a mean value $E_m \leqslant 100$ kV/cm.
We obtain now :

$$P_{max} = 0.132 \text{ e Å}.$$

In view of the interionic distance in MgO (2,1 Å), this result seems astonishingly small. It may indicate that the effect of the trapping of the hole on one oxygen is counterbalanced by the lattice distorsion around the vacancy.

5. References.

(1) L.E. Halliburton, D.L. Cowan, W.B.J. Blake and J.E. Wertz, 1974, Phys. Rev. B., 8, 1610.
(2) G. Rius, 1974, Thèse d'Etat, Université de Grenoble, n° A.O.8419
(3) G. Rius and A. Hervé, 1974, Solid State Communications, 15, 2, 421.
(4) W.P. Unruh, Y. Chen and M.M. Abraham, 1973, Phys. Rev. Letters, 30, 446.
(5) J.E. Wertz, G. Saville, P. Auzins and J.W. Orton, 1959, Discussion of the Faraday Society, 28, 136.

THE BEHAVIOR OF POSITIVE MUONS IN "MUONIC" U_2-CENTERS

P.F. Meier and A. Schenck
Schweizerisches Institut für Nuklearforschung, SIN
CH-5234 Villigen, Switzerland

Abstract: Positive muons implanted in nonconducting solids form hydrogenlike muonium atoms (μ^+e^-) with properties similar to those of U_2- centers, e.g. the occurrence of a superhyperfine interaction with neighbor nuclei. The resulting change of the residual μ^+-polarization in longitudinal magnetic fields and the occurring characteristic precession frequencies are investigated.

1. Introduction. Currently extensive studies are made on the interaction of muonium with matter (Schenck and Crowe, 1974). The depolarization of initially fully polarized positive muons in solids can be detected with the help of the anisotropic angular distribution of the emitted positrons. In most nonmetallic solids muonium atoms (μ^+e^-) are formed where 50 % are in a stationary state (muon and electron spin parallel). The other 50 % are in a state which is a superposition of two eigenstates resulting in a rapid precession of the muon spin in the contact field. This precession is too fast to be detected experimentally. Thus the muons forming a muonium atom with antiparallel spins appear depolarized and the residual polarization is 1/2. In magnetic fields parallel to the initial muon spin this depolarization is quenched (Ferrell and Chaos, 1957).
Experiments in general confirm this picture. In weak fields, however, the polarization is often less than 1/2 and approaches the value of the muonium theory, after a steep rise, only if the field is increased (Buhler et al., 1965).
In this work we consider the contact interaction of the electron in the muonium atom with surrounding nuclei with spin different from zero. This interaction is well known from ESR and ENDOR investigations of paramagnetic impurity centers and is called superhyperfine interaction. One can argue that paramagnetic muonium is most probably formed at an interstitial lattice site, thus resembling very much interstitial paramagnetic hydrogen centers which are called U_2-centers. The latter have been most extensively studied in alkali halides (Spaeth and Sturm, 1970), demonstrating the significance of superhyperfine interactions. Thus it is to be expected that muonium in alkali halides will also experience a sizeable superhyperfine interaction. This may be the reason for the unusual quenching behaviour obtained in KCl (Ivanter et al., 1972).

2. Model. We consider the spin system given by the contact interaction between muon and electron and the corresponding Zeemann terms (i.e. the Breit-Rabi-Hamiltonian describing muonium)plus the contact interaction between electron and a spin I. I stands for the sum of spins of equivalent nuclei in e.g. the first neighbor shell. For the sake of simplicity other spin species are neglected and it is assumed that the superhyperfine interaction is isotropic and that the Zeemann energies of the nuclei can be neglected.
The Hamiltonian reads

$$H = A\,\vec{S}_\mu \cdot \vec{S}_e + a\,\vec{S}_e \cdot \vec{I} + H_Z$$

with the Zeemann energy

$$H_Z = g_\mu\,\mu_0^\mu\,BS_\mu^z + g_e\mu_0^e BS_e{}^z .$$

The muonium interaction energy A leads to a hyperfine frequency ν_0 = 4463 MHz in vacuum. The values for the superhyperfine frequency (a/h) in alkali halides vary between about 10 and 200 MHz.

3. μ^+-Polarization in zero field. In the absence of a magnetic field the eigenvalues of the Hamiltonian H are found to be given by (if $I \geq 1$)

$$\lambda_1 = A/4 + Ia/2$$
$$\lambda_3 = A/4 - (I+1)a/2$$
$$\lambda_{2,4} = -A/4 - a/4 \pm \left[(A-a/2)^2 + I(I+1)a^2\right]^{1/2}/2.$$

The corresponding transition frequencies ω_{14}, ω_{24}, and ω_{34} are of the order of ω_0 and are usually not resolvable in experiments. An approximate expression (valid for large values of I) for the observable μ^+-polarization is then given by

$$P_\mu^{obs}(t) \simeq \frac{1}{6}\left\{1 + \cos\omega_{12}t + \cos\omega_{23}t\right\}.$$

An inspection of results from ENDOR measurements on U_2-centers in alkali halides shows that the modulations of the polarization with frequencies ω_{12} and ω_{23} should be visible, provided that enough events are recorded to guarantee the necessary statistics (10^7 events or more). This explains the fact that the experiments performed on KCl and NaCl (Myasishcheva et al., 1967) did not reveal any modulation of the polarization in zero or small transverse fields. Compared to the muonium case the constant term is reduced from 1/2 down to 1/6. Such a reduction of the observed polarization below 1/2 is indeed seen in many experiments. Moreover, by including further superhyperfine interactions with other nuclei (e.g. in the second neighbor shell or different species in the first neighbor shell) one gets a still smaller constant term and further oscillating terms.

4. Quenching of depolarization in longitudinal fields. In longitudinal fields the depolarization is quenched. The μ^+-polarization has been evaluated numerically for various values of the superhyperfine interaction a. Typical results for the time independent term are presented in Fig. 1. With increasing field the muonium system first is decoupled from the nuclei (Paschen-Back effect for (μe)-I in fields where the Zeemann energy H_Z exceeds the interaction energy a). This leads to a steep rise of the polarization. If the field is further increased the curve approaches that of muonium and finally exhibits the Paschen-Back effect for μ-e.
This behavior of the polarization in longitudinal fields has indeed been seen in various substances. The most pronounced examples are SiO_2 and Al_2O_3 (Minaichev et al., 1970) where chemical reactions of muonium are expected to be negligible. The depolarization in SiO_2 agrees with the one calculated for muonium. The depolarization in Al_2O_3, however, shows exactly the behavior which is expected for a superhyperfine interaction of the electron in the muonium atom. This different behavior of the muon in SiO_2 and Al_2O_3 confirms our theory since in these substances only the Al^{27} nuclei have a spin different from zero (except for Si^{29} with abundance of 4.7 %). The calculated time dependence of the polarization in both longitudinal and transverse fields shows a superposition of various oscillating terms with weights depending on the field strength. Some of these oscillations should be detectable in measurements with high statistics.

5. Conclusion. In conclusion, the consideration of the superhyperfine interaction of the electron in the muonium atom, accounts for the observed high depolarization in weak magnetic fields. Furthermore, we predict modulations of the time dependence of the polarization which should be detectable in careful experiments. The analysis of such field dependent transition frequencies then gives information on the electron g-value, the main hyperfine interaction and the various occurring superhyperfine interactions and on their isotropic and anisotropic properties. These informations may be used to check on electronic and structural properties of the crystals. Hydrogenic U_2-centers have been observed only in a relatively small number of crystals. In contrast, muonic U_2-centers can be expected to be formed in almost every kind of nonconducting crystals, thus promising to be a valuable probe in solid state physics.

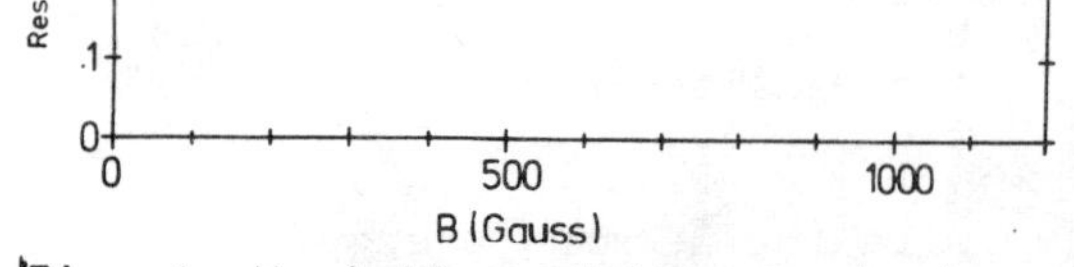

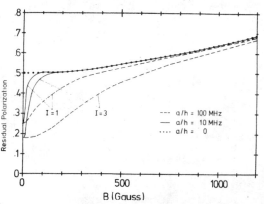

Fig. 1: Variation of the constant term of the muon polarization P_μ^Z in longitudinal field for superhyperfine interaction with spins I = 1 and I = 3 and frequencies a/h = 10 MHz and 100 MHz. The dotted curve represents the polarization for muonium.

6. References.

Buhler A. et al., 1965, Nuovo Cimento 39, 812
Ferrel R.A. and F. Chaos, 1957, Phys. Rev. 107, 1322
Ivanter I.G. et al., 1972, Zh. Eksp. Theor. Fiz. 62, 14 [Sov.Phys. JETP 35, 9]
Minaichev E.V. et al., 1970, Zh.Eksp, Teor.Fiz.58, 1586 [Sov.Phys.JETP 31,849]
Myasishcheva G.G., et al., 1967, Khim. Vysok En. 1, 394 [Soviet High Energy Chemistry 1, 343]
Schenck A. and K.M. Crowe, 1974, Proceedings of the Topical Meeting on Intermediate Energy Physics, Cern Report 74-8
Spaeth J.M. and M. Sturm, 1970, phys. stat. sol. 47, 739

ESR STUDY OF PARAMAGNETIC IRRADIATION CENTERS IN ThO_2 SINGLE CRYSTALS

I.Ursu, S.V.Nistor and S.A.Marshall+
Institute for Atomic Physics, P.O.B.35, Bucharest, Romania
+Argonne National Laboratory, Argonne, Ill.60439, U.S.A.

Abstract. We report the results of an ESR study of ThO_2 single crystals which were previously γ-ray irradiated. New paramagnetic centers have been observed in crystals with higher concentration of impurities.

1. Introduction. Along its history the ESR method proved to be one of the most powerfull experimental methods in understanding the structure of many paramagnetic irradiation centers (Ursu,1968). Though a big deal of interest was shown concerning the defects in semiconductors and ionic crystals, mainly alkali and alkali earth halides, a few results were published about ThO_2 (Neeley et al,1967),(Röhrig and Schneider, 1969), (Kolopus, Finch and Abraham, 1970) and (Rodine and Land,1971). The study of this highly refractory material which seems to have an important future in nuclear power generation is also of scientific interest. Its relatively simple structure of fluorite type has large interstitial sites which can accomodate high concentrations of impurities which could determine the formation of a large variety of defects.

2. Experiment. Two sets of ThO_2 single crystals from different sources were employed in our study. One set of crystals displaying an amber coloration were obtained from Norton Company and were previously employed in ESR studies on iron (Marshall and Nistor,1972) and gadolinium ions (Marshall et al,1973). A chemical analysis indicated also the presence of other paramagnetic impurities in concentrations of a few ppm as well as other cationic impurities some of which reached comparatively high concentrations (up to a few hundreds ppm). The other set of small colourless crystals with characteristic octahedral habit was grown at Oak Ridge National Laboratory. The very good quality of these crystals and the low content of impurities was proved by the presence of very narrow , weak Gd^{3+} ESR absorptions. These lines did not exhibit the anomalous broadening below 60°K which could be observed for the other set of crystals.

The ESR measurements were performed using either the equipment already described in a previous paper (Marshall and Nistor,1972) or a standard E-12 Varian spectrometer with homodyne detection. For both spectrometers gas tight microwave cavities of TE_{011} mode were employed. The γ-ray irradiations were made using a ^{60}Co source of about 8000 Curie in which a dewar with liquid nitrogen containing the cavity could be inserted.

3. Experimental results. After γ-ray irradiation at room temperature (RT)

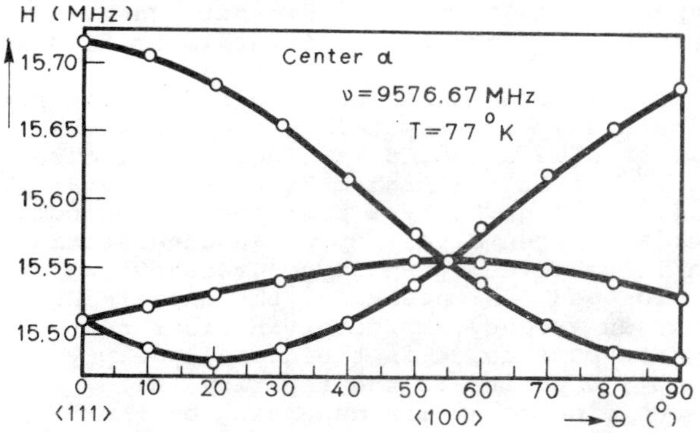

Fig.1. Isofrequency plot of the α-center spectra.

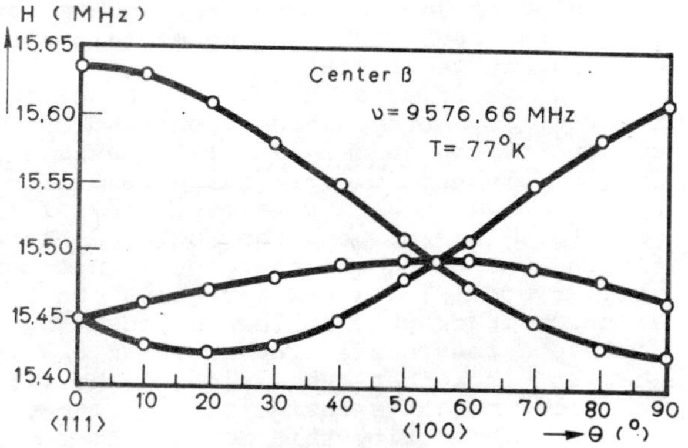

Fig.2. Isofrequency plot of the β-center spectra.

both sets of crystals did not exhibit any new ESR absorption. Irradiating at liquid nitrogen temperature (LNT) no changes could be observed in the ESR spectrum of crystals from ORNL. The samples grown by Norton Company did exhibit relatively strong anisotropic ESR signals. An isofrequency plot of these new lines for which the Zeeman vector is rotated in a (110) plane is given in Figs.1 and 2. Each set of lines may be interpreted as being the superposition of four equivalent spectra due to centers at sites with axial symmetry, the local axis being directed along one of the four ⟨111⟩ directions. The principal characteristics of these centers called α respectively β are given in table 1. No hyperfine structures could be observed for any orientation or temperature down to 1.3°K. The identification of these ESR absorptions as being due to different paramagnetic centers is also supported by different growing and annealing kinetics of the corresponding absorptions. Irradiating the crystal for consecutive periodes of 1/2 hours the α -center lines intensity reaches the saturation after about 5 hours. In the case of β-center it is necessary only 2 hours. With H∥⟨111⟩ the final intensity ratio is $I_\alpha : I_\beta = 5:1$ both centers displaying the same linewidths. From thermal bleaching experiments it was found that the α -center bleaches faster at RT (half-life of 15 minutes) than at LNT (half-life of 2 days). For periods of weeks no changes of the β -center intensity could be observed. Heating the samples at higher temperatures (up to 200°C) the β -center can be also annealed but a new relatively broader line grows in the neighbourhood of high -field component of Fe^+ spectrum. This isotropic line cannot be annealed even by heating the crystal up to 1000°C. Moreover the spin-lattice relaxation time measurements have shown a change in the relaxation process of Fe^+ perhaps through a spin-diffusion mechanism involving the new paramagnetic center.

Center	$g_\parallel$	$g_\perp$	Spin	Symmetry	Thermal stability	Associated charge
α	1.8541	1.8821	1/2	axial⟨111⟩	unstable at 77°C	-e
β	1.8633	1.8888	1/2	axial⟨111⟩	stable at RT	-e

Table 1. Characteristic data of the irradiation centers in ThO_2

Pulse recovery experiments were performed on both α and β -centers between LNT and 1.4°K . In this range of temperatures the recovery traces of the ESR absorption peaks did not exhibit for any orientation a simple exponential time dependence. However, some average relaxation times could be determined for different slopes. Comparing with Gd^{3+} relaxation, both centers yield longer relaxation times, the longest belonging to the β -center (up to 8o m sec at 1.4°K).

4. Discussion. Like in the case of other paramagnetic irradiation centers observed in ThO_2, it is difficult to propose definite structural models, due to the absence of any hyperfine structure. As both thorium and oxygen does not exhibit nuclear moments (excepting the ^{17}O isotope with an abundance of 0.037% and consequently its effect would be undetectable) the absence of hyperfine structures supports the idea that observed paramagnetic centers are associated with intrinsic defects of the lattice. However there are strong arguments that impurities has to be considered proposing structural models of the electron traps; the saturation of ESR absorption intensities at relatively low doses of irradiation, the presence of paramagnetic centers in samples with important level of impurities, the relatively large number of paramagnetic centers reported until now.

Concerning the β -center, a comparison with the + 174 center reported by Tint (1970) in samples UV-irradiated at LNT indicates similar structure. The fact that Tint did not see the α -center in his samples and also that we did not observe other ESR absorptions can be explained by the existence of different impurities in the respective crystals.

More information about the structure of the defects can be obtained considering their C_{3v} symmetry. This symmetry occurs naturally for any defect complex whose components are joined by a ⟨111⟩ axis. ESR studies on doped ThO_2 and CeO_2 indicated the presence of rare earth ions like Yb^{3+} or Eu^{2+} (Abraham et al,1965,1969) situated at sites with axial symmetry of ⟨111⟩ type given by a positive charge compensator. This compensator could be either an oxygen divacancy or an interstitial cation. Supposing the charge compensation is achieved by an oxygen vacancy, a simpler model for α and β centers can be proposed. In this model the unpaired electron is trapped at the compensating oxygen vacancy. Moreover in order to achieve a reasonable trap depth an O^{2-} ion is considered in the normally empty nearest - neighbor cubic site of the vacancy, opposite to the impurity. With an excess negative charge of only one or two electron units depending on the valence of impurity this model seems more realistic. The differences in g values and trap depth for α and β centers can be explained as being due to different impurity ions participating in the formation of the trap.

5. References

Abraham M.M., Finch C.B., Reynolds R.W. and Zeldes H., 1969, Phys.Rev.187, 451.
Abraham M.M., Weeks R.A., Clark G.W. and Finch C.B., 1965, Phys.Rev.137, 138.
Kolopus J.L., Finch C.B. and Abraham M.M., 1970, Phys.Rev.B2, 2040.
Neeley V.I., Gruber J.B. and Gray W.J., 1967, Phys.Rev.158, 809.
Marshall S.A. and Nistor S.V., 1972, Phys.Rev.B6, 24.
Marshall S.A., Nistor S.V., Huang C.Y. and Marshall T., 1973, Phys.Stat.Sol.b59,55.
Rodine E.T. and Land P.L., 1971, Phys.Rev.B4, 2701.
Röhrig R. and Schneider J., 1969, Phys.Lett.A30, 371.
Tint G.S., 1970, Defect structures in single crystal ThO_2. Thesis, Temple Univ.
Ursu I., 1968, La Résonance Paramagnétique Electronique, Dunod, Paris.

SECTION Q

Free Radicals

ENDOR IDENTIFICATION OF A R-NH_2 RADICAL IN GAMMA IRRADIATED ANTIFERROELECTRIC NH_4 H_2 AsO_4 SINGLE CRYSTALS.

B. Lamotte, Section de Résonance Magnétique, Département de Recherche Fondamentale, Centre d'Etudes Nucléaires de Grenoble, BP 85 Centre de Tri, 38041 Grenoble-Cedex, France.

Abstract. Identification is made, by ENDOR at 1.5°K, of a free radical species, created by gamma irradiation at room temperature of crystals of NH_4 H_2 AsO_4, whose EPR lines lie around g = 2, and which shows hyperfine interactions with two nearly equivalent protons and one nitrogen. The radical, called R-NH_2, is of the π type character on the NH_2 fragment, and the spin density in the nitrogen $2p_\pi$ orbital is about 0.6.

1. Introduction. We have undertaken the identification of paramagnetic defects created by gamma irradiation in the compounds of the KDP (K H_2 PO_4) family, in order to try to use these species as probes of the hydrogen bonds and of the ferroelectric or antiferroelectric properties which are present in these compounds (1-3).

In the course of this program, we present here the identification of a new radical R-NH_2, which has been detected in single crystals of ADA (i.e. the ammonium dihydrogen arsenate : NH_4 H_2 AsO_4) irradiated at room temperature. We have already mentionned (1) that the EPR spectrum taken at room temperature shows, in addition to the lines corresponding to the $(AsO_4)^{4-}$ and $(AsO_3)^{2-}$ radicals, a group of lines centered close to g = 2, in which one may recognize the spectrum of a rotating NH_3^+ radical. This group of lines is in fact more complex and non identified lines appear to be closely mixed with those corresponding to the NH_3^+ radical. In the present study, ENDOR has been used to identify the species which is responsible of their lines. ENDOR was made at 1.5°K (i.e. in the antiferroelectric phase) using already published experimental techniques (2).

2. Results. We have detected, in the range from 20 to 65 MHz, several groups of ENDOR lines whose frequency variations with the magnetic field direction exploring the crystallographic planes are presented in Fig. 1. These curves follow the orthorhombic P $2_1 2_1 2_1$ symmetry of the crystal. But since the crystal contain two types of domains corresponding to the exchange of their a and b axes, any exploration of an ac plane contain simultaneously the curves corresponding to the bc plane, as it can be seen in Fig. 1.(see below).

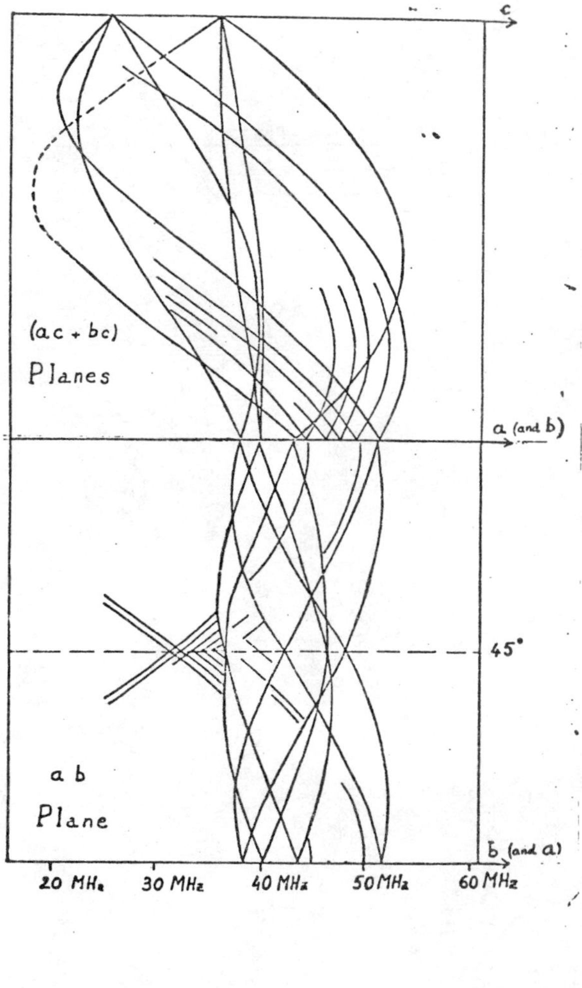

These ENDOR lines correspond to two protons and one nitrogen. For the two protons we have here their complete curves for all the sites for the ENDOR high frequency transition, but their low frequency transition could not be detected owing to the poor sensitivity. For the same reason, the ENDOR lines corresponding to the nitrogen are far more lacunar ; however they were easily recognized to be due to nitrogen by their weak splittings coming from the Zeeman nuclear, and quadrupolar terms.

From the experimental data, it was possible to deduce with a correct precision (the ENDOR line widths were about 0.5 MHz) the hyperfine tensors of the two protons. The hyperfine tensor of the nitrogen could be much more roughly evaluated with the little portions of curves experimentally obtained.

For the three tensors, since high and low frequency ENDOR transitions could not be obtained both, relative signs of their principal values, and of the director cosines of their principal directions could not be deduced.

However the obtained tensors are such that they can be, quite plausibly, organised in a coherent model, which, in turn, determine the relative signs of their principal values. In effect, the tensors of the two protons are typically of the aromatic type and they are nearly equivalent. The tensor of the nitrogen is nearly axial. Moreover, the principal directions of the two protons which would, in the aromatic model, correspond to the π direction, are nearly parallel and also parallel to the axis of the nitrogen tensor.

These results can be seen in Table 1, where the tensors are given with the signs corresponding to the model.

	Isotropic couplings	Dipolar tensors: Principal values.	Principal directions : their direction cosines on a	b	c
Proton 1 (a)	- 46.8 MHz	- 1.2 MHz -34.9 MHz +36.2 MHz	± 0.935 ± 0.240 ± 0.264	± 0.309 ± 0.913 ± 0.265	± 0.177 ± 0.329 ± 0.928
Proton 2 (a)	- 46.8 MHz	- 3.4 MHz -35.2 MHz +38.6 MHz	± 0.954 ± 0.265 ± 0.137	± 0.120 ± 0.759 ± 0.640	± 0.273 ± 0.594 ± 0.756
Nitrogen (b)	50 MHz	+50 MHz -25 MHz -25 MHz	± 0.94	0.0	± 0.33

Table 1

(a) Results are given with ± 0.3 MHz (b) Results are given within ± 3 MHz

They show that the NH_2 fragment is of aromatic character and that the $2p_\pi$ orbital of the nitrogen atom bears about sixty per cent of the spin density. The π axis is close to the a axis while the two N-H directions are near to the bc plane, the $N-H_1$ direction being near to the c axis and the $N-H_2$ direction at about 120° from it.

3. References.

(1) Gaillard, J., Constantinescu, O., and Lamotte, B., 1971, J. Chem. Physics, 55, 5447.
(2) Gaillard, J., Gloux, P., and Lamotte, B., 1974, Molecular Physics, 27, 6, 1441.
(3) Lamotte, B., Gaillard, J., and Constantinescu, O., 1972, J. Chem. Physics, 57, 3319.

RESONANCE FIELD SHIFTS IN THE PARAMAGNETIC AND ANTIFERROMAGNETIC STATES OF TYPICAL ORGANIC RADICALS

H. Ohya-Nishiguchi and O. Takizawa
Department of Chemistry, Faculty of Science, Kyoto University, Kyoto 606, Japan

Abstract. Paramagnetic and antiferromagnetic resonances from 1.3 to 300 K of single crystals of di-p-anisylnitroxide and DPPH solvent complexes have been investigated. The resonance field shifts observed are discussed in terms of a dipolar field due to the neighbouring magnetic moments induced by the applied field and/or a small uniaxial anisotropic field.

1. Introduction. Measurements of static magnetic susceptibilities (χ) of some paramagnetic organic free radicals have indicated a ferromagnetic or antiferromagnetic transition at low temperature. In this paper we describe some effects of the magnetization of the crystals on the electron resonance conditions, appeared typically in the para- or antiferro-magnetic state of some organic free radicals. According to the results of the chemical analyses, the proton magnetic resonance (PMR) and χ, DPPH solvent complexes are classified into three categories represented by the following crystals; a) solvent free DPPH crystallized from diethylether, b) DPPH·benzene (1:1) complex (hereafter abreviated as DPPH·Bz), and c) DPPH·CCl_4 (4:1) complex. These compounds have exhibited the traits of the χ-T curves individually different, although they are all paramagnetic above 1 K (Fujito, et al, 1973). On the other hand, χ and PMR of di-p-anysilnitroxide (DANO) have clearly shown a magnetic phase transition into the antiferromagnetic state at 1.67 K, as is shown in Fig. 1 (Takizawa, 1974). In the paramagnetic region above the temperature a broad maximum of the susceptibility has been observed at about 5 K, which is shown to be consistent with nearest neighbour antiferromagnetic Heisenberg exchange of a quadratic layer lattice (Duffy, et al, 1969). In these single crystals appreciable resonance field shifts (δH) which depend on the temperature and the direction of the applied field have been observed at low temperatures. We wish to examine here some mechanisms of the shifts in connection with the magnetic properties described above.

2. Experiments. Single crystals of DPPH and its solvent complexes were grown from the corresponding solutions by the spontaneous evapolation of solvent after twice or three times recrystallizations. DANO single crystals were obtained by the recrystallization from the ethyl alcohol solution. The hyperfine lines of Mn(II) ion in MgO were used for the measurements of δH, which is expressed by

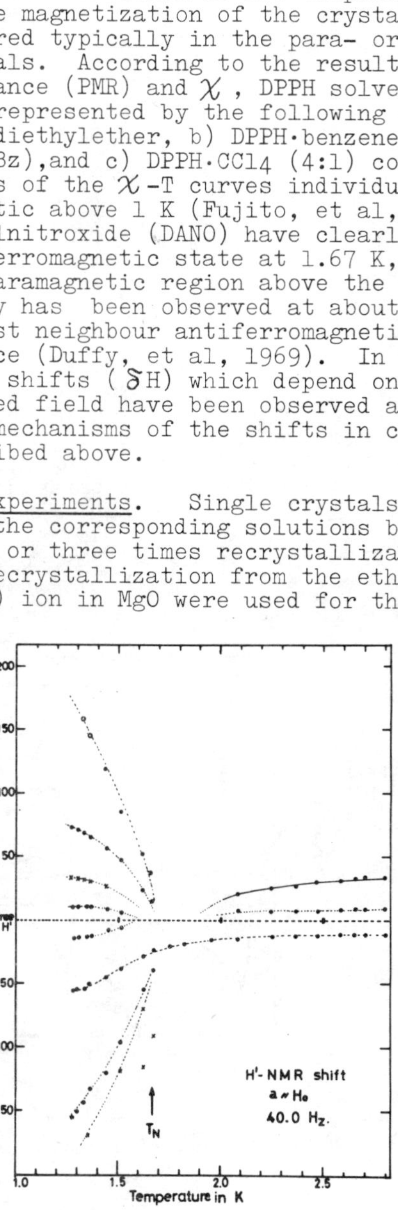

Fig. 1. Temperature dependence of the proton magnetic resonance shifts of DANO single crystal (H_0// a-axis). T_n indicates the antiferromagnetic transition.

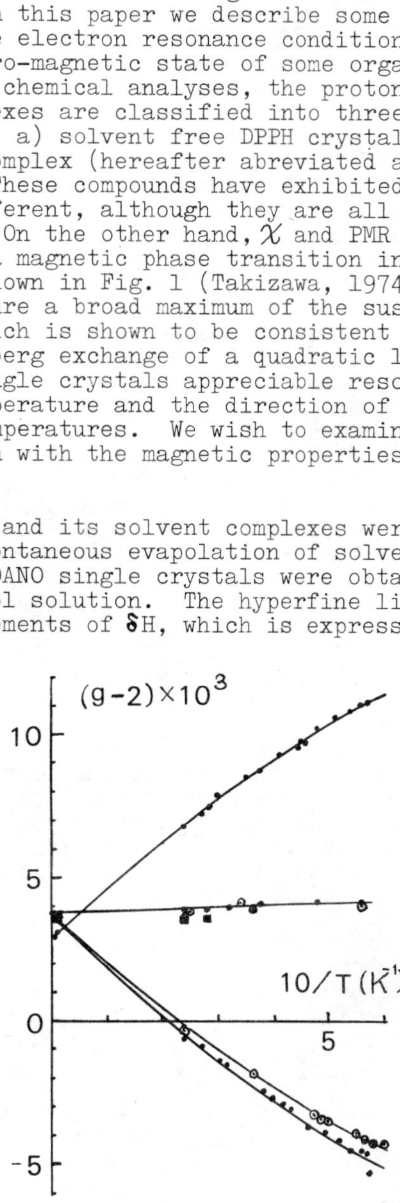

Fig. 2. Temperature dependence of g_{ap} of DPPH and its solvent complexes. The solid lines shown correspond to $g=g_{rt}+A/(T+0.45)$, where A is a parameter to fit the data.

the difference between the resonance field (H_{res}) and that at room temperature (H_{rt}), or by an apprent g-value;

$$\delta H = H_{res} - H_{rt} \qquad (1)$$

$$g_{ap} = \frac{H_{rt}}{H_{rt} + \delta H} \cdot g_{rt} \qquad (2)$$

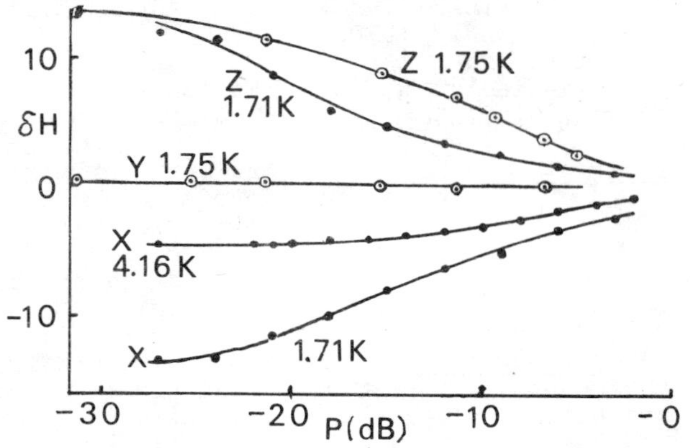

Fig. 3. Resonance field shift as a function of microwave power.

3. Results and Discussion.

a) δH in the paramagnetic region. Temperature dependence of g_{ap} of DPPH and its solvent complexes are shown in Fig. 2, where the symbols X, Y, and Z signify the directions of the applied magnetic field parallel to X, Y, and Z axis respectively. X axis is parallel to the long edge of the crystals, Z axis perpendicular to both the broader face and to the X axis, and Y axis perpendicular to both Z and X axis. The g_{ap} along the Z axis ($g_{ap}(Z)$) of DPPH·Bz decreases rapidly with decreasing temperature. On the other hand DPPH·$\frac{1}{4}CCl_4$ shows an increase of g_{ap} along the X axis which is in agreement with the results reported by Singer and Kikuchi (Singer, 1955). The g_{ap} is independent of the resonance frequency (X- and Q-band), and can be expressed by a formula $g_{rt}+A/(T+\Theta)$. These results suggest that δH is proportional to χ times the applied field. In order to clarify further the mechanism of δH we have measured the resonance field shift as a function of the microwave power along some typical directions at 4.2 and 1.7 K (see Fig. 3). It should be noted that δH decreases with increasing microwave power and converge to the non-shift position at room temperature. The power dependence of δH is in agreement with that of the saturation factor $1/(1+\gamma^2 H_1^2 T_1 T_2)$ in the experimental errors. These results can be reasonably interpreted by the mechanism that the magnetizationof the neighbouring molecules induced by the applied field, which is proportional to χH in the paramagnetic state, leads to a dipolar field (H_d) at the position of the resonant moment. Similar resonance field shifts have been observed in the case of DANO single crystal. The EPR spectra of the solvent free DPPH have shown an isotropic g-value independent of the temperature variation from 3.5 to the room temperature (g=2.0035).

b) δH near the phase transition temperature of DANO. δH in DANO single crystal expressed by g_{ap} showed much complicated temperature dependence (solid lines in Fig. 4), which can be simplified successfully by the subtraction of the H_d described above from the resonance field (dotted lines in Fig. 4). One of the characteristic features in the figure is that at the transition temperature (T_n) neither the abrupt change of the resonance field nor the divergence of the line width has been observed. Above T_n δH depends linearly on the resonance frequency, while below T_n δH deviates from the linearity (X- and Q-band). These phenomena can be interpreted by existence of an small anisotropic field below T_n assumig a condition

$$H_{res} \gg (H_E H_A)^{\frac{1}{2}} \qquad (3)$$

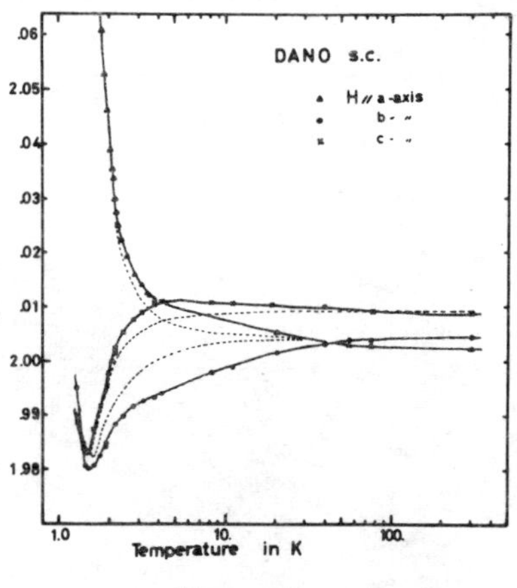

Fig. 4

Another feature in DANO is the existence of unique axis of the resonance field shift along the a-axis which may be related to the direction of the π-orbitals in DANO molecule.

4. References.

Duffy, W., Strandburg, D. L., and Deck, J. F., 1969, Phys. Rev., 183, 567.
Fujito, T., et al, 1972, Chem. Letters, 1972, 557.
Singer, L. S., and Kikuchi, C., 1955, J. Chem. Phys., 23, 1738.
Takizawa, O., 1974, to be published.

THE TEMPERATURE DEPENDENCE OF THE ESR SPECTRUM OF BIS-(TRIFLUORO-METHYL)-NITROXIDE AND DI-T-BUTYLNITROXIDE.

Tjeerd J. Schaafsma,
Laboratory of Molecular Physics, Agricultural University,De Dreyen6,Wageningen, The Netherlands.

Abstract. As is evident from the temperature dependence of their ESR spectra, di-t-butylnitroxide (I) and bis-(trifluormethyl)-nitroxide (II) show quite different association-behaviour. At room temperature I exhibits a spectrum consisting of two structureless lines, one with a linewidth of a few Gauss, the other having a width of more than hundred G. By lowering the temperature, the broad component disappears with a simultaneous increase in the intensity of the narrow line. The broad transition is ascribed to I in the gasphase, the other to pure liquid radical. While a single narrow line persists when the temperature is lowered down to $4^{\circ}K$, the linewidth is strongly temperature dependent. In contrast, II dimerizes to a diamagnetic compound if T is lowered. One subsequently observes the elimination of spin exchange, quenching of radical motion (and -isotropy) and the rotation of $-CF_3$ groups, when II is cooled down to $77^{\circ}K$. The results are discussed in terms of simple models.

1. Introduction. Nitroxides are widely used in ESR studies of structure and motion of biological macromolecules (Swartz,1972).To interpret lineshapes under various conditions, it is necessary to have a clear idea about the various interactions of the electron spin with its environnement. Therefore, we have studied the temperature dependence of the ESR spectrum of two simple nitroxides, di-t-butylnitroxide(DTBN,I) and bis-(trifluoromethyl)nitroxide (BFMN,II), both in their pure state, in the range 4-$400^{\circ}K$.

2. Experimental. Synthesis of I (Hoffman,1964) was followed by purification by vacuum-distillation and VPC; II was a gift from Dr.Russell Reinhard at Texaco Inc.,Beacon,U.S.A. ESR measurements were done with a Varian V 4502 spectrometer on degassed I in a sealed quartz-tube and on II condensed into a thickwalled quartztube free of oxygen (Schaafsma 1968). The viscosity of I was measured at various temperatures using an Ostwald viscosimeter.

3. Results. Fig.1 shows the ESR spectrum of I at $T>300^{\circ}K$. At $433^{\circ}K$ a broad unresolved line with a peak-to-peak derivative linewidth $\Delta H_{p-p} = 150$ G is observed together with a much narrower line of ≈ 7G wide. The broad line is identified as the gasphase spectrum by its disappearance on lowering T, whereas the central component, due to liquid I, grows at the same time. By extrapolating to zero vaporpressure, one obtains the unmodulated spin-rotational width $\Delta H_{SR}(0)$ by subtracting the hfs contribution ($2a_N \approx 30$G) from the observed linewidth.

From (Kivelson,1957 1960)

$$\Delta H_{SR}(0) = C\gamma \hbar^{-1} [8kT. \tilde{I}]^{\frac{1}{2}} \qquad (1)$$

where the average moment of inertia $\tilde{I} \approx 5\times10^{-38}$ gcm^2, $\gamma = 5.68\times10^{-8}$G rad^{-1} sec and $k,\hbar,T$ have their usual meaning, we calculate the r.m.s. spin-rotation interaction constant $C = 1.4\times10^{7}$ rad sec^{-1} or 0.82 G which is about 5 times smaller than for II (Schaafsma 1968)

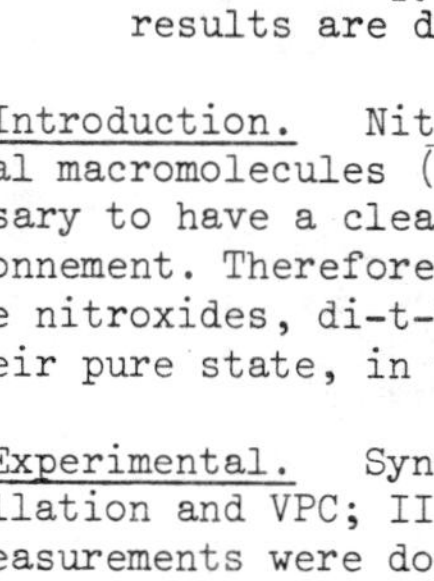

Fig.1. T dependence of the ESR(X-band) spectrum of pure DTBN.

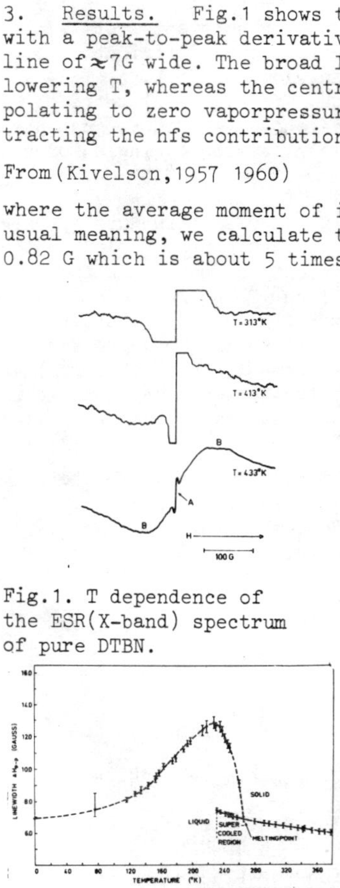

Fig.2. Peak-to-peak derivative linewidth ΔH_{p-p} vs.T of pure DTBN.

Although the linewidth ΔH of the central line of fig.1 is strongly dependent on T, as shown in fig.2, the susceptibility follows a simple Curie-Weiss law down to $4^{0}K$. Below that temperature, no reliable data could be obtained due to instrumental effects on the lineshape, which is Lorentzian from 4 to $380^{0}K$. The exchangefrequency ν_{ex} in the solid is calculated to be $\approx 2\times10^{11}sec^{-1}$ at $4^{0}K$, estimating the rigid lattice linewidth, arising from dipolar interaction and the anisotropy in g and a to be 90 G.
The increase in ΔH with T is probably due to the increase in the r.m.s. distance between nearest neighbours, thereby reducing the exchange interaction J(r). At melting ($267^{0}K$), ΔH is determined by two main processes (Krüger,1972): exchange modulation and rotational modulation. The first reduces the dipolar electron-electron and electron-nuclear interactions, as well as the g-value anisotropy and the isotropic hyperfine interaction. Except for the latter, rotational motion in the liquid can also contribute to line-narrowing. A plot of ΔH vs. T/η, where η is the viscosity of pure liquid DTBN, does not yield a straight line in the region 230-$380^{0}K$, demonstrating that the model of translational diffusion for dilute solutions (Kivelson,1957,1969,1972; Plachy,1967), cannot be applied to the pure liquid radical.
We estimate the exchange frequency ν_e by calculating $\bar{v}zr^{-1}$, where $\bar{v}$ is the r.m.s. speed, z is the number of nearest neighbours on a sphere with r.m.s. radius r, given by the liquid density $\rho = MN^{-1}r^{3}$, where M is the molecular weight and N Avogadro's number. Assuming that every molecular encounter leads to spin exchange and a molecular radius of 3.2 Å (Plachy,1967), we obtain

$$\nu_e^{-1} \approx 1\times10^{-12} \text{ sec.}$$

This is somewhat smaller than in the solid, and an order of magnitude smaller than the rotational correlation time, which is calculated to be 5×10^{-10}sec. at T = 267^{0}K from a hydrodynamic model. We thus conclude that the drop in ΔH upon melting is largely due to an increase in the rate of spin exchange and not to the onset of molecular reorientation. Bis(trifluoromethyl)nitroxide (Blackley & Russell,1965) shows a very different behaviour, as is shown in fig.3. On lowering T of a sample tube with 1 atm. BFMN at room temperature, one detects at 222^{0}K a broad gasphase spectrum, due to spinrotation interaction (Schaafsma & Kivelson,1968) and an exchange-narrowed line due to the liquid. Continuing, the gasphase spectrum disappears, as with DTBN, but now the radical concentration has also decreased. Since we observe spin exchange at rather low concentration in the solid (m.p. 203^{0}K) there is migration of the radical monomer through the solid by a displacement reaction with its nearest neighbours, as has been reported previously for solutions of II (Chin & Weissman,1970). Thus, there is no sharp break in the linewidth due to the quenching of molecular motion on solidification, as with DTBN. At 130^{0}K we find a superposition of an isotropic spectrum with $a_N \approx a_F$ = 8.9 G and an anisotropic spectrum which we tentatively describe with

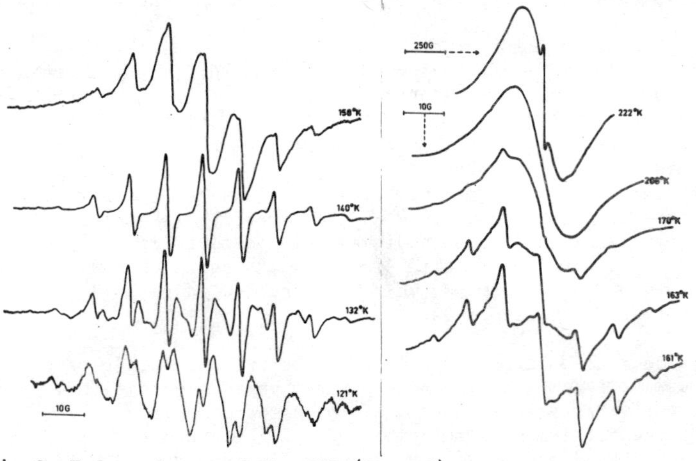

Fig.3. T dependence of the ESR (X-band) spectrum of pure BFMN. Note the different magnetic field scale of the spectrum at 222^{0}K.

g_1 = 2.003(2), g_2 = 2.007(4), g_3 = 2.010(6), a_1 = 9.4($\pm$0.1 G), a_2 = 8.9($\pm$0.3)G, a_3 = 9.2($\pm$0.1)G.

Note that $g_2 - g_1 = g_3 - g_2$ and $a_1 \approx a_2 \approx a_3$, leading to near-inversion symmetry around the spectral center. Since the spectrum can be interpreted with one set of hyperfine constants, we conclude that all six ^{19}F nuclei are still equivalent and have anisotropic coupling constants very similar to those of ^{14}N. At 77^{0}K the $-CF_3$ groups become inequivalent and the resulting distribution of anisotropic ^{19}F couplings wipes out all structure within the Gaussian width of 26 G. A high-gain search in other field regions did not reveal any parallel transitions. The highly symmetrical Gaussian lineshape at 77^{0}K is in accordance with the assumptions made above. At 208^{0}K, just above the melting point, ν_{ex} is large enough to span the total hyperfine width causing a single Lorentzian transition with a width of 13.6 G. From

$$\Delta H_{SI} = \frac{1}{\pi\sqrt{3}} \gamma \nu_{ex}^{-1} \sum_i \frac{1}{3} I_i (I_i + 1) a_i \qquad (2)$$

where I is the nuclear spin, we obtain $\nu_{ex} = 2.4 \times 10^8 \sec^{-1}$. At 160^{0}K hyperfinestructure appears, implying that ν_{ex} just comes in the range of the hyperfinesplitting constant, or $\nu_{ex} < 10^8 \sec^{-1}$. At 113^{0}K, all magnetic anisotropy is averaged out since ν_{ex} is sufficient to span the g factor anisotropy $\Delta g = g_i - g_0$(i=1,2,3). Thus, $\nu_{ex}(113^0K) \approx 3 \times 10^7 \sec^{-1}$. At still lower T the rate of migration is too slow to average out the g anisotropy, but it may cause the hyperfine coupling to appear isotropic for both types of nuclei at 130^{0}K. Cooling to 77^{0}K is required to let the hyperfine anisotropies broaden the spectrum again. The exchange rate may then be so small, that other kinetic processes take over, such as freezing out the $-CF_3$ rotation. Without further data we cannot distinguish between quenching of orientational averaging by exchange and freezing out of real rotational tunneling. Similar phenomena have been observed for bicyclic mitroxides (Mendenhall,1973).

4. Acknowledgement. The author is indebted to Dr.D. Kivelson of the Chemistry Department at UCLA, where part of the experiments were carried out, for his advice and encouragement. Dr. Richard Chui (UCLA,Phys.Dept.) made the measurements at liquid He temperature.

5. References.

Blackley,W.D., Reinhard,R.R., 1965, J.Am.Chem.Soc. 87, 802
Chin, S.A., Weissman,S.I., 1970, J.Chem.Phys. 53, 841
Hoffman,A.K., 1964, J.Am.Chem.Soc. 86, 641
Kivelson,D., 1957, J.Chem.Phys. 27, 1087
Kivelson,D., 1960, J.Chem.Phys. 33, 1094
Kivelson,D., 1972, Electron Spin Relaxation in Liquids, ed. L.T. Muus, P.W. Atkins, Plenum Press, New York
Krüger,G.J., 1972, Adv.Mol.Relaxation Processes, Elsevier Publ.Co., Amsterdam, p. 235-251
Mendenhall,G.D., Ingold,K.U., 1973,J.A.M.Chem.Soc. 95, 6390
Plachy,W., Kivelson,D., 1967, J.Chem.Phys. 47, 3312
Schaafsma,T.J., Kivelson,D., 1968, J.Chem.Phys. 49, 5235
Swartz,H.M., Bolton,J.R., Borg,D.C., eds. 1972, Biological Applications of Electron Spin Resonance, Wiley Interscience, New York.

ELECTRON NUCLEAR TRIPLE RESONANCE OF FREE RADICALS IN SOLUTION

K.P. Dinse, R. Biehl, K. Möbius
Institut für Molekülphysik, Freie Universität Berlin, 1 Berlin 33, Boltzmann-strasse 20, Germany.

Abstract. Electron Nuclear Triple Resonance (TRIPLE) experiments on free radicals in solution give a considerable signal-to-noise improvement as compared to ENDOR. In addition the observed linewidth is decreased and information about the number of protons contributing to a particular hfs constant is obtained. The experimental set up is described and the results are discussed in terms of a phenomenological theory.

Electron Nuclear Double Resonance (ENDOR) has proved already to be a powerful experimental technique for the analysis of complex ESR spectra, organic radicals in solution are an impressive example (Kwiram, 1971). ENDOR measurements not only revealed hyperfine splitting constants (hfsc) of large molecules of low symmetry but also gave information about structural properties through the elucidation of relaxation mechanisms (van Willigen et al., 1973). These results have been obtained in spite of certain drawbacks of the ENDOR method. Some of these are small obtainable ENDOR signals, the requirement of large NMR fields and partial loss of the important information about the number of protons belonging to a particular hfsc. Two of these obstacles can be overcome to a certain extent by performing Electron Nuclear Triple Resonance (TRIPLE). This extension of the ENDOR experiment was proposed by Freed in 1968 but up to now had not been performed on radicals in solution. In contrast to an ENDOR experiment, in which one of the two possible NMR transitions belonging to a set of equivalent nuclei with frequencies $\omega_n = |\gamma_n B_0 \pm 1/2\ \gamma_e a_n|$ is excited to desaturate the ESR transition, in TRIPLE resonance both nuclear transitions are excited simultaneously. To produce these two NMR fields at the sample, which may differ in frequency by up to 10 MHz, we used a cavity/solenoid broadband-arrangement, which was excited by a 400 W distributed amplifier. With 400 W rf power it was possible to produce simultaneously the two NMR fields with an amplitude of 10 Gauss (rot.) each. Experimental details are discussed elsewhere (Dinse et al., 1974).

It is quite easy to understand qualitatively the behaviour of a S = I = 1/2 spin system irradiated with two or three resonant fields. Neglecting coherence effects, a simple electrical circuit diagram description is sufficient to predict the ratio of TRIPLE to ENDOR enhancements as a function of microwave and NMR field strengths and of the various relaxation times (T_{2e}, T_{1e}, T_{2n}, T_{1n}) of the system (Freed, 1972), if the discussion is restricted to proton-TRIPLE and -ENDOR where cross relaxation effects can be usually ignored (Freed, 1967). In this simple picture one identifies lattice induced transition rates with electrical conductances and looks for the change of an "ESR conductance" while short circuiting NMR transitions. For the relaxation rates being in the limit of $W_e \gg W_n$ - a situation frequently met with organic radicals in solution - it is easy to show that to first order in W_n/W_e the limiting ENDOR enhancement is given by

$$E\,[\%] = 1/2 \cdot W_n/W_e \ .$$

The limitation due to the small nuclear relaxation rate can be overcome in the TRIPLE experiment. Simple estimates show an improvement of a factor of 2 compared to ENDOR, if the rf induced transition rates are equal to the lattice induced rates W_n, and a factor of 6 is gained if induced transition rates are 4 times W_n. By the same method it can be shown, that the number of protons belonging to a particular hfsc can be obtained approximately from high power TRIPLE line intensity ratios (Dinse et al., 1974).

Investigations have been performed on several radicals. Fig. 1 shows ENDOR and TRIPLE spectra of the tetraphenyl-pyrene radical anion. This example shows clearly a considerable increase in the signal-to-noise ratio as compared to ENDOR. In addition, the TRIPLE lines give information about the number of protons contributing to a particular hfs constant, an information which is lost in high power ENDOR. Furthermore we have shown (Dinse et al., 1974), that the saturation broadening of the NMR transitions - inherent in both ENDOR and TRIPLE experiments - is reduced by roughly a factor of 2 in the TRIPLE experiment.

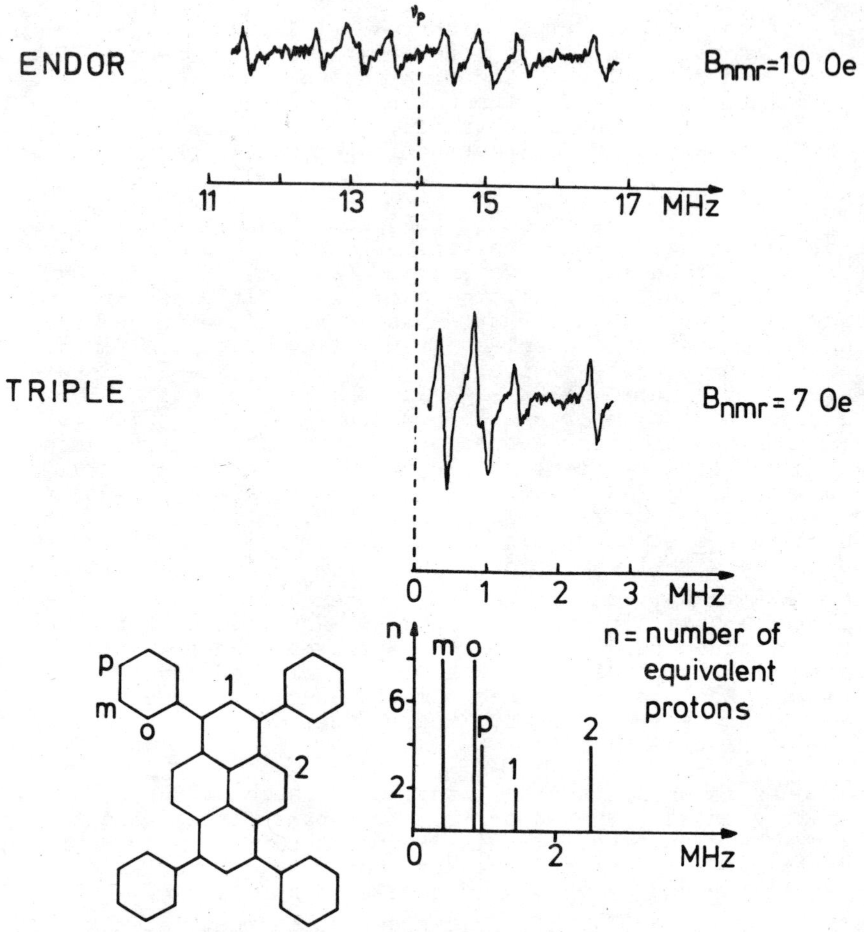

Fig. 1 ENDOR and TRIPLE spectra of tetra-phenyl-pyrene$^{(-)}$ at 215°K

References

Dinse, K.P., Biehl, R., a. Möbius, K., 1974, J.Chem.Phys. in press.
Freed, J.H., 1967, J.Phys.Chem. 71, 38.
Freed, J.H., 1972, in: Electron Spin Relaxation in Liquids, L.T. Muus and P.W. Atkins Edts., Plenum Press, N.Y.
Kwiram, A.L., 1971, Ann.Rev.Phys.Chem. 22, 133.
van Willigen, H., Plato, M., Biehl, R., Dinse, K.P., a. Möbius, K., 1973, Mol.Phys. 26, 793.

Low Temperature Magnetic Properties of DPPH.$(C_6D_6)_x$

P. Grobet*, R. Verlinden** and L. Van Gerven
Laboratorium voor Vaste Stof-Fysika en Magnetisme, University of Leuven, 3030 Leuven (Belgium)

Abstract. NMR and low field EPR measurements on powdered DPPH.$(C_6D_6)_x (0 \leq x \leq 1)$ are performed in order to get more detailed information about the hyperfine coupling, the proton spin resonance spectrum and the properties of the electron spin system at liquid helium temperatures. In pure DPPH the formation of dimers is clearly observed.

1. Introduction. Important information about the complex hyperfine interaction of the unpaired electron spin with the different proton spins in DPPH, as well as on the properties of the electron spin system itself, can be obtained from proton magnetic resonance (PMR) at low temperatures. Some work has been done along this line (Gutowsky et al., 1959; Karimov, 1961; Anderson et al., 1960). In the low temperature PMR spectrum of most of our samples we observed (Grobet, 1973), like other investigators, besides several lines, paramagnetically shifted by hyperfine coupling, a strong unshifted, central, diamagnetic line, indicating that many of the protons in those samples are not hyperfine coupled to unpaired electrons. Anderson et al. (1960), who investigated DPPH.C_6H_6 and DPPH.C_6D_6, ascribe this unshifted line to protons of solvent molecules, crystallized into the solid. However, already from our PMR spectra of DPPH.$(CDCl_3)_{1/3}$ and DPPH.$(CCl_4)_x$, where we observed a strong central line (P. Grobet et al., 1973), we could invalidate somewhat Anderson's explanation: the unshifted line is probably not only due to solvent protons. In order to verify this preliminary conclusion, and in order to get more detailed data about the hyperfine coupling in DPPH, about the origin of the unshifted PMR line and about the electron spin system itself at low temperatures, we concentrated our research on solid DPPH.$(C_6D_6)_x (0 \leq x \leq 1)$, using PMR as well as EPR.

2. Samples and experimentals. In our experiments on DPPH.$(C_6D_6)_x$ three different samples have been investigated, all powdered and ground: DPPH.C_6D_6 (the saturated sample), DPPH.$(C_6D_6)_{0.02}$ (the dried sample) and DPPH (the pure sample). The saturated sample is crystallized from a very diluted solution of DPPH powder, supplied by Eastman Kodak Co, in C_6D_6 (99.5 % purity). Pumping on the powder is restricted to a minimum. The dried sample is prepared from our saturated sample by a heat treatment at 65°C in vacuo during 48 hours. The pure sample is prepared from a very diluted solution of DPPH powder in ether and purified in a chromatographic column (T. Laederlich et al., 1962). Afterwards it is crystallized and dried in vacuo. All PMR spectra are taken from a broad line NMR spectrometer with marginal oscillator working at 40 MHz. The temperature of the samples is varied, in a conventional liquid helium dewar between 4.2 K and 1.4 K.

3. Experimental results and discussion. a) Saturated DPPH. The PMR spectrum of saturated DPPH shows at 4.2 K a quite weak unshifted line and four shifted lines (fig. 1a). The experimental hyperfine coupling constants a_H calculated from the paramagnetic shifts of the four principal shifted lines, agree very well with those found by Anderson et al. (1960) for DPPH.C_6D_6 (Verlinden et al., 1974). The shifts ΔH as a function of temperature, do not follow a Curie-Weiss law down to the lowest temperature 1.4 K. Assuming ΔH to be strictly proportional to the electron spin susceptibility χ, the deviation corresponds to an electron spin susceptibility lower than given by a Curie-Weiss or Langevin law. This result is in disagreement with the observation of Anderson et al. (1960), who found for ΔH a Curie law behaviour down to 1.5 K. There is, at least qualitative, agreement between our results of ΔH vs. $1/T$ and the experimental χ vs. $1/T$ data obtained on DPPH.C_6D_6 by T. Fujito et al. (1972). Assuming the formation, below a given temperature, of molecule pairs (dimers), which give rise to a thermally accessible pair state susceptibility for the electron spin system, given by $\chi = C/T\,[3 + \exp(J/kT)]$ (C = the Curie constant; J = the energy separation between the singlet ground level and the triplet excited level of the dimer), Fujito et al. (1972) deduce from their reults a value for J/k in DPPH.C_6H_6 of 1.04 K. Our PMR shift data fit with a value of J/k = 0.82 K (fig. 2). When lowering the temperature or increasing the frequency, no change of resolution nor of overall line shape of the hyperfine PMR spectrum is observed. This

* "Onderzoeksleider" of the Belgian N.F.W.O.
** "Navorser" of the Belgian I.I.K.W.

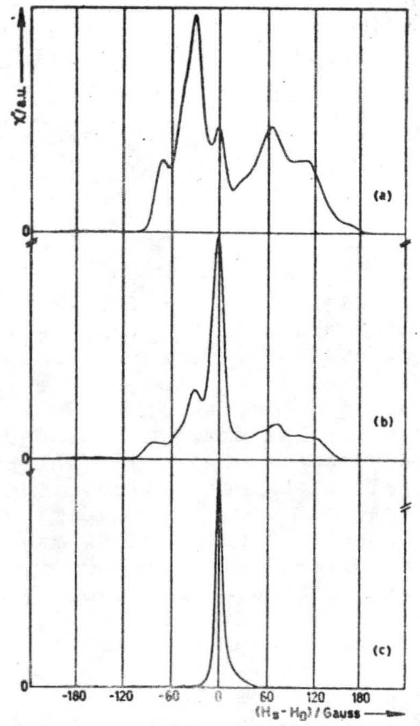

Fig. 1 The proton magnetic resonance spectrum of DPPH. $(C_6D_6)_x$ powder (39.47 MHz; 4.2 K). a) satureted DPPH (x=1). b) dried DPPH(x≙0.02) c) pure DPPH(x<0); T=1.3K.

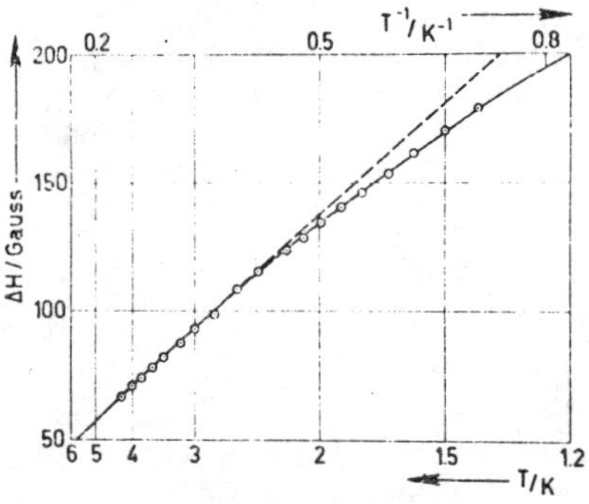

Fig. 2 The paramagnetic shift of one of the principal lines (saturated DPPH); ———: theoretical curve based on pair state susceptibility (J/k = 0.82 K); ---: theoretical curve based on Langevin type susceptibility.

indicates that the hyperfine splitting is higly anisotropic.

b) Dried DPPH. The position of the shifted lines in the PMR spectrum of dried DPPH is the same as in saturated DPPH (fig. 1b). However measurements of the shifts as a function of temperature are difficult, because of the much lower intensity of these lines vs. the unshifted one, and also because of broadening of the lines with decreasing temperature. Most strikingly, dried DPPH shows a very strong unshifted line. This line, for sure, is not due to solvent molecules: the solvent used is deuterated for 99.5 % and, moreover, the sample is dried thoroughly. Diamagnetic impurities, possibly introduced during the drying process, cannot be the cause either. Indeed, saturating again the dried sample with deuterated benzene reduces the unshifted line to the intensity of the one in saturated DPPH. We may conclude that the statement of Anderson et al. (1960) - that the central line is principally due to solvent molecules - is incorrect.

c) Pure DPPH. The pure sample provides a PMR spectrum which is completely different from the two others (fig. 1c). Around and below 2 K it contains an almost symmetric single unshifted line with a width between inflexion points of 5.6 Gauss, independent of temperature and magnetic field. Above 2 K some gradual splitting of the line occurs. Around 10 K the unshifted line seems to disappear and the PMR spectrum becomes predominantly paramagnetic, although weakly resolved. The building up of an unshifted line below 10 K could be very well connected to the rather drastic decrease of χ with decreasing temperature below 10 K, found by Fujito et al. (1972). Both facts could be explained, again, by the presence of a thermally accessible pair state, similar to the one, appearing at much lower temperatures, in saturated DPPH. An antiferromagnetic transition is excluded, since it cannot produce a clear, narrow, unshifted line (Wittekoek et al., 1966). The presence of such a pair state (dimer) in pure DPPH around 4 K is proved by the observation in the EPR spectrum (57 MHz) of the typical g = 4 line (Verlinden et al., 1974).

We may conclude that the most characteristic PMR spectra are the one of saturated DPPH.C_6D_6 (fig. 1a) and the one of pure DPPH (fig. 1c). Solvent concentrations in between ($0<x<1$) produce a mixture of both these characteristic spectra. In respect to samples crystallized in other solutions than benzene, we may conclude that an unshifted line shows up whenever not all free radical molecules are separated by solvent molecules. In these circumstances - provided the temperature is low enough - here and there molecule pairs are formed, which have a strongly reduced pair state electron spin susceptibility.

References.

Anderson M.E., Pake G.E. and Tuttle Jr. T.E. 1960, J. Chem. Phys. 33, 1581.
Fujito T., Enoki T., Ohya-Nishiguchi H. and Deguchi Y. 1972, Chem. Lett. 7,557.
Grobet P., Verlinden R. and Van Gerven L. 1973, Magnetic Resonance and Related Phenomena, 413, V. Hovi, North Holland, Amsterdam.
Gutowsky H.S., Kusomoto H., Brown T.H. and Anderson D.H. 1959, J. Chem. Phys. 30, 860.
Karimov Y. and Shehgolev I. 1961, Soviet Phys. JETP 40, 3.
Laederlich t. and Traynard P. 1962, Comp. Rend. 254, 1826.
Verlinden R., Grobet P. and Van Gerven L. submitted to Chem. Phys. Letters.
Wittekoek S. and Poulis N.J. 1966, Physica 32, 693.

ENDOR SPECTROSCOPY OF THE FREE RADICALS OF SOME CONJUGATED HYDROCARBONS IN LIQUID SOLUTION

F.W. Heineken and T.C. Christidis
Department of Physics, American University of Beirut, Lebanon.

Abstract. ENDOR spectra have been obtained for phenanthrene, pyrene, perylene, pentacene, 1,2,3,4-dibenzanthracene and 9,10-diphenylanthracene in solution. For the last two radicals no ESR data are known. Hyperfine coupling constants are determined with an accuracy of ± 10 kHz and the assignments are compared to the results of HMO and McLachlan type calculations.

1. Introduction. Several authors (Lagendijk et al, 1970; Dinse et al, 1972; Biehl et al, 1973) have reported Electron Nuclear Double Resonance (ENDOR) studies of conjugated hydrocarbon free radicals in liquid solution. Their investigations covered the molecules naphtalene, anthracene. tetracene, p-terphenyl, m-terphenyl and 9-phenylanthracene. It has been possible to extend this work and in this paper ENDOR spectra of phenanthrene, pyrene, perylene, pentacene, 1,2,3,4-dibenzanthracene and 9,10-diphenylanthracene are reported. The spectra of pyrene, perylene and pentacene did not yield any new data, compared to the existing ESR data. However the measurements of the ENDOR spectrum of phenanthrene make it possible to favour the smaller coupling constant of the two contradicting values, reported in the ESR literature (McLachlan, 1959; Colpa et al, 1961). No ESR has been reported previously for the last two radicals. This is most probably due to the complex and unresolved nature of the ESR spectra (Fig.1). Hückel molecular orbital as well as McLachlan type calculations have been performed. The assignments of the experimental splitting constants are compared to the results of the above calculations.

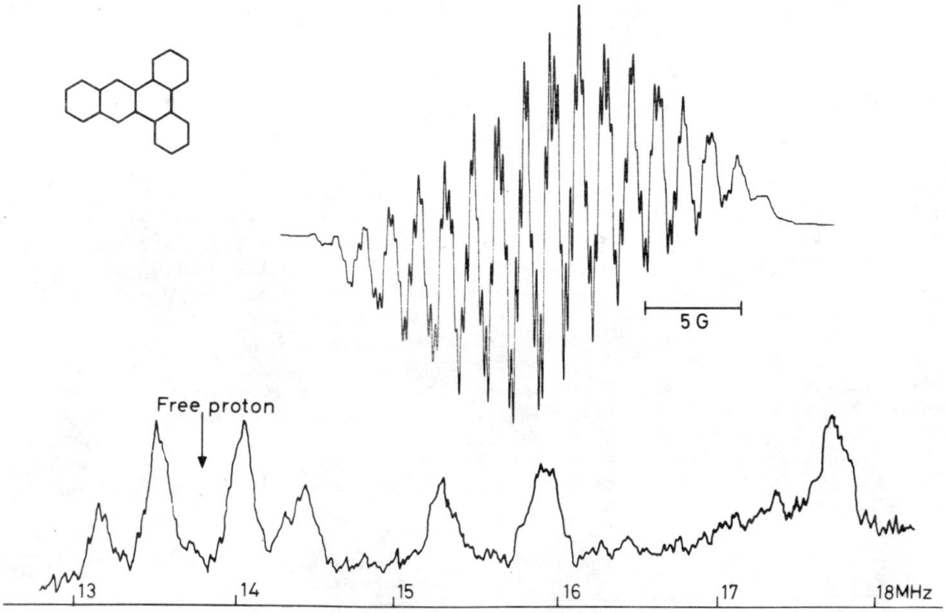

Fig.1. ESR and ENDOR spectrum of 1,2,3,4-dibenzanthracene.

2. Experimental Technique. The chemical compounds were obtained commercially and were used without any further purification. The free radicals are prepared by reduction of the hydrocarbons with a Na or K mirror under high vacuum in 1,2-dimethoxyethane (DME). The radical concentration is about 10^{-3} M. The optimum temperature for recording the ENDOR spectra is just above the freezing point of the solvent (about -60 °C). The ENDOR spectra are recorded with an ENDOR spectrometer built in our department. A special designed rectangular ENDOR cavity (Christidis et al, 1973) was used, in which RF fields of 8 G in the rotating frame were generated.

3. Results and Discussion. The experimental hyperfine structure constants of phenanthrene, pyrene, perylene and pentacene have been collected in Table 1, together with the values reported in the ESR literature. The spectra of pyrene, perylene and pentacene did not yield any new data. However for phenanthrene two different splitting constants have been reported in the literature. Both authors claim respectively a difference of 100 and 200 mG between the two smallest splitting constants a_3 and a_5. This difference could not be observed in the ENDOR spectrum. This may be due to the fact that this difference is within the ENDOR linewidth for this particular free radical (about 230 kHz or 165 mG). However our ENDOR measurements definitely favours the smaller difference and the splitting constants given by Colpa et al.
No ESR has been reported previously for the last two free radicals, 1,2,3,4-dibenzanthracene and 9,10-diphenylanthracene. The ENDOR signal-to-noise ratio is relatively small for these free radicals. This is partly due to their small ESR signal-to-noise ratio. Increasing the

Table 1

Experimental hyperfine splitting constants in Gauss.

		a_1	a_2	a_3	a_4	a_5	a_6
phenanthrene 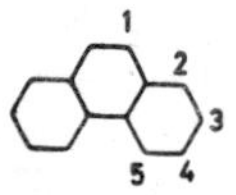	ENDOR	4.58	3.83	0.55	2.97	0.55	--
	ESR+	4.75	3.84	1.02	3.02	0.91	--
	ESR*	4.43	3.71	0.43	2.88	0.63	--
pyrene 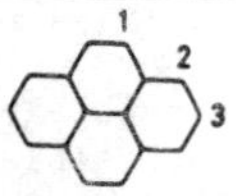	ENDOR	2.13	4.81	1.04	--	--	--
	ESR**	2.08	4.75	1.09	--	--	--
perylene 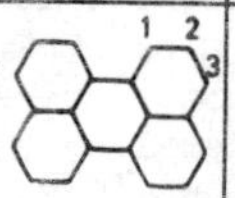	ENDOR	3.06	0.43	3.51	--	--	--
	ESR*	3.09	0.46	3.53	--	--	--
pentacene 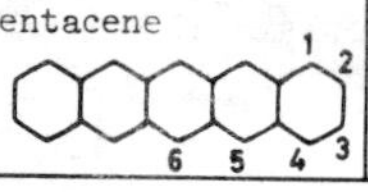	ENDOR	0.89	0.89	--	--	3.03	4.27
	ESR*	0.88	0.88	--	--	3.01	4.27

(+McLachlan, 1959; *Colpa et al, 1963; **Hoijtink et al, 1961)

concentration of the sample did not give any improvement. Since the signal-to-noise ratio is small and the ENDOR signal amplitude is not directly proportional to the number of equivalent protons, it was rather difficult to discern overlapping lines. However the total linewidths of the ESR spectra could easily be obtained and these were very useful in the final analysis. Hückel and McLachlan molecular orbital calculations were performed on these two molecules and the assignment of splitting constants is made on the basis of these calculations. In Table 2 the ENDOR measurements have been collected together with the calculated spin densities. The agreement is reasonably well in the case of 1,2,3,4-dibenzanthracene, most probably due to the fact that the molecule is planar. Only three ENDOR lines are observed for 9,10-diphenylanthracene, while five lines are to be expected. However the low frequency ENDOR line has an amplitude, which is three times as large as the others. This may be due to overlap of lines. The assignment of splitting constants as given in Table 2 is made with the help of the molecular orbital calculations and is based on the fact that this assignment is the only one that gives the correct total ESR spectrum width. The agreement with the calculations turned out to be poor. This is not unreasonable, because of the twist angle of the phenyl groups with the plane of anthracene (Biehl et al, 1973), which violates the assumption of $\sigma - \pi$ delocalization in the molecular orbital calculations.

Table 2

Experimental hyperfine structure in Gauss
and theoretical spindensities

		a_1	a_2	a_3	a_4	a_5	a_6	a_7
1,2,3,4 - dibenzanthracene 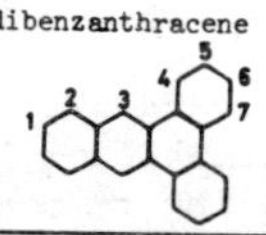	ENDOR	1.49	2.79	5.30	0.18	1.06	0.46	0.18
	HMO	0.051	0.114	0.179	0.004	0.026	0.020	0.008
	McLachlan	0.046	0.142	0.241	-0.005	0.016	0.014	0.000
9,10 - diphenylanthracene 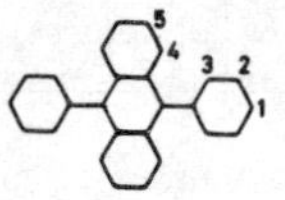	ENDOR	0.23	0.23	1.45	2.62	0.23	--	--
	HMO	0.047	0.001	0.043	0.058	0.034	--	--
	McLachlan	0.059	-0.019	0.056	0.068	0.027	--	--

4. References.

Biehl, R., Plato, M., Mobius, K., Dinse, K.P., Congress Ampere 17 (1973) 423.
Christidis, T.C., Heineken, F.W., J. Phys. E: Sci. Instr. 6 (1973) 432.
Christidis, T.C., Heineken, F.W., Chem. Phys. 2 (1973) 239.
Colpa, J.P., Bolton, J.R., Mol. Phys. 6 (1963) 273.
Dinse, K.P., Biehl, R., Mobius, K., Plato, M., J. Magn. Res. 6 (1972) 444.
Hoijtink, G.J., Townsend, J., Weissman, S.I., J. Chem. Phys. 34 (1961) 507.
Lagendijk, A., Tromp, N.F.M., Glasbeek, M., Van Voorst, J.P.W., Chem. Phys. Letters 6 (1970) 152.
McLachlan, A.D., Mol. Phys. 3 (1959) 233.

PAIRWISE TRAPPING OF RADICALS IN IRRADIATED SOLIDS

M.C.R. Symons
Department of Chemistry, The University, Leicester LE1 7RH.

Abstract. The phenomenon of pairwise trapping of radicals in irradiated solids is briefly reviewed, and the form of the resulting e.s.r. spectra discussed. Details of an e.s.r. spectrum assigned to radical pairs are presented and discussed.

In the field of solid-state transition-metal ions there are two common circumstances in which spin-spin interactions between pairs of paramagnetic ions occur. One is for diamagnetic host crystals containing moderate concentrations of a paramagnetic dopant such that there is an appreciable statistical chance that neighbouring sites be occupied by the dopant. The other common situation is that two paramagnetic cations are bonded to a common ligand so that they are held close together.

It is less well appreciated that a similar situation can arise in photolytic and radiation studies. In both, there are often two paramagnetic primary products formed from a single energy quantum. These may separate, in which case e.s.r. spectroscopy will detect isolated doublet species, or they may be constrained to remain together. In that case, combination or disproportionation may destroy the radical products directly, or if they are prevented from moving by the rigid matrix, a pseudo triplet-state species may result. If the separation between the radicals within each pair is spread over a range, then these triplets will probably not be detected by e.s.r. If, however, they are precisely separated, then good e.s.r. spectra will result.

In the absence of hyperfine coupling and large g-value variation, the g = 2 spectrum has two $M_S = 1$ transitions whose powder spectrum will take the form shown in Fig. 1. The separation (2D) between the outer, parallel features is directly linked to the mean separation between the electrons (R) (on the point-dipole approximation) by $D = g^2\beta^2R^{-3}\,(3\cos^2\theta - 1)$ and hence a good estimate of R can be obtained. There will also be a g = 4 transition which can be crudely described as $M_S = 2$. If hyperfine coupling is present in the parent radicals, this may appear as if the two unpaired electrons are completely delocalised across the two species. Alternatively, if spin-exchange is small (because orbital overlap is poor) then the hyperfine coupling will remain the same as that in the parent radicals.

The first e.s.r. observation of such "pair-trapping" was for photolysed $K_2S_2O_8$ crystals (Atkins et al., 1963). The primary act is thought to give two $\cdot SO_4^-$ radicals and in order to explain the large separation of 15.8 Å, it was necessary to postulate a side-ways reaction with neighbouring $S_2O_8^{2-}$ ions, the final process being to give (Barnes and Symons, 1966)

$$O_3S\dot{O}O^- + O_3SOSO_3^{2-} + O_3SOSO_3^{2-} + \dot{O}OSO_3^- .$$
$$|\leftarrow\text{---------}\ 15.8\ \text{Å}\ \text{---------}\rightarrow|$$

Since then there have been many observations of this phenomenon (Box, 1971; Ledeber, 1969). One particularly clear cut and significant result occurs for the molecule $(CH_3)_2C(CN)N{=}NC(CN)(CH_3)_2$, which is widely used as a source of free radicals. On photolysis of the solid at 77 K, a complex e.s.r. spectrum results (Ayscough et al., 1964) which was later analysed in terms of two $(CH_3)_2\dot{C}CN$ radicals trapped ca. 6 Å apart (Symons, 1967). Thus the process $RN = NR \xrightarrow{h\nu} R\cdot + N_2 + \cdot R$ occurs efficiently at 77 K.

Radiolysis gives pair-trapping far less frequently, and indeed it is perhaps surprising that it occurs at all. One example that seems to be closely similar to photolysis in that it probably stems from the decomposition of an electronically excited molecule (McRae and Symons, 1968), is $PhOCOOPh \rightarrow PhO\cdot + CO_2 + \cdot OPh$ (PhO· ... ·OPh separated by 5.8 Å).

However, the primary act in γ-radiolysis is generally electron ejection, and I have suggested elsewhere (Symons, 1967) that pair-trapping may occasionally occur because one or more molecules adjacent to the ejecting molecule may have a very high electron-capture cross-section. This concept has been modified by Iwasaki et al. (Iwasaki et al. 1969) and Box (Box, 1971) who suggest that the ejected electron is initially captured at some distance away from the radical-cation, which then transfers a proton to a neighbouring molecule. On being thermalised, the electron may return under the influence of the positive charge, but will then react with the protonated neighbour: $SH \rightarrow \dot{S}H^+ + e_t^-$; $\dot{S}H^+ + SH \rightarrow S\cdot + SH_2^+$; $SH_2^+ + e^- \rightarrow S\cdot + H_2$.

We have recently discovered a most unusual and well defined example of pair-trapping which illustrates yet another possible mode of interaction (Neilson and Symons). Exposure of bromomaleic acid ($HO_2CCH{=}C(Br)CO_2H$) to ^{60}Co γ-rays at 77 K results in an extremely

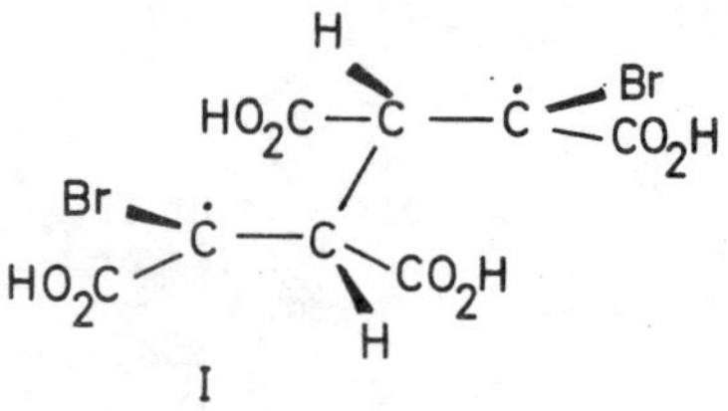

I

complex powder e.s.r. spectrum (Fig. 2) which we have interpreted in terms of the bi-radical I. This interpretation explains satisfactorily the estimated separation of 5.5 Å, the hyperfine coupling to ^{71}Br and ^{81}Br (exchange is weak so only a single bromine nucleus is detected) and the directions of the various tensor components. We suggest that an excited state molecule bonds specifically to its nearest neighbour (only one is conveniently placed) to give this radical in one step. It is not easy to see why this dimerisation should occur unless an electronically excited molecule is involved. This is an example of a reaction that would only occur in the solid-state provided orientations and separations are correct in the first instance. In this case this is true for specific pairs of radicals: if, however, monomer molecules are properly placed in chains, then solid-state chain-polymerisation can result.

When pair-trapping is observed in high energy radiolyses, it is generally a relatively minor occurrence, the major act being the formation of the components as well separated species. These need not be equivalent, although they frequently are. A good example of this is the formation of the d^7 ion $Fe(CN)_5NO^{4-}$ [or $Fe(CN)_4NO^{3-}$] by γ-irradiation of sodium nitroprusside (McNeil et al., 1966; Goodman et al., 1966). The major product at room temperature are well separated d^7 ions, but pairs are present in concentrations that are thousands of times too high for purely statistical trapping. These are arbitrarily formed from all of the many types of nearest neighbours.

In contrast, pair-trapping can sometimes be remarkably specific, with only very low relative concentrations of the trapped separate components. This is, of course, normal for photolytic pairs (Atkins et al., 1963; Barnes and Symons, 1966), or for pairs formed from excited state species (Box, 1971), as is the case for bromomaleic acid pairs (Neilson and Symons). It seems, however, to be rare in most radiation experiments.

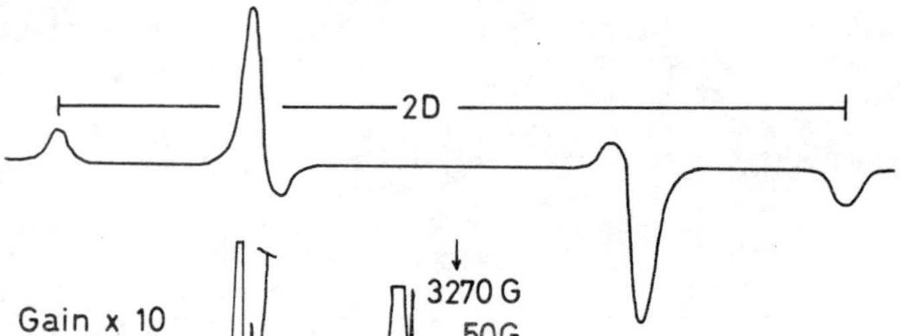

Fig. 1 - Hypothetical first derivative e.s.r. spectrum for radical-pairs having g = 2 and axial symmetry.

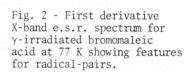

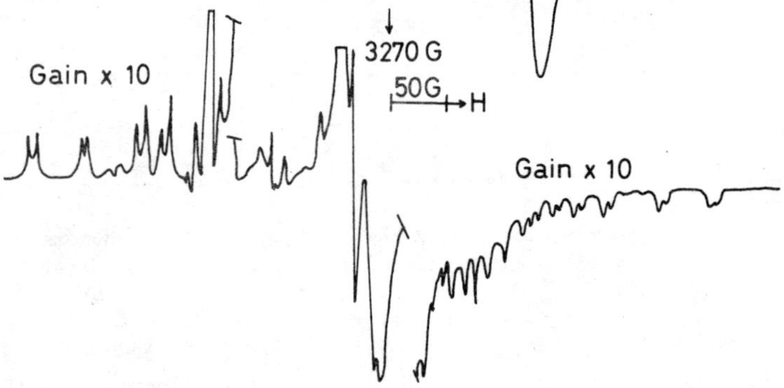

Fig. 2 - First derivative X-band e.s.r. spectrum for γ-irradiated bromomaleic acid at 77 K showing features for radical-pairs.

References

Atkins, P.W., Symons, M.C.R. and Trevalion, P.A., 1963, Proc. Chem. Soc., 222.
Ayscough, P.B., Brooks, B.R. and Evans, H.E., 1964, J. Phys. Chem., 68, 3889.
Barnes, S.B. and Symons, M.C.R., 1966, J. Chem. Soc., 66.
Box, H.C., 1971, J. Phys. Chem., 75, 3426.
Goodman, B.A., McNeil, D.A.C., Raynor, J.B. and Symons, M.C.R., 1966, J. Chem. Soc. (A), 1547.
Iwasaki, M., Ichikawa, T. and Ohmori, T., 1969, J. Chem. Phys., 50, 1991.
Ledeber, Y.S., 1969, Radiation Effects, 1, 213.
McNeil, D.A.C., Raynor, J.B. and Symons, M.C.R., 1966, Mol. Phys., 10, 297.
McRae, J.A. and Symons, M.C.R., 1968, J. Chem. Soc. (B), 428.
Neilson, G.W., and Symons, M.C.R., unpublished results.
Symons, M.C.R., 1967, Nature, 213, 1226.
Symons, M.C.R., 1967, in "Radiation Research", North-Holland Publishing Co., Amsterdam, p. 131.

ETUDES PAR R.P.E. DE SEMI-COKES D'ANTHRACENE

C.Simon, H.Estrade, D.Tchoubar et J.Conard
Centre de Recherche sur les Solides à Organisation Cristalline Imparfaite, Rue de la Férollerie, Orléans (Loiret), France

Abstract. The results presented in this paper show the relationship between structural properties and the characteristics of the E.P.R. lines in a soft carbon obtained from very pure anthracene charred at 400°C and heat-treated between 400°C and 1200°C.

1. Introduction. On sait qu'il est possible d'obtenir des microcristaux de graphite à partir de corps organiques carbonisés puis traités thermiquement jusqu'à environ 3000°C. L'évolution vers la structure tridimensionnelle du graphite pose encore de nombreux problèmes aux chercheurs qui veulenten comprendre les mécanismes, mais il semble qu'elle soit conditionnée par les tout premiers stades de carbonisation et de cokéfaction. En particulier, la qualité cristallographique du produit final dépend du produit d'origine (Edström et al.1969). Le but de l'étude présentée ici était d'étudier par rayons X et R.P.E., ces premiers stades de transformation d'échantillons extrêmement purs dont on sait qu'ils donnent après traitement thermique un excellent graphite (d_{002} = 3,354).

2. Préparation des échantillons. Les échantillons fournis par le Groupe Français d'Etude du Carbone sont obtenus par carbonisation vers 400°C de cristaux d'anthracène. Nous avons donc au départ un semi-coke qui ne contient que du carbone et de l'hydrogène avec un taux de cendres de 10ppm. Ce produit est ensuite retraité thermiquement en atmosphère neutre aux températures désirées (H.T.T.) (Simon 1974).

3. Données cristallographiques. Les semi-cokes d'anthracène présentent dès 400°C une véritable structure et il a été possible par des études fines des formes des diagrammes de diffraction des R.X. (bandes et modulations), de définir les dimensions moléculaires conformes au modèle proposé par Kinney (1956).
A 400°C, le produit contient un mélange de molécules d'anthracène et de bisanthène (diamètre moyen observé 7 Å) (Fig 1) et quelques rubans de bisanthènes associés. On observe que, en moyenne, 15 de ces molécules s'empilent parallèlement les unes aux autres avec un espacement moyen de 3,40±0,01 Å.

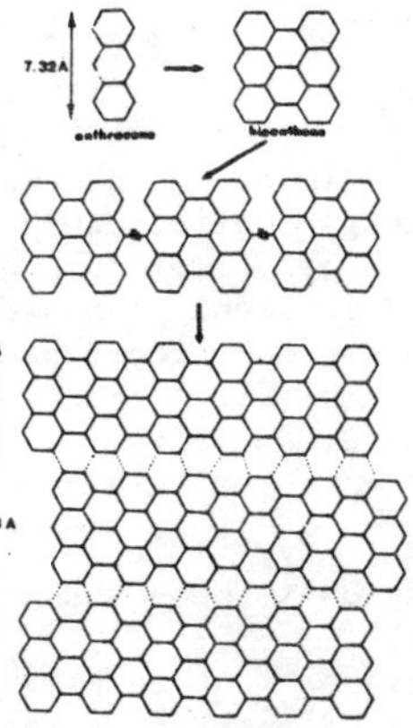

Fig 1

A 600°C, les rubans de 3 bisanthènes reliés sont nettement majoritaires. Les dimensions des motifs observées sont de 23 sur 7 Å mais avec un très net effet mosaïque, dû à la forme gauche des rubans : les liaisons entre bisanthènes ne sont pas rigides et le ruban n'est pas plan. Cette non-planéité s'accompagne d'une désorientation des rubans les uns par rapport aux autres : leur empilement est moins régulier avec un espacement moyen de 3,42±0,01 Å et un nombre moyen de couches dans un empilement de 11 donc plus faible qu'à 400°C.
A 800°C, les liaisons internes aux rubans se sont renforcées; ces rubans s'associent entre eux (mais encore faiblement) 3 par 3 pour former des macromolécules bidimensionnelles isotropes mais gauches : les diamètres observés sont de 27 Å. L'empilement est encore plus mauvais qu'à 600°C : il est limité à 6 couches d'espacement moyen 3,42±0,01 Å.
Aux températures de traitement plus élevées, les macromolécules coalescent pour former de véritables couches graphitiques parallèles, désorientées les unes par rapport aux autres, c.à.d. la structure turbostratique classique.

4. Résonance paramagnétique électronique. Comme tous les semi-cokes, les semi-cokes d'anthracène présentent des raies de R.P.E. très intenses dès 400°C.
a/ Les largeurs de raies sont relativement faibles (fig 2) surtout vers 650-700°C avec un effet oxygène (élargissement des raies en présence d'air) très important à ces températures de traitement où la raie passe de 20 à 0,1 G par simple dégazage. Cet effet est certainement lié à une porosité maximale ouverte, formée par les défauts d'empilement observés aux R.X.
b/ La susceptibilité paramagnétique passe par un maximum pour une H.T.T. de 700°C environ : le nombre de centres paramagnétiques mesurés est de 4.10^{20} par gramme, ce qui est pratiquement égal au nombre de molécules de bisanthènes qu'on aurait si toute la substance était sous cette forme : chaque motif bisanthène est donc radicalaire (radical bisanthryl) (fig 3).
Pour des H.T.T. supérieures à 700°C, il y a diminution du nombre de centres paramagnétiques ce qui correspond à la formation des rubans puis à la coalescence de ces motifs mais chacun possède encore plusieurs électrons qui se délocalisent.
c/ Cette délocalisation est confirmée par l'évolution des largeurs de raies et les temps de relaxation avec la H.T.T. (fig 4). Vers 800°C, on a $T_1 = T_2$, et tous deux décroissent quand la H.T.T. croît c.à.d. quand la délocalisation augmente par croissance des motifs et apparition du caractère conducteur.

Ce phénomène est confirmé par la transition observée (Carmona et al. 1974) dans des échantillons équivalents, sur la conductivité et l'effet Hall, transition interprétée par les auteurs comme une transition semi-conducteur-métal vers 800°C.

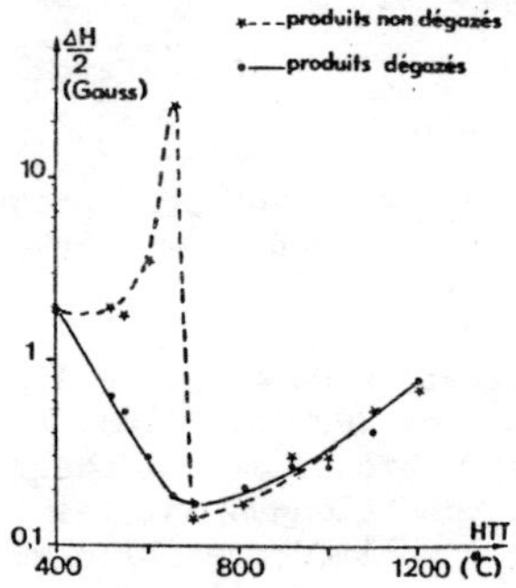

Fig 2

Largeurs de raie de R.P.E. en fonction de la température de traitement (H.T.T.)

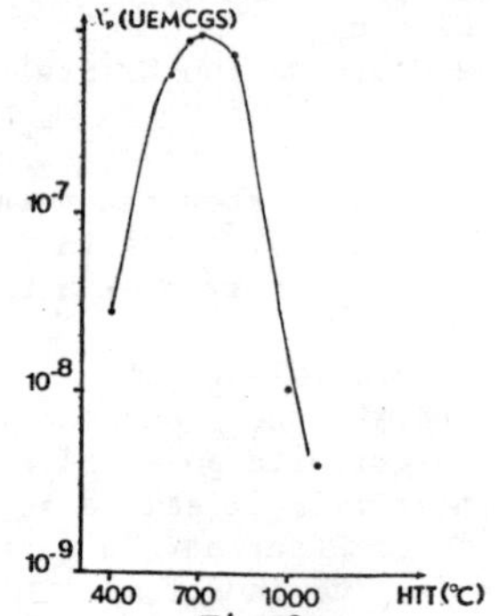

Fig 3

Susceptibilité paramagnétique en fonction de la température de traitement (H.T.T.)

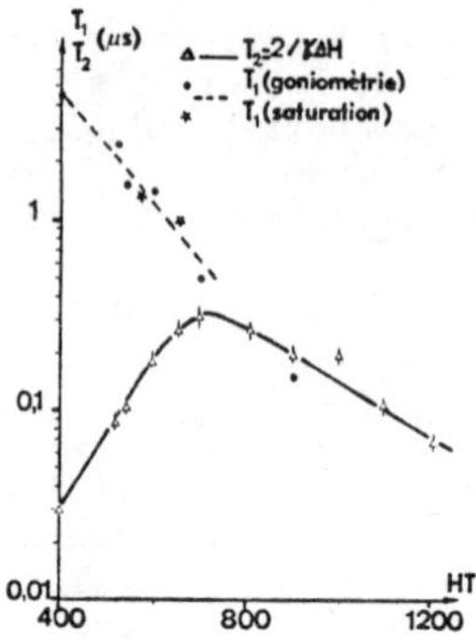

Fig 4

Temps de relaxation électronique en fonction de la température de traitement (H.T.T.)

d/ Les largeurs de raies très faibles observées dès 600°C laissent apparaître une anisotropie du facteur de Landé généralement masquée dans les autres semi-cokes. Les formes de raies ont pu être interprétées comme dues à une poudre présentant une anisotropie axiale de "g" : on peut donc mesurer $g_{//c}$ et $g_{\perp c}$ pour des orientations du champ directeur parallèle et perpendiculaire à l'axe c des cristallites.

En fonction de la H.T.T. (fig 5), $g_{\perp c}$ varie peu et reste proche du g_ℓ de l'électron libre : le couplage spin-orbite reste faible puisque relatif à des orbites plan à plan toujours difficiles. On peut cependant remarquer que, on a $g_{\perp c}$ des cokes d'anthracène supérieur au même $g_{\perp c}$ du graphite : ceci peut s'expliquer d'une part par le très bon couplage entre plans (d_{002} très proche de celle du graphite), d'autre part par l'existence probable de couplages par le bord des motifs, couplages qui se font d'autant plus sentir que les dimensions de ces motifs sont faibles. Le facteur de Landé $g_{//c}$ varie beaucoup plus vite avec la H.T.T. pour rejoindre la valeur du graphite, suivant en cela l'augmentation des dimensions planes des macromolécules et donc la délocalisation des électrons dans le plan graphitique. Une évolution semblable s'observe sur le diamagnétisme de ces matériaux (Simon 1974).

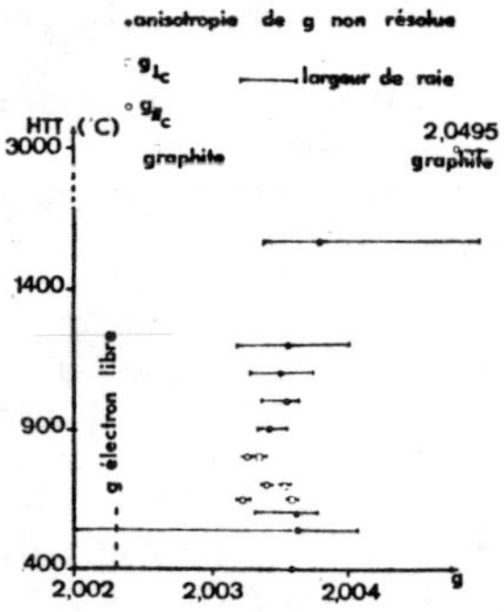

Fig 5

Facteurs de Landé pour des orientations perpendiculaire et parallèle à l'axe c des cristallites.

En champ élevé (13000 G) le coke traité à 700°C présente une légère différence entre g_{xx} et $g_{yy} \sim g_{\perp c}$ due sans doute à l'anisotropie des dimensions du tribisanthène.

5. Conclusion. Par ces études corrélées de diffraction des rayons X et de R.P.E., nous pensons avoir mis en évidence les premiers stades d'organisation des cokes d'anthracène, organisation remarquable pour de si basses températures de traitement et qui facilite sans doute l'évolution ultérieure vers un graphite très bien cristallisé.

6. Bibliographie.

Carmona, F., Delhaes, P., Kenyer, G., Manceau, J.P., Solid State Com. à paraitre
Edström, T., Lewis, I.C., Carbon 7, 1969,85
Kinney, C.R., Proc.1st and 2nd Conf.Carbon, 1956,83
Simon, C., Thèse 3ème cycle, 1974, Orléans

SECTION R

The Liquid State

SPIN-LATTICE RELAXATION OF ^{119}Sn NUCLEI IN LIQUIDS

J. Puskar, T. Saluvere, E. Lippmaa, A. B. Permin* and V. S. Petrosyan*
Department of Physics, Institute of Cybernetics, Academy of Sciences of the Estonian SSR, Tallinn 200001, USSR.
* Moscow State University, Moscow 117234, USSR.

Abstract. The temperature dependence of ^{119}Sn and ^{13}C spin-lattice relaxation times T_1 of a series of liquid alkyltin halides have been studied. Both spin-rotation and dipole-dipole interactions govern the ^{119}Sn relaxation, while the latter modulated by segmental motion determines the ^{13}C T_1.

1. Introduction. The spin-lattice relaxation time T_1 measurements of heteronuclei provide useful information about the molecular and segmental motion in liquids and the molecular structure.

2. Results and Discussion. We have investigated the temperature dependence of the ^{13}C and ^{119}Sn spin-lattice relaxation time T_1 in a series of alkyltin halides and their $CDCl_3$ solutions over the entire liquid range using a previously reported (Puskar et al., 1974) experimental technique. Only a few ^{119}Sn T_1 measurements are reported previously (Sharp, 1972; Saluvere and Puskar, 1973). All samples were carefully purified, degassed with argon gas and sealed into 14 mm (1.2 ml) o.d. spherical ampoules. Accuracy of the measured T_1 values was ± 5 to 10 %, the sample temperature was controlled within ±1°C.

The results of T_1 measurements are given in Figs. 1 and 2. The data were analyzed in terms of three relaxation mechanisms: spin-rotation interaction, dipole-dipole coupling with protons and scalar coupling with rapidly relaxing halogen nuclei. This procedure has been described in detail (Puskar et al., 1974).

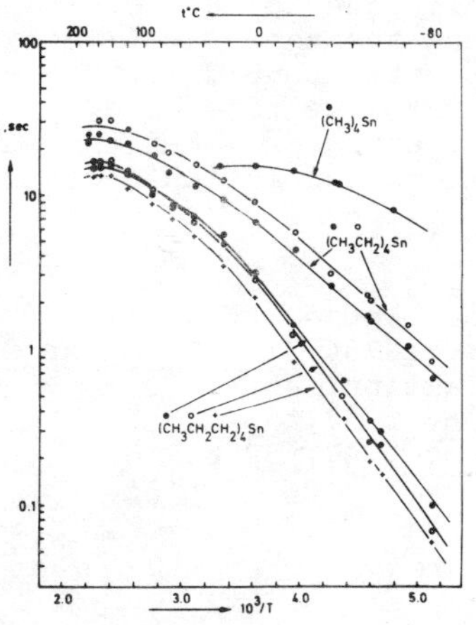

Fig.1. T_1 temperature dependence of ^{13}C nuclei in tetramethyl-, tetraethyl- and tetrapropyltin.

In Me_4Sn, Et_4Sn and Pr_4Sn dipole-dipole contribution to the ^{13}C spin-lattice relaxation dominates over the entire investigated temperature range with some admixture of spin-rotation contribution at higher temperatures (Fig.1). In Et_4Sn and Pr_4Sn for the ^{119}Sn nuclei (Fig.2a) the dipole-dipole contribution dominates at lower temperatures, whereas at higher temperatures only the spin-rotation contribution to the spin-lattice relaxation is significant. At the maximum of T_1, $T_1(sr) \cong T_1(dd,H)$. On the spin-lattice relaxation of the ^{119}Sn nuclei in Me_4Sn, Me_3SnCl, Me_3SnBr, Me_2SnCl_2, Me_2SnBr_2, $MeSnCl_3$, $MeSnBr_3$, Et_2SnCl_2 and Et_2SnBr_2 (Figs.2a,b,c) spin-rotation interaction dominates over the entire liquid range. From these compounds only the solutions of Me_3SnCl and Me_3SnBr in $CDCl_3$ (molar concentration 1:5) (Fig.2d) at a lower temperature the contribution from the dipole-dipole interaction with protons is detectable from the nuclear Overhauser effect measurements. In Me_3SnI (Fig.2c) near the melting point the spin-rotation interaction and scalar coupling with the rapidly relaxing ^{127}I nuclei, and some admixture of the dipole-dipole interaction with protons (only ~20% from scalar contribution) govern the ^{119}Sn spin-lattice relaxation.

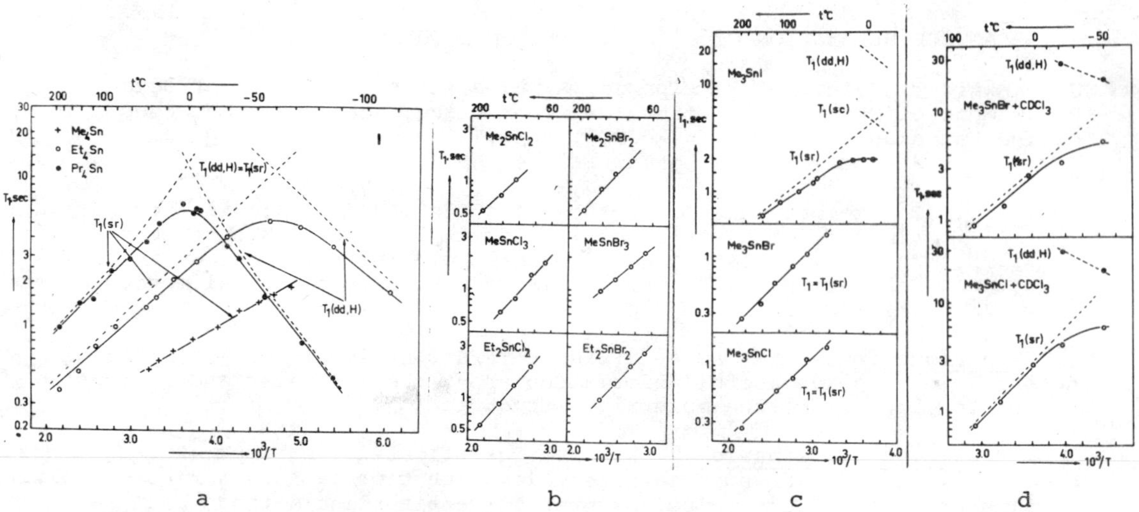

Fig.2. T_1 temperature dependence of ^{119}Sn nuclei in some tin compounds.

From the separated spin-lattice relaxation mechanisms some useful information about the molecular and segmental motion is obtained. Comparison of the diffusion constants for a molecule as a whole and for a α-carbon in Et_4Sn and Pr_4Sn, calculated from the molecular structure and the dipole-dipole relaxation data (Wallach, 1967) of the ^{13}C and ^{119}Sn nuclei shows that α-carbons in the alkyl groups reorient about ten times more rapidly than the molecule as a whole. In the isotropic diffusion limit of a rigid molecule for the dipole-dipole relaxation time $T_1(dd,H)$ of the ^{13}C nuclei in ethyl group $T_1(\alpha)/T_1(\beta) = 1.5$, which due to the more rapid reorientation of the methyl group differs from the averaged experimental value of ~1.3. In Pr_4Sn at $t < 0°C$ the γ-carbon reorients about 2.7 times more rapidly as compared to the α-carbon, where at $t > 120°C$ the rate is about 1.7, respectively. This indicates that at higher temperatures, when the molecule as a whole reorients more rapidly, the role of segmental motion decreases.

3. References

Puskar, J., Saluvere, T. and Lippmaa, E., 1974, Proc. XVIII Congress AMPERE.

Saluvere, T. and Puskar, J., 1973, XIth European Congress on Molecular Spectroscopy, Tallinn, abstr.83.

Sharp, R.R., 1972, J.Chem.Phys. 57, 5321.

Wallach, D., 1967, J.Chem.Phys. 47, 5258.

SELF-DIFFUSION, SPIN-LATTICE RELAXATION AND MOLECULAR MOTION IN LIQUID PHOSPHINE

K. Krynicki, D. W. Sawyer and J. G. Powles
The Physics Laboratories, The University, Canterbury, U.K.

Abstract. The self-diffusion coefficient has been measured for liquid PH_3 and PD_3 using spin echoes for protons, deuterons and ^{31}P nuclei. The results are discussed and used in the re-interpretation of the nuclear spin-lattice relaxation data. The molecules in liquid phosphine move by relatively large reorientational and translational steps. Inertial effects seem important in the molecular reorientation, and the reduced correlation times fit between the 'J' and 'M'-diffusion models.

1. Introduction. A previous n.m.r. study of phosphine [16] indicated that the molecules in the liquid can move relatively freely. However, the analysis given in [16] for the liquid was incomplete due to lack of any experimental data on the self-diffusion and spin-rotation coupling. These parameters have now become available, and a brief description is given here of an attempt [8,9] to obtain a more complete picture of the molecular motion in this model compound.

2. Self-diffusion. We have measured the self-diffusion coefficient, D, for pure liquid phosphines, PH_3 and PD_3 along the coexistence curves, from the triple point (t.p.), 139.4K, to room temperature [8], using spin echoes of protons, deuterons and ^{31}P nuclei at 4.4MHz. A calibrated linear field gradient [7] of 6 to 12 $Oe\,cm^{-1}$ was used, and the results were corrected for the usual non-linearity in the diode detection. Within the limits of experimental error, ranging from 5 to 10%, the values of D for all three nuclear species have the same magnitude and temperature dependence in both liquids. This behaviour does not correlate with the viscosity difference of about 10% [8] in contrast with a rather good correlation between viscosities and self-diffusion in liquid NH_3 and ND_3 [14]. For phosphines between the t.p. and 200K an approximate relation $D = 5.18 \times 10^{-4}\exp(-413/T)cm^2s^{-1}$ holds, but above this region D rises faster with temperature T to reach 15×10^{-5} cm^2s^{-1} at 293K. Attempts to account for the behaviour of D, using our viscosity data [8] in the modified Stokes-Einstein formula [18], 'cubic cell' model [5], and Dullien's relationship [4] have failed except at the lowest temperatures. The deuteration independence of our D values, within the stated accuracy, would be compatible with the phosphine molecules interacting with a central potential [12], e.g. of the Lennard-Jones form [15] for which the ratio, $D(PH_3)/D(PD_3) = [M(PD_3)/M(PH_3)]^{\frac{1}{2}} = 1.044$, M being the molecular weight. This potential is consistent with the molecular radius a = 1.99 Å [15] used below.

3. Spin-lattice relaxation and reorientational correlation times. We shall refer to the previous paper [16] for the data on nuclear spin-lattice relaxation times, T_1's, used here. From the above analysis of D we shall expect little relation between viscosity and molecular reorientation in liquid phosphine, and we shall try an inertial treatment as used for ammonia [2]. The molecular reorientation is expected to be approximately isotropic, in accord with rather small differences between the main tensor components of the moments of inertia [10] (the mean values are: $I_0(PH_3) \equiv I_{OH} = 6.57 \times 10^{-40}$ and $I_0(PD_3) \equiv I_{OD} = 12.83 \times 10^{-40} g\,cm^2$) and the spin-rotation coupling [3] (the r.m.s. values are $<c^2>^{\frac{1}{2}} = 115.31$ kHz for ^{31}P and $<c^2>^{\frac{1}{2}} = 7.90$ kHz for protons in PH_3).

The molecular reorientation correlation times, τ_{QD} for PD_3 are calculated from the measured deuteron T_1 [16], whereas τ_{QH} for PH_3 are deduced from the assumed relation, $\tau_{QH} = \tau_{QD}(I_{OH}/I_{OD})^{\frac{1}{2}}$. Thus the reduced correlation times for molecular reorientation become equal, as $\tau^*_{QH} = (kT/I_{OH})^{\frac{1}{2}}\tau_{QH} = (kT/I_{OD})^{\frac{1}{2}}\tau_{QD} \equiv \tau^*_{QD}$, k being Boltzmann's constant. Our values of τ_{QH} agree very well in magnitude and temperature dependence with those obtained from the Raman band shapes in the liquid [17], the apparent activation energies being 0.52 and 0.50 kcal $mole^{-1}$, respectively.

The phosphorus T_1's in both PH_3 and PD_3 are controlled by the spin-rotation interactions [16]. The spin-rotation correlation times, τ_{srH} for PH_3 can then be readily evaluated from the measured values of $T_1(^{31}P)$ in liquid PH_3, using $<c^2>^{\frac{1}{2}} = 115.31$ kHz for this nucleus. (Incidentally, in [16] there was an overlooked error of a factor of 3 arising from the erroneous use of $(2\pi)^{-1}(2c_\perp^2 + c_\parallel^2)^{\frac{1}{2}}$ as $<c^2>^{\frac{1}{2}}$ kHz instead of $(2\pi)^{-1}[(2c_\perp^2 + c_\parallel^2)/3]^{\frac{1}{2}}$.) Finally, since for isotopic species $<c^2>^{\frac{1}{2}}$ is approximately proportional to I_0, we calculate $<c^2>^{\frac{1}{2}} = 59.1$ kHz for ^{31}P nuclei in PD_3 and hence deduce τ_{srD} from $T_1(^{31}P)$ in liquid PD_3. The values of the reduced spin-rotation correlation times, $\tau^*_{sr} = (kT/I_0)^{\frac{1}{2}}\tau_{sr}$, for liquid PH_3 and PD_3 agree to within less than 8%. This supports our assumptions and shows that the molecular reorientation, which is very fast in the liquid ($\tau_{QH} = 2.1 \times 10^{-13}$s at the t.p.), is dominated by the inertial effects. We find that τ_{sr} is proportional to $T^{5/2}$ and that above 210 K the ratio τ_{srH}/τ_{QH} becomes greater than one attaining the value of 3 at 293 K.

When we plot log τ_Q^* against log τ_{sr}^* and compare our graph with the extended rotational diffusion models of McClung [11], our values fit between the 'J' and 'M'- diffusion models as observed for liquid HCℓ and HBr [6]. On the other hand, if viscosity is used to deduce τ_{QH} from $\tau_{QH}^{'}$ then the data would favour approximately the M-diffusion model which seems less plausible for liquids.

The r.m.s. angle of molecular reorientation for liquid PH_3, estimated classically as $<\theta^2>^{\frac{1}{2}} = 3^{\frac{1}{2}}\tau_{sr}^*$, varies from 32° at the t.p. to 160° at 293K.

4. Proton $T_{1\,inter-d}^{-1}$ and self-diffusion. Using the values of τ_{QH} and τ_{srH} discussed above together with other relevant molecular data, we can calculate the proton spin-rotational and intramolecular dipolar contributions [1] and subtract them from the proton experimental T_{1e}^{-1} [16] in liquid PH_3. A careful estimate shows that the remaining relaxation rate must be predominantly due to the intermolecular magnetic proton-proton and proton-^{31}P nucleus interactions which we call $T_{1\,inter-d}^{-1}$. This contribution can also be estimated [1],using our values of D, the molecular radius a = 1.99 Å, and the measured density for PH_3 [8]. The following three corrections derived by Muller [13] have been made: (i) for nuclear spin off-centre effect (~5%); (ii) for non-uniform distribution of neighbours around a given molecule (13 to 20%); (iii) to allow molecular jumps of arbitrary size, r. If the translational diffusion equation holds, the correction (iii), expressed usually as $<r^2>/4a^2$, is the only variable in comparing the calculated and experimental $T_{1\,inter-d}^{-1}$. The r.m.s. flight distance $<r^2>^{\frac{1}{2}}/2a$, estimated from our results for liquid phosphine, varies from varies from unity at the t.p. to about 2 at 293K.

We conclude that the molecules in liquid phosphine are relatively free to move by large reorientational and translational steps.

A detailed account of this work will be published elsewhere [8,9].

5. References

1. Abragam, A., 1960, The Principles of Nuclear Magnetism, Oxford University Press, London.
2. Atkins, P.W. et al., 1969, Mol. Phys., 17, 329.
3. Davies, P.B. et al., 1971, J. Chem. Phys., 55, 3564.
4. Dullien, F.A.L., 1963, Trans. Faraday Soc., 59, 856.
5. Houghton, G., 1964, J. Chem. Phys., 40, 1628.
6. Krynicki, K., J. Magn. Res., to be published.
7. Krynicki, K., Rahkamaa, E.J. and Powles, J.G., 1974, Mol. Phys., 28, 853.
8. Krynicki, K. and Sawyer, D.W., Mol. Phys., to be submitted for publication.
9. Krynicki, K., Sawyer, D.W. and Powles, J.G., Mol. Phys., to be submitted for publication.
10. Maki, A.G. et al., 1973, J. Chem. Phys., 58, 4502.
11. McClung, R.E.D., 1969, J. Chem. Phys., 51, 3842; 1972, ibid., 57, 5478.
12. McLaughlin, E., 1960, Physica, 26, 650.
13. Muller, B.H., 1966, Phys. Lett., 22, 123; Harmon, J. and Muller, B.H., 1969, Phys. Rev., 182, 400.
14. O'Reilly, D.E. et al., 1973, J. Chem. Phys., 58, 4072.
15. Reid, R.C. and Sherwood, T.K., 1966, The Properties of Gases and Liquids, McGraw-Hill, New York.
16. Sawyer, D.W. and Powles, J.G., 1971, Mol. Phys., 21, 83.
17. Schwartz, M. and Wang, C.H., 1974, Chem. Phys. Lett., 25, 26.
18. Sutherland, G.B.B.M., 1905, Phil. Mag., 9, 781.

PHASE SHIFTED PULSE SEQUENCE FOR THE MEASUREMENT OF SPIN-LATTICE RELAXATION IN HIGH RESOLUTION NMR SPECTRA OF COMPLEX SYSTEMS §

D.E. Demco*, P. Van Hecke** and J.S. Waugh
Department of Chemistry, Massachusetts Institute of Technology, Cambridge, Massachusetts 02139

Abstract. A phase modification of the "non-ideal" inversion (saturation) recovery pulse sequence is presented, which eliminates for the interference effects of transverse magnetization in the case of T_1 measurements in high resolution complex spectra.

Inversion and saturation recovery methods are routinely used for the determination of spin-lattice relaxation times, either in single or - after Fourier transform - in complex liquid systems (Freeman et al., 1969, 1970, 1971; Freeman er al., 1972; Freeman et al., 1969; McDonald et al., 1973; Markley et al., 1971; Vold et al., 1968). The observed recovery of the magnetization (of the individual lines) towards thermal equilibrium depends on the non-idealities in the pulse sequence used, and so does T_1, which value is extracted from this recovery.

We analyzed (Demco et al., 1974) the effects of such non-idealities for the case of phase modifications in the inversion recovery pulse sequence, $180^\circ_0 - t - 90^\circ_\phi$ for an arbitrary ratio of T_1/T_2^*. A new method for the measurement of T_1 is then proposed, which compensates for the effects of those non-idealities on the apparent recovery of the magnetization.

The pulses are supposed to be narrow and intense. The non-ideal character of the pulse is related to the frequently encountered case of an inhomogeneity ΔH_1 in the amplitude of the r.f. field. Furthemore, these pulses are applied at a frequency ω_0, off-resonance with respect to the frequency ω of the different lines in the spectrum ($\Delta\omega = \omega - \omega_0$). The effect of such non-ideal pulse, described by a propagator, is then calculated using density matrix formalism, where for simplicity, a spin system characterized by a single absorption line is assumed. The magnetization recovery after the 90°_ϕ pulse, detected along an axis, at an angle α with the x-axis, is given by

$$\frac{M_0\,(t_{90^\circ}) - M\,(t_{180^\circ} - t - t_{90^\circ})}{M_0(t_{90^\circ})} = C(\alpha,\ \phi,\ \theta,\ t)\exp(-t/T_1)$$

where $\theta = \tan^{-1}\,\omega_1/\Delta\omega$ and $\omega_1 = \gamma\ (\bar{H}_1 + \Delta H_1)$

The factor $C(\alpha, \phi, \theta, t)$ takes into account the effects of non-idealities in the inversion recovery pulse sequence. This recovery is non-exponential, depending on T_1/T_2^*, the relative pulse phase, the off-resonance frequency, the observation direction and the r.f. field inhomogeneity.

For $T_2^* \ll T_1$, the magnetization recovery is exponential, independent of pulse phases; a well-known result.

When $T_1 \sim T_2^*$, the recovery is non exponential, due to the interference effects produced by the transverse magnetization. In such cases, one has to produce a static field gradient between the pulses, in order to artificially shorten T_2^* (Vold et al., 1968). Our analysis however, shows that for a $180^\circ_0 - t - 90^\circ_0$ and a $180^\circ_0 - t - 90^\circ_{180}$ inversion recovery sequence, the non-exponential effects on the magnetization recovery, after the 90° observation pulse, exactly cancel. Consequently, successive use of a phase and antiphase sequence and averaging of the magnetization recoveries, yields a purely exponential recovery with the real T_1 value.

Compared to the use of the pulsed field gradient, this technique is simple and allows a maximum resolution of the partly relaxed spectrum. The method can be applied even well to any selective or non-selective pulse excitation in inversion- or saturation recovery pulse sequence.

References.

§ Supported in part by the National Science Foundation and in part by the National Institutes of Health.

* Permanent address: Institute for Atomic Physics, P.O. Box 35, Bucharest, Romania.

**"Aangesteld Navorser" of the Belgian N.F.W.O.
Permanent address: Universiteit Leuven, Dept. Natuurkunde, 3030 Heverlee, Belgium.

D.E. Demco, P. Van Hecke and J.S. Waugh, 1974, J. Mag. Res., to be published.

R. Freeman and H.D. Hill, 1969, J. Chem. Phys. 51, 3140; 1970, ibid. 53, 4103; 1971, ibid 54, 3367.

R. Freeman, H.D. Hill and R. Kaptein, 1972, J. Mag. Res. 7,82.

R. Freeman and S. Wittekoek, 1969, J. Mag. Res. 1, 238.

G.G. McDonald and J.S. Leigh Jr., 1973, J. Mag. Res. 9, 358.

J.L. Markley, W.J. Horsley and M.P. Klein, 1971, J. Chem. Phys. 55, 3604.

R.L. Vold, J.S. Waugh, M.P. Klein and D.E. Phelps, 1968, J. Chem. Phys. 48, 3831.

SELFDIFFUSION OF PHOSPHATE AND POLYPHOSPHATE ANIONS IN AQUEOUS SOLUTION

H.S.Kielman and J.C.Leyte
Gorlaeus Laboratory, Dept. of Phys.Chem., University of Leiden, Leiden, the Netherlands.

Abstract. Selfdiffusioncoefficients of phosphate and polyphosphate anions were measured with the method of the NMR spin echo technique in the presence of a pulsed field gradient. Using the theory of Brownian motion of ellipsoides the selfdiffusioncoefficients for the polyphosphates can satisfactory be described.

1. Introduction. By measuring the magnetization (M) at the time of the spin echo (2τ) as a function of the field gradient (g), the selfdiffusioncoefficient (D) was obtained using the equation of Stejskal and Tanner 6):

$$(1) \quad \ln(M(2\tau)/M(o)) = -D(\gamma g \delta)^2 \Delta$$

with δ = time of the gradient pulse
Δ = diffusiontime

The unknown value of g is found by calibration with respect to a sample with a known self-diffusioncoefficient from tracer measurements 4). Perrin 5) described the Brownian motion of an ellipsoide in solution and calculated the translational diffusioncoefficient as a function of the length of the three half axes a, b and c:

$$(2) \quad D = \frac{kT}{12\pi\eta_o} \int_0^\infty \frac{ds}{\sqrt{(a^2+s)(b^2+s)(c^2+s)}}$$

with k = constant of Boltzmann
η_o = viscosity of the solvent (H_2O)
T = absolute temperature

In the limit of a sphere with radius r equation (2) reduces to the relation of Stokes-Einstein:

$$(3) \quad D = \frac{kT}{6\pi\eta_o r}$$

This equation will be used for phosphoric acid. For ellipsoides with equal minor axes (a=b=r) and varying major one (c) equation (2) yields:

$$(4) \quad D = \frac{kT}{6\pi\eta_o r} \cdot \frac{\ln(p+\sqrt{p^2-1})}{\sqrt{p^2-1}} \quad \text{with } p = \frac{c}{r}.$$

Describing a polyion as a sequence of monomeric units linked together linearly we can calculate p as a function of the degree of polymerization. Equation (4) will be used in calculating self-diffusioncoefficients for polyphosphates.

2. Results. We measured the selfdiffusiondata in phosphoric acid and polyphosphates by the ^{31}P nucleus at 24.29 MHz and 25°C with a Bruker pulse spectrometer. Only the data mentioned in table I were measured at other temperatures. In figure 1 the results are compared with diffusion-coefficients obtained with a concentration gradient by Edwards and Huffmann 1) by plotting our results in a graph from their paper. At 0.5 eq.l.$^{-1}$ our result agrees reasonably well with the diffusioncoefficient for H_3PO_4 calculated by Edwards and Huffmann from their data (7.2×10^{-6} cm^{2}sec^{-1}). Table I shows measurements in phosphoric acid (1.00 eq.l.$^{-1}$) as a function of temperature demonstrating the linear dependence of the selfdiffusioncoefficient on η^{-1}. From this table a radius for the H_3PO_4 molecule of 3.2 ± 0.4 Å. is calculated. From this value it is concluded that one hydration shell is attached to the phosphoric acid molecule.

Sodiumpolyphosphates, $(NaPO_3)_n$, were prepared with varying degree of polymerization (3-350) and the selfdiffusioncoefficients were measured. These data are shown in figure 2. It is seen that the measured data can well be described by the calculated values from equation (4), represented by the solid line, where we take the radius equal to that of the monomeric ion (orthophosphate) and calculate the major axes from molecular dimensions, assuming the polyion to consist of a linear string of PO_4 tetraeders. The viscosity used in the calculation is that of the solvent (H_2O). Comparing our results on the polyionselfdiffusion with diffusion data obtained by nonequilibrium methods 2)3) it is seen that the pulsed gradient method yields much lower diffusion constants at high molecular weights. It is suggested that nonequilibrium data are largely determined by the counterion mobilities.

3. References.

1) Edwards O.W. and Huffmann E.O., J.Am.Chem.Soc. 63,1830 (1959)
2) Katchman B.J. and Smith H.E., Arch.Biochem. and Biophys. 75,396 (1958)
3) Kern E., Ph.D.Thesis, Michigan State University (1966)
4) Mills R., J.Am.Chem.Soc. 77,6116 (1955)
5) Perrin F., J.Phys.Radium 7,1 (1936)
6) Stejskal E.O. and Tanner J.E., J.Chem.Phys. 42,288 (1965)

TABLE I

selfdiffusiondata of 1.00 eq.l^{-1} phosphoric acid in H_2O as a function of temperature.

T(K)	D ($10^{-6}cm^2sec^{-1}$)	η_o(CP) (H_2O)	$\frac{kT}{6\pi\eta_o D}$ (Å)
278	4.1	1.52	3.3
283	5.5	1.31	2.8
288	6.9	1.14	2.9
293	6.8	1.01	3.1
298	7.0	0.89	2.9
303	8.7	0.80	3.2
308	8.7	0.72	3.6
313	10.9	0.66	3.2

r=3.2 ± 0.4 Å

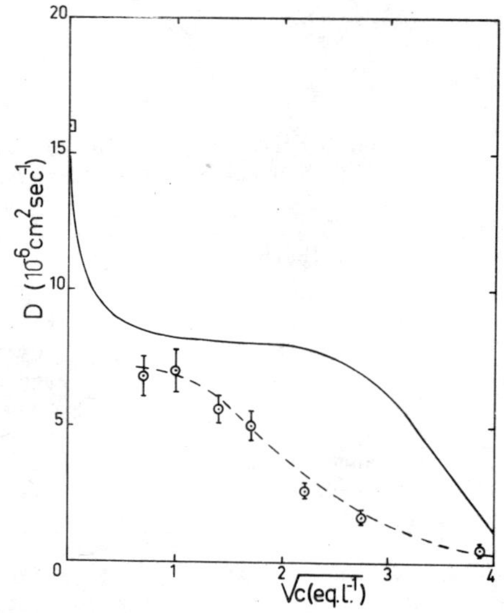

Fig. 1.
Selfdiffusion- and diffusioncoefficients for phosphoric acid at 25°C as a function of concentration.

⊙ observed values (error bars 10%)
⊡ Nernst limiting value
—— diffusiondata by reference 1)

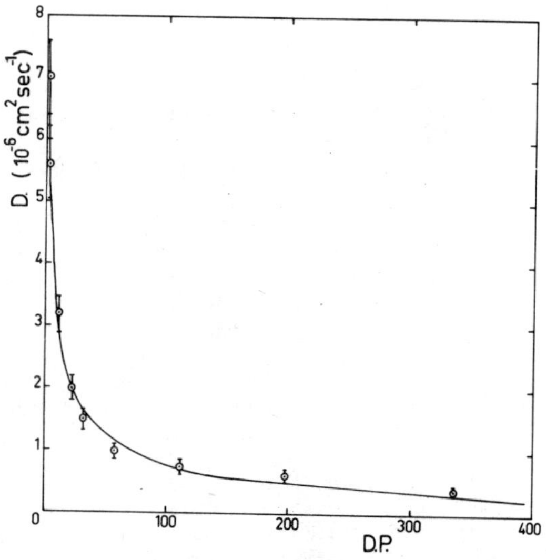

Fig. 2.
Selfdiffusioncoefficients of the polyions in 1.00 eq.l^{-1} sodiumpolyphosphates as a function of the degree of polymerization (DP)

⊙ observed values
—— calculated line with equation (4) and l=0.8 r with r=3.2 Å where l=length of one monomeric unit.

SPIN LATTICE RELAXATION OF ^{14}N AND ^{15}N NUCLEI IN LIQUID NITROGEN

K. Krynicki, E. J. Rahkamaa* and J. G. Powles
The Physics Laboratories, The University, Canterbury, U.K.
*Department of Physics, University of Helsinki, Finland.

Abstract. The measured spin-lattice relaxation times for liquid nitrogen along the coexistence curve are used to obtain independent values of the molecular-reorientation correlation times. It is concluded that the molecular reorientation is unusually free but is frictionally rather than collisionally hindered.

Measurements are reported of the nitrogen nucleus spin-lattice relaxation time, T_1, in liquid nitrogen $^{14}N_2$ and liquid nitrogen $^{15}N_2$ along the liquid-vapour coexistence line from 77.3 K up to the critical temperature and for the fluid at the critical density up to 145 K. Our results for $^{14}N_2$ agree quite well with those of Armstrong and Speight (1970) but less well with those of De Reggi *et al* (1969) where the comparison is possible. No previous results for $^{15}N_2$ have been reported. $T_1(^{14}N_2)$ is of order milliseconds and is dominated by quadrupole interactions whereas $T_1(^{15}N_2)$ is of order seconds and is due to spin-rotation interaction.

Values of the molecular reorientational correlation time, τ_Q, and the molecular angular momentum correlation time, τ_{sr}, have been determined from $T_1(^{14}N_2)$ and $T_1(^{15}N_2)$ respectively.

It is found that at the normal boiling point temperature (77K) τ_Q and τ_{sr} are virtually equal. τ_Q falls slowly with increasing temperature corresponding to an activation energy of 0.73 kJ mole^{-1}. τ_{sr} rises rapidly with temperature and near the critical temperature is some nine times as long as τ_Q. The very short correlation times, of order a few times 10^{-13}s, suggest that the molecules are rather free to reorient and indeed the times are of order $(kT/I_o)^{\frac{1}{2}}$ where I_o is the molecular moment of inertia. The reduced correlation times, $\tau^* = (kT/I_o)^{\frac{1}{2}}\tau$, correspond roughly with the Hubbard relation, $\tau_Q^*\tau_{sr}^* = 1/6$ (1963) at the lowest temperature but depart from it at higher temperatures, somewhat in the manner predicted by Gordon (1966) for linear molecules and by McClurg (1969) for spherical molecules. However the results agree rather better with an extension of Steele's (1963) reorientational model as extended by Kluk and Powles (1974) which is based on a frictional type resistance to reorientational motion rather than gas-like collisions. The root mean square angle of molecular reorientation is shown to increase rapidly from some 30° at the triple point to some several hundred degrees at the critical point.

The values of τ_Q and τ_{sr} are also compared with those predicted by the computer simulation method (Barojas *et al*, 1973; Cheung and Powles, 1974). The simulation values are within a factor of about 1.5 of the experimental values and show the same trend of variation with temperature at the low temperatures, for which they are so far available.

It is concluded that molecular reorientation in liquid nitrogen is relatively free but is not gas-like and is better described by a frictional model even at temperatures and densities approaching the critical point.

A detailed account of this work will be published elsewhere (Krynicki *et al*, 1974).

References

Armstrong, R.L. and Speight, P.A., 1970, J. magn. Res. **2**, 141.
Barojas, J. et al, 1973, Phys. Rev. A **7**, 1092.
Cheung, P.S.Y. and Powles, J.G., 1974, Mol. Phys., to be published.
De Reggi, A.S. et al, 1969, J. magn. Res. **1**, 144.
Gordon, R.G., 1966, J. chem. Phys., **44**, 1870.
Hubbard, P.S., 1963, Phys. Rev., **131**, 1155.
Kluk, E. and Powles, J.G., 1974, Mol. Phys., to be published.

Krynicki, K., Rahkamaa, E.J. and Powles, J.G., 1974, Mol. Phys., to be published.
McClung, R.E.P., 1969, J. chem. Phys., **51**, 3842.
Steele, W.A., 1962, J. chem. Phys., **38**, 2411.

SPIN-LATTICE RELAXATION OF ^{29}Si NUCLEI IN ORGANIC COMPOUNDS

J. Puskar, T. Saluvere and E. Lippmaa
Department of Physics, Institute of Cybernetics, Academy of Sciences of the Estonian SSR, Tallinn 200001, USSR.

Abstract. The temperature dependence of ^{29}Si spin-lattice relaxation time T_1 in $SiCl_4$, $MeSiCl_3$, Me_2SiCl_2 and $MeSiCl_3$ has been studied over the entire liquid range. Spin-rotation interaction is predominant at higher temperatures with significant dipole-dipole contribution only near the freezing point.

1. Introduction. Small negative magnetogyric ratio and low natural abundance of ^{29}Si nuclei both contribute to the very low intensity of ^{29}Si NMR signals. Only a few ^{29}Si spin-lattice relaxation time T_1 measurements are reported (Hunter and Reeves, 1968; Levy et al., 1972, 1973; Harris and Kimber, 1973).

2. Results and Discussion. We have investigated the temperature dependence of ^{29}Si spin-lattice relaxation time T_1 in compounds of the general formula Me_nSiCl_{4-n} (n=0,1,2,3) over the entire liquid range. The T_1 measurements were carried out at a magnetic field of 14.1 kg with a universal NMR spectrometer (Lippmaa et al., 1966), using the Fourier transform technique with a NIC 1083 computer and a π-τ-$\pi/2$ rf pulse sequence. All samples were degassed with pure argon gas and sealed into 14 mm o.d. (1.2 ml) spherical ampoules. Accuracy of the measured T_1 values is ± 5 to 10 %. The observed spin-lattice relaxation time T_1 is given by the simple expression:

$$T_1^{-1} = T_1^{-1}(dd,H) + T_1^{-1}(sr) + T_1^{-1}(sc) + T_1^{-1}(\sigma) \qquad (1)$$

Contributions from the scalar coupling with the rapidly relaxing ^{35,37}Cl nuclei, $T_1^{-1}(sc)$ and chemical shift anisotropy $T_1^{-1}(\sigma)$ to ^{29}Si spin-lattice relaxation are negligible in the compounds investigated. The measured relaxation times T_1 as well as the spin-rotational $T_1(sr)$ and dipole-dipole contributions $T_1(dd,H)$ to it are shown in Fig.1.

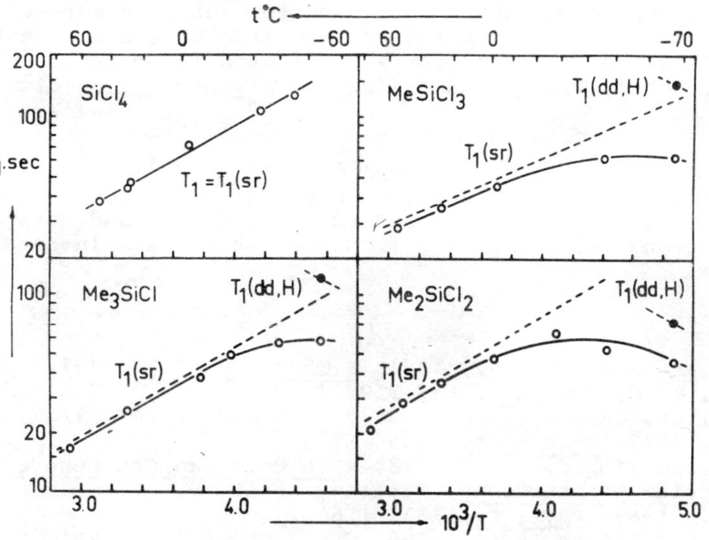

Fig. 1. The temperature dependence of the spin-lattice relaxation time T_1 in silicon compounds.

The share of the dipole-dipole interaction was determined from the nuclear Overhauser effect under the influence of the decoupling rf field according to

$$T_1^{-1}(dd,H) = 2\gamma_{Si}(I_z - I_o)\cdot(\gamma_H T_1 I_o)^{-1} = KT_1^{-1} \qquad (2)$$

where I_z and I_o are the total integrated intensities of the ^{29}Si signal under double resonance (^{29}Si - {H}) and single resonance conditions, respectively.

The spin-rotation contribution to T_1 is obtained from the opposite temperature dependence of $T_1(dd,H)$ and $T_1(sr)$. At the maximum in T_1,

$T_1(sr) \cong 2T_1$ (Rhodes et al., 1968).

In $SiCl_4$ the spin-rotation contribution to T_1 dominates over the entire liquid range. In $MeSiCl_3$, Me_2SiCl_2 and Me_3SiCl the contribution from the dipole-dipole interaction with protons and spin-rotation contribution to T_1 of ^{29}Si nuclei are of comparable magnitude near the melting point with the spin-rotation contribution dominating at higher temperatures. The results are in good agreement with previous T_1 temperature dependence measurements of ^{29}Si nuclei in Me_4Si (Levy et al., 1973).

The significance of the spin-rotation interaction as a relaxation mechanism in the investigated compounds makes it attractive to calculate ^{29}Si spin-rotation constants from the relaxation data. From the molecular structure and the dipole-dipole contribution $T_1^{-1}(dd,H)$ data we get the effective molecular reorientation time τ_c. Then according to (Hubbard, 1963) for a spherical molecule

$$T_1^{-1}(sr) = (2\pi IC/h)^2 \cdot (3\tau_c)^{-1} \qquad (3)$$

where C and I are the averaged spin-rotation constant and the calculated from structural parameters molecular moment of inertia, respectively. The spin--rotation constants C calculated from the experimental data are given in Table 1.

Table 1. Relaxation times and spin-rotation constants for some silicon compounds.

Compound	t°C	T_1 (sec)	T_1(dd,H) (sec)	T_1(sr) (sec)	$I \cdot 10^{40}$ $g \cdot cm^2$	$\bar{C}$ kHz
Me_4Si	-50	42*	214*	52*	270	-1.26
Me_3SiCl	-54	58.3	122	112	390	-0.91
Me_2SiCl_2	-68	45.6	73.5	120	510	-1.06
$MeSiCl_3$	-68	64.1	145	115	627	-0.89

* G.C.Levy et al., 1973

Deverell, Gillen and Sharp have used the relation between the spin--rotation interaction constant and the magnetic shielding constant to obtain absolute chemical shift scales for ^{19}F, ^{31}P and ^{119}Sn nuclei (Deverell, 1970; Gillen, 1972; Sharp, 1972). According to (Flygare and Goodisman, 1968)

$$\sigma_{Av} = \sigma_d' + \sigma_p' = \sigma_d' + \frac{e^2}{2mc^2} \cdot \frac{\pi CI}{M\mu_N\gamma_I} \qquad (4)$$

where Deverell's notation is used.

For Me_4Si, the chemical shifts standard in ^{29}Si NMR, and using $\sigma_d' = 878.7$ ppm (Bonham and Strand, 1964), we get from the estimated spin-rotation constant $\sigma_{Av} \cong 550$ ppm.

3. References

Bonham, R.A. and Strand, T.G., 1964, J.Chem.Phys. 40, 3447.
Deverell, C., 1970, Mol.Phys. 18, 319.
Flygare, W.H. and Goodisman, J., 1968, J.Chem.Phys. 49, 3122.
Gillen, K.T., 1972, J.Chem.Phys. 56, 1573.
Harris, R.K. and Kimber, B.J., 1973, XIth European Congress on Molecular Spectroscopy, Tallinn, abstr.75.
Hubbard, P.S., 1963, Phys.Rev. 131, 1155.
Hunter, B.K. and Reeves, R.L., 1968, Can.J.Chem. 46, 1399.
Levy, G.C., 1972, J.Amer.Chem.Soc. 94, 4793.
Levy, G.C., Cargioli, J.D., Juliano, P.C. and Mitchell, T.D., 1972, J.Mag.Res. 8, 399.
Levy, G.C., Cargioli, J.D., Juliano, P.C. and Mitchell, T.D., 1973, J.Amer.Chem.Soc. 95, 3445.
Lippmaa, E., Past, J., Olivson, A. and Saluvere, T., 1966, Eesti NSV Tead.Akad.Toim., Füüs.-Matem. 15, 58.
Rhodes, M., Aksnes, D.W. and Strange, J.H., 1968, Mol.Phys. 15, 541.
Sharp, R., 1972, J.Chem.Phys. 57, 5321.

^{15}N AND ^{13}C SPIN LATTICE RELAXATION IN NEAT LIQUIDS AT HIGH MAGNETIC FIELDS

D. Schweitzer and H.W. Spiess
Department of Molecular Physics, Max-Planck-Institute, 6900 Heidelberg, Jahn-Straße 29, F.R.G.

Abstract. General features of ^{15}N spin lattice relaxation in nitrobenzene, pyridine and triethanolamine are discussed and compared with ^{13}C relaxation mechanisms.

General features of ^{15}N spin lattice relaxation are probably best discussed in comparison with ^{13}C, where a lot of experimental material has been collected in the past few years. For ^{13}C usually two relaxation mechanisms are predominant: intramolecular dipole dipole interaction (DDI) between the nuclear spins of the carbon and directly bound protons and spin rotation interaction (SRI), since spin rotation constants for ^{13}C are relatively large. On the other hand, relaxation due to anisotropic chemical shift (ACS) could be demonstrated for ^{13}C only in a few cases [1.] and intermolecular DDI can be neglected almost completely.

For ^{15}N the first two relaxation mechanisms are also important as was shown by Lippmaa et al. [2.], e.g., in aniline and nitrobenzene. We shall demonstrate that for ^{15}N relaxation due to ACS and intermolecular DDI can also be significant by combining ^{15}N and ^{13}C spin lattice relaxation rate measurements in the same molecule as a function of frequency and temperature. The relaxation measurements provide information about correlation times for both nuclei as well as values for the anisotropy of the chemical shift and the spin rotation constants for ^{15}N [3.] These values can be compared with those obtained independently from the powder spectra of the frozen liquids.

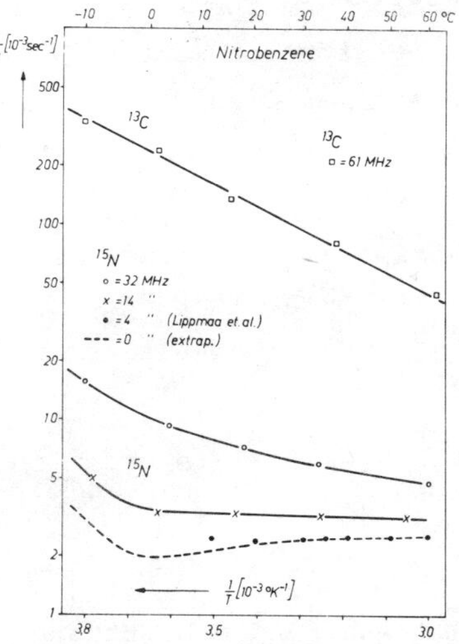

fig. 1

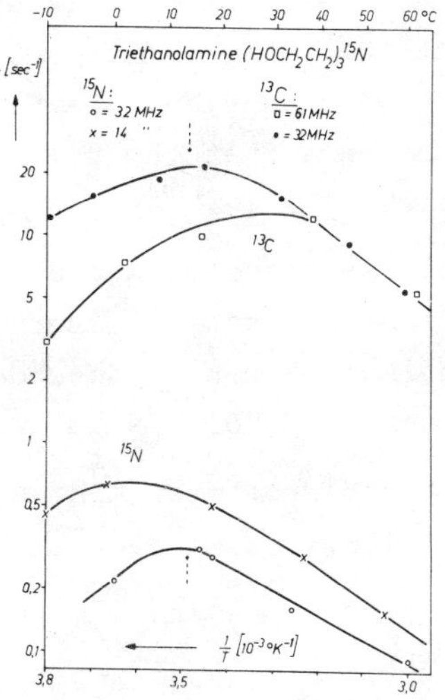

fig. 2

In fig. 1 the relaxation rates $1/T_1$ of ^{15}N and ^{13}C in nitrobenzene are shown. It can be seen that at 32 MHz (equivalent to a magnetic field of about 75 kG) the relaxation due to ACS is predominant and even at + 60° C still accounts for about half of the total relaxation rate of the ^{15}N. With correlation times τ_c obtained from the ^{13}C relaxation rates (fig. 1) (^{13}C relaxation rates for aromatic ring carbons are solely determined in a good approximation by intramolecular DDI in this temperature range) we could determine a value $|\Delta\sigma| = |\sigma_{\parallel} - \sigma_{\perp}| = 400$ ppm for the ACS of the ^{15}N from the T_1 measurements. We also measured $\Delta\sigma$ directly in the powder by FT-NMR at - 105° C. The $\hat{\sigma}$-tensor is found to be almost axially symmetric with $\Delta\sigma = -398 \pm 20$ ppm [4.)]. From the shift components measured in the solid one can calculate the ^{15}N spin rotation constants: $C_{xx} = 11.4 \pm 1.5$ KHz, $C_{yy} = 1.6 \pm 0.5$ KHz, and $C_{zz} = 1.1 \pm 0.5$ KHz. Here the x-direction coincides with the twofold symmetry axis of nitrobenzene. Spin rotation constants we obtained indepedently from the relaxation measurements in the liquid agree remarkably well with those given above [4.)].

Similarly for pyridine a significant frequency dependence of the ^{15}N relaxation rate was observed [3.)]. The frequency independent part of the relaxation rate is dominated by SRI at high temperatures (above 0° C). Below -30° C the major contribution to this rate is given by DDI, which is about 90 o/o due to intermolecular DDI [3.)]. The reason is that the non-bonding orbital of the nitrogen can interact with protons of neighbouring molecules. Again $|\Delta\sigma|$ can be obtained from our T_1 data giving $|\Delta\sigma| = 710$ ppm. From the powder spectrum we obtained for ^{15}N: $\Delta\sigma = 672 \pm 20$ ppm, $C_{xx} = 16.5 \pm 1$ KHz, $C_{yy} = 11.5 \pm 1$ KHz, and $C_{zz} = 0.6 \pm 1$ KHz. Here the z-direction coincides with the direction perpendicular to the molecular plane of the pyridine.

In triethanolamine $(HOCH_2CH_2)_3N$ the nitrogen atom has a non-bonding orbital as in the case of pyridine. From a molecular model, however, it can be seen that in this case the nitrogen atom is more or less shielded by the surrounding atoms of the molecule [5.)]. Therefore, relaxation due to intermolecular DDI might be small. The results of the relaxation rate measurements for ^{15}N and ^{13}C are shown in fig. 2. Here the frequency dependence is opposite to that of ^{15}N in nitrobenzene (fig. 1) and pyridine because in triethanolamine the correlation times are much longer so that $\omega\tau_c \approx 1$. From the maximum of the relaxation rate we get the correlation times τ_c. With those τ_c the relaxation rates for the ^{13}C as well as for the ^{15}N can be completely accounted for by intramolecular DDI [5.)]. The nature of the motion of the carbons and the nitrogen must be the same because we find the relaxation rate maxima at 32 MHz for C and N at the same temperature. This also means that the relaxation is caused by reorientation of the molecule as a whole.

These examples show that for ^{15}N intramolecular as well as intermolecular DDI, SRI and ACS can be important relaxation mechanisms depending on the system and experimental conditions. Consequently, a considerable amount of information can be gathered from ^{15}N relaxation studies.

We thank Professor K.H. Hausser for his special interest in this work and for helpful discussions.

References.

1.)a) H.W. Spiess, D. Schweitzer, U. Haeberlen, and K.H. Hausser, J.Mag.Res. 5, 101 (1971),
b) H.W. Spiess and H. Mahnke, Ber.d.Bunsenges.f.Phys.Chem. 76, 990 (1972),
c) G.C. Levy, J.D. Cargioli, F.A.L. Anet, J.Am.Chem.Soc. 95, 1527 (1973),
d) H.W. Spiess, D. Schweitzer, and U. Haeberlen, J.Mag.Res. 9, 444 (1973),
2.) E. Lippmaa, T. Saluvere, and S. Laisaar, Chem.Phys.Lett. 11, 120 (1971),
3.) D. Schweitzer and H.W. Spiess, J.Mag.Res. 15, Sept. 1974, in press,
4.) D. Schweitzer and H.W. Spiess, J.Mag.Res. 15, Nov. 1974, in press,
5.) D. Schweitzer and H.W. Spiess, to be published.

DEUTERON MAGNETIC RELAXATION OF POLY(METHACRYLIC ACID) IN AQUEOUS SOLUTIONS.

J.Schriever and J.C.Leyte
Gorlaeus Laboratories, Department of Physical Chemistry, University of Leiden, Leiden, the Netherlands.

Abstract. Deuteron relaxation rates in aqueous solutions of selectively deuterated poly(methacrylic acid) were studied at 80°C as a function of the Larmor frequency.

1. Introduction. Since more than 25 years the influence of varying the total charge of the macromolecule on the time averaged geometric form of such molecules- e.g. the conformation of weak polyacids such as PMA$(-CH_2-CCH_3COOH)_n$ or PAA$(-CH_2CHCOOH)_n$- has been studied with various methods. These weak polyacids have been shown to excist as flexible more or less densely coiled chains when bearing no charge. Upon charging such polyelectrolytes- which can be done by adding a strong base to the polyacid solution- more extended conformations are induced as a consequence of the increasing Coulomb repulsion between the negatively charged groups along the polymer chain (COO^-).

Variation of the charge (degree of neutralization, α) on the polyelectrolyte molecules is therefore a convenient way to introduce conformational changes in these polymers.

Using solutions of PMA in which the methylene groups are deuterated- $(-CD_2-CCH_3COOH/D)_n$ - one can measure the deuteron relaxation rates of respectively the CD_2-deuterons and of the $COOD/D_2O$-deuterons. Because of the dominance of quadrupole relaxation as a relaxation mechanism for the deuterons these investigations may yield information on the influence of conformational changes on internal rotations of the PMA molecule.

For the relaxation of the $COOD/D_2O$-deuterons we will use the same approach as applied earlier to PAA which gives 1)

$$(T_{1,2}^{-1})_{obsvd} = \frac{2X_1}{2X_1+X_2}\,\frac{D_o}{D}\,(T_{1,2}^{-1})_o + \frac{X_2}{2X_1+X_2}\,(1-\alpha)\,(T_{1,2}^{-1})_{COOD} \quad \cdots \quad (1)$$

where the index o indicates pure solvent, D the selfdiffusion coefficient and X_1 and X_2 the molar fractions of D_2O molecules and PMA monomeric units respectively.

By using the Huntress 2) formalism for anisotropic diffusional reorientation of the deuteron bond axis one calculates - assuming a cylindrical symmetry of the electrondistribution in the deuteron bond - for the relaxation rates:

$$T_1^{-1} = \frac{3}{8}\left[\frac{eQeq}{\hbar}\right]^2 \frac{1}{5} \sum_{i=-2}^{+2} C_i\,\tau_i\,[(1+\omega_o^2\tau_i^2)^{-1} + 4\,(1+4\omega_o^2\tau_i^2)^{-1}] \ldots (2)$$

$$T_2^{-1} = \frac{3}{8}\left[\frac{eQeq}{\hbar}\right]^2 \frac{1}{5} \sum_{i=-2}^{+2} C_i\,\tau_i\,[\frac{3}{2} + \frac{5}{2}(1+\omega_o^2\tau_i^2)^{-1} + (1+4\omega_o^2\tau_i^2)^{-1}] \ldots (3)$$

where $eQeq/\hbar$ is the deuteron quadrupole coupling constant and ω_o the Larmor frequency. The coefficients C_i and the correlation times τ_i are given as function of the principal values of the rotational diffusion tensor and the average orientation of the bond axis with respect to diffusion tensor principal axes in Table I of reference 1).

II Experimental. Deuterated PMA with a DP of 1.5×10^3 was prepared from the selectively deuterated methylester of which the synthesis will be published elsewhere; the samples with different α were prepared in the same way as in reference 1). The PMA monomolarfraction in H_2O was 0.0107 and in D_2O 0.0108. The temperature of the sample was controlled at 80°C by a Bruker temperature unit. Relaxation rates were measured in the same way as in ref. 1), with the exception of the T_2 of the CD_2-deuterons which was measured from the decay after a $\frac{\pi}{2}$ puls; this decay was averaged with a Varian CAT. Diffusion coefficients were measured in the same way as in ref. 1).

III Results. Using selfdiffusion measurements on the D_2O solutions at 80°C and a measured value for $(T_{1,2}^{-1})_o$ equal to 0.755^{-1} we have calculated the $(T_1^{-1})_{COOD}$ with formulae (1). In fig. 1 we have plotted these relaxation rates as a function of α together with $(T_1^{-1})_{CD_2}$ and $(T_2^{-1})_{CD_2}$; the relaxation rates of both types of deuterons were measured at 4.6 MHz.
A steep decrease in the relaxation rates is observed upon charging the polymers (the same effect has been noted in earlier proton studies 3)). From eqns.(2) and (3) it is seen that changes in the correlation times and / or changes in the orientation of the bond axes with respect to the diffusion tensor principal axes may be responsible for this behavior. Both effects may occur as a result of the uncoiling of the polymer.

The same techniques have been used to calculate the COOD-relaxation rates for α=0.05; α=0.15 and α=0.31 at 4 other frequencies - 9.2 MHz; 8.0 MHz; 6.0 MHz and 4.0 MHz. Samples with the same α were measured in H_2O at these frequencies to evaluate the CD_2- relaxation rates. From these experimental results we have calculated the relative spectral density $J(\omega)/J(o)$ for the 3 different α's and the 2 kinds of deuterons and we have plotted the results in fig. 2 and 3.

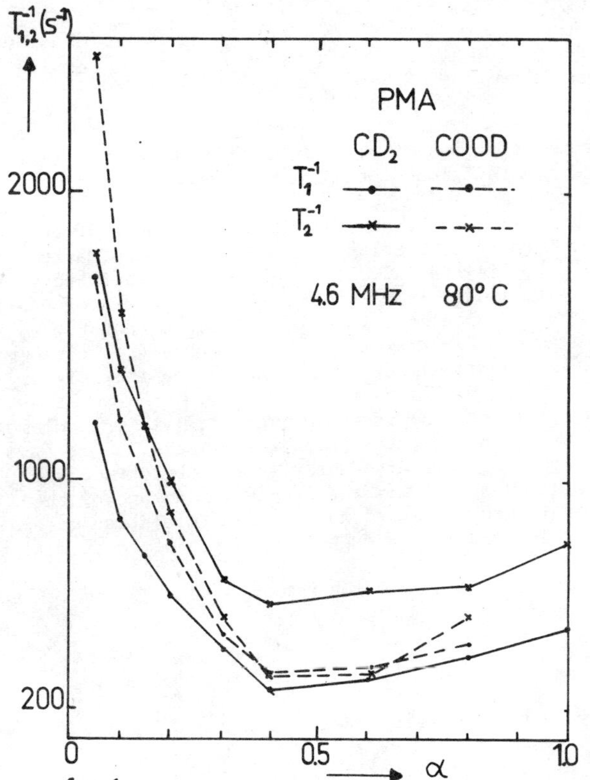

fig. 1

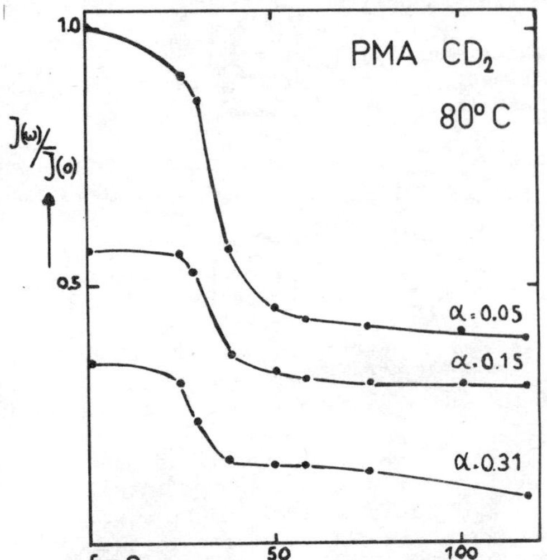

fig. 2

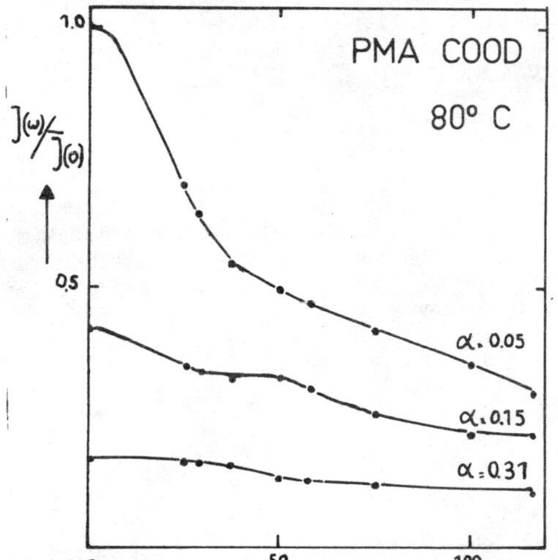

fig. 3

In fig. 2 and 3 J(ω) is displayed relative to J(o) for the smalles α-value (=0.05) in each figure. The frequency dependence of the spectral densities indicate anisotropic rotation and the isotropic limits of eqns. 2 and 3 cannot therefore be applied.

An interesting difference in the changes of the spectral densities for the $-CD_2-$ and the -COOD groups emerges from these figures. The shape of the spectraldensities is more or less the same for the different α-values in the case of $-CD_2-$ (indicating no dramatic changes in the larger correlation times). In contrast, for -COOD the spectral density flattens out nearly completly upon charging the polymer.

This indicates that only relatively short correlation times contribute to the relaxation behavior of the -COOD deuterons in the partially extended macromolecule.

References:

1) J.J.van der Klink,J.Schriever and J.C.Leyte, Ber.Bunsenges.Physik.Chem. 78, 369 (1971)
2) W.T.Huntress,Jr., Adv.Magn.Res. 4, 1 (1970)
3) J.Schriever and J.C.Leyte, Abstracts Iupac Intern.Symp. on Macromol. Leiden 1970, 473 (1970).

MÖSSBAUER SCATTERING FROM HIGHLY VISCOUS GLYCEROL

M. SOLTWISCH AND D. QUITMANN
Freie Universität Berlin
Institut für Atom- und Festkörperphysik, D-1000 Berlin 33, Boltzmannstr. 20, Fed. Rep. Germany

Abstract. The quasielastic scattering from glycerol of 14.4 keV γ-rays has been studied between -30°C and $+20^{\circ}$C at the first diffraction maximum ($2\Theta=11^{\circ}$) and at $2\Theta=20^{\circ}$. From the linewidth observed the diffusion constant was derived. Comparison with NMR measurements is made. The mean square displacements due to "fast" motions $\langle x^2\rangle_f$ appears to be related to the "slow" movement by $\langle x^2\rangle_f \propto \Gamma^{1/2}$.

Introduction: Among the highly viscous glass forming liquids, glycerol is the most thoroughly studied example. The majority of measurements working at the atomic scale have been done on the spin relaxation times of the glycerol protons. Scattering with reciprocal momentum transfer corresponding to atomic and molecular separations directly scans the atomic and/or molecular motion. Energy resolved photon scattering samples the movement of the molecular backbone, while NMR follows the movement of individual protons in glycerol (as does neutron scattering). The necessary energy resolution for keV-photons is, however, available only through the Mössbauer effect and corresponds to the high viscosity regime.
As in neutron scattering, the energy distribution of the scattered intensity has a quasielastic line due to the low frequency ("slow") movements, i.e. the average diffusional velocity. The high frequency components ($\Delta E \gtrsim 10^{-6}$ eV) are observed integrally.
So far, Mössbauer scattering experiments on glycerol were performed by Elliot, Hall and Bunbury (EllH66), and Champeney and Woodhams (ChaW 68). However, the results were not very precise.

Experiment and Results: In the scattering experiment we use a source of 250 mCi ^{57}Co diffused into a 12 μm Rh foil over 2x3 mm^2 (from Amersham Radiochemical Center); it is mounted on a electromechanical drive system. A 5cm long graded collimator of 2x3 mm^2 opening is placed in front of the source. The scattering chamber is a 3x5x0.5 cm^3 temperature regulated metal frame with 0.5mm Be windows for thermal homogeneity, enclosed in a dry air-chamber. Temperature was stable to $\pm 0.5^{\circ}$C and homogeneous to better than 0.2°C over the sample. A 3x10 mm collimator of 2 cm length was placed in front of the detector, a Si(Li) diode. Source-scatterer-detector distances were about 10 cm both. The source was mounted on the movable arm of an X-ray goniometer, whereas the detector was fixed.
The sample was obtained from 98% glycerol (Merck DAB 7) by heating to 110°C for two days under vacuum. To check the water content, the melting point was determined and found to be $(18.3\pm.1)^{\circ}$C which corresponds to less than 0.2% water content. The thickness of 0.5 cm was chosen as the optimum caused by photoelectric absorption. Measurements were taken at $2\Theta=11^{\circ}$ and at $2\Theta=20^{\circ}$; the angular spread was $\pm 3^{\circ}$. The background including Compton scattering was separately measured. Depending on the width of the quasielastic line, we used thick absorbers of either $Li_3FeF_6+(NH_4)_3FeF_6$ or stainless steel which had 1.7mm/s and 0.6mm/s FWHM, respectively; their line shapes were approximated as sums of Lorentz curves. The calibration of the f-factor of the source-absorber-combination was performed using the (200) Bragg-reflection from LiF.
For the quasielastic scattering, a simple Lorentzian broadening was assumed with characteristic parameters FWHM Γ and relative intensity f. Good fits were obtained in every case.
The results for Γ and f are presented in figs. 1 and 2 as D vs 1/T and $\Gamma^{1/2}$ vs. ln f/k^2 where $D=\Gamma/2\hbar k^2$ and $k=4\pi \sin\Theta/\lambda$ have been used.

Discussion: For the interpretation, we start from the calculated energy distribution of Mössbauer-γ-radiation emitted from a diffusing necleus (SinS60). We use the limit of small step diffusion which leads to the Lorentzian distribution with $=2\hbar k^2 D$. The diffusion constant D obtained from the first diffraction maximum, fig. 1, can be well described in the temperature region considered by

$$D = D_o \exp(-E/kT) \text{ with } D_o = 22(5)\ \mathrm{cm}^2/\mathrm{s}\ ;\ E=12.5(5)\ \mathrm{kcal/mol\ for\ } 240\mathrm{K} \lesssim T \lesssim 290\ \mathrm{K}.$$

The errors quoted are rough estimates.

The diffusion constants derived from various experiments are compared in a more sensitive plot in fig. 3. Note that in (AleB 69), a sample with only 98% glycerol was used.
When doing a comparison between the different techniques it has to be kept in mind that they are sensitive to different aspects of the atomic motion:

a) quasielastic Mössbauer scattering in the first maximum reflects the time modulation of the O- and C-density correlation over about 5 Å.
b) neutron scattering and pulse gradient NMR are determined by the motion of individual protons.
c) proton relaxation times measure the modulation of the intermolecular proton-proton distances (with strongest weighing at the lower limit ≈2Å)

Starting from the model of translational small step diffusion, a very satisfactory discussion had been presented by Fiorito and Meister (FioM 72); this description included the temperature dependence of the appearent thermal activation energy E. Within this picture, the satisfactory agreement between NMR and our data is taken as a sign that the assumption of individual protons

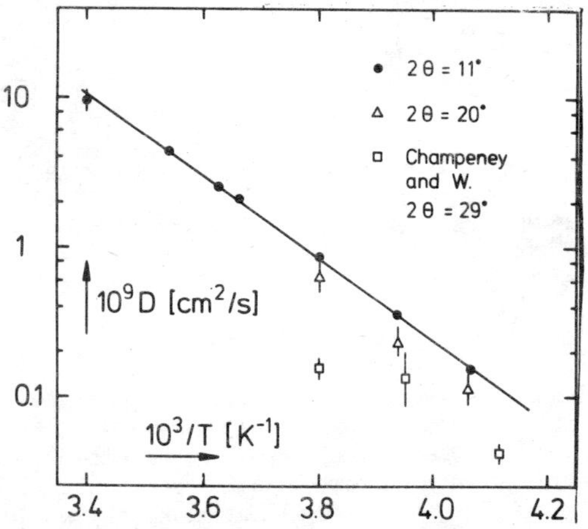

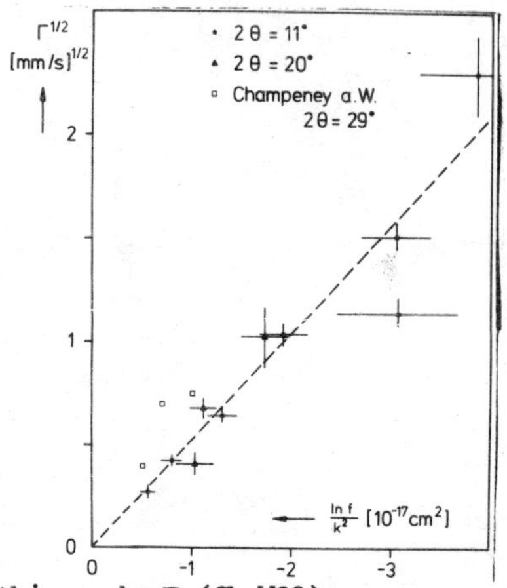

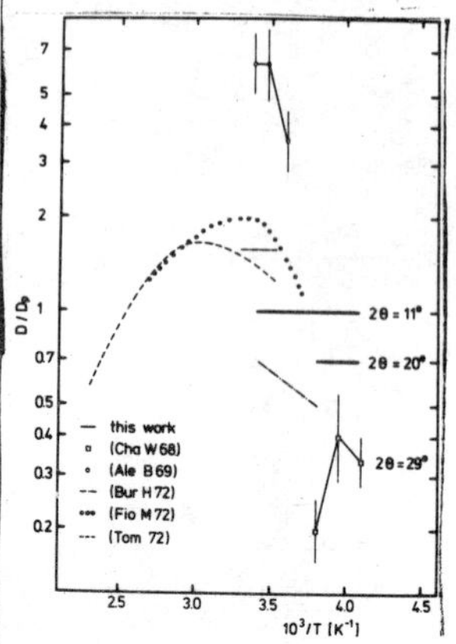

Fig. 1 Diffusion constant of glycerol, • Δ this work; □ (ChaW68)

Fig. 2 Relation between width Γ and relative intensity f of the quasielastic line; plotted is $\Gamma^{1/2}$ vs. ln f/k^2.

Fig. 3 Comparison of diffusion constants for glycerol as obtained by different methods. (ChaW68) and this work: Mössbauer scattering; (AleB69): neutron scattering; (Tom72): pulse gradient NMR; (Fio M72), (Bur H72): proton ralaxation NMR. All data have been normalized to D_p=21.8cm²/s·exp(-6308K/T). For most of the data, only representative lines are drawn.

riding with their molecules is correct. However, an unexpected increase of the minimum proton-proton (intermolecular) distance was obtained in (FioM 72). Recently, the whole model of pure translational diffusion has been questioned on the basis of new relaxation measurements in deuterated glycerol (KinZ 73).
For the further discussion of atomic and molecular motion in glycerol we would like to draw attention to two points which follow from our measurements and go beyond the simple diffusional picture:

In the first diffraction maximum (corresponding to ≈ 5Å), inter- and intramolecular correlations contribute to the width of the quasielastic line. At high momentum transfer (2Θ=20°,29°) the linewidth is determined by density modulations over smaller distances, i.e. especially intramolecular correlations. Here one observes a slower modulation (see Fig. 1). This is the same direction of deviation as is observed in neutron scattering on water where Γ falls below $2\hbar k^2 D$ at larger k (see e.g.(Spr 72)).
The NMR relaxation times depend on the "slow" motions ($\omega \lesssim 10^8$/sec), while the "fast" motions are essentially averaged out (see, however, (KinZ73)). In the scattering experiment with good energy resolution, the latter integrally produce a reduction of the intensity I_q of the quasielastic line, $I_q \propto \exp(-k^2\langle x^2\rangle_f)$. This serves to derive the "fast" mean square displacement $\langle x^2\rangle_f$ from the data. One obtains a drastic increase of $\langle x^2\rangle_f$ with T. A relation $\Gamma \propto \langle x^2\rangle_f$ has been suggested by Jensen (Jen71) from a simple formula for combining diffusional and oscillatory motion. We find that $\Gamma \propto (\langle x^2\rangle_f)^2$ represents the data of this experiment and of (ChaW68) definitely better; this plot is presented in fig.2. We have no explanation or model for this observation. However, the rapid growth and narrowing of the quasielastic line at the expense of the high frequency movements reminds one of the central peak with is observed when diffusional disorder freezes at phase transitions as in ferroelectrics. At any rate, a pure scaling of the frequency spectrum by a single temperature dependent parameter like the correlation time may not be sufficient.

References:

1. (Ale B 69) B. Alefeld, M. Birr und A. Heidemann: Naturwiss. 56 (69) 410
2. (Bur H 72) L.J. Burnett and J.F. Harmon: J. Chem. Phys. 57 (72) 1293
3. (Cha W 68) D.C. Champeney and F.W.D. Woodhams: J. Phys. B2 (68) 620
4. (Ell H 66) J.A. Elliott, H.E. Hall and D.S.P. Bunbury: Proc. Phys. Soc. 89 (66) 595
5. (Fio M 72) R.B. Fiorito and R. Meister: J. Chem. Phys. 56 (72) 4605; see also G. Preissing et al.: Z. Physik 246 (71) 84
6. (Jen 71) J.H. Jensen: Phys. kond. Mat. 13 (71) 273
7. (Kin Z 73) J.P. Kintzinger and M.D. Zeidler: Ber. Bunsenges. Phys. Chem. 77 (73) 95
8. (Sin S 60) K.S. Singwi and A. Sjölander: Phys. Rev. 120 (60) 1093
9. (Spr 72) T. Springer: Ergebn. Exakt. Naturw. 64 (72) 1
10. (Tom 72) D.J. Tomlinson: Mol. Phys. 25 (72) 735

INTERPRETATION OF THE J_{X-H} NUCLEAR SPIN-SPIN COUPLING CONSTANTS IN HYDRIDES XH_n WITH A HULTHÉN POTENTIAL LCAO MODEL

P. Pyykkö [+‡] and J. Jokisaari[*]
+ Department of Physics, University of Jyväskylä, SF-40720 Jyväskylä, Finland
* Department of Physics, University of Oulu, SF-90100 Oulu, Finland

Abstract. The indirect nuclear spin-spin coupling in the hydrides CH_4 to PbH_4 is considered using a simple model for the X-H fragment. The basis consists of bound and continuum solutions in a Hulthén screened Coulomb potential, and a hydrogen atomic orbital. The coupling constant, J, is an integral over all momenta, k, of a spectral density function. The increase of coupling with increasing Z is successfully interpreted, if a relativistic correction is included. The absolute values are about half of the experimental.

1. Introduction. Quantitative calculation of nuclear spin-spin coupling constants, J, in molecules has proved difficult. Perhaps the first and only such calculation is that of Kowalewski et al. (1974) for H_2. They found that a large basis set was necessary and that correlation effects amount to about 20 per cent of the observed coupling.

In the present work we try to obtain a physical feeling for what parts of the momentum, or energy space are important in determining the value of J in the particular case of J_{X-H} in hydrides XH_n. We also attempt to interpret the increase of the reduced coupling constant $K_{XH} = J_{XH}/(g_X\, g_H\, \beta_n^2)$ in the series C-H to Pb-H.

2. The LCAO model. We describe the screened Coulomb field of the heavy atom by a Hulthén potential

$$V(r) = -Zq/\left[\exp(qr) - 1\right], \quad (1)$$

where Z is the nuclear charge and q a screening parameter. Analytic solutions for the bound s states χ_{ns} and continuum s states χ' in this potential are known (Lindhard and Winther 1971). The value of q is fixed by demanding that the valence χ_{ns} be equal to a Hermann-Skillman wave function at origin. The X-H chemical bond is described by a simple LCAO MO

$$\Phi = N(\chi_{ns} + a\,\chi_H) \quad (2)$$

where χ_H is a hydrogen AO having $\zeta = 1.2$ and $a = 0.3$. The excited electronic state is described by χ', orthogonalized against Φ:

$$\Phi' = N'\left[\chi' - Na\, S'\, \Phi\right], \quad S' = \langle \chi_H \mid \chi' \rangle \quad (3)$$

The singlet ground state Ψ_0 and the triplet excited state Ψ' are Slater determinants

$$\Psi_0 = \mid \Phi(1)\alpha(1)\Phi(2)\beta(2) \mid, \quad \Psi' = \mid \Phi(1)\alpha(1)\,\Phi'(2)\alpha(2) \mid \quad (4)$$

Only the Fermi contact contribution is considered. Defining a spectral density function for the nuclear spin-spin coupling,

$$\Phi(k) = \langle \Psi_0 | \delta_H | \Psi'(k) \rangle \langle \Psi'(k) | \delta_X | \Psi_0 \rangle / \left[E_0 - E(k)\right] \quad (5)$$

we express the reduced coupling constant as

$$K_{XH} = C \int_0^\infty \Phi(k)dk\,, \quad C = 256\,\pi^2 B(n,z)\beta^2/9. \quad (6)$$

k is the momentum of the intermediate state. E_0 is approximated by the Hulthén binding energy and $E(k) = k^2/2$. At the chosen values of q there are no excited bound states so the entire spin-spin coupling arises from the continuum. Relativistic effects are important for the heavier atoms and are described by the hydrogen-like relativistic corrections B(n,z) (Pyykkö et al. 1973).

3. Results. The calculated and observed reduced coupling constants K_{XH} (in $10^{20} cm^{-3}$) are shown in the following table. R is the X-H distance (in a.u.).

‡ Present address: Department of Physical Chemistry, Åbo Akademi, SF-20500 Åbo (Turku), Finland

X	q	R	B(n, Z)	K_{XH} (calc.)	K_{XH} (exp.)
C	2.3408	2.067	1.004	+ 21.7	41.8
Si	2.8531	2.797	1.023	+ 32.2	84.9
Ge	3.8314	2.881	1.125	+ 101	233
Sn	3.8859	3.214	1.348	+ 166	431
Pb	4.4752	3.3	2.592	+ 492	(947)

The experimental values are taken from Dalling and Gutowsky (1971). The spectral density functions $\Phi(k)$ for CH_4 and PbH_4 are shown in the figure.

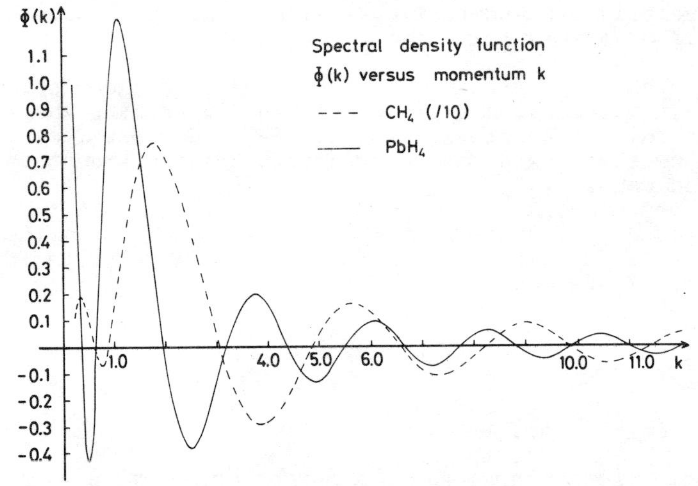

4. Discussion. The 23-fold increase of K_{X-H} from C-H to Pb-H is successfully interpreted by the model. The absolute values of K are too small by a factor of 2, if the Hermann-Skillman valence AO spin densities are used. The following reasons may account for this: (a) The continuum orbital χ' is spherically symmetrical while the true excited state is enhanced at the proton, (b) correlation effects are neglected and, (c) the bond model is only qualitative, and the χ_{ns} is not very good at larger distances. The maxima of $\Phi(k)$ occur at $(E_o - E(k))$ of about 50 eV for CH_4 and about 17 eV for PbH_4. This casts some doubt at the traditional effective energy denominator of 15 eV for any molecule. A more detailed manuscript of this work is under preparation.

5. Acknowledgements. We thank Dr. H. Jansen and Mr. E. Latvamaa for helpful discussions. This work was begun while P.P. was at CECAM, Université de Paris XI, France.

6. References

Dalling, D.K. and Gutowsky, H.S., 1971, J. Chem. Phys. 55, 4959.

Kowalewski, J., Roos, B., Siegbahn, P. and Vestin, R., 1974, Chem. Phys. 3, 70.

Lindhard, J. and Winther, A., 1971, Nucl. Phys. A166, 413.

Pyykkö, P., Pajanne, E. and Inokuti, M., 1973, Internat. J. Quantum Chem. 7, 785.

SPIN DENSITY DISTRIBUTION IN THE CARBONYL GROUP FROM THE PROTON MAGNETIC RESONANCE OF PARAMAGNETIC COMPLEXES

K.Jackowski and Z.Kęcki
Laboratory of Intermolecular Interactions, Institute of Fundamental Problems of Chemistry, Warsaw University, ul. Pasteura 1, 02-093 Warszawa, Poland.

Abstract. PMR contact shifts were measured for the octahedral Ni/II/ and Co/II/ complexes of ketones. Up field shifts were observed for the methyl and methylene groups which were directly bonded to the carbonyl group. All the other shifts were down field. It is explained in terms of the characteristic spin density distribution within the carbonyl group.

1. Introduction. In paramagnetic complexes some of the metal ion unpaired electrons can be delocalized onto ligands and the isotropic contact coupling of unpaired electron and nucleus spin magnetic moments can be observed as NMR contact shifts. The expression used to evaluate the coupling constant A_s[Hz] from the NMR contact shifts is

$$\frac{\Delta H}{H_o} = -A_s \cdot \frac{\bar{g}\mu_B S(S+1)}{(\gamma_N/2\pi)3kT} \qquad /1/$$

where $\bar{g}$ is the averaged g factor for the complex, μ_B is the absolute value of the Bohr magneton, γ_N is the nuclear magnetogyric ratio, S is the total spin number of the electrons which are delocalized onto ligands, ΔH is the contact shift at an applied magnetic field H_o[2]. The A_s constant is proportional to the density of unpaired electrons /the spin density/ occupying s-orbitals. The mechanism by which the spin density reaches the nuclei on the ligands is closely related to the geometry of the complex. Octahedral complexes especially afford an excellent opportunity for the elucidation of the spin delocalization mechanism because of the strict separation of π- and σ-bonding; the t_{2g} metal d-orbitals can transmit spin to the ligand via π-bonding only and the e_g metal orbitals via σ-bonding only. Finally, the spin density distribution associated with the ligands provides an insight into the electronic structure of the ligand molecule.

2. Experimental. Ni/II/, Co/II/ and Zn/II/ perchlorates were obtained from dry chlorides and $AgClO_4$ directly in the solutions of ketones. $Co[CH_3CO \cdot H]_6[InCl_4]_2$ was synthesized in the nitromethane solution[1]. All the reagents were carefully purified and dehydrated. PMR spectra were recorded on JNM-3H-60 /JEOL Co., Tokyo/ spectrometer of high resolution at 60 MHz. Shift measurements were carried out using the sideband technique. TMS was applied as the internal standard. The solutions of zinc perchlorate were used to estimate the diamagnetic contributions of the paramagnetic shifts. The pseudocontact contributions were neglected.

3. Results and Discussion. We have examined the PMR contact shifts of solvated ketones to the paramagnetic cations: Ni^{2+} and Co^{2+}. The significant result of this study can be summarized:

1/ the up field shifts were observed only for the methyl or methylene protons next to the carbonyl group /$A_s<0$, the negative spin density on the protons/;
2/ all the other shifts were down field /$A_s>0$, the positive spin density on the protons/;
3/ the shifts of the nickel and cobalt solvates were similar.

As an illustration the contact shifts of acetone and methylethylketone obtained for six-coordinated complexes are given in Tab.I. The contact shifts were determined from the averaged shifts observed for the ketone solutions of nickel

Tab.I. Contact Shifts and Coupling Constants for NiL_6^{2+} and CoL_6^{2+} Complexes.

Complex	Signal	$\frac{\Delta H}{H_o}$ [ppm]	A_s [Hz]
$Ni/CH_3 \cdot CO \cdot CH_3/_6^{2+}$	$-CH_3$	+3.6	$-4.1 \cdot 10^4$
$Co/CH_3 \cdot CO \cdot CH_3/_6^{2+}$	$-CH_3$	+2.9	$-3.1 \cdot 10^4$
$Ni/CH_3^{(M)} \cdot CO \cdot CH_2 \cdot CH_3^{(E)}/_6^{2+}$	$-CH_3^{(M)}$	+7.0	$-8.1 \cdot 10^4$
	$-CH_2-$	+1.1	$-1.3 \cdot 10^4$
	$-CH_3^{(E)}$	-0.9	$+1.0 \cdot 10^4$
$Co/CH_3^{(M)} \cdot CO \cdot CH_2 \cdot CH_3^{(E)}/_6^{2+}$	$-CH_3^{(M)}$	+7.7	$-8.2 \cdot 10^4$
	$-CH_2-$	-1.7	$+1.8 \cdot 10^4$
	$-CH_3^{(E)}$	-0.4	$+4.0 \cdot 10^3$

and cobalt perchlorates. According to Driessen and Groeneveld[1]) the Ni^{2+} and Co^{2+} cations form with acetone the octahedral complexes and the coordination occurs via the oxygen atom of the acetone. The observed contact shifts can be explained in terms of the spin density delocalization in the octahedral complexes as follows:

1/ the primary effect is the delocalization of two e_g unpaired electrons /the positive spin density/ in the σ-bonding system;
2/ the σ-delocalization causes the spin polarization of the carbonyl π orbital by inducing the positive spin density at the oxygen atom and the negative spin density at the carbon atom as shown in Fig.1.;
3/ the negative spin density is transferred directly from the π-orbital to the 1s hydrogen orbitals of alkyl groups by hyperconjugation;
4/ the contact shifts for the nickel and cobalt complexes are similar because in the cobalt complexes only the two e_g electrons are delocalized like in the nickel complexes and the delocalization of the t_{2g} unpaired electron does not occur.

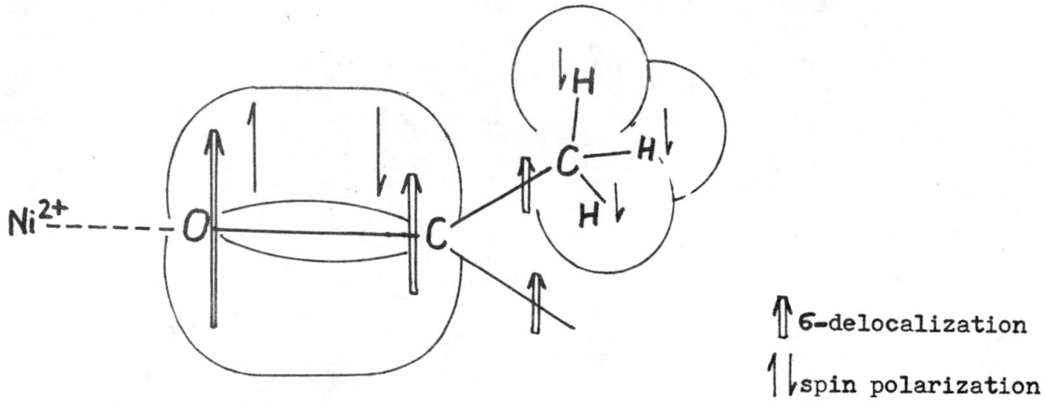

Fig.1. The spin density distribution within the carbonyl group.

This mechanism gives the proper spin densities on the protons due to the characteristic distribution of spin density within the carbonyl group, Fig.1. We hope that the mechanism is valid also for aldehydes and esters on condition that their six molecules are coordinated to the Ni^{2+} or Co^{2+} ion the carbonyl oxygen atoms. On the ground of this assumption it can be expected for the complex of acetaldehyde: the large positive spin density on the aldehyde proton because of the σ-delocalization and the negative spin density on the methyl protons because the spin polarization and hyperconjugation are more effective at the methyl protons than the σ-delocalization. We were able to measure the relative spin densities for the complexed acetaldehyde when the complex - $Co[CH_3\cdot CO\cdot H]_6[InCl_4]_2$ was dissolved in nitromethane and the ratio of the spin densities methyl protons/aldehyde proton was -0.03/+1.00. The large positive spin density on the aldehyde proton is no doubt due to the σ-delocalization of the e_g electrons. This result is consistent with the proposed mechanism of spin delocalization. In order to test if the hyperconjugation is important as the part of the mechanism we carried out LCAO MO INDO calculations for some model radicals[3])/i.e. molecules formaldehyde, acetaldehyde or acetone with one excess β electron /m_s= -1/2 / on the lowest π^*antibonding orbital/. The calculations showed that remarkable negative spin density on the methyl hydrogens and only small positive spin density on the aldehyde hydrogen was produced, e.g. for the acetaldehyde model -0.051 and +0.004, respectively. The large negative spin density on the 1s methyl hydrogen orbitals /-0.051/ proves that spin density may be easily transferred from the π-type carbonyl orbital to the 1s hydrogen orbitals of the neighbouring alkyl group due to the hyperconjugation. And it is the reason that the delocalization of spin density within carbonyl molecules occurs in the way described above.

4. References

1. W.L.Driessen, W.L.Groeneveld, Rec.Trav.Chim., 88,977/1969/; 90,87/1971/.
2. G.A.Webb, in E.F.Mooney/ed./"Annual Reports on NMR Spectroscopy", Academic Press, London 1970, pp.211-259.
3. K.Jackowski, Z.Kęcki, to be published.

LONG RANGE H-H COUPLING CONSTANTS IN SOME 2-SUBSTITUTED OXETANES

Jukka Jokisaari
Department of Physics, University of Oulu, SF-90100 Oulu 10, Finland

Abstract. The long range coupling constants between the methine proton and the phenyl protons in some substituted 2-phenyloxetanes have been calculated at the INDO and CNDO/2 level of approximation using the finite perturbation method. Furthermore, the effect of temperature on these coupling constants has been studied in 2-(2-chlorophenyl)oxetane.

1. Introduction. Recently oxetane molecule has been a subject of many works (cf. Jokisaari, 1974). The ring-puckering vibration of that molecule has been studied by far-infrared spectroscopy (Jokisaari et al. 1973 and references therein) and also nuclear magnetic resonance has been applied using nematic solvents (Cole et al. 1972, Khetrapal et al. 1973). The investigations show that the oxetane ring is planar. However, it seems that the 2-substitution leads to a non-planar structure with puckering angle varying between 3 and 7 degrees (Fomichev et al. 1972, Jokisaari, 1974).

The aim of this work has been to study the rotation about the C-C bond between the oxetane and phenyl rings in 2-phenyloxetane, 2-(4-fluorophenyl)oxetane, 2-(2-chlorophenyl)oxetane and 2-(3-chlorophenyl)oxetane. This investigation has been realized by semiempirical INDO and CNDO/2 calculations and by studying the temperature dependence of the long range coupling constants.

2. Methods. In the calculations the QCPE 141 program with three more subprograms has been used. The calculation of the spin-spin coupling constants has been based on the finite perturbation method (Pople et al. 1968) and the convergence has been tested by the diagonal elements of the spin density matrix.

The 60 MHz PMR spectra of 2-(2-chlorophenyl)oxetane have been recorded by the Varian A60 spectrometer and the temperature has been controlled by the V-6040 unit. The analyses of the spectra have been carried out using the program LAOCN3. The computations have been performed in both cases with the UNIVAC 1108 system.

3. Results. The short notations J^o, J^m and J^p for the coupling constants between the methine proton and the ortho, meta and para (relative to the position of the oxetane part) positioned phenyl protons, respectively, have been used. Labelling of the protons and the definition of the rotational angle θ is given in Fig. 1a.

When the total energies of the studied systems are plotted at different conformations (see Fig. 1b) it can be observed two equal energy minima for 2-phenyloxetane, 2-(4-fluorophenyl)oxetane and 2-(3-chlorophenyl)oxetane at $\theta=135^o$ and $\theta=315^o$ and one absolute minimum for 2-(2-chlorophenyl)oxetane at $\theta=135^o$. These θ values correspond to the conformations where the oxygen atom of the oxetane ring is situated close to the plane of the phenyl ring. The mean values produced by the states $\theta=135^o$ and $\theta=315^o$ for J^o's are -0.563 Hz in 2-phenyloxetane and -0.572 Hz in 2-(4-fluorophenyl)oxetane, the experimental values being

a) 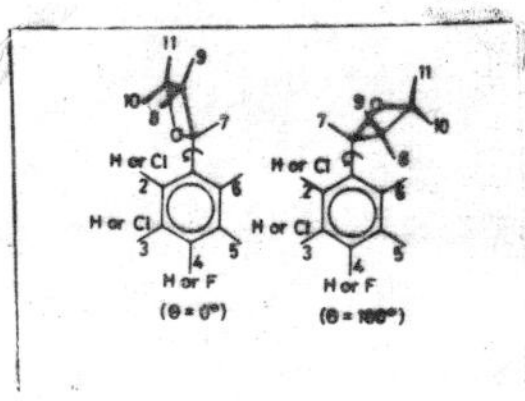

b) 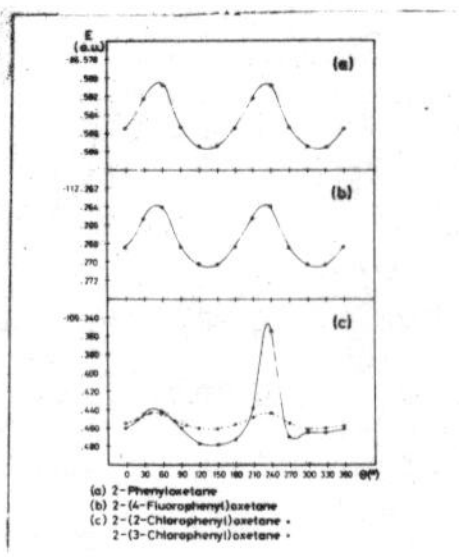

Figure 1. a)Labelling of the protons and definition of the rotational angle.
b)Calculated total energies (in a.u.) at different conformations.

-0.64 Hz and -0.63 Hz, respectively. However, it appears that the same values are obtained by assuming a free rotation about the C-C bond.

The CNDO/2 calculations show that the meta coupling constants in 2-(2-chlorophenyl)- and 2-(3-chlorophenyl)oxetane are more dependent on conformation than the ortho ones. Therefore, they are better suited for conformational studies. For instance the meta coupling constant J^m_{57} in 2-(2-chlorophenyl)-

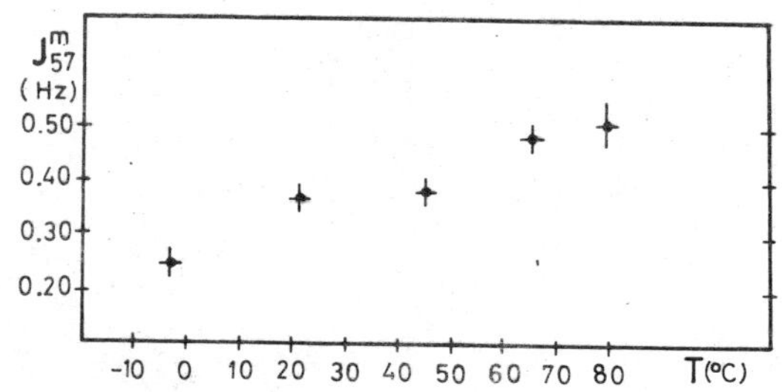

Figure 2. Experimental meta coupling constant J^m_{57} of 2-(2-chlorophenyl)-oxetane as a function of temperature.

oxetane is well reproduced by the equation

(1) $$J^m_{57}(\text{Hz}) = 0.299 + 0.300 \cos\theta + 0.241 \cos 2\theta.$$

According to the above equation the difference in J^m_{57} at $\theta=0^{\circ}$ and at $\theta=180^{\circ}$ is 0.6 Hz. When studying the experimental values given in Fig. 2. it can be found that at T=-3 °C J^m_{57}=0.26 Hz while at T=+80 °C J^m_{57}=0.51 Hz. On the contrary, the other meta coupling constant, J^m_{37}, and the ortho coupling constant remain constant within the error limits. It seems that the change of J^m_{57} can be related, at least partly, to the rotation or vibration about the C-C bond.

4. References.

Cole, K.C., and Gilson, D.F.R., 1972, J.Chem.Phys. 56, 4363
Fomichev, A.A., Kostyanovskii, R.G., Zon, I.A., Gella, I.M., Zakharov, K.S., and Markov, V.I., 1972, Dokl.Akad.Nauk., USSR, 204, 644
Jokisaari, J., and Kauppinen, J., 1973, J.Chem.Phys. 59, 2260
Jokisaari, J., 1974, Ph.D.thesis, in press
Khetrapal, C.L., Kunwar, A.C., and Saupe, A., 1973, Mol.Phys. 25, 1405
Pople, J.A., McIver, J.W., and Ostlund, N.S., 1968, J.Chem.Phys. 49, 2965

INDIRECT DETERMINATION OF J AND T_1 IN URANIUM HEXAFLUORIDE BY PULSED NMR

I.Ursu, D.E.Demco, V.Simplaceanu, N.Vâlcu and N.Ilie[x]
Institute for Atomic Physics, P.O.Box 35, Bucharest, Romania
[x]Student Trainee 1974, University of Bucharest, Bucharest, Romania

Abstract. Measurements of T_2^{-1} and T_1^{-1} nuclear magnetic relaxation rates of ^{19}F nuclei in liquid UF_6 samples with known enrichments in ^{235}U have been used in order to estimate the scalar coupling constant J_{F-U} and the relaxation time T_{1U} of the ^{235}U nuclei.

1. Introduction. Pulsed NMR methods have been used for determining the scalar coupling constant J by rotating frame relaxation measurements (Sears, 1972), transverse and longitudinal relaxation in the laboratory frame (Shoup, 1972) or by measuring the transverse relaxation time with various pulse spacings in the Carr-Purcell sequence (Boden et al.,1966). In order to succeed, either a knowledge of the relaxation time of the S spin species is necessary or this relaxation time must be long enough in order to avoid the necessity of using prohibitively large r.f. magnetic fields.

The X-ray studies (Hoard and Stroupe,1944) showed that in the condensed state the molecule of UF_6 is a tetragonally distorted octahedron with the axial bonds slightly longer than the equatorial ones. NMR spectra (Blinc et al.,1973; Rigny 1965; Blinc et al.,1966) confirmed the existence of non-equivalent fluorine positions at low temperatures. Above 20°C the spectrum is motionally narrowed and consists of a single resonance line. The anisotropy of the chemical shift modulated by the molecular motions is the main relaxation mechanism in solid UF_6 (Blinc et al.,1965; 1971), but the spin-rotational interaction is dominant in the liquid state (Rigny, 1967; Blinc and Lahajnar, 1968) of the UF_6 with natural (0.712%) ^{235}U abundancy. Recently, a strong dependence of the T_2 of the ^{19}F nuclei on the enrichment in ^{235}U of the liquid UF_6 was reported (Ursu et al. 1974) and explained by considering the quadrupole relaxation of the ^{235}U nuclei (S=7/2). In this work a preliminary estimation is made of the scalar coupling constant J_{F-U} and of the relaxation time of the ^{235}U nuclei in liquid UF_6.

2. Experiment and Results. Five UF_6 samples with 0.5%, 4.399%, 10%, 49% and respectively 93.29% atomic enrichments were purchased from British Nuclear Fuels Ltd. Measurements were performed at 60 MHz and 36 MHz on a commercial Bruker B-KR 322s spectrometer at 338°K, 348°K and 356°K. T_1 was measured by the usual inversion recovery method and T_2 was measured with the Carr-Purcell-Meiboom-Gill sequence using the relaxation time counter B-KR 300z16.

A noticeable difference was observed between the longitudinal and transverse relaxation rates. The difference is larger at higher enrichments, where it becomes an order of magnitude. The analysis of the relaxation mechanisms (Ursu et al.,1974) showed that the quadrupole relaxation of the ^{235}U nuclei is responsible for this difference, via the indirect scalar coupling between ^{19}F and ^{235}U nuclei (Abragam, 1960). Assuming the extreme narrowing condition for ^{235}U nuclei, $T_{1U} = T_{2U}$ and the contributions to the ^{19}F relaxation rates are (Ursu et al.,1974):

$$(T_1^{-1})_Q = 11.5\ J^2_{F-U}\ T_{1U}\ I_U/|1 + (\omega_{oF} - \omega_{oU})^2\ T^2_{1U}| \qquad (1)$$

$$(T_2^{-1})_Q = 5.75\ J^2_{F-U}\ T_{1U}\ I_U\{1 + 1/|1 + (\omega_{oF} - \omega_{oU})^2\ T^2_{1U}|\} \qquad (2)$$

where I_U stands for the enrichment in ^{235}U. These contributions are negligible at natural abundancy, but become dominant in T_2^{-1} at higher abundancies. As we do not know the value of T_{1U}, the simple method of using the difference between the transverse and longitudinal relaxation rates (Abragam, 1960) is not feasible. Instead we use the difference between the slopes of the relaxation rates plotted vs. the enrichment. These slopes are given by Eqs. 1 and 2 when divided by I_U. Denoting them by R_{1Q} and R_{2Q}, we have:

$$2R_{2Q} - R_{1Q} = 11.5\ J^2_{F-U}\ T_{1U} \qquad (3)$$

and from Eq. 1 divided by I_U we get:

$$T_{1U} = (2R_{2Q} - 2R_{1Q})^{1/2}/(\omega_{oF} - \omega_{oU}) \qquad (4)$$

and from Eq. 3:

$$J_{F-U}(Hz) = (1/2\pi)|11.5\ (2R_{2Q} - R_{1Q})/T_{1U}|^{1/2} \qquad (5)$$

The results are given below.

Table 1. The values of J_{F-U} and T_{1U} in liquid UF_6

f_{oF}(MHz)	T(^oK)	$R_{1Q}(sec^{-1})$	$R_{2Q}(sec^{-1})$	$T_{1U}.10^9 sec$	J_{F-U}(Hz)
60	338	2.75 ± 0.1	88.6 ± 1.3	21.3 ± 1	4240 ± 183
	348	2.26 ± 0.5	104.7 ±13.6	25.8 ± 7.2	4207 ± 1300
	356	2.89 ± 0.05	113.4 ± 2.3	23.7 ± 0.5	4566 ± 130
36	338	2.61 ± 0.53	76.5 ± 4.2	34.0 ± 8	3124 ± 700
	348	2.98 ± 0.39	82.4 ± 9.5	33.0 ± 6	3290 ± 711
	356	2.80 ± 0.89	96.5 ± 9.6	37.0 ±13	3376 ± 1205

These values are preliminary and the differences among them require further investigations which are under way in our laboratory.

This work was supported in part by the International Atomic Energy Agency under contract no.1121/RB and the State Committee for Nuclear Energy of Romania under contract no. 570.

3. References

Abragam, A., 1960, Principles of Nuclear Magnetism, Oxford Univ.Press, London

Blinc, R., Marinkovic, V., Pirkmajer, E., Zupancic,I., 1963, J.Chem.Phys. 38,2474

Blinc, R., Pirkmajer,E., Zupancic,I., Rigny, P., 1965, J.Chem.Phys. 43, 3417

Blinc, R., Pirkmajer,E., Slivnik,J., Zupancic,I., 1966, J.Chem.Phys. 45, 1488

Blinc, R., and Lahajnar,G., 1968, NIJS Report R-530

Blinc, R., Lahajnar,G., Slivnik,J., 1971, Chem.Phys.Lett.11, 344

Hoard,J.L., and Stroupe,J.D., 1944, Cornell Report A-1296

Rigny,P., 1965, CEA France Report R-2827

Rigny,P., and Virlet,J., 1967, J.Chem.Phys.47, 4645

Ursu,I., Demco,D.E.,Simplaceanu,V.,Vâlcu,N., 1974, Rev.Roum.Phys. 19, 605

COMPARATIVE ESR AND DIELECTRIC ABSORPTION MEASUREMENTS ON SPECIES CAUSING BROAD BACKGROUND ESR SIGNALS OF MANGANESE(II) IN SOLUTIONS

M.Stockhausen and M.Strassmann
Institut für Physik, Universität Mainz, D 65 Mainz

Abstract. Background, due to ion pair species, becomes broadened and powder like with increasing dielectric correlation time, indicating rotational tumbling as relevant magnetic relaxation process.

Solutions of Mn(II) salts in organic solvents only in a few cases show uniform ESR spectra of the shape expected theoretically; broad background signals (BBS) are a general feature (e.g. Stockhausen 1973). Several examples are given in Fig.1. To systematize the BBS somewhat arbitrarily, they are called type I if approximately lorentzian with $g \approx 2$; type II if markedly broadened in the wings but with $g \approx 2$; type III if asymmetric and centered at $g > 2$. Intermediate shapes between I and II are known. Type III, since only observed in rare cases (e.g. chloro complexes), will not be referred to in this communication.

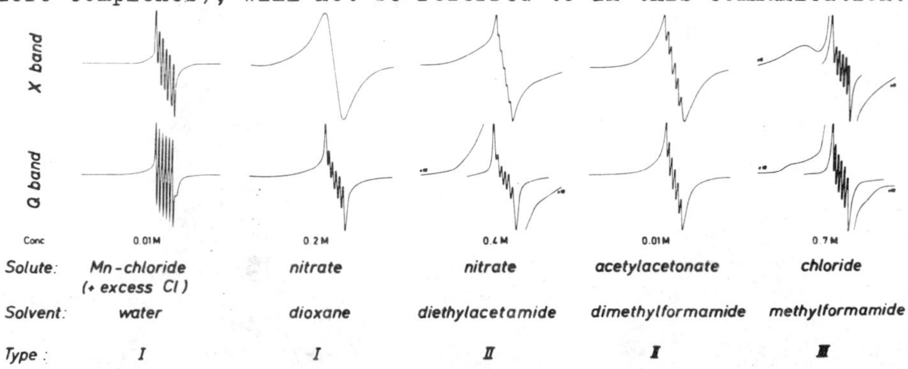

Fig.1 BBS examples

The appearance of BBS in a given solvent is anion dependent. Often a resolved main part of the spectrum is observed, having the same linewidth as with other anions. Since obviously due to free ions (meaning ions not involved in complexes with counterions), it will not be considered here. The remaining BBS intensity generally increases with decreasing conductivity. From this and other, e.g. optical arguments, BBS may be ascribed to unspecified ion pair complexes.

Relaxation of Mn(II) is governed by stochastic modulation of zfs as caused by the ligand field of the complex. The underlying dynamical process may be fluctuations in the ligand shell, or rotational tumbling of certain entities. Type II BBS, resembling powder spectra, were discussed by Burlamacchi (1973) and coworkers (e.g. Martini et al. 1974) in terms of the latter process, assuming random ligand field sites. Considering BBS as due to ion pair species, which generally will exhibit an electric dipole moment, the opportunity is offered for investigating their dynamical behaviour independently by dielectric absorption spectroscopy. In the following some representative examples are selected showing clear BBS. All measurements are at room temperature.

Type I is purely found in some solutions of low conductivity, e.g. Mn-nitrate in ethylacetate, tetrahydrofuran and dioxane. ESR spectra are uniform but tend to type II in the most viscous solvent (dioxane). From X and Q band spectra one gets a correlation time τ_{xQ} = 14 ps in THF, 18 ps in dioxane. Dielectric absorption (Fig.2a) shows two relaxation regions. The high frequency region is due to water stemming from the hydrated salt, as indicated by Fig.2b; the dielectrically ineffective fraction of water will be bound in the Mn-complex. Its relaxation region is the low frequency one, which, by comparison with other dielectric data, may be ascribed to a definite species in the order of first (maximum second) coordination sphere, relaxing by rotational tumbling. The correlation time as derived from the absorption maximum is τ_D = 100 ps (THF) resp. 320 ps (dioxane). Comparison of $\tau_D/3$ with τ_{XQ} seems to exclude the rotational process for magnetic relaxation, but taking into account the possibility of a spectral density function flattened in any way, rotational tumbling is likely to dominate the magnetic relaxation, too (Stockhausen 1974).

Type II BBS (e.g. Mn-nitrate in diethylacetamide) may formally be apportioned into a lorentzian part with medium linewidth, and additional broad wings. In comparison to free ion spectra, hfs splitting is slightly increased, g slightly decreased. It is not clear if those spectra belong to distinct species. Dielectric absorption shows qualitatively the same behaviour as type I, but the rotational correlation time of the complex is remarkably longer (τ_D = 580 ps). An obvious parallelism between increasing BBS and increasing τ_D is illustrated

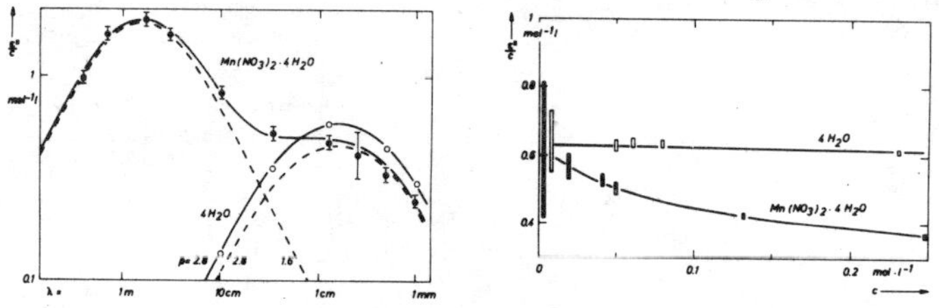

a Fig.2 b

Dielectric absorption, Mn-nitrate in dioxane (0,05 M). a) Salt absorption separated into 2 regions (dashed); p = Fröhlich parameter b) Salt and corresponding water absorption at high frequency maximum (λ = 8 mm) vs. conc.

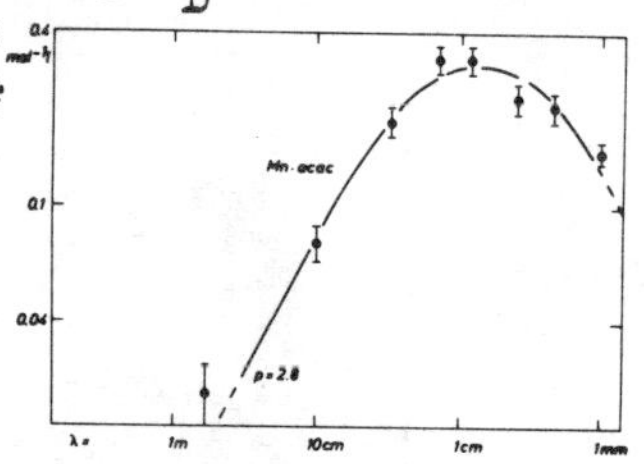

Fig.3

Dielectric absorption, Mn-acetylacetonate in benzene

by comparison with a homologous solvent (dimethylacetamide: weak BBS, τ_D = 370 ps). This confirms the general explanation of those BBS as given by Burlamacchi et al., as mentioned above.

Since BBS as discussed so far seem to be caused by reorientation of species with quasistatic zfs, the question arises if ligand fluctuations may contribute to magnetic relaxation. Proper times τ_{XQ} are expected from free ion results to be in the 5 ps range. Indeed, an example exhibiting both relaxation processes is the chelate complex Mn-acetylacetonate. In several solvents, BBS is observed as well as resolved hfs lines, the latter (with $\tau_{XQ} \approx 5$ ps) being ascribed to the complex itself in this case. Dielectric absorption is unique in showing a high frequency region, τ_D = 4 ps ($\approx \tau_{XQ}$!), but no reorientational region (Fig.3), presumably because of the complex symmetry (like similar complexes, Di Carlo et al. 1973). The correspondence: Narrow ESR lines - fast dielectric (probably fluctuation) process, and: BBS - slow (rotational) dielectric process (here ineffective), supports the conclusions drawn on ion pair species.

References:
Burlamacchi L and Romanelli M 1973, J Chem Phys 58, 3609
Di Carlo E N, Watson E, Varga C E and Chamberlain W J 1973, J Phys Chem 77,1073
Martini G, Romanelli M and Burlamacchi L 1974, in: Molecular Motions in Liquids (Reidel Dordrecht)
Stockhausen M 1973, Ber Bunsenges physik Chem 77, 338
- 1974, Conf on Molecular Dynamics of Complexes in Solution (Liège)

SECTION S

High Resolution NMR in Solids

EFFECTS OF LATTICE DYNAMICS IN MULTIPLE PULSE NMR

C. R. Dybowski and R. W. Vaughan
Division of Chemistry and Chemical Engineering, California Institute of Technology, Pasadena, California 91109

Abstract. The use of the multiple pulse NMR techniques for study of restricted and anisotropic motion is discussed using preliminary results from studies of two polymers and dipalmitoyl lecithin.

1. Introduction. The preponderance of experimental results obtained with the multiple pulse NMR techniques (see Vaughan, 1974, for a recent review) has been for systems in which the spins are fixed rather rigidly to the lattice, and one can consider the secular terms in the dipolar Hamiltonian to have components only at zero frequency. When molecular motion is present, it can interfere with the averaging effects of the pulse sequence (Haeberlen and Waugh, 1969). However, if the molecular motion is not isotropic, substantial benefit can accure from the use of the line narrowing pulse sequences. The lineshape observed in the case of restricted or anisotropic motion can be a complicated function of the effects of the molecular motion on the several contributing Hamiltonians, and by performing several separate but related multiple pulse NMR measurements, it is possible to separate and determine the individual contributions to the lineshape. This is of considerable advantage in studies of molecular dynamics in motionally ordered systems such as polymers, liquid crystals, and ordered biological material, as well as molecules moving on surfaces.

2. Experimental Details. The spectrometer used for the measurements reported here has been described by Vaughan, et al. (1972) and a previously discussed eight pulse cycle (Rhim, et al., 1973) was used for the multiple pulse measurements. The eight pulse cycle was modified by introducing a phase error in the P_x pulse in order to introduce a zeroth order Hamiltonian (see Table II, Rhim, et al., 1974) which serves to second average any small off-resonance Hamiltonians, and thus removes chemical shift and magnetic field inhomogeneity effects. The use of this phase shifted eight pulse cycle and details on the preparation and characterization of the two cis-polyisoprene samples are described in more detail by Dybowski and Vaughan (1974).

3. Results and Discussion.

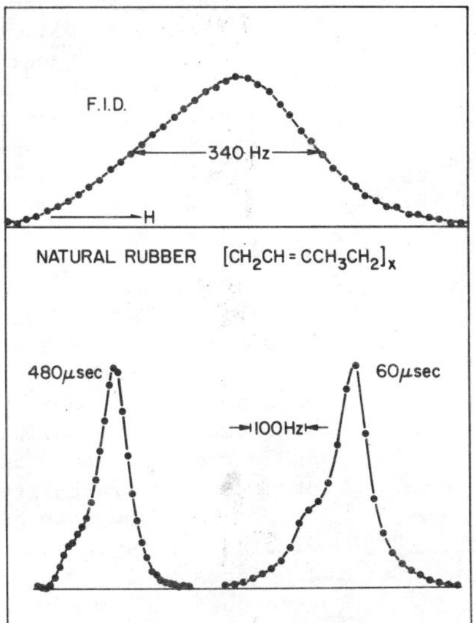

Fig. 1

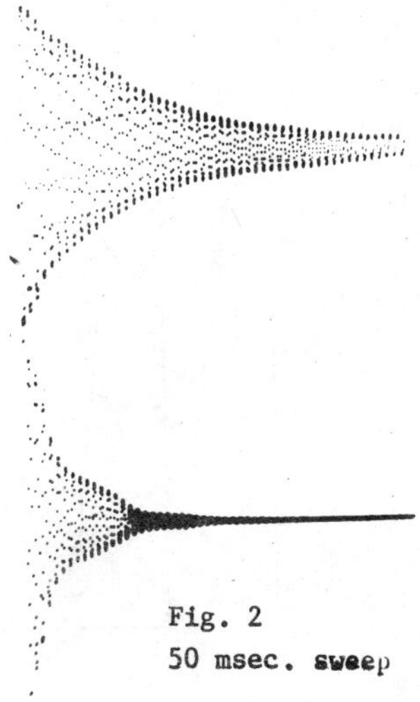

Fig. 2
50 msec. sweep

The room temperature free induction decay and multiple pulse spectra of cis-polyisoprene at two cycle times are shown in Fig. 1. The free induction decay spectrum is slightly asymmetric indicating the presence of fine structure on the line, and its width of 340 Hz indicates that a sizeable amount of molecular motion is presented. A two-peak spectrum is observed for the multiple pulse spectrum of the natural rubber in which the small shoulder peak can be associated with the chemically shifted proton bonded to the allyl carbon. The lineshape of the multiple pulse spectrum contains contributions from the chemical shift Hamiltonians, relaxation phenomena, and instrumental effects. In order to further separate these, the phase modified form of the eight pulse cycle described above was applied. In Fig. 2 is a comparison of the decay observed with the normal eight pulse cycle (bottom) and the phase modified eight pulse cycle (top). The normal eight pulse spectrum was taken 2 kHz off resonance and Fourier transforms to the spectrum observed in Fig. 1, while the phase shifted eight pulse experiment was done at resonance and the observed beat frequency is a measure of the phase shift introduced in the P_x pulse.

A plot of the decay envelope for the phase shifted eight pulse experiment is exponential with a time constant of 17 msec. This decay is due to relaxation phenomena associated with the complex interaction of the dipolar and rf Hamiltonian, modulated by the molecular motion present and to instrumental effects. Since the eight pulse and phase altered eight pulse measurement were made under similar experimental conditions, it is possible to deconvolute the line associated with the phase altered eight pulse measurement from the normal eight pulse spectrum observed in Fig. 1, and isolate the contribution due to the chemical shift Hamiltonians. Such an analysis gives values near $(1.9\text{ppm})^2$ for the second moment of the superposition of partially motionally averaged chemical shift powder patterns from the chemically inequivalent protons in the sample. Further analysis of this quantity coupled with similar data at other temperatures will clearly furnish additional information on the nature of the motional processes present.

The contributions to the lineshape of the free induction decay spectrum given in Fig. 1 are due to secular dipolar, chemical shifts, and relaxation effects. The relaxation contribution is shown to be small in this case from measured values of the spin-lattice relaxation time (determined for this sample by standard 180-90 pulse measurements to be 95 msec.). The experimentally determined second moment of the free induction decay spectrum can be separated into chemical shift and dipolar contributions since the chemical shift contribution is known from the above analysis of multiple pulse data. The second moment of the free induction decay spectrum in Fig. 1 is near 23,700 Hz^2 and when corrected for the chemical shift contribution one obtains a secular dipolar contribution of 12,500 Hz^2. Information on the anisotropic and restricted nature of the molecular motion present in the sample can be obtained from the size of the secular dipolar contribution and the fact that it can be removed by application of the eight pulse cycle. The ability to remove the dipolar contribution implies that over the time relevant to eight pulse cycles used (see Fig. 1), it is an effectively static quantity and is thus due to anisotropic or restricted motion. One can, therefore, define an order parameter for this system as the square root of the ratio of the second moment of secular dipolar contribution to the rigid lattice second moment and obtain a value of 0.006 for natural rubber at room temperature. This order parameter is a measure of the freedom of the spins on the polymer to move in the hindering potential of the polymer network.

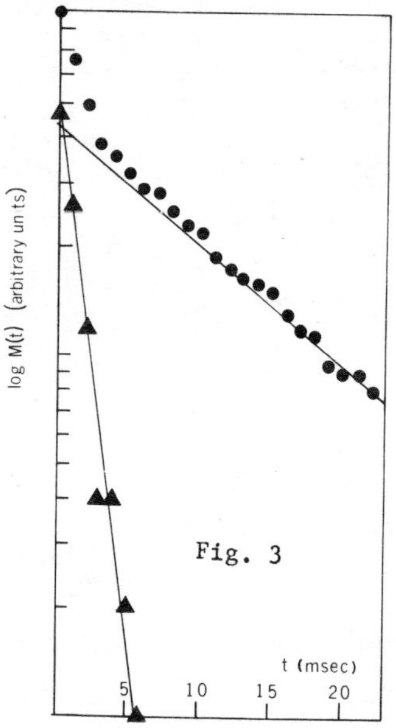

Fig. 3

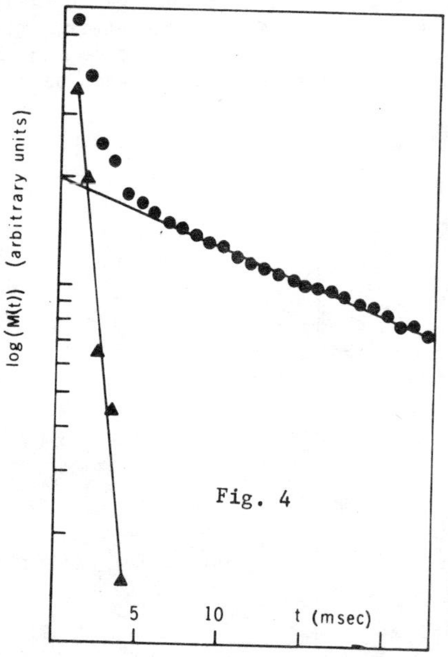

Fig. 4

Similar experimental data has been obtained on a carbon-filled sulfur vulcanized cis-polyisoprene sample and on samples of dipalmitoyl lecitin (see Dybowski and Vaughan, 1974 and Dybowski, et al., to be published). The results of the phase altered eight pulse cycle measurements on these two materials exhibit, however, quite different behavior. Figs. 3 and 4 exhibit the non-Lorentzian nature of the decay observed in these two cases and also indicate that it is possible to fit the observed decay with a sum of two Lorentzians in both cases. Decay constants of 1.5 and 13.5 msec. were observed for the filled rubber and correspond to 60 and 40% of the signal intensity, respectively. In the case of the diaplmitoyl lecithin decay constants of 1 and 22 msec. were observed with intensities of 80 and 20%, respectively. In both cases the relative intensities correspond to the relative density of methyl and backbone protons, i.e. in the filled rubber 60/40 observed 62.5/37.5 calculated and in the lecithin 80/20 observed and 77/23 calculated. Thus, it appears that the methylene protons exhibit an order of magnitude shorter relaxation time than the methyl protons in both of these samples. A quantitative treatment of this behavior is in preparation.

4. Acknowledgement. This work was supported in part by the Petroleum Research Fund of the American Chemical Society (#6915-AC5,6) and the National Science Foundation (#32517).

5. References

Dybowski, C. R. and R. W. Vaughan, 1974, submitted to Macromolecules.
Dybowski, C. R., R. W. Vaughan, C. H. Seiter and S. I. Chan, to be published.
Haeberlen, U. and J. S. Waugh, 1969, Phys. Rev., 185, 420.
Rhim, W-K., D. D. Elleman and R. W. Vaughan, 1973, J. Chem. Phys., 58, 1772; 59, 3740.
Rhim, W-K., D. D. Elleman, L. B. Schrieber and R. W. Vaughan, 1974, J. Chem. Phys., 60, 4595.
Vaughan, R. W., 1974, Annual Review of Materials Science, Vol. 4.
Vaughan, R. W., D. D. Elleman, L. M. Stacey, W-K. Rhim and J. W. Lee, 1972, Rev. Sci. Instr., 43, 1356.

IMPROVED RESOLUTION IN MULTIPLE PULSE NMR

A.N. Garroway, P. Mansfield and D.C. Stalker
Department of Physics, Nottingham University, Nottingham NG7 2RD

Abstract. A requirement of multiple pulse sequences is that the line-narrowing efficiency should not vary significantly over the breadth of the line to be observed; otherwise, the resultant lineshape will be distorted. In this report we identify three resonance offset broadening mechanisms. New multiple pulse sequences which successfully reduce the effect of these sources of broadening are demonstrated.

1. Introduction. Figure 1 shows the line-narrowing results as a function of resonance offset in CaF_2 and in liquid trifluoro-acetic acid, in which the dipolar interactions are motionally averaged, for the {132;1$\bar{3}$2} sequence (Mansfield, 1971) with τ=6.4μs. Three salient features of Fig. 1 are: (i) In the liquid the linewidth near resonance reflects the limit imposed by the static field inhomogeneity $\langle\delta\nu_0^2\rangle^{\frac{1}{2}}$ and then deteriorates linearly with increasing resonance offset. (ii) In the solid at resonance there is significant broadening which improves more or less symmetrically to yield the highest resolution at an offset of 0.5 kHz. (iii) Beyond 0.5 kHz in the solid the linewidth deteriorates with increasing offset; note also that the slope of the solid data is greater than for the liquid in this region.

To explain these observations and to design improved sequences we examine terms in the effective Hamiltonian which produce this behaviour.

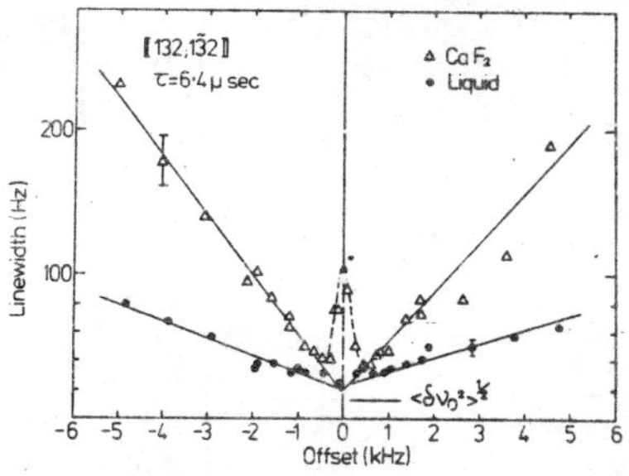

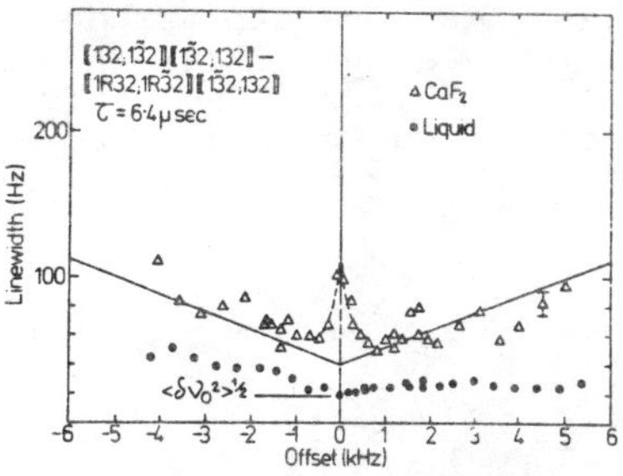

Fig. 1 and 2. ^{19}F linewidth versus offset for (1) the {132;1$\bar{3}$2} sequence and (2) the {132;1$\bar{3}$2}{1$\bar{3}$2;132}{1R32;1R$\bar{3}$2}{1$\bar{3}$2;132} sequence, τ=6.4μs, in CaF_2 and trifluoro-acetic acid.

2. Damping on and off resonance

2.1 Rf inhomogeneity. For the {132;1$\bar{3}$2} cycle we have previously shown (Mansfield et al, 1973) that inhomogeneity of the transmitter field leads to a linewidth which, well away from resonance, depends linearly on resonance offset; the slope of the liquid data in Fig. 1 is explained by this effect. Rather than constructing a more homogeneous coil, we have devised a pulse sequence which compensates for rf inhomogeneity damping. The sequence is represented by {132;1$\bar{3}$2}{1$\bar{3}$2;132}{1R32;1R$\bar{3}$2}{1$\bar{3}$2;132}; the R indicates the point in the cycle where a $\pm 270^0$ pulse has replaced a $\mp 90^0$ pulse. Figure 2 shows the results for this sequence with τ=6.4 μs and we see that the linewidths in the liquid are essentially independent of resonance offset, indicating that the compensation scheme has succeeded. Although the results for the solid have improved somewhat, the linewidth is still greater than in the liquid and, further, has a non-zero slope. We conclude that rf inhomogeneity compensation or improved coil design does not materially improve the resonance offset dependence in the solid. We now examine damping terms arising only in the solid.

2.2 Damping in a solid. For the sequences discussed here, the average Hamiltonian $\bar{H}^{(0)}$ contains only the rf damping term discussed above. The next higher term $\bar{H}^{(1)}$ in the effective Hamiltonian does not contribute to damping. The important term is $\bar{H}^{(2)}$ which we write as

$$\bar{H}^{(2)} = \bar{H}^{(2)}_{d^3} + \bar{H}^{(2)}_{d^2c} + \bar{H}^{(2)}_{dc^2} + \bar{H}^{(2)}_{c^3} \qquad (1)$$

where $d^n c^{3-n}$ indicates the product of n dipolar and (3-n) chemical shift (resonance offset) terms. We examine the damping of these in turn.

If the fractional loss of signal over each signal cycle is constant (exponential approximation), then the linewidths for the separate contributions are

$$W_{d^n c^{3-n}} = \frac{3\,\tau^4}{\sqrt{2}\,(36)^2}\,K_{2n}|\Delta\omega|^{(5-2n)}, \quad n=1,2,3 \qquad (2)$$

where K_{2n} has the character of a $(2n)^{th}$ moment. The term in $\bar{H}^{(2)}_{c^3}$ alters only the scale factor and does not affect the damping. In particular note that for the pure dipolar term (n=3), the linewidth varies inversely with $|\Delta\omega|$.

If the leading terms in the expression for the signal amplitude are regarded instead as a Gaussian expansion, then we predict (Gaussian approximation)

$$W_{d^n c^{3-n}} = \frac{\sqrt{(2\ln 2)}\;\tau^2}{\pi\;36}\,(K_{2n})^{\frac{1}{2}}|\Delta\omega|^{3-n}, \quad n= 1,2,3 \qquad (3)$$

Here there is no offset dependence of the pure dipolar linewidth.

We find experimentally for the {132;1$\bar{3}$2} sequence that near resonance the linewidth decreases as we move away from resonance, as predicted in Eq. (2) for the pure dipolar case. At the point of optimal narrowing (0.5 kHz) the linewidth increases as τ^4 for $6.4 \le \tau \le 12.8$ µs. We ascribe this dependence to the growth of the pure dipolar term which rapidly dominates any dipolar resonance offset term with a weaker τ dependence. However, at 1.4 kHz from resonance the linewidth exhibits roughly a τ^2 behaviour and we suggest that at this point the higher order terms add to produce a Gaussian rather than Lorentzian lineshape, accounting for the τ^2 dependence of Eq. (3). Rhim et al (1973) have observed a similar dependence, but offer a different explanation.

It has been shown that for a fully permuted symmetry cycle the pure dipolar term $\bar{H}^{(2)}_{d^3}$ vanishes identically on resonance (Mansfield, 1971). For technical reasons we have not tested such a sequence. However, there is some advantage (Garroway et al, 1974) in using a partially permuted symmetry cycle, for example {132;1$\bar{3}$2}{123;1$\bar{2}$3}. Over this cycle $\bar{H}^{(2)}_{d^3}$ is reduced by a factor of two from its corresponding value in the parent {132;1$\bar{3}$2} cycle.

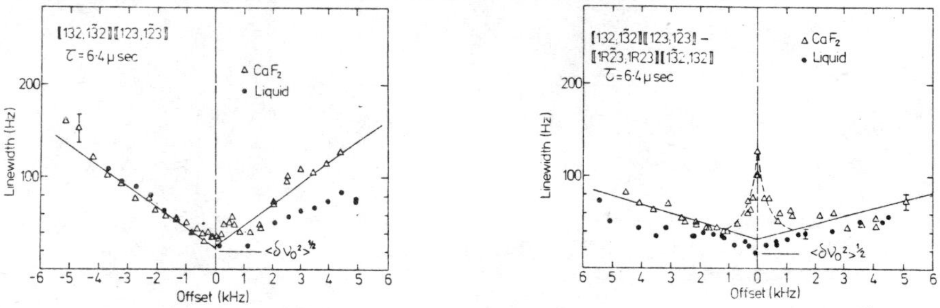

Fig. 3 and 4. ^{19}F linewidth versus offset for (3) the {132;1$\bar{3}$2}{123;1$\bar{2}$3} sequence and (4) the {132;1$\bar{3}$2}{1$\bar{2}$3;123}{1R23;1R$\bar{2}$3}{1$\bar{3}$2;132}sequence, τ=6.4 µs, in CaF_2 and trifluoro-acetic acid.

Figure 3 shows the line-narrowing results for the new {132;1$\bar{3}$2}{123;1$\bar{2}$3} sequence, with τ=6.4 µs. The peak near resonance has diminished, as expected from the decrease in $\bar{H}^{(2)}_{d^3}$.

Having demonstrated some control over rf inhomogeneity and the pure dipolar damping terms, we may then construct a cycle which compensates for both effects. Line-narrowing results for this sequence,{132;1$\bar{3}$2}{1$\bar{2}$3;123}{1R23;1R$\bar{2}$3}{1$\bar{3}$2;132}; are shown in Fig. 4. Notice that the linewidths in the liquid and solid are almost comparable beyond 1 kHz from resonance. Unfortunately, there is a significant damping term on resonance which we ascribe to a subtle spectrometer misadjustment and which we do not think is endemic to the sequence.

3. Conclusions. We have shown that by compensating for rf inhomogeneity and employing partially permuted symmetry cycles, the line-narrowing ability of multiple pulse sequences becomes less sensitive to resonance offset. We emphasize that this resonance offset dependence may in practice be more important than that the optimal resolution achievable.

We acknowledge the Science Research Council for a postdoctoral research assistantship (ANG) and a research studentship (DCS).

4. References.

Garroway A.N., Mansfield P. and Stalker D.C. 1974 (to be published).

Mansfield P. 1971 J. Phys. C. 4 1444.

Mansfield P.,Grannell P.K., Garroway A.N. and Stalker, D.C. 1973. Proc. First Specialized Colloque Ampere, Inst. of Nuc. Phys., Krakow, J.W. Hennel, ed.

Rhim W-K., Elleman D.D. and Vaughan R.W. 1973 J. Chem. Phys. 59 3740.

Proton Multiple Pulse Single Crystal Study of Potassium Hydrogen Maleate.

A.M. Achlama (Chmelnick), U. Kohlschütter, and U. Haeberlen.

Department of Molecular Physics, Max-Planck-Institute Heidelberg, Germany.

The magnetic shielding tensors of all magnetically-inequivalent protons in the unit cell could be independently determined. Two of these protons are carboxylic, forming hydrogen bonds. The relevance of the knowledge of the full shielding tensors to the accurate positioning of protons in hydrogen bonds is demonstrated.

The other protons in the unit cell are olefinic. This study contains the first direct determination of the full magnetic shielding tensor of olefinic protons. This is the first type of protons whose magnetic shielding tensor is not even approximately axially-symmetric, the principal values being - 2.4, - 5.1, - 7.3 $\pm$ 0.5 ppm (from adamantane). The principal directions of the shielding tensors of the olefinic protons constitute a unique case: they reflect all characteristic directions of the carbon-carbon double bond, while the C-H direction is of no importance.

Our measurements also indicate that protons in general distinguish themselves from other nuclei belonging to double bond systems in that the <u>least</u> shielded direction is that of the lone 2p carbon orbitals forming the π system.

NUCLEAR MAGNETIC SHIELDING TENSORS FOR ^{1}H, ^{13}C, AND ^{15}N IN ORGANIC SOLIDS

H.W. Spiess, U. Haeberlen, J. Kempf, and D. Schweitzer
Department of Molecular Physics, Max-Planck-Institute, 6900 Heidelberg, Jahn-Straße 29, F.R.G.

Abstract. A survey will be given of nuclear magnetic shielding tensors obtained by multiple pulse techniques for ^{1}H in carboxylic acids and by high field NMR for ^{13}C in carbonyl- and carboxyl groups of aromatic compounds and for ^{15}N in pyridine and nitrobenzene.

Solid state NMR for $I = 1/2$ nuclei has received considerable attention in the last years due to the development of new techniques by which an increase of resolution of solid state spectra can be achieved. This allows the determination of nuclear magnetic shielding tensors $\hat{\sigma}$ containing considerable more information than the isotropic shifts that can be measured in liquids. We have studied $\hat{\sigma}$ tensors for protons in a number of carboxylic acids by multiple pulse techniques at 90 MHz and for ^{13}C and ^{15}N in organic solids by high field NMR at 61 and 32 MHz, respectively. Some of the general features of the shielding tensors have become apparent regarding the size of the anisotropy as well as the orientation of the principal axes system of $\hat{\sigma}$ relative to the molecule.

For protons the largest values for the anisotropy of $\hat{\sigma}$ reported so far are found for protons in hydrogen bonds, the shielding tensors being almost axially symmetric. The values for the shielding anisotropy $\Delta\sigma = \sigma_{\parallel} - \sigma_{\perp}$ are typically about 20 ppm in carboxylic acids. From single crystal studies (see, e.g. Haeberlen et al., 1974) it follows that the unique axis of the $\hat{\sigma}$ tensor which is also the most shielded direction, lies along the hydrogen bond. We have shown that these features of the $\hat{\sigma}$ tensor for protons in hydrogen bonds are due mainly to the diamagnetic effect (Haeberlen et al., 1974), which can be handled quantitatively as described by Gierke and Flygare, 1972.

^{13}C shielding tensors were studied for carbons involved in carbonyl and carboxyl groups (Kempf et al., 1974) as shown in Table 1. The most shielded direction is perpendicular to the sp^2 plane. From these data and the values for the corresponding aliphatic compounds (Waugh et al., 1972) it follows that both the components of largest and smallest shielding, σ_{zz} and σ_{xx} respectively, vary relatively little in this series, whereas the value for the intermediate shielding can be anywhere between -71 and +78 ppm (see also Table 1). For ^{13}C the main features of the shielding anisotropy are governed by the paramagnetic contribution. The much larger variation of σ_{yy} compared with σ_{xx} and σ_{zz} in this series can be understood if one assumes that the y-axis generally is close to the C = O bond as observed in single crystals of benzophenone and benzoic acid (Kempf et al., 1972, 1974). This was further supported by studying the ^{13}C

Compound	σ_{xx} (ppm)	σ_{yy} (ppm)	σ_{zz} (ppm)
Benzophenone	-79	-36	+94
Benzoic Acid	-38	+ 5	+90
Silver Benzoate	-53	+18	+101
Benzoic Acid Anhydride	-42	+50	+114

Table 1: Principal elements of ^{13}C shielding tensors relative to CS_2, estimated accuracy ± 4 ppm.

shielding anisotropy in thiobenzophenone so that the shielding components for a C = O and for a C = S group could be correlated with the corresponding optical spectra (Kempf et al., 1974).

The determination of ^{15}N shielding anisotropies allows a comparison of the values for $\Delta\sigma$ for ^{13}C and ^{15}N in isoelectronic compounds (Table 2).

Isoelectronic Pair	$\Delta\sigma$ [ppm] ^{13}C	^{15}N	References
benzene/pyridine	180[a]	672	[a]Pines, A., Gibby, M.G., Waugh, J.S., 1972, Chem.Phys.Letters, 15, 373.
Ag Benzoate/nitrobenzene	-112	-398	[b]Lauterbur, P.C., 1958, Phys.Rev.Letters, 1, 343.
CO_3^{2-}/NO_3^-	75[b]	210[c]	[c]Gibby, M.G., Griffin, R.G., Pines, A., Waugh, J.S., 1972, Chem.Phys.Letters, 17, 80 [d]Ozier, I., Crapo, K.M., Ramsey, N.F., 1968, J.Chem.Phys., 49, 2314.
CO/N_2	401[d]	635[e]	[e]Baker, M.R., Anderson, C.H., Ramsey, N.F., 1964, Phys.Rev. 133 A, 1533.

Table 2: Anisotropy of the shielding for ^{13}C and ^{15}N in isoelectronic systems. For details see Schweitzer and Spiess, 1974.

Despite of the limited material available now, it seems to be clear that the shielding anisotropy in general is substantially larger for ^{15}N (Schweitzer and Spiess, 1974). The large increase in $\Delta\sigma$ for ^{15}N compared with ^{13}C for the first three pairs (Table 2) seems to be especially interesting. For the pyridine/benzene pair one might be tempted to attribute this increase to the existence of the non-bonding orbital at the nitrogen atom in pyridine in contrast to benzene. The data for nitrobenzene and the nitrate ion show, however, that the increase in $\Delta\sigma$ is almost as big in these cases where no such simple explanation is obvious.

These results show that by studying shielding tensors for different nuclei and for series of compounds some general features can be stablished so that from the experimental determination of shielding components one will get meaningful information about the electronic structure of molecules.

References.

Gierke, T.D., Flygare, W.H., 1972, J.Am.Chem. Soc., 94, 7277.

Haeberlen, U., Kohlschütter, U., Kempf, J., Spiess, H.W., Zimmermann, H., 1974, Chem.Phys. 3, 248.

Kempf, J., Spiess, H.W., Haeberlen, U., Zimmermann, H., 1972, Chem.Phys.Letters, 17, 39;—, 1974, Chem.Phys. 4, 269.

Schweitzer, D., Spiess, H.W., 1974, J.Magn.Res. -, in print.

Waugh, J.S., Gibby, M.G., Kaplan, S., Pines, A., 1972, Proc. XVIIth Colloque Ampère, Turku.

NMR-"LINE NARROWING" OF TWO-SPIN SYSTEMS BY MEANS OF MULTIPLE PULSE EXPERIMENTS

U. Haubenreißer and B. Schnabel
Sektion Physik der Friedrich-Schiller-Universität, DDR-69 Jena, Max-Wien-Platz 1

Abstract: The process of removing the direct magnetic dipole-dipole interaction (DIA) between the two equal nuclei of a two-spin system by means of different multiple pulse experiments is investigated and discussed as a function of pulse distance τ. The theoretical time dependence of the NMR signals are compared with the results of proton magnetic resonance experiments of Kieserit $MgSO_4 \cdot H_2O$.

Using a number of different multiple pulse NMR experiments (for example MW4: Ostroff and Waugh, 1966 / Mansfield and Ware, 1968; WHH4 and LG6: Haeberlen and Waugh, 1968) it is possible, to remove the influence of the direct magnetic dipole-dipole interaction (DIA) on the NMR signal of solids to a high degree. In the case, that there are weakly interacting systems of two equal nuclear spins in the solid, this process exhibits some special properties. The transverse nuclear magnetization of one isolated two-spin system does not decay in time but is purely periodical. According to this, the signal envelope during a multiple pulse NMR experiment is to be expected to be time-independent, too, and it cannot be used to characterize the reduction of DIA. Therefore, in order to analyse the process of removing the Pake splitting the Fourier transform of the time-dependent NMR signal as a whole is performed.

As a first case, fig. 1 shows the theoretical time-dependent magnetization during a MW4 experiment (δ-Pulses). Fig. 2 shows the corresponding experimental result which was measured using Kieserit $MgSO_4 \cdot H_2O$ (Haubenreißer and Schnabel, 1973) with the Pake splitting $2\omega_p = 2\pi \cdot 84$ kHz and pulse distance $2\tau = 36\,\mu s$. The weak envelope decay is due to the residual DIA between the two-spin systems of Kieserit.

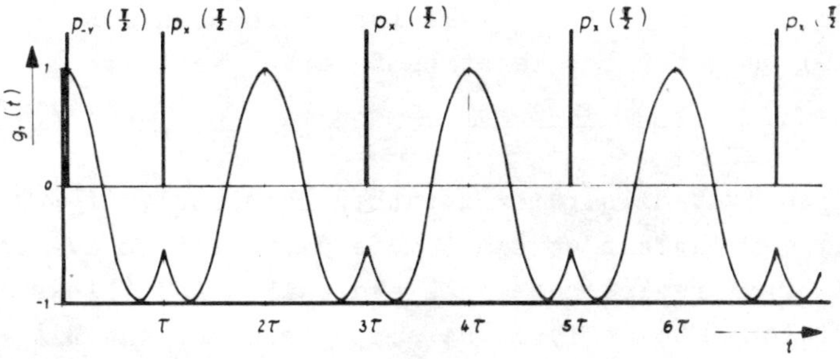

Fig. 1

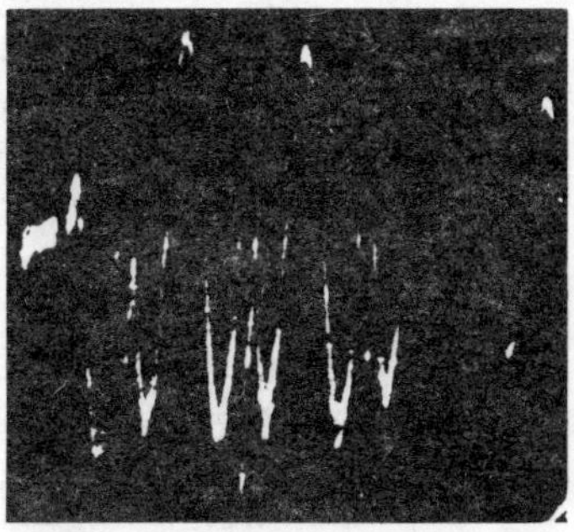

Fig. 2

With the simplifying assumptions, that the NMR signal extends symmetrically with respect to t = 0 from t = $-\infty$ to t = $+\infty$ we get the relaxation function $g_1(t)$:

$$g_1(t) = \frac{a_0}{2} + \sum_{k=1}^{\infty} a_k \cos\frac{k\pi}{\tau} t \quad , \qquad (1)$$

the spectrum of which is given by the Fourier amplitudes:

$$a_k = 2(-1)^k \frac{\sin \omega_p \tau}{\omega_p \tau} \cdot \frac{1}{1 - \left(\frac{k\pi}{\omega_p \tau}\right)^2} \quad . \qquad (2)$$

If $\omega_p\tau \gg 1$ is valid there are side band lines in the surrounding of the Pake frequency ω_p (fig.3), the amplitudes of which are given by (3) within the limiting condition k >> 1:

$$a_{k\pm j} \approx (-1)^j \frac{\sin \varepsilon\pi}{(\varepsilon \mp j)\pi} \quad . \qquad (3)$$

The distance of the side band lines is $\Delta\omega_k \approx \omega_p / k$, and their amplitude is maximum in the surrounding of $\omega' = \omega_p$ and nearly zero in the neighbourhood of $\omega' = 0$ (at resonance). In opposition to this, if $\omega_p\tau < 1$ is valid, there is a rather strong line at resonance $\omega' = 0$, that means the DIA between the two spins is removed rather completely and the whole system behaves like a system of rather isolated nuclear spins. The condition $\omega_p\tau < 1$ is of the same kind as derived formerly for solids in a different way (Mansfield and Ware, 1968).

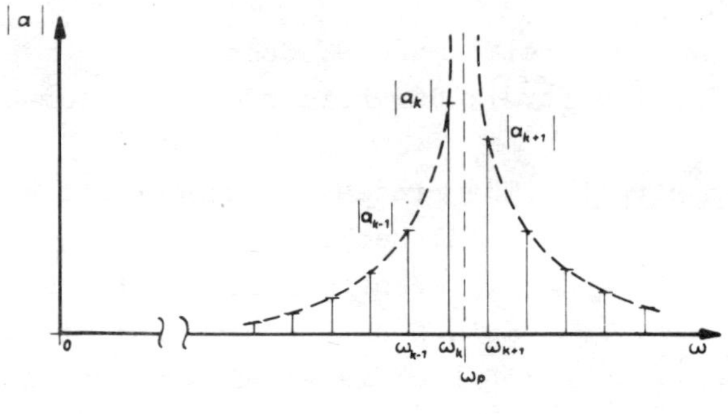

Fig. 3

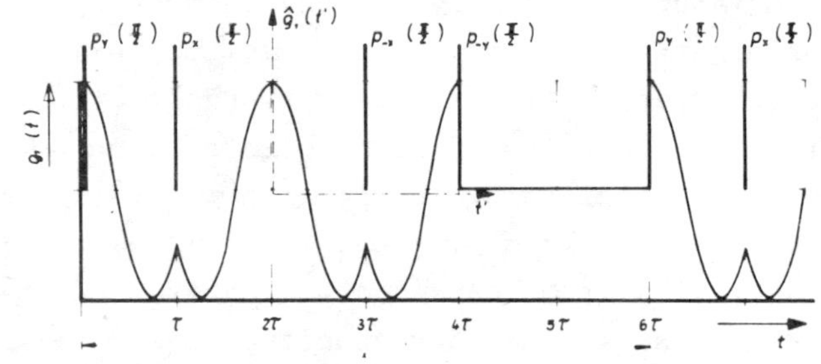

Fig. 4

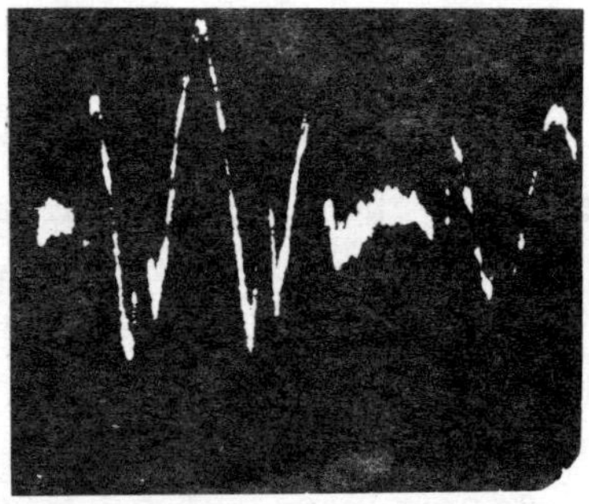

Fig. 5

Fig. 4 shows the theoretical time-dependent magnetization during a WHH4 experiment (δ-pulses). Fig. 5 shows the corresponding experimental result of Kieserit with the pake splitting $2\omega_p = 2\pi \cdot 25$ kHz and pulse distance $\tau = 18\,\mu$s. Under the same conditions as above (a new scale $t' = t + 2\tau$ ist introduced for symmetry with respect to $t' = 0$) we get the relaxation function $g_1(t')$:

$$g_1(t') = \frac{2}{3}\frac{\sin\omega_p\tau}{\omega_p\tau} + \frac{4}{3}\sum_{k=1}^{\infty}\left(\frac{\sin\omega_p\tau}{\omega_p\tau}\cdot\frac{\cos\frac{\pi}{3}k}{1-\left(\frac{\pi k}{3\omega_p\tau}\right)^2} + \frac{1}{1-\left(\frac{3\omega_p\tau}{\pi k}\right)^2}\cdot\frac{\sin\frac{2\pi}{3}k}{\frac{2\pi}{3}k}\right)\cos\frac{k\pi}{3\tau}t' \; . \qquad (4)$$

Again the DIA within the two-spin system is removed rather completely if $\omega_p\tau < 1$. The same is true for the LG6 experiment but somewhat more difficult to evaluate. The given procedure is capable of a precise numerical evaluation and allows, within the cited limiting conditions, to analyse the suppression of the DIA between the two spins depending on the cycle time t_c and the Pake splitting. This dependence is somewhat different for each multiple pulse experiment.

References:

Haubenreißer, U. and Schnabel, B., 1973, Proc. "1. Spec. Coll. Ampère", Kraków, p. 140

Mansfield, P. and Ware, D., 1968, Phys. Rev., 168, 318

Ostroff, E.D. and Waugh, J.S., 1966, Phys. Rev. Lett., 16, 1097

Haeberlen, U. and Waugh, J.S., 1968, Phys. Rev., 175, 453

CHEMICAL SHIELDING TENSOR OF $^{13}COO^-$ IN SINGLE CRYSTAL L-ALANINE[1,2]

J. J. Chang and A. Pines[3]

University of California, Berkeley, California 94720 USA

Employing proton-enhanced nmr, we have obtained ^{13}C spectra in single crystals of L-alanine, one of the simplest amino acids (CH_3-$CHNH_3^+$-COO^-). The center carbon peaks show resolved splittings due to dipolar coupling with the ^{14}N nucleus. The unit cell is orthorhombic (space group $P2_12_12_1$) and contains four molecules not related by centers of symmetry. Thus, the analysis of crystal rotation data yields four chemical shielding tensors σ for the $^{13}COO^-$ groups, related to each other by the crystal symmetry operations. The average values of the principal elements for the $^{13}COO^-$ shielding tensors are:

$$\sigma_{11} = -115.9; \; \sigma_{22} = -55.2; \; \sigma_{33} = 19.1$$

in ppm (±2) relative to an external reference of liquid benzene. The average orientation of the principal axes of σ relative to the local structure of the carboxyl group is similar to the orientation we have found previously for other carboxyl groups, with the most shielded element σ_{33} perpendicular to the $^{13}COO^-$ plane.[4] The average orientation of σ_{11} and σ_{22} is shown in the figure below. The different ^{13}C-O bond lengths,[5] and slight rotation of the principal axes σ_{11} and σ_{22} from ideal C_{2v} symmetry (dotted lines) may reflect the effects of hydrogen bonding.[5]

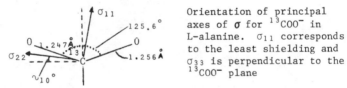

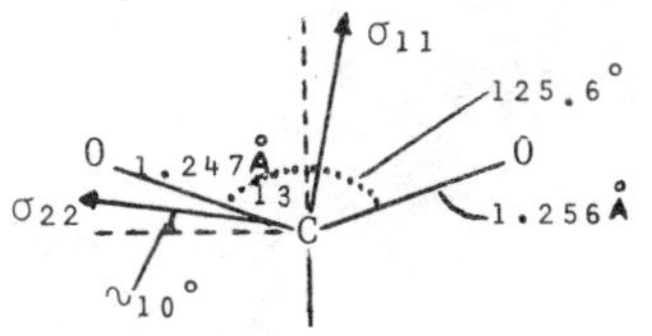

Orientation of principal axes of σ for $^{13}COO^-$ in L-alanine. σ_{11} corresponds to the least shielding and σ_{33} is perpendicular to the $^{13}COO^-$ plane

1. Supported by the National Science Foundation and Research Corporation and by the U. S. Atomic Energy Commission through the Inorganic Materials Research Division of the Lawrence Berkeley Laboratory.

2. To be presented at the 18th Ampere Congress, Nottingham, England, September, 1974.

3. Alfred P. Sloan Fellow.

4. A. Pines, J. J. Chang and R. G. Griffin, J. Chem. Phys., in press.

5. H. J. Simpson, Jr., and R. E. Marsh, Act. Cryst., 20, 550 (1966).

CROSS POLARIZATION DYNAMICS

D. E. Demco,* J. Tegenfeldt† and J. S. Waugh
Department of Chemistry, Massachusetts Institute of Technology, Cambridge, Massachusetts 02139, U.S.A.

Abstract. The theory of memory functions has been used to calculate the rate of transfer of energy between abundant and rare spins in solids under double-resonance conditions. The results reproduce previous measurements on CaF_2 under a variety of experimental conditions.

1. The Problem. In proton-enhanced nuclear induction spectroscopy (Pines, et al., 1973), abundant spins I in a solid are brought to a condition of low effective spin temperature and then put into thermal contact with rare spins S by various rotating-frame double resonance techniques (Hartmann and Hahn, 1962; Lurie and Slichter, 1964; MacArthur, et al., 1969). The resultant cooling of the S spins is manifested in an enhanced magnetization, $\overline{M_S}$, which is used as the initial condition for an S-spin Bloch decay occurring after the thermal contact is broken and while the I spins are strongly decoupled by resonant irradiation. The thermal mixing and decay may be repeated frequently for rapid-averaging purposes until the I spins have been appreciably heated, after which the I spins are repolarized by spin-lattice relaxation or by thermal contact with a third species (Demco, et al., in press). The resulting enhanced sensitivity may be further optimized by appropriate inhomogeneous echo methods.

The enhancement of S magnetization over it's Curie value in each thermal mixing, for spin-1/2 species with a large abundance ratio, is given under the usual experimental conditions by

$$\frac{M_S}{M_{oS}} = \frac{H_{eS}}{(H_{1I}^2 + H_{LI}'^2)^{1/2}}$$

where H_{eS} is the effective Zeeman field in the rotating frame for the S spins during thermal mixing and the other symbols have their usual meanings. Clearly maximum magnetization is favored by large H_{eS}: however, the rate of approach to this final situation is governed critically by a time constant T_{IS} which depends upon H_{eS}. Two common situations arise: (a) mixing is carried out with spin-locking of the I spins and both species irradiated at resonance (Hartmann and Hahn, 1962); then T_{IS} lengthens as $\omega_{1S} = \gamma_S H_{1S}$ is increased beyond the Hartmann-Hahn condition $\omega_{1I} = \omega_{1S}$. (b) The I spins are demagnetized in the rotating frame (MacArthur, et al., 1969), and T_{IS} decreases as ω_{1S} increases. The dependence of T_{IS}^{-1} on ω_{1S} has been called a dipolar fluctuation spectrum by MacArthur, et al., who observed it to be exponential. The surprising nature of this result, together with the importance of T_{IS} in the design of double resonance experiments of maximum sensitivity, led us to construct a theory of cross relaxation spectra.

2. Theory. As usual, the I-S part $\mathcal{H}_{IS}^{0}$ of the appropriate truncated dipolar interaction is treated as a perturbation which couples I and S reservoirs. The calculation of T_{IS} is tantamount to calculating a correlation function $C(\tau)$, whose form depends on whether the I species are locked (SL) or demagnetized (ADRF). To evaluate $C(\tau)$ we have introduced a memory function K (Mori, 1965):

$$\frac{d}{dt} C(\tau) = - \int_0^\infty K(\tau-\tau')C(\tau')d\tau'$$

It may be argued on quite general grounds that K may be assumed gaussian with reasonable accuracy, although $C(\tau)$ can take on a variety of functional dependencies. We have made this assumption and reinforced it by calculation of moments of K: for CaF_2 in various orientations and under both SL and ADRF conditions, M_4/M_2^2 is close to the value of 3 expected for a gaussian, although the corresponding moment ratios for $C(\tau)$ deviate widely from this value.

3. Experimental Tests and Predictions. Cross-relaxation spectra $T_{IS}^{-1}(\omega_{1S})$ and $T_{IS}^{-1}(\omega_{1S}-\omega_{1I})$ have been calculated by Fourier inversion of $C(\tau)$ computed by the memory-function method, for both ADRF and SL cases, for $I = {}^{19}F$, $S = {}^{43}Ca$ in CaF_2 in various orientations. All lattice sums were evaluated by computer, and the resulting spectra have no adjustable parameters of any kind. MacArthur et al. (1969) have measured such spectra in the ADRF case for the [111] and [110] orientations. Our calculations agree with their measurements in every case within experimental error, both for the long time behavior and for the transient oscillations of $C(\tau)$ occurring at short times: a representative comparison is shown in the figure.

Our calculations for the SL case predict considerably different behavior, with generally larger values of T_{IS}. As one goes from $\vec{H}_o$ along [111] to [110] and [100], $T_{IS}^{-1}(\omega)$ becomes less and less nearly exponential for the ADRF case and less and less nearly gaussian for the SL case.

A detailed account of this work, including a discussion of other theoretical approaches, has been submitted for publication elsewhere (Demco, et al., to be published).

4. References.

Demco, D. E., Kaplan, S., Pausak, S. and Waugh, J. S., Chem. Phys. Letters, in press.

Demco, D. E., Tegenfeldt, J. and Waugh, J. S., submitted to Phys. Rev.

Hartmann, S. R. and Hahn, E. L., Phys. Rev. 128, 2042 (1962).

Lurie, F. M. and Slichter, C. P., Phys. Rev. 133A, 1108 (1964).

MacArthur, D. A., Hahn, E. L. and Walstedt, R. E., Phys. Rev. 188, 609 (1969).

Mori, H., Progr. Theoret. Phys. 33, 423 (1965).

Pines, A., Gibby, M. G. and Waugh, J. S., J. Chem. Phys. 59, 569 (1973).

* On leave from the Institute of Atomic Physics, Bucharest.

† On leave from the University of Uppsala.

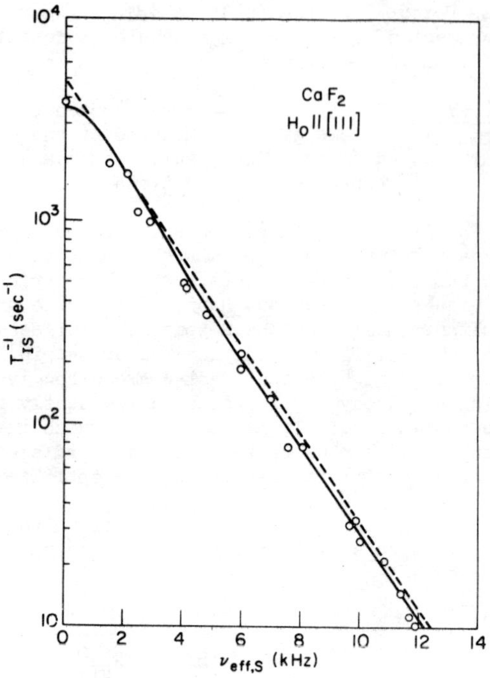

Cross relaxation spectrum for ^{43}Ca-^{19}F in a CaF_2 crystal. Circles are experimental points of MacArthur et al. (1969) and the dashed line is their fit to an exponential spectrum. The solid line is calculated by the present theory.

THEORETICAL TREATMENT OF THE INFLUENCE OF H_I-FELD INHOMOGENEITY ON THE NMR NARROWING BEHAVIOUR OF THE SIMPLE AND A PULSED LEE-GOLDBURG EXPERIMENT

U. Haubenreisser and B. Schnabel
Sektion Physik der Friedrich-Schiller-Universität, DDR-69 Jena, Max-Wien-Platz I

Abstract: In order to calculate the limiting influence of the rf field inhomogeneity, a theoretical treatment is given generally valid for NMR line narrowing experiments affecting the spin operator part of the dipolar interaction. Applied to three selected cases the results are discussed as a function of the sample dimensions.

By a number of special NMR experiments affecting the spin operator part of the direct magnetic dipole-dipole interaction, it is possible to reduce the NMR line width of solids. Especially because of the rather high rf field strength necessary, this line narrowing effect may be limited by the effective rf field inhomogeneity depending on the size of the sample in a given coil. A calculation was given before (Lan, Pfeifer and Schmiedel, I973) assuming a Gauß distribution of the rf field. However, it is very important to relate the properties of the rf coil and the size of the sample on one hand to the spatial dependence of the line narrowing behaviour on the other hand. To do this, we calculate the zero-order average dipolar Hamiltonian depending on the field distribution in the rf coil, which may be represented as

$$H_{Ix} = H_{Ix}(0,0) + \Delta H_{Ix}(x,\rho) = H_{Ix}(0,0) + \frac{I}{2!}\left(x^2 - \frac{\rho^2}{4}\right) H_{1x}^{(2)}(0,0) + \cdots$$
$$H_{1y} = \Delta H_{1y}(x,\rho) = \tfrac{1}{2}\rho\, H_{1x}^{(2)}(0,0) + \cdots \qquad (I)$$

where x, ρ are cylindrical coordinates, assuming the rf coil to be cylindrically symmetric and x to be parallel to the axis of the coil. The relation to the geometrical dimensions of the single-layer solenoid is given by

$$H_{Ix}(0,0) = I\bar{\nu}A(A^2+R^2)^{-1/2}$$
$$H_{Ix}^{(2)}(0,0) = -I\bar{\nu}\, 3AR^2(R^2+A^2)^{-5/2}, \qquad (2)$$

where I is the rf current in the coil, $\bar{\nu}$ the turning number density, 2A the length and 2R the diameter of the coil. Calculating the second moment M_2' of the narrowed NMR line by the well known formula and integrating it over the sample volume, we can define the narrowing factor V according to

$$\bar{M}_2' = V^{-2} M_2, \qquad (3)$$

where M_2 is the ordinary NMR second moment of the solid.
In this way, the influence of the effective field in the rotating frame H_e, inhomogeneons in size and direction, was calculated for the LG (Lee and Goldburg, I965) and the LG6 (Haeberlen and Waugh, I968; sequence of I20° pulses combined with simultanions d.c. pulses in a tilted rf coil) experiments, assuming the angle between $\vec{H}_e$ and $\vec{H}_o$ to be 54.7° in the center of the coil. In the case of LG6, the inclination of the coil axis was chosen as 45°. δ- pulses were assumed, as in the case of WHH4 (Waugh, Huber and Haeberlen, I969), which was calculated for comparison. In the three cases given we get the effective reduced second moments of the sample as (H_{eo} is the value of H_e at the center of the coil)

$$\text{LG:}\ \bar{M}_2' = \left(\frac{r}{R}\right)^4 \left\{I + \left(\frac{A}{R}\right)^2\right\}^{-4} \left\{\frac{2}{5}\left(\frac{a}{r}\right)^2 - \frac{1}{6}\left(\frac{a}{r}\right)^2 + \frac{1}{24}\right\} M_2 \left(1 + \tfrac{1}{2}\frac{\gamma^2 H_{eo}^2}{M_2}\right)$$

$$\text{LG6:}\ \bar{M}_2' = \left(\frac{r}{R}\right)^4 \left\{I + \left(\frac{A}{R}\right)^2\right\}^{-4} \left\{0.333\left(\frac{a}{r}\right)^4 + 0.0432\left(\frac{a}{r}\right)^2 + 0.0344\right\} M_2$$

$$\text{WHH4:}\ \bar{M}_2' = I.85\left(\frac{r}{R}\right)^4 \left\{I + \left(\frac{A}{R}\right)^2\right\}^{-4} \left\{\frac{I}{5}\left(\frac{a}{r}\right)^4 - \frac{I}{I2}\left(\frac{a}{r}\right)^2 + \frac{I}{48}\right\} M_2 \qquad (4)$$

As a consequence of the rf field distribution regarded only approximately up to the terms given in (I), these formulas are valid within the limits $A/R > 2$, $a/r < 2$ and $r/R < 3/4$ which are usually sufficient for line narrowing experiments (2a and 2r are maximum sample length and sample diameter, respectively). Assuming certain coil dimensions (2A = I6 mm, 2R = 8 mm) the narrowing factor V was calculated and plotted as a function of sample length at constant sample diameter (Fig. I, 2 and 3).

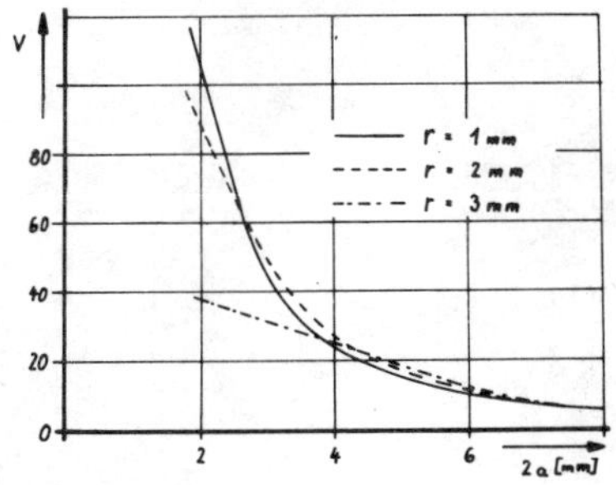

Fig. I. LG-experiment ($\gamma H_{eo} M_2^{-I/2} = 5$)

Fig. 2. LG6-experiment

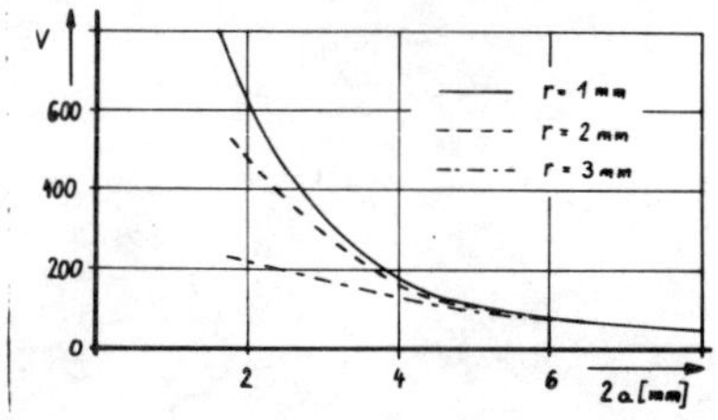

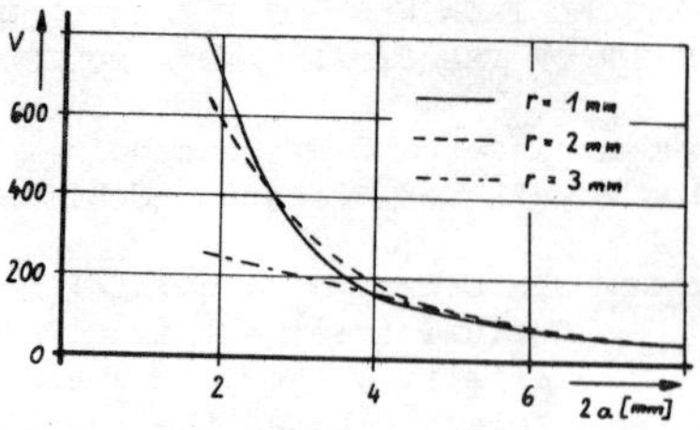

Fig.3. LG6-experiment

As can be seen, the narrowing factor V is rather small in the case of LG as a consequence of the rf field inhomogeneity, as this influence is governing the narrowing effect for $H_e M_2^{1/2}$ where the narrowing factor without rf field inhomogeneity is rather high. The narrowing behaviour of LG6 and WHH4 is by no means essentially different. Using a sample with 2a=2r=4 mm, we have a narrowing factor of about 150 in both cases. If 2a 4mm. there is no essential dependence of the narrowing factor on r in all three cases. This means the rf field decay in x-direction is dominant over that one in ρ-direction. As the dependence of different terms in the formulas (4) on a/r is different, the narrowing factor V is maximum at a certain value of a/r. Using, for example, a cylindrical sample of given volume in the case of LG6, we get

$$(a/r)_{opt} = 0.45, \tag{5}$$

which is within the limiting conditions given above for the validity of the procedure.

References

Haeberlen, U. and Waugh, J.S., 1968, Phys.Rev. 175,453
Lan, N.Q., Pfeifer, H. and Schmiedel, H., 1973, Proc. I. Spec. Coll. Ampere, Krakow, 116.
Lee, M. and Goldburg, W.I., 1965, Phys.Rev. 140, A 1261
Waugh, J.S., Huber, L.M. and Haeberlen, U., 1968, Phys.Rev.Lett. 20,180

^{1}H-NMR INVESTIGATIONS ON SINGLE CRYSTALS OF KH_2PO_4 AND $(NH_4)H_2PO_4$ (KDP AND ADP) BY VARIOUS MULTIPLE PULSE METHODS INCLUDING DOUBLE RESONANCE EXPERIMENTS

R. Willsch, U. Burghoff, R. Müller, H. Rosenberger, G. Scheler, M. Pettig and B. Schnabel
Sektion Physik der Friedrich-Schiller-Universität, DDR-69 Jena, Max-Wien-Platz 1

Abstract: By a combined application of the multiple pulse line narrowing experiments WHH4 and MW4 including special double resonance experiments single crystals of KDP and ADP were investigated. As a result we obtained the second moments due to ^1H-^1H homonuclear and ^1H-^{31}P heteronuclear dipole interaction separately, the proton chemical shift tensors and ^1H-^{31}P heteronuclear three-spin splittings.

In the last years NMR multiple pulse methods for line narrowing (MW4: Ostroff and Waugh, 1966 / Mansfield and Ware, 1966) and high resolution (WHH4: Waugh et al., 1968) in solids were developed. If high resolution is hindered by a large heteronuclear interaction, one can apply a WHH4 experiment with simultaneous heteronuclear spin decoupling (WHH4-DR: Mehring et al., 1971). In analogous way the heteronuclear interaction can be re-established by a MW4-DR (double resonance) experiment, the chemical shift being removed (Müller and Willsch, 1974).

We investigated single crystals of KDP and ADP by ^1H-resonance and ^1H-^{31}P double resonance at room temperature, using a double resonance pulse spectrometer developed in our laboratory. The resonance frequencies of ^1H and ^{31}P are 60,0 and 24,3 MHz, respectively. The cycle time is 30 μs for WHH4, and the pulse distance for MW4 is 2τ = 30 μs. Resonance offset in the WHH4-experiments was 9,75 kHz for protons.

The effective decay time T_{2e} of transversal magnetization in the MW4- and MW4-DR experiment was measured in dependence on orientation of the single crystals in the static magnetic field H_0 (Fig. 1). Using the well-known Van Vleck formulas the second moments $M_{2II} = \langle\Delta\nu^2\rangle_{II}$ of ^1H-^1H dipolar interaction and $M_{2IS} = \langle\Delta\nu^2\rangle_{IS}$ of ^1H-^{31}P dipolar interaction were calculated for KDP from the crystal data (Fig. 2). By comparing the KDP experimental values T_{2e} and T_{2e}^{DR} in Fig. 1 and the computed second moments we obtain the following expressions for relaxation rates in the MW4 and MW4-DR experiments: $(T_{2e})^{-1} = \alpha \cdot M_{2II}^{3/2} + \beta \cdot M_{2IS}^{3/2}$ and $(T_{2e}^{DR})^{-1} = \alpha \cdot M_{2II}^{3/2} + \gamma \cdot M_{2IS}^{3/2}$, where α and β are of the same order and $\gamma \gg \alpha, \beta$ (1). Using these relations, the ADP second moments M_{2II} and M_{2IS} (Fig. 2) were derived from experimental values of T_{2e} and T_{2e}^{DR} (Fig. 1). A good agreement with theore-

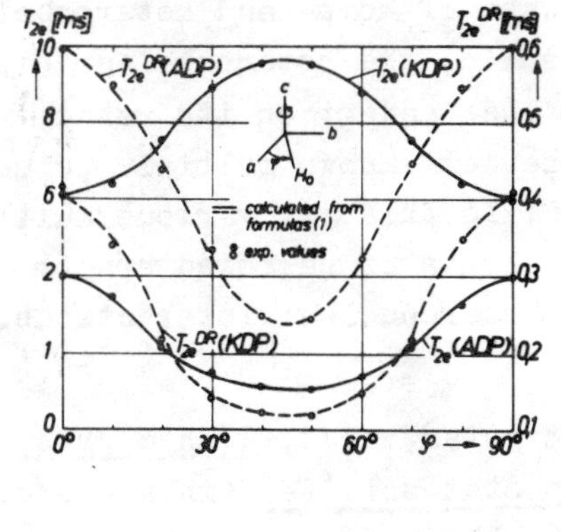

Fig. 1

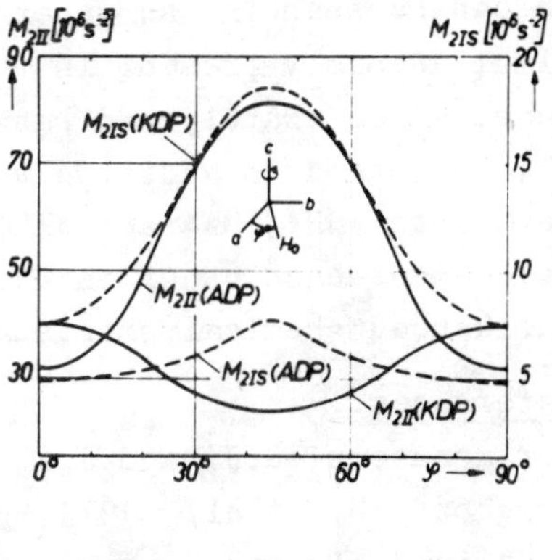

Fig. 2

tical total second moments of ADP (Adriaenssens and Bjorkstam, 1971) requires a value of α only slightly different from that of KDP.

In the MW4-DR experiment the ^{1}H-^{31}P interaction is re-established nearly completely by 180° rf pulses to the ^{31}P spins, whereas the ^{1}H-^{1}H dipolar interaction is suppressed effectively by the MW4 pulse train to ^{1}H spins. Thus in KDP one can observe splittings caused by three-spin dipolar interaction between protons and their two neighbouring phosphorus nuclei in an orientation of large M_{2IS} (Fig. 2, $\varphi = 45^{\circ}$). The computer Fourier transform of the MW4-DR induction decay shows a symmetrical spectrum with four lines (±630 Hz and ±3,70 kHz, corrected by the MW4-DR scaling factor $\sqrt{2}$). This agrees with the corresponding WHH4 spectrum (±620 Hz and ±3,75 kHz, corrected by the WHH4 scaling factor $\sqrt{3}$) and with the line positions of ±700 Hz and ±3,8 kHz calculated from geometrical structure by the average Hamiltonian method. In other orientations the WHH4 spectra are more complicated and one can find up to eight lines due to heteronuclear three-spin interaction and anisotropic chemical shifts of the protons in KDP.

The proton chemical shift tensors of KDP and ADP were measured by the WHH4-DR method. Depending on orientation the phosphorus-decoupled proton resonance spectra of KDP show only one doublet caused by anisotropic chemical shift (Fig. 3). These two lines are related to both the chemical shift tensors different in direction due to the two different O-H···O hydrogen bond directions in the crystal. The tensors are equal in size, the main values are $\sigma_{\parallel} = 9 \pm 2$ ppm and $\sigma_{\perp} = -28 \pm 4$ ppm relative to the single line of adamantane (Burghoff et al., 1974).

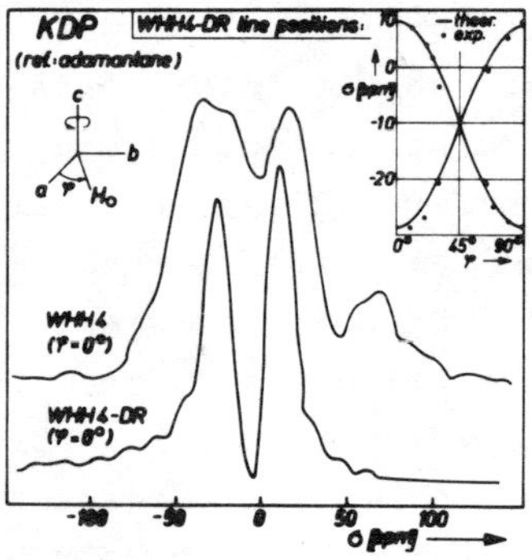

Fig. 3

The WHH4 spectra of an ADP single crystal show one broad line ($\Delta\nu_{1/2} \sim 1$ kHz). Only by means of the WHH4-DR experiment it was possible to resolve the different lines of the chemically shifted protons. The corresponding tensors are equal in size, too, with $\sigma_{\parallel} = 6 \pm 2$ ppm and $\sigma_{\perp} = -30 \pm 4$ ppm. A third line fixed at $-8{,}5 \pm 2$ ppm (relative to adamantane) is related to the protons of the reorientating ammonium groups in ADP.

As can be seen in the general case of homo- and heteronuclear dipole interaction simultaneously present in the solid, the heteronuclear dipole interaction can be measured separately and rather independent on its relative size by applying the MW4-DR method in addition to the well-known multiple pulse line narrowing experiments WHH4, MW4 and WHH4-DR. In this way various multiple pulse methods complement each other in the analysis of narrowed spectra of solids, which are influenced by chemical shifts and heteronuclear interactions.

References:

Adriaenssens, G.J. and J.L. Bjorkstam, 1971, J.chem.phys. 55, 1137
Burghoff, U. et al., 1974, phys.stat.sol. (a) (in the press)
Mansfield, P. and D. Ware, 1966, Phys.Lett. 23, 421
Mehring, M. et al., 1971, J. chem. phys. 54, 3239
Müller, R. and R. Willsch, 1974, submitted to J. magn. res.
Ostroff, E.D. and J.S. Waugh, 1966, Phys. Rev. Lett. 16, 1097
Waugh, J.S. et al., 1968, Phys. Rev. Lett. 20, 180

^{13}C CHEMICAL SHIELDING IN AMMONIUM HYDROGEN OXALATE HEMIHYDRATE

R. G. Griffin, D. J. Ruben, and L. J. Neuringer
Francis Bitter National Magnet Laboratory,† Mass. Institute of Technology, Cambridge, MA 02139 USA

Abstract. The ^{13}C chemical shielding tensors in ammonium hydrogen oxalate hemihydrate, $(NH_4)H\ C_2O_4 \cdot \frac{1}{2}H_2O$, have been determined by a single crystal study. The two types of carboxyl groups in the molecule -- e.g. ionized and protonated -- possess distinctly different chemical shielding tensors. The principal values and orientations of the tensors are discussed and compared with ^{13}C shielding tensors in other oxalate molecules.

I. Introduction. Using the direct observation double resonance experiment introduced by Pines et al. (1973), we have determined the ^{13}C chemical shift tensors in ammonium hydrogen oxalate hemihydrate, $(NH_4)H\ C_2O_4 \cdot \frac{1}{2}H_2O$ (AHOX), from single crystal studies. We chose to study this oxalate molecule because of its chemical simplicity and because it allows us to observe an ionized and protonated carboxyl group and thus determine the differences in their chemical shift tensors. In addition, the crystal structures of a number of other oxalates are known and, consequently, shift tensor studies of these compounds allow one to assess the effects of chemical and structural perturbations, such as changes in conformation, on ^{13}C chemical shift tensors. As we shall see below, we have detected changes in ^{13}C shift tensors due to protonation. The change we observe in a carboxyl tensor due to protonation is different from the effect of the simple conformation change produced by the twisting of carboxyl planes relative to one another, such as occurs in diammonium oxalate (Griffin et al. to be published).

II. Ammonium Hydrogen Oxalate Hemihydrate. The crystal structure of AHOX has recently been determined (Kuppers, 1973) and was found to be orthorhombic (space group Pnma) with eight molecules per unit cell. The symmetry of the AHOX crystal, which includes a number of inversion centers and twofold screw axes, allows only two distinct chemical shift tensors which we assign to the ionized and protonated carboxyl carbons. The principal values of these two tensors are listed in Table 1 and their orientation is illustrated in Figure 1. Table 1 also includes the principal values of the ^{13}C shift tensors in three other oxalate molecules -- diammonium oxalate and hydrated and anhydrous oxalic acid -- whose tensors have been examined in powder or single crystal studies. We first discuss the ^{13}C tensors in AHOX and then compare them with those found in diammonium oxalate and the oxalic acids.

In Table 1 are listed the principal values for the two ^{13}C tensors in AHOX which belong to the ionized and protonated carboxyl groups. We see that the σ_{11} and σ_{33} components of both tensors are very nearly identical, being -120 ppm and + 17.3 ppm with respect to liquid C_6H_6, respectively. However, for the protonated carboxyl the σ_{22} component is moved in the direction of increased shielding (upfield) by 7.9 ppm relative to σ_{22} for the ionic carboxyl tensor. Concurrent with this change in σ_{22} is a change in the tensor orientation which is illustrated in Figure 1.

Figure 1 is a schematic representation of AHOX illustrating the flat carbon oxygen frame. The view is along the perpendicular to the COO planes which coincides with the direction of σ_{33}. For the ionic COO one can surmise that σ_{11} and σ_{22} lie in the plane of the paper with σ_{11} almost along the C-C bond direction. The reason that σ_{11} is not exactly parallel to this direction is due to the fact that the carboxyl group does not possess C_{2v} symmetry (C1 - 01 = 1.236 Å and C1 - 02 = 1.257 Å). As we see from Figure 1, σ_{11} is actually 6.8 ± 1.5° away from the C-C vector in the direction of the (shorter) C1-02 bond (Chang et al., 1974).

Table 1

^{13}C oxalate shift tensor components, trace of the tensor, and isotropic shift in aqueous solution. All relative to liquid C_6H_6

Compound	σ_{11}	σ_{22}	σ_{33}	$\frac{1}{3}Tr\,\sigma_{ii}$	$\bar{\sigma}_{iso}$
$NH_4HC_2O_4 \cdot \frac{1}{2}H_2O$					
COO^-	-120.5	-14.4	17.3	-39.2	-38.2
COOH	-123.9	-6.5	17.3	-37.7	-38.2
$(NH_4)_2\ C_2O_4$					
$\cdot\ H_2O$	-116.6	-29.9	19.3	-42.4	-44.0
$(COOH)_2$					
$\cdot\ 2H_2O$	-119.8	-3.2	20.1	-34.3	-32.9
$(COOH)_2$	-121.8	5.6	16.8	-33.2	-32.9

Figure 1 also illustrates the effect of protonation on the orientation of the carboxyl tensor. As with the ionized carboxyl group the σ_{33} component is perpendicular to the COOH plane; however, the σ_{11} and σ_{22} components assume a different orientation. As we see from Figure 1, σ_{11} and σ_{22} are now rotated about σ_{33} by $24.4 \pm 1.7°$ so that σ_{22} is $\sim 7.4°$ from the C = 0 direction. Thus, protonation of the carboxyl group produces two effects: first it moves σ_{22} in the direction of increased shilding and secondly it rotates σ_{11} and σ_{22} about σ_{33} towards the C = 0 direction. We note that in the case of the symmetrical ketone, benzophenone, σ_{22} lies along the C = 0 direction (Kempf et al., 1972).

III. Other Oxalate Tensors. Chemical shift tensors in other oxalate compounds are similar, but not exactly the same, as those found in AHOX. In diammonium oxalate monohydrate (Griffin et al., to be publ.), $(NH_4)_2\ C_2O_4 \cdot H_2O$ (DAOX), the principal values of σ_{11} and σ_{22} are close to those for AHOX but the σ_{22} component is shifted in the direction of decreased shielding (see Table 1) by about 15 ppm relative to the ionic group in AHOX. In addition, the orientation of the tensors is somewhat different. DAOX is unique among the oxalates in that the COO planes are twisted by 28° with respect to one another in the crystal lattice (x-ray studies have shown all other oxalate ions to be flat). Apparently the twisting of the COO groups also produces a 10° rotation of σ_{22} and σ_{33} tensor components about the C-C bond-σ_{11} direction towards the <u>longer</u> C-O bond. The rotation of the shift tensors together with the orientation of the carboxyl groups is illustrated in Figure 2 which is a view of the DAOX molecule looking down the C-C bond. To date, this is the only carboxyl shift tensor which shows an appreciable tilt of σ_{33} away from the perpendicular to the COO plane (Chang et al., 1974; Kempf et al, 1972; Chang et al., to be published).

^{13}C shift tensors in two other oxalate molecules have also been examined, specifically, the hydrated and anhydrous forms of oxalic acid. In both of these compounds the carboxyl groups are protonated and, thus, one might expect the σ_{22} components of their tensors to be moved in the direction of increased shielding. Inspection of Table 1 shows that, indeed, this is the case. The orientation of these tensors will be discussed further elsewhere (Griffin et al., to be published).

IV. Conclusions. The chemical shift tensors in AHOX have been determined and compared to shift tensors in other oxalate molecules. The effect of protonation on a carboxyl tensor is first to move σ_{22} in the direction of increased shielding and second to rotate σ_{11} and σ_{22} about σ_{33} so that σ_{22} moves nearer to the shorter of the CO bonds. In contrast, the twisting of the carboxyl planes relative to one another in DAOX produces a rotation of σ_{22} and σ_{33} about σ_{11} so that σ_{33} moves towards the longer of the two C-O bonds. The overall orientation of these tensors is in general agreement with the tensor orientations observed for carboxyl groups in other molecules (Chang et al., 1974; Kempf et al., 1972; Chang et al., to be published).

V. References.

†Supported by the National Science Foundation. Parts of this work, specifically that on diammonium oxalate and the oxalic acids, were performed by one of us (R. G. G.) in collaboration with A. Pines, S. Pausak, and J. S. Waugh.

Chang, J. J., Griffin, R. G., and Pines, A., J. Chem. Phys., 60, 2561 (1974).

Chang, J. J., Pines, A., and Griffin, R. G., to be published; Chang, J. J. and Pines, A., proceedings of this conference.

Griffin, R. G., Pausak, S., Pines, A., and Waugh, J. S., to be published.

Kempf, J., Speiss, H. W., Haekeilen, U., and Zimmerman, H., Chem. Phys. Letters, 17, 39 (1972).

Kuppers, H., Acta Cryst. B24, 318 (1973).

Pines, A., Gibby, M. G., and Waugh, J. S., J. Chem. Phys., 59, 569 (1973).

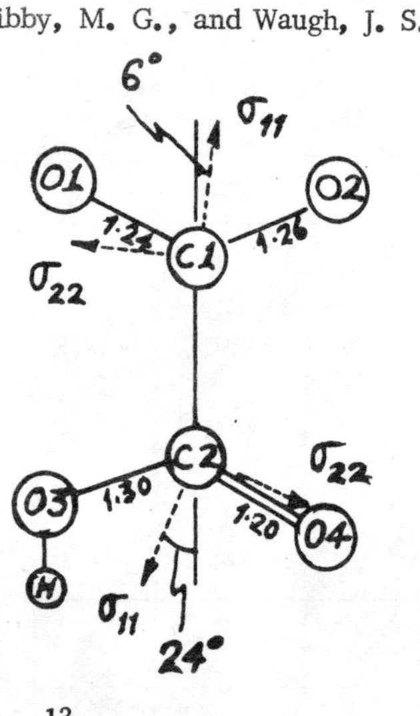

Figure 1: ^{13}C tensors in AHOX -- view ⊥ to COO planes.

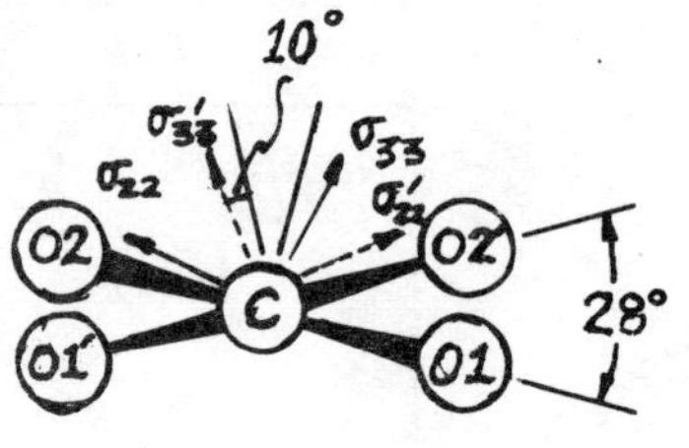

Figure 2: ^{13}C tensor in DAOX -- view ∥ to the C-C bond.

TRANSIENT OSCILLATIONS IN NMR CROSS-POLARIZATION EXPERIMENTS IN SOLIDS

Luciano Müller, Anil Kumar, Thomas Baumann and Richard R. Ernst
Laboratorium für Physikalische Chemie, Eidg. Technische Hochschule, CH-8006 Zürich

Abstract. A quantitative analysis of transient oscillations in cross-polarization experiments is given based on a simple theoretical treatment.

I. Introduction. In the course of proton-carbon-13 cross-polarization experiments in ferrocene, utilizing the Hartmann-Hahn condition, it has been found that the magnetization transfer depends in an oscillatory manner on the cross-polarization time (Müller et al,1974). Similar phenomena have also been observed in cross-polarization experiments in liquids (Hartmann and Hahn,1962). This is in contrast to the common assumption that cross-polarization can be described by an exponential process. The observed phenomenon is caused by a particularly strong dipolar interaction of the ^{13}C spin with the directly bonded proton. The damping of the oscillation is determined by the further interactions of the two spins with the surrounding protons.

2. Theory. The following treatment singles out one ^{13}C spin S and the directly bonded proton spin I and treats the interactions to the remaining proton spins I_r by a perturbation treatment of second order.The secular part of the Hamiltonian $\mathcal{H}_0$ for one particular S spin is given in a frame rotating with the two applied rf frequencies ω_I and ω_S . $\mathcal{H}_0$ is divided into three parts, $\mathcal{H}_0^{IS}$ for the particular spins I and S, $\mathcal{H}_0^{II}$ for the surrounding proton bath and $\mathcal{H}_1$ describes the interaction between the two:

$$\mathcal{H}_0 = \mathcal{H}_0^{IS} + \mathcal{H}_0^{II} + \mathcal{H}_1 \qquad (1)$$

Here, $$\mathcal{H}_0^{IS}/\hbar = \omega_{1I} I_z + \omega_{es} S_z + b' (I_x S_x + I_y S_y)/2 , \qquad (2)$$

where $\omega_{1I} = - \gamma_I H_{1I}$, and $\omega_{es} = (\omega_{1S}^2 + \Delta^2)^{1/2}$ are the effective fields for I and S spins respectively (assuming on-resonance irradiation for I and off-resonance irradiation with offset Δ for S). The z-axis for the I and S spins are taken to be the directions of the respective effective fields in the rotating frame; the reduced dipolar coupling constant is $b' = \sin\Phi\, \gamma_I \gamma_S \hbar (3\cos^2\theta - 1)/(2r_{IS}^3)$ in which $\tan\Phi = \omega_{1S}/\Delta$ and θ is the angle between static magnetic field and rotation axis of the ferrocene molecule.

$$\mathcal{H}_0^{II}/\hbar = \sum_r \omega_{1I} I_{zr} - \sum_{r,s} a_{rs}(I_{zr} I_{zs} - (I_{xr} I_{xs} + I_{yr} I_{ys})/2)/2 \qquad (3)$$

with the motion-averaged dipolar coupling constants a_{rs}.

$$\mathcal{H}_1 = \sum_j b_j' (I_{xj} S_x + I_{yj} S_y)/2 - \sum_k a_k(I_{zk} I_z - (I_{xk} I_x + I_{yk} I_y)/2)/2 . \qquad (4)$$

Spin diffusion among the surrounding spins is assumed to be sufficiently fast such that they remain in internal equilibrium. It is then possible to write the density operator as a product $\rho = \sigma \rho_{II}$, where σ is the density operator of the (I,S) subsystem and satisfies, up to second order in $\mathcal{H}_1$, the differential equation:

$$\begin{aligned}\dot{\sigma} &= - (i/\hbar)[\mathcal{H}_0^{IS},\sigma] - (1/\hbar^2)\mathrm{Tr}\{\int_0^t d\tau [\mathcal{H}_1,[\mathcal{H}_1(-\tau), \sigma(t) - \sigma_o]] \\ &= - (i/\hbar)[\mathcal{H}_0^{IS},\sigma] - \Gamma (\sigma(t) - \sigma_o)\end{aligned} \qquad (5)$$

where, $\mathcal{H}_1(-\tau) = \exp(-\tau i(\mathcal{H}_o^{IS}+\mathcal{H}_o^{II}))\,\mathcal{H}_1\exp(+i\tau(\mathcal{H}_o^{IS}+\mathcal{H}_o^{II}))$. (6)

Straightforward calculations give for Γ: (7)

$$\Gamma(\sigma) = [I_z,[I_z,\sigma]]R_I^z + ([I_x,[I_x,\sigma]]+[I_y,[I_y,\sigma]])R_I + ([S_x,[S_x,\sigma]]+[S_y,[S_y,\sigma]])R_S \; .$$

In this expression, terms which do not affect the S magnetization under Hartmann-Hahn condition and two small terms have been omitted. The coefficients are given by

$$R_I^z = (1/9)\langle\Delta\omega^2\rangle_{II}\int_0^t d\tau\cos^2(b'\tau/4)\; g(\sum_k a_k I_{zk})$$
$$R_I = (1/36)\langle\Delta\omega^2\rangle_{II}\int_0^t d\tau\cos(b'\tau/4)\; g(\sum_k a_k I_{xk}) \qquad (8)$$
$$R_S = (1/4)\langle\Delta\omega^2\rangle_{IS}\sin^2\Phi\int_0^t d\tau\cos(b'\tau/4)\; g(\sum_j b_j' I_{xj})$$

where $g(\sum_k a_k I_{xk}) = \mathrm{Tr}\{\sum_k a_k I_{xk}\exp(-i\mathcal{H}_o^D\tau)\sum_j a_j I_{xj}\exp(+i\mathcal{H}_o^D\tau)\}/\mathrm{Tr}\{(\sum_k a_k I_{xk})^2\}$ is the normalized correlation for the $\sum a_k I_{xk}$, and $\mathcal{H}_o^D$ is the dipolar part of $\mathcal{H}_o^{II}$. $\langle\Delta\omega^2\rangle_{IS}$ is the second moment of a single line of the doublet of the undecoupled ^{13}C spectrum and $\langle\Delta\omega^2\rangle_{II}$ is the second moment of the proton spectrum.

The integration limits are now extended to infinity. Then the R's become relaxation rates and Eq.(5) can be solved using the initial condition

$$\sigma(0) = (\,1 - \beta\hbar\omega_{oI}I_z\,)/\mathrm{Tr}(1) \; . \qquad (9)$$

One finds for the time dependence of the observed signal amplitude $\langle S_z\rangle(t)$:

$$\langle S_z\rangle(t) = \mathrm{Tr}\{\sigma(t)S_z\} = (1/4)\beta\hbar\,\omega_{oI}\{1 - [1+4(R_I-R_S)R_I^z/b'^2](1/2)\exp[-(R_I+R_S)t] -$$
$$-[\cos(b't/2)+((R_I+R_S)2+R_I^z)(1/b')\sin(b't/2)-(R_I-R_S)R_I^z(4/b'^2)\cos(b't/2)]\,\cdot \qquad (10)$$
$$\cdot\,(1/2)\exp[-(R_I+R_S+R_I^z/2)t\,]\,\} \; .$$

This equation reduces to the result of the earlier simplified treatment given by (Müller et al,1974) for the special case when $R_S=0$, $R_I^z = R_I$ and $b' \gg R_I^z$.

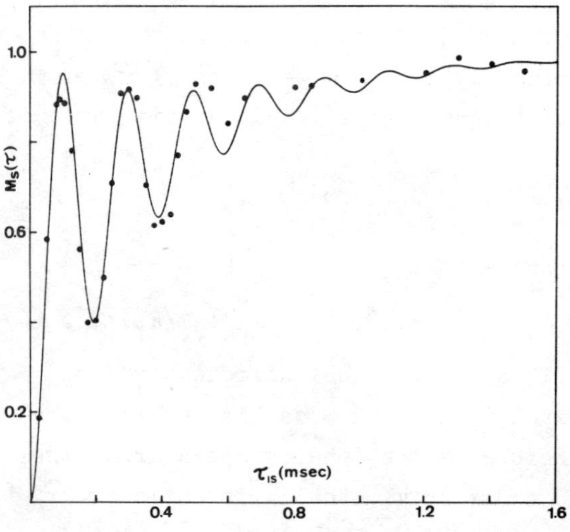

Figure. It shows a set of experimental measurements together with a theoretical plot based on eq.(10). The parameter values for the theoretical curve are: $b'/2\pi$ = 10.19 kHz, $R_I/2\pi$ = 1.7 kHz, $R_S/2\pi$ = 0.3 kHz, $R_I^z/2\pi$ = 2.55 kHz. The rf field amplitudes are $\omega_{1I}/2\pi = \omega_{eS}/2\pi$ = 17.4 kHz.

3. References.

S.R.Hartmann and E.L.Hahn,Phys.Rev.,128, 2042,(1962)

L.Müller,A.Kumar,T.Baumann and R.R.Ernst,Phys.Rev.Lett.,32, 1402,(1974).

SECTION T

Jahn–Teller Ions

EXPERIMENTAL TECHNIQUES FOR THERMAL DETECTION OF STRONGLY-COUPLED IONS

W. S. Moore, T. M. Al-Sharbati*, I. A. Clark and A. P. Knowles
Department of Physics, University of Nottingham, University Park, Nottingham NG7 2RD, ENGLAND.
*Department of Physics, College of Science, University of Basrah, Basrah, IRAQ.

Abstract. Broad highly-asymmetric absorptions are typically detected from strongly-coupled non-Kramers ions when electric-field-induced thermally-detected EPR techniques are employed The angle and temperature dependence of linewidth and lineshape have given much new information on the couplings of such ions to their microscopic environment and to each other.

Introduction. Gorter (1936) and Schmidt and Solomon (1966) have stated that it is possible to detect paramagnetic ions in a specimen by their resonant absorption of electromagnetic energy in a magnetic field if the resulting rise in temperature following magnetic relaxation of a thermally isolated specimen is observed. We have successfully adapted the idea for the specific case of di electric crystals containing a wide variety of paramagnetic impurity ions (Moore et al., 1973a,b, Moore, 1974). The experiments have been performed at liquid helium temperatures so that full advantage can be taken of the extremely small heat capacity possessed by such crystals at these low temperatures. Under such conditions even a very small heat leak from the thermally isolated specimen leads to a reasonable thermal time constant τ for the crystal and hence to large temperature rises when a small power is dissipated in the crystal by magnetic relaxation, typically 10^5 degree watt^{-1}, with $\tau = 1$ sec.

Experimental Details. We have used very small carbon film resistance thermometers to measure the crystal temperature rise. Figure 1 shows the evacuable resonator we currently use, where the single crystal quartz rod is used to transport the heat generated in the crystal to the thermometer This rod adds little to the heat capacity of the combined crystal plus thermometer, but is very necessary to avoid direct heating of the thermometer by the eddy currents which would be induced by incident microwaves. The heat leak is via the electrical leads to the thermometer rather than via the thin 'Mylar' plastic supports for the rod, thus allowing control of τ. In order that $\tau \sim 1$ sec, we have used 1 x 0.1 x 0.001 cm aluminium foil electrical leads. The thermometer resistance is measured by a conventional 1 kHz AC Wheatstone bridge. Provision is made to pass DC current through the carbon film thermometer whilst still measuring its resistance with AC. This raises the temperature of the crystal + rod + thermometer above the temperature of the helium bath which surrounds the resonator. The thermometer resistance is calibrated for different temperatures by pumping over the helium bath, and thus we are able to measure the absolute power delivered by the relaxing paramagnetic ions as a function of temperature under the influence of known microwave fields. We believe that the crystal, rod and thermometer are in thermal equilibrium with each other at all times because of the excellent thermal conductivity of quartz, and the known rapid rate with which the monochromatic phonons generated in the crystal by relaxation come into equilibrium. Thus in order to observe spectra, the bridge out-of-balance signal, proportional to the resistance change, is plotted as a function of magnetic field, when a decrease in resistance indicates crystal heating. Figure 2 shows a typical spectrum and illustrates two further points. The first is that the base line is not straight, due to thermometer magnetoresistance, and the second is that quite large field-independent heating is found in some specimens. We can however allow for both of these effects if necessary by drawing a baseline without incident microwave power if we ensure that the crystal temperature is kept the same in the absence of microwave power by passing the appropriate DC current through the resistance thermometer.

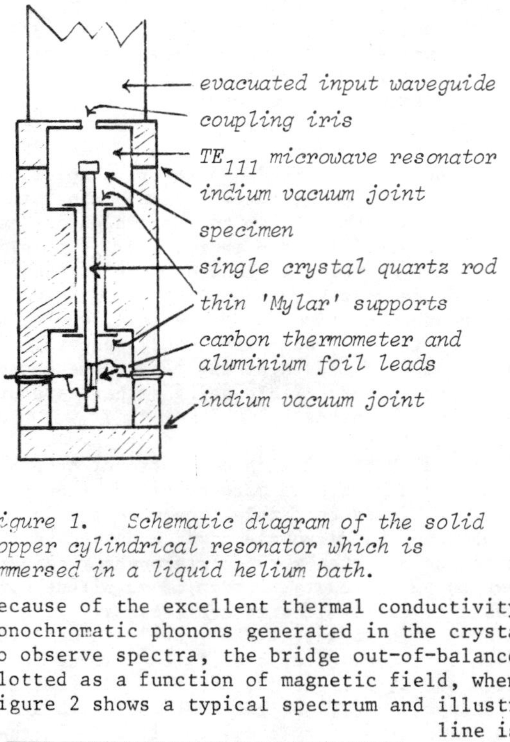

Figure 1. Schematic diagram of the solid copper cylindrical resonator which is immersed in a liquid helium bath.

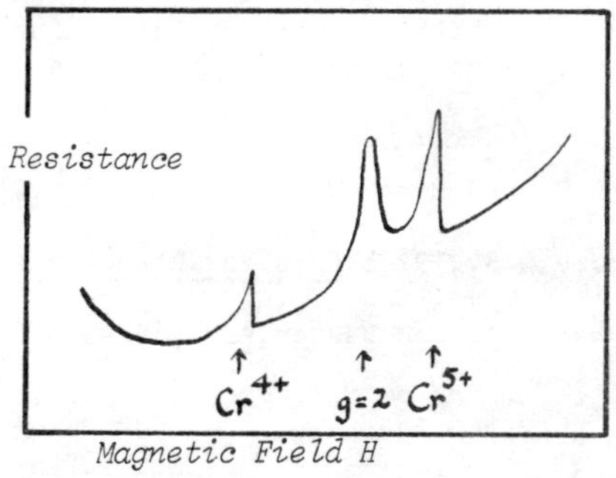

Figure 2. X-band TD EPR spectrum of Al_2O_3:Mg at 3 K, H // c.

Transition Probabilities. Perhaps the most important fact that the technique has produced is that for non-Kramers ions in non-centrosymmetric sites, the RF electric field e_{RF} will induce transitions which are normally highly forbidden ($g_\perp = 0$, $\Delta J_{eff} = 0,1,2$) for the RF magnetic field h_{RF}. In fact all the absorption lines of such ions which we have so far seen in Al_2O_3 by thermal detection (TD) have been electric-field-induced EPR transitions (EF-EPR), and appear to be identical to the corresponding APR spectra, where available (Moore et al., 1973b). In Figure 2 for instance the line at g = 2 corresponds to transitions induced by h_{RF} when perpendicular to the c-axis, while the lines from Cr^{4+} and Cr^{5+} are due to transitions induced by e_{RF} when perpendicular to the c-axis. To determine the relative importance of these two mechanisms, we first make measurements with the crystal placed almost on the end wall of the cylindrical resonator where h_{RF} is maximum and then with the crystal in the centre of the resonator where e_{RF} is maximum (as shown in Figure 1). In each of these two positions, we make use of the bimodal property of the cylindrical TE_{111} resonator to enable the RF fields to be turned through exactly 90° by a small change in frequency from one mode resonance to the other. The frequencies and axes of the two non-degenerate normal modes of the resonator can be determined either by the shape anisotropy or dielectric anisotropy of the specimen, or by pieces of dielectric material suitably placed within the resonator. To obtain the spectrum of Figure 2, the specimen was placed in an intermediate position and neither normal mode axis was aligned with the c-axis of the Al_2O_3 single crystal, thus allowing both e_{RF} and h_{RF} transitions to be seen simultaneously. Because of the existence of EF-EPR we can now explain Kramers/non-Kramers ion coupling in general. This has been somewhat of a mystery since magnetic coupling between such pairs of ions is ruled out by the non-Kramers ion having $g_\perp = 0$. We now think (Bates et al., 1974) that it is due to EF-EPR transitions of the non-Kramers ion caused by the fluctuating electric field produced when the quadrupole moment of a Kramers ion with J' > 1 is changed by a transition.

Lineshapes and Linewidths. As the lattice coupling of a series of ions such as Ti^{3+}, V^{4+}, Cr^{5+} ($3d^1$) increases, they become increasingly sensitive to lattice defects, strain and the random microscopic electric fields due to charged defects (henceforth all called 'strain'). Thus broad and usually asymmetric absorption lines are usual as shown on Figure 2, and the observed linewidths of $3d^1$ ions are in the order $Cr^{5+} > V^{4+} > Ti^{3+}$. We have examined the ions Fe^{2+}, Fe^{3+}, Cr^{2+}, Cr^{3+}, Cr^{4+}, Cr^{5+}, Ti^{3+}, Ni^{3+} in Al_2O_3 and found a variety of widths and shapes, where the characteristic properties of the unstrained site often give rise to discontinuities in the absorption-field spectrum. One of the most common shapes found is exhibited by both Cr^{4+} and Cr^{5+} in Figure 2 where there is a sharp high-field edge corresponding to zero strain and a low-field tail produced by the strain distribution. The energy separation between doublets, $|\pm 1\rangle$ for Cr^{4+} and $|\pm\frac{1}{2}\rangle$ for Cr^{5+} depends quadratically on strain and if a Gaussian distribution is assumed, an exponential lineshape ($\sim \exp(h_E - h)$) is expected and found in many cases, where h_E is the magnetic field for the edge and h is the difference between the actual field and h_E. In a similar way we hope to find the detailed distribution of strain by its extreme effect on the Ni^{3+} ion EF-EPR absorption spectrum in Al_2O_3 (Abou-Ghantous et al., 1974a,b). Towards this end we have constructed a simple apparatus to measure the zero-field lineshape for this ion. The apparatus is identical to that which has been described before except that there is no resonator and the crystal + rod + thermometer is merely inserted into a piece of waveguide where it is exposed to the travelling electromagnetic waves generated by a swept-frequency BWO. So far we have found the zero strain edge of the Ni^{3+} absorption at the predicted frequency of 14.1 GHz followed by an enormous high-frequency tail, and we are currently studying the effects of steady magnetic fields on the absorption-frequency spectra. This latter non-resonant method has the great advantage that it yields a straight baseline because there is no magnetoresistance change in the thermometer, the magnetic field being either zero or constant. It is consequently much easier to find the true lineshapes.

Sensitivity. It is difficult to compare the sensitivity of TD EF-EPR with conventional EF-EPR and APR. The apparatus for TD is certainly much simpler and it can be much more sensitive than a conventional high frequency modulation EPR spectrometer. This is certainly true when the ions of interest have broad lines and short relaxation times, because the microwave power can then be increased to levels where it would be impossible to balance and use even a superheterodyne receiver to detect the very broad lines. The TD method also seems to be at least an order of magnitude more sensitive than APR, although it suffers from being unable to detect ions in cubic sites. Even though our thermometers are far from optimal, we have been able to detect Fe^{2+}, Cr^{4+} and Cr^{2+} in Al_2O_3 down to concentrations of 1 in 10^8 which corresponds in the latter case to 10^{13} spins in an inhomogeneous line at least 1 kG wide. In the spectrum shown in Figure 2, chromium was present only in its natural abundance and was measured to be < 3 ppm. TD is of course inferior to conventional detection when the lines are narrow and the relaxation times very long, but it is nevertheless still a useful technique under these circumstances because the height of the saturated lines depends on $1/T_1$. Thus resonant cross-relaxation can and has been studied, and it may even be possible to detect ENDOR thermally.

References

Abou-Ghantous, A., Bates, C.A., Clark, I.A., Fletcher, J.R., Jaussaud, P.C. and Moore, W.S., 1974a, J.Phys.C, 7, 2707.

Abou-Ghantous, M., Jaussaud, P.C., Bates, C.A., Fletcher, J.R. and Moore, W.S., 1974b, Phys.Rev.Letters, 33, 530.

Bates, C.A., Moore, W.S., Al-Sharbati, T.M., Steggles, P., Gavaix, A., Vasson, A. and Vasson, A.M., 1974, J.Phys.C, 7, L83.

Gorter, C.J., 1936, Physica, 3, 995.

Moore, W.S. and Al-Sharbati, T.M., 1973a, J.Phys.D, 6, 367.

Moore, W.S., Bates, C.A. and Al-Sharbati, T.M., 1973b, J.Phys.C, 6, L209.

Moore, W.S., Proc.Fifth Internat. Symposium on Magnetic Resonance, to be published in Pure and Appl. Chem., 40, Nos 1 and 2, 1974/1975.

Schmidt, J. and Solomon, I., 1966, J.Appl.Phys., 37, 3719.

THE SUPPRESSION OF THE TRIGONAL CRYSTAL FIELD SPLITTING OF Ti^{3+} IN ALUM

A. Jesion, Y.H. Shing and D. Walsh
Department of Physics, Eaton Electronics Laboratory, McGill University, P.O. Box 6070, Montreal, Quebec, Canada H3C 3G1

Abstract. The ground state of Ti^{3+} in CsAl alum is a quasi Γ_8 state with a very small trigonal splitting due to the presence of a dynamic Jahn-Teller effect. The model also explains the low temperature properties of CsTi alum.

The ground state of Ti^{3+} as a dilute substitutional impurity for Al in CsAl alum was found recently (Shing and Walsh, 1974) using EPR techniques below 3K to be a quasi Γ_8 quartet with a very small trigonal splitting. A spin-Hamiltonian was derived for this quasi Γ_8 state, which described well the unusual EPR spectrum and anisotropy. The angular EPR spectra of Figs. 1 and 2 measured in {110} and {111} planes respectively confirmed the presence of four symmetry-equivalent sites with the trigonal symmetry axis along <111> directions, which is in agreement with this alum structure (Cromer et al., 1966). Three strain-broadened closely-spaced EPR lines were detected corresponding to the three allowed $\Delta M = \pm 1$ transitions. The generalized spin-Hamiltonian, which includes terms linear in H, for a quasi Γ_8 state in a trigonal lattice has the form

$$\begin{aligned} H_s = {} & D_{20}T_{20}(\vec{S}) + g_{\parallel}\mu_B H_z T_{10}(\vec{S}) + \frac{1}{\sqrt{2}} g_{\perp}\mu_B H_x[T_{1-1}(\vec{S}) - T_{11}(\vec{S})] \\ & + f_{\parallel}\mu_B H_z T_{30}(\vec{S}) + \frac{1}{\sqrt{2}} f_{\perp}\mu_B H_x[T_{3-1}(\vec{S}) - T_{31}(\vec{S})] \\ & + h\mu_B\{H_z[T_{33}(\vec{S}) - T_{3-3}(\vec{S})] - \sqrt{\frac{3}{2}}\, H_x[T_{32}(\vec{S}) + T_{3-2}(\vec{S})]\} \quad , \end{aligned}$$

where the parallel direction is along the trigonal [111] axis and $T_{\ell m}(\vec{S})$ are tensor operator equivalents of rank ℓ expressed in terms of the spin angular momentum $\vec{S}$. These operator equivalents are related to the Stevens' operator equivalents and are defined as irreducible tensors which possess the convenient transformation properties (Buckmaster et al. 1972). The spin-Hamiltonian parameters were determined from the angular spectra to be:

$g_{\parallel} = 1.1937 \pm 0.001$, $\quad g_{\perp} = 0.6673 \pm 0.005$,
$u_{\parallel} = -0.0144 \pm 0.002$, $\quad u_{\perp} = -0.0868 \pm 0.005$,
$D = 39.3 \pm 1.0$MHz, $\quad$ where $(u_{\parallel}, u_{\perp}) = (\sqrt{\frac{5}{2}})\,(f_{\parallel}, f_{\perp})$.

The $\Delta M = \pm 1$ EPR transitions were calculated using these spin-Hamiltonian parameters and were found in very good agreement with the corresponding experimental data.

The unique properties of a quasi Γ_8 state with a very small zero field splitting D arises as follows. The dominant crystalline electric field in CsAl alum at the Al site due to the surrounding octahedron of water molecules is cubic. Thus to first order the ground orbital state of Ti^{3+} in this alum is Γ_{5g}. From the previous results of Ti^{3+} in CH_3NH_3Al alum (Shing et al., 1974) this Γ_{5g} state will also be strongly coupled to the Γ_{3g} vibrational mode. Under these circumstances the effective trigonal Hamiltonian becomes (Abou-Ghantous et al., 1974):

$$H_{trig} = \frac{1}{3}\,[\gamma v - \frac{(v')^2}{\Delta}]\,[3\ell_z^2 - \ell(\ell+1)] \quad ,$$

where v is the trigonal splitting of the Γ_{5g} state to first order, γ is the vibronic reduction factor, v' is the admixture of the excited Γ_{3g} and the ground state by the trigonal field, Δ is the cubic field splitting and the effective orbital angular momentum $\ell = 1$. Since the first term within the first square bracket contains the vibronic reduction factor γ but the second term is unaltered by the Jahn-Teller coupling, γv will be comparable in magnitude with $\frac{(v')^2}{\Delta}$. The parameter v is positive for Ti^{3+} in CsAl alum which leads in this case to an almost complete cancellation of the effective trigonal field. Thus the ground Γ_8 spin-orbit state will be slightly split by the residual trigonal field into doublets giving the unusual EPR properties of a quasi Γ_8 state. Since the zero field splitting D given above is very small one can choose $\gamma \simeq \frac{(v')^2}{v}$. By evaluating directly the matrix elements of $T_{20}(\vec{L})$ one finds the ratio $v'/v = \sqrt{2}$. The parameter v can be derived indirectly by comparing the experimental zero field splittings of Cr^{3+} in both Al_2O_3 (Laurance and Lambe, 1963) and CsAl alum (Danilov et al., 1973) and using the accepted value (Macfarlane et al., 1968) for Ti^{3+} in Al_2O_3 of $v = \frac{2}{3}(1000)cm^{-1}$. Thus we estimate for Ti^{3+} in CsAl alum $v \simeq \frac{2}{9}(1000)cm^{-1}$. Choosing $\Delta = 18{,}000cm^{-1}$ we obtain $\gamma = 0.03$, which is considerably less than the previous estimate for Ti^{3+} in CH_3NH_3Al alum (Shing et al., 1974). With $\gamma = 0.03$ the predicted excited Γ_7 state is at $6cm^{-1}$ which can be confirmed for instance by spin-lattice relaxation via the Orbach process to that state. With regard to the vibronic reduction factors for Ti^{3+} in methylammonium alum it is important to note that this crystal undergoes a low-temperature ferroelectric phase transition (only the high temperature form is β-alum). Below T_c the symmetry of the Al site is lowered from $\bar{3}$ to 1 (Fletcher and Steeple, 1964), although the EPR spectrum does show approximately trigonal symmetry. Excellent agreement was also found between the experimental g-values and the calculated values using the experimentally determined vibronic reduction factors. Nevertheless, these reduction factors are probably unreliable, since deviations from the quasi symmetry $\bar{3}$ will alter the energy levels from which the parameters are deduced.

EPR of single crystal CsTi alum was first reported by Bleaney et al. (1955) in the temperature range 2.5K-4.2K. The spectrum indicated that the crystal had the usual alum structure with four ions in the unit cell. The g-values were $g_{\parallel} = 1.25 \pm 0.02$ and $g_{\perp} = 1.14 \pm 0.02$, and the line width was constant over the temperature range given above at 250 ± 50 gauss. In a recent preliminary report these g-values were confirmed by Harrowfield (1971), who also found the EPR of deuterated CsTi alum to be approximately isotropic with $g = 1.31 \pm 0.02$. Crystal field theory has failed to account satisfactorily for the magnetic properties of CsTi alum and there is a well-known disagreement between the low temperature theoretical spin-lattice relaxation time and the experimental results (Van Vleck, 1958).

The complete description of the ground state of CsTi alum by a quasi Γ_8 quartet has not been attempted to date. However, all the above anomalies can be accounted for qualitatively by a quasi Γ_8 description. The trigonal distortion of CsTi alum will be close but not identical to that in CsAl alum, and from the EPR results we assume that the trigonal field cancellation in the former is more complete than that in the latter. Consequently, only one broad, fast-relaxing transition will be detected, while the small zero field splitting of the quasi Γ_8 quartet will provide a large specific heat below 0.01°K, in agreement with previous experimental results (Benzie and Cooke, 1951). Since the Ti^{3+} in CsAl alum spectrum shows three closely-spaced, strain-broadened lines, the line width of CsTi alum will be primarily due to strain and not exchange as was previously assumed. This also eliminates a longstanding theoretical difficulty - the experimental Curie-Weiss constant indicated negligible exchange interactions, as we would expect from nearest neighbour Ti distances in CsTi alum.

References

Abou-Ghantous, M., Bates, C.A., and Stevens, K.W.H., 1974, J. Phys. C:Solid State Phys., 1, 325-338.
Benzie, J., and Cooke, A.H., 1951, Proc. Roy. Soc., A209, 269-278.
Bleaney, B., Bogle, G.S., Cooke, A.H., Duffus, R.J., O'Brien, M.C.M., and Stevens, K.W.H., 1955, Proc. Phys. Soc., A68, 57-58.
Buckmaster, H.A., Chatterjee, R., and Shing, Y.H., 1972, Phys. Stat. Sol., 13a, 9-50.
Cromer, D.T., Kay, M.J., and Larson, A.C., 1966, Acta Cryst., 21, 383-389.
Danilov, A.G., Vial, J.C., and Manoogian, A., 1973, Phys. Rev. B8, 3124-3133.
Fletcher, R.O.W., and Steeple, H., 1964, Acta. Cryst., 17, 290-294.
Harrowfield, B.V., 1971, 8th Australian Spectroscopy Conference, Clayton, Australia, Aug. 16-20. (1972, Phys. Abstr., 75, 594, No. 10288)
Laurence, N., and Lambe, J., 1963, Phys. Rev. 132, 1029-1036.
Macfarlane, R.M., Wong, J.Y., and Storge, M.D., 1968, Phys. Rev. 166, 250-258.
Shing, Y.H., Vincent, C., and Walsh, D., 1974, Phys. Rev. B9, 340-341.
Shing, Y.H., and Walsh, D., 1974, Phys. Rev. Letters, (submitted)
VanVleck, J.H., 1958, Disc. Far. Soc., 26, 96-102.

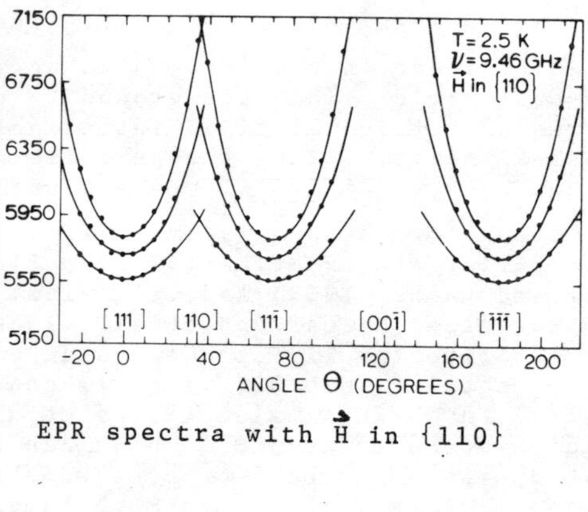

EPR spectra with $\vec{H}$ in {110}

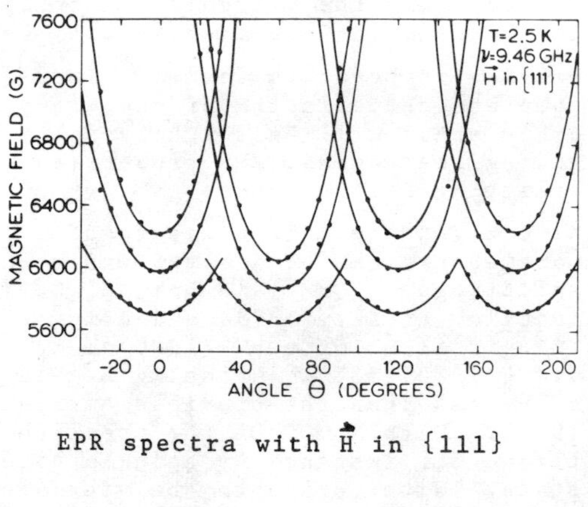

EPR spectra with $\vec{H}$ in {111}

INFLUENCE OF THE JAHN-TELLER EFFECT ON THE PARAMAGNETIC PROPERTIES OF THE GROUND-STATE OF ZnS/Cu^{2+}

B. Clerjaud and A. Gelineau
Laboratoire de Luminescence II, Université Paris VI, 4 pl. Jussieu, 75230 Paris Cedex 05, France
Equipe de recherche associée au C. N. R. S.

Abstract. The weak gyromagnetic factor $|g(\Gamma_7)| = 0.71$ of the ground-state $\Gamma_7\,(^2T_2)$ as well as the absence of E. P. R. spectrum corresponding to $g(\Gamma_7)$ are explained as being due to a dynamic Jahn-Teller effect within the ground-state. The coupling occurs with E phonons and the Jahn-Teller energy is 900 cm^{-1}.

1. Introduction. Electron paramagnetic resonance (E. P. R.) spectrum of a Cu^{2+} center in zinc sulfide having cubic symmetry has never been observed in spite of research in this field (Holton et al., 1969). On the other hand, the gyromagnetic factor $g(\Gamma_7)$ of the ground-state $\Gamma_7\,(^2T_2)$ has been measured from Zeeman effect experiments in the near infrared absorption spectrum (de Wit, 1969 - Broser et al., 1970) or emission spectrum (Wöhlecke, 1974) and has been found to be $|g(\Gamma_7)| = 0.71$.

The aim of this communication is to explain this weak gyromagnetic factor, as well as the absence of E. P. R. spectrum as being due to a dynamic Jahn-Teller effect within the ground-state $^2T_2$.

Up to now, several authors (Broser and Maier, 1966 - Bates and Chandler, 1971 a, b - Yamaguchi and Kamimura, 1972 - Maier and Scherz, 1974 - Sauer et al., 1974) have considered the possibility of the influence of a Jahn-Teller effect in Cu^{2+} in ZnS, but they failed in the explanation of the weak gyromagnetic factor.

The investigation of a Jahn-Teller effect in an orbital triplet state is a complicated problem, because in a site of tetrahedral symmetry, this state can be coupled, apart from the A_1 total symmetrical mode, to one E and to two T_2 vibrational modes (Struge, 1967), but fortunately optical uniaxial stress experiments (Maier, 1972) have shown that the $^2T_2$ level is coupled predominantly with E modes of vibration. Such a coupling has permit to interpret the position of the narrow lines in the optical spectra as well as the ratio between their intensities (Clerjaud and Gelineau, 1974).

2. Gyromagnetic Factor. We have performed a second order calculation, considering the spin-orbit and Zeeman interactions as perturbations to the Jahn-Teller problem. This calculation involving the $^2T_2$ ground-state and the ^2E excited state vibronic levels leads to (Clerjaud and Gelineau, 1974) :

$$g(\Gamma_7) = -\frac{2}{3}\left\{1 + 2e^{-3S_E/2} - \frac{2\lambda}{\hbar\omega} e^{-3S_E}\left[G(3S_E) + G(\tfrac{3}{2}S_E)\right] + \frac{4\lambda}{\Delta}(1 - e^{-3S_E/2})\right\}$$

where $S_E = E_{JT}/\hbar\omega$ is the Jahn-Teller energy in units of E phonons of frequency ω, λ is the spin-orbit constant, Δ is the splitting between the ^2E and the $^2T_2$ states and $G(x) = \sum_{n=1}^{\infty} \frac{x^n}{n.\,n!}$

The important lattice modes has been shown (Abou-Ghantous et al., 1974) to be those with wavelengths of the same order as the diameter of the electron orbit, i. e. optical phonons in our case ; this led us to take $\hbar\omega = 300$ cm^{-1}. If we take $\lambda = -600$ cm^{-1}, $\Delta = 6900$ cm^{-1} and $S_E = 3$, value which fits the optical data (Clerjaud and Gelineau, 1974), we get $g(\Gamma_7) = -0.8$ (without taking into account the reduction in the orbital part of the Zeeman interaction due to covalency which should diminish $|g(\Gamma_7)|$ of about 0.1). It is to be noted that the contribution of the ^2E excited state is important (0.23).

3. Spin Lattice Coupling. The Hamiltonian describing the effect of strain is :

$$\mathcal{H} = V'_{A_1}\,\mathcal{J}\,\frac{1}{3}(e_{11}+e_{22}+e_{33}) + V'_E(\varepsilon_\theta e_\theta + \varepsilon_\varepsilon e_\varepsilon) + (V'_{T_{2a}} + V'_{T_{2b}})(\tau_\xi e_{23} + \tau_\eta e_{31} + \tau_\zeta e_{12})$$

where $\mathcal{J}$ is the unit matrix, ε_θ, ε_ε, τ_ξ, τ_η, τ_ζ, are given in (Ham, 1965, 1972), e_{ij} is a component of the strain tensor, e_θ and e_ε are defined in (Ham, 1965, 1972). There is a close connection between the strain coefficients V' and the linear Jahn-Teller coupling coefficients V (Ham, 1968). If we assume the cluster model, we get for tetrahedral coordination :

$$V'_{A_1} = 2RV_{A_1}, \quad V'_E = \frac{2\sqrt{2}}{3}RV_E, \quad V'_{T_{2a}} = \frac{4}{3}RV_{T_{2a}}, \quad V'_{T_{2b}} = -\frac{2\sqrt{2}}{3}RV_{T_{2b}}$$

where R is the nearest neighbor distance. We assume E mode coupling ; we can thus neglect $V_{T_{2a}}$ and $V_{T_{2b}}$; on the other hand, we have no experimental evaluation of V_{A_1}. V_E can be deduced from the Jahn-Teller energy : $E_{JT} = V_E^2/2\mu\omega^2$, where μ is the effective mass of the mode ; if we take for μ, the sulfur mass (Ham, 1968, 1972), we get : $|V'_E| = 27\,800$ cm^{-1}.

The effect of strains on the ground-state $\Gamma_7\,(^2T_2)$ is described by the following term in the Spin-Hamiltonian (Tucker, 1966 - Tucker and Rampton, 1972) : $\mathcal{H}_{S-L} = \mu_B \sum_{ij} h_{ij} H_i S_j$. To first order in strain, h_{ij} can be expressed by the following tensor expression : $h_{ij} = \sum_{k\ell} F_{ijk\ell}\, e_{k\ell}$ where the fourth-rank tensor F is the spin-lattice coupling tensor. h_{ij} is not necessarely symmetric, however it is in our high symmetry case (Tucker and Rampton, 1972 - Black and Donoho, 1968) ; it is thus possible to use the Voigt notation and to represent the fourth-rank F tensor by a six dimensional square matrix which has the same properties as the photoelastic matrix of the $\bar{4}3m$ class (Nye, 1957). This matrix has three independent elements F_{11}, F_{12} and F_{44} that we shall evaluate now. We consider the Zeeman-plus-strain hamiltonian as a perturbation to the Jahn-Teller-plus-spin-orbit hamiltonian. The strain hamiltonian has no effect within the $\Gamma_7\,(^2T_2)$ ground-state, it is thus necessary to consider the effect of excited states in the frame of a second order perturbation calculation. The first excited state is the $\Gamma_8\,(^2T_2)$ state which lies very close to $\Gamma_7\,(^2T_2)$ because of the Jahn-Teller reduction of the spin-orbit coupling coefficient. It is to be noted that the total symmetrical part of the strain hamiltonian is inactive and that only the e_θ component of the E strain has an influence. A straightforward calculation leads to :

$$F_{11} = -2F_{12} = \frac{8V'_E}{9}\left\{\frac{2+e^{-\frac{3}{2}S_E}}{\lambda'} + \frac{e^{-\frac{3}{2}S_E}}{\hbar\omega}\left[G(3S_E) - \frac{1}{2}G\left(\frac{3}{2}S_E\right)\right] + \frac{1}{\Delta}\left[4e^{3/2 S_E} - 1\right]\right\}$$

where we have taken for simplicity $E(\Gamma_8) - E(\Gamma_7) = \frac{3}{2}\lambda'$; λ' being the reduced spin-orbit constant : $\lambda' = -\lambda e^{-\frac{3}{2}S_E}$. F_{44} is equal to zero because we assume that there is no coupling to T_2 modes of vibrations. Taking $S_E = 3$, $\Delta = 6900\ \mathrm{cm}^{-1}$, $\hbar\omega = 300\ \mathrm{cm}^{-1}$ and $\lambda = -600\ \mathrm{cm}^{-1}$, we get :

$$|F_{11}| = 2|F_{12}| \simeq 10\,000 / \text{unit strain}\ , \quad F_{44} = 0.$$

These very large values explain that no Electron Paramagnetic Resonance spectrum corresponding to g (Γ_7) has been observed : because of the presence of internal strains in the crystals and of the great sensitivity of g (Γ_7) to strains, the E. P. R. spectrum should be very broad and as it is impossible to have a doping of isolated copper greater than 40 ppm (Suslina and Fedorov, 1974) in ZnS crystals, the E. P. R. spectrum of Cu^{2+} is inobservable.

Acknowledgments. We are grateful to Dr. Maier, Professor Scherz and Dr. Wöhlecke for making available their results prior to publication.

4. References.

Abou-Ghantous, M., Bates, C. A., Chandler, P. E. and Stevens, K. W. H., 1974, J. Phys. C 7, 309

Bates, C. A. and Chandler, P. E., 1971 a, Proceedings of the Sixteenth Congress AMPERE (Publishing house of the Academy of the Socialist Republic of Romania, Bucharest), p. 249

Bates, C. A. and Chandler, P. E., 1971 b, J. Phys. C 4, 2713

Black, T. D. and Donoho, P. L., 1968, Phys. Rev. 170, 462

Broser, I. and Maier, H., 1966, J. Phys. Soc. Japan 21, suppl., 254

Broser, I., Scherz, U. and Wöhlecke, M., 1970, J. Lumin. 1, 2, 39

Clerjaud, B. and Gelineau, A., 1974, Phys. Rev. B 9, 2832

Ham, F. S., 1965, Phys. Rev. 138, A 1727

Ham, F. S., 1968, Phys. Rev. 166, 307

Ham, F. S., 1972, Electron Paramagnetic Resonance, ed. by S. Geschwind (Plenum, New York) p. 1

Holton, W. C., de Wit, M., Watts, R. K., Estle, T. L. and Schneider, J., 1969, J. Phys. Chem. Solids 30, 963

Maier, H., 1972, Abstracts of the Intern. Conf. on Luminescence, Leningrad (unpublished), p. 108, and private communication

Maier, H. and Scherz, U., 1974, Phys. Stat. Sol. (b), 62, 153

Nye, J. F., 1957, Physical Properties of Crystals (Clarendon, Oxford)

Sauer, U., Scherz, U. and Maier, H., 1974, Phys. Stat. Sol. (b), 62, K 71

Sturge, M. D., 1967, Solid State Physics, vol. 20 (Academic, New York), p. 91

Suslina, L. G. and Fedorov, D. L., 1974, Phys. Stat. Sol. (a), 21, 389

Tucker, E. B., 1966, Physical Acoustics, vol. IV A, ed. by W. P. Mason (Academic, New York) p. 47

Tucker, J. W. and Rampton, V. W., 1972, Microwave Ultrasonics in Solid State Physics (North-Holland, Amsterdam)

de Wit, M., 1969, Phys. Rev. 177, 441

Wöhlecke, M., 1974, (to be published)

Yamaguchi, T. and Kamimura, H., 1972, J. Phys. Soc. Japan 33, 953

THE MULTIMODE MODEL OF THE JAHN-TELLER EFFECT IN AN ORBITAL DOUBLET

M. Abou-Ghantous[x], C.A. Bates[xx], I.A. Clark[xx], J.R. Fletcher[xx], P.C.Jaussaud[x] and W.S.Moore[xx]

x Service B.T., Centre d'Etudes Nucléaires de Grenoble, BP 85 Centre de Tri, 38041 Grenoble-Cedex France.

xx Department of Physics, University of Nottingham, University Park, Nottingham NG7 2RD, England.

Abstract. Experimental resonance data from Ni^{3+} : Al_2O_3 are explained with a Jahn-Teller model for Ni^{3+} ion coupled to the whole lattice. Ionic parameters are deduced from the results such as the optical excited levels ..etc.

Ni^{3+} ion in Al_2O_3 has always been thought to be a very simple ion. From the acoustic loss experiments of Sturge et al.(1967) and E.P.R. experiments by Geschwind and Remeika (1962), it was believed that the Ni^{3+} ion was undergoing a very pronounced static Jahn-Teller effect. These authors tried to explain their results by involving a large anharmonic term in the energy of lattice vibrations.

We would like to report on some acoustic paramagnetic resonance (APR) experiments, as well as some electric-field-induced EPR (EF-EPR) experiments. These experiments must be interpreted in terms of a dynamic Jahn-Teller effect of intermediate strengh.

Fig.(1) shows the spectrum obtained both in APR and EF-EPR. As shown on the figure, the spectrum can be interpreted as the sum of three main lines. The dependance of the features A, B, C, D with the angle χ between the c-axis of the sample and magnetic field H is shown on Fig.2. Experiments suggest that the site symmetry of Ni^{3+} ions is C_{3v}.

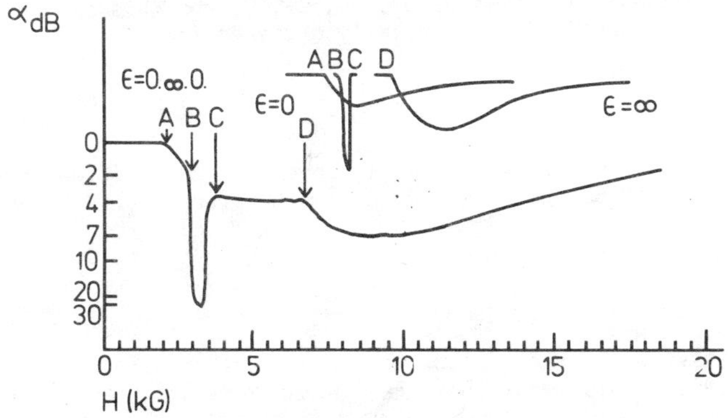

FIG 1

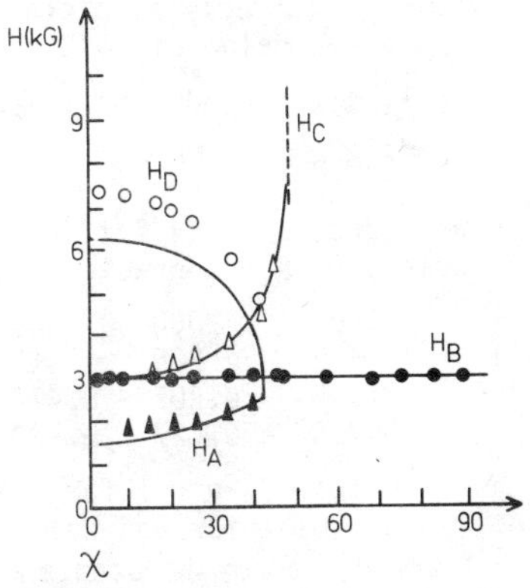

FIG 2

The $(3d)^7$ Ni^{3+} ion is known to be in the strong field region. Its ground state in C_{3v} symmetry is a ^2E state. Its properties may then be described by the following hamiltonian (Abou-Ghantous et al.1974) :

$$\mathcal{H} = apT_3S_Z + (2 + \Delta g)\beta \underset{\sim}{H}.\underset{\sim}{S} + qk\beta(T_+O_- + T_-O_+) + pf\beta T_3H_Z + j\beta H_ZS_Z + 1/2qV_E(Q_+T_- + Q_-T_+) \quad (1)$$

with $|ap| = 0.94 \pm 0.01\ cm^{-1}$ $\Delta g = 0.150 \pm 0.005$

$|fp| = 0.02 \pm 0.005$ (a and f also have the same sign).

In zero magnetic field, we are left with the first and last terms only. The last term describes the effect of strains and will be ignored. The first gives a splitting of the two Kramers doublets between which one should be able to excite directly the EF-EPR transitions at the correct frequency. This experiment has now been done and preliminary results indicate the presence of a strong microwave absorption from 0.47 cm^{-1} ($= \frac{1}{2}$ | ap|) to higher energies.

We must now relate the phenomenological parameters introduced in (1) and deduced by experiment with a physical model for the system. We thus proceed as follows.

(i) coefficients a, Δg, k, f, j are calculated in a static crystalline field model by second and third order perturbation theory.

(ii) coefficients for a dynamic lattice can be deduced from (i) by changing

a → ap, f → fp, k → kq, where p, q are Ham's reduction factors due to overlapping of vibrational functions (p reduces A_2-type orbital operators and q reduces E-type orbital operators). This change has already been incorporated in equation (1). We thus get the following equations.

../..

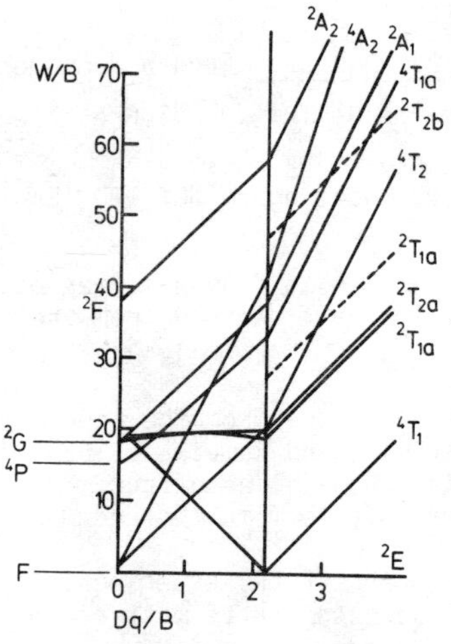

FIG 3

$$|ap| = \left|\left(\frac{4}{3}\zeta^2 \ \Delta_4^{-2} \ v + 2 \ \sqrt{2}\ v' \ \ \zeta \ D_1\right)p\right| = 0.94 \ \ldots (2)$$

$$|fp| = |\ 2\sqrt{2}\ v' \ D_2 p\ | = 0.02 \ \ldots (3)$$

$$\Delta g = 4\ \zeta^2\ \Delta_4^{-2} + 2\ \zeta\ D_3 = 0.150 \ \ldots (4)$$

$$kq = 2\ \zeta\ D_4 q = 0.051 \ \ldots (5)$$

where ζ, v and v' are the spin-orbit coupling constant and the diagonal and off-diagonal trigonal field parameters, and where Δ_4, D_1, D_2, D_3, D_4 are expressions involving the relative energies of the excited states in the cubic crystal field (Fig.3)

We first estimate ζ from Table 1 obtained from values published in Abragam and Bleaney (1970).

TABLE 1

		Ti	V	Cr	Mn	Fe	Co	Ni	Cu
ζ (solutions)	2+	120	168	236	335	404	528	644	829
ζ (Al_2O_3)	3+	159	219	292	380	486			
Ratio (3+/2+)		1.32	1.3	1.2	1.1	1.2			

We thus assume the value for Ni^{3+} to be 770 $\pm$ 80 cm^{-1}

We then assume Dq/B = 2.2 as a starting value. This gives us D_1, D_2, D_3, D_4 and thus Δ_4, which in turn gives a new value of Dq/B etc. Self consistency is readily obtained for

Δ_4 = 5500 cm^{-1}, Δ_{1a} = 24000 cm^{-1}, Δ_{2a} = 25000 cm^{-1}

Δ_{1b}= 34500 cm^{-1}, Δ_{2b} = 54500 cm^{-1}.

From (5), we deduce q = 0.467 ; from (3) v'p = 59 cm^{-1} and from (1) vp = -207 cm^{-1}. If we know v, we can calculate p. A reasonable range for v is 500 cm^{-1} to 1500 cm^{-1}. Therefore :

-v (cm^{-1})	1500	1000	500
v' (cm^{-1})	428	285	142
p	0.138	0.207	0.414
2q-p	0.796	0.727	0.520

We can thus see that 2q-p < 1 for all cases considered. Following various authors (Englman and Halperin, (1973), Gauthier and Walker (1973), this means that the cluster model of the dynamic Jahn-Teller effect cannot be used and that we have to couple the Ni^{3+} ion in the whole lattice. Furthermore, this shows that we are in an intermediate coupling case as in strong or weak coupling, the cluster and lattice models are indistinguishable (Fletcher, 1972).
We would emphasize the fact that from properties of the ground state, we have been able to determine the energy levels of the excited states, which may be used to explain optical spectra.

References

Abou-Ghantous M., Bates C.A., Clark I.A., Fletcher J.R., Jaussaud P.C. and Moore W.S. 1974, J. Phys.C.Sol.State Phys. 7, 2707 and Phys.Rev.Letters, Vol.33, p. 530

Abragam A. and Bleaney B., 1970, Electron Paramagnetic Resonance of transition ions, Clarendon Press, Oxford.

Fletcher J.R., 1970, J.Phys.C : Sol.Stat.Phys.,5, 852

Gauthier N., and Walker M.B., 1973, Phys.Rev.Letters, 31, 1211.

Geschwind S. and Remeika J.P., 1962, J.Appl.Phys.Suppl. to 33, 370.

Halperin B., and Englman R., 1973, Phys.Rev.Letters, 31, 1052.

Sturge M.D., Krause J.I, Gyorgy E.M., Lecraw R.C. and Merritt F.R., 1967, Phys.Rev. 155, 218-24.

SECTION U

Theoretical Aspects

NON-LINEAR EFFECTS IN PARAMAGNETIC RESONANCE

L. Van Gerven and J. Accou
Laboratorium voor Vaste Stof-Fysika en Magnetisme, Universiteit Leuven, Leuven, Belgium.

Abstract New terms are added to Bloch's equation, which take into account changes in magnetization, negligible in strong external static fields H_s but important in low fields. The HF field of frequency ω_0 is supposed to be linear and to have amplitudes, up to values of the order of H_s and of ω_0/γ. This gives rise to a typical non-linear situation, quite different from non-linear effects due to saturation, and hence to Fourier components in the magnetization of higher order in ω_0. The modified Bloch equation is solved numerically for this quite general case. One of the striking results is that 2 or 3 Fourier components are sufficient to describe accurately the behaviour of the spin system, even for $H_1 \underline{\approx} \omega_0/\gamma$. In order to verify the theoretical results experiments on DPPH in low static and high HF fields are made, using a simple magnetic dispersion spectrometer of novel design.

The original Bloch equation (Bloch, 1946), aimed at describing the magnetodynamic behaviour of a paramagnetic spin system, does not provide in many cases an accurate description of this behaviour, e.g.:1. when the relaxation function is not exponential; 2. in the presence of saturation: when $p^2 \not\ll qr$; 3. in low static fields: when $q \not\ll 1$ or $q/\eta \not\ll 1$; 4. in strong HF fields: when $p \not\ll 1$ or $p/\eta \not\ll 1$.
$p \equiv H_1/H_0$; $q \equiv \delta/H_0$; $r \equiv \zeta/H_0$; $\eta \equiv H_s/H_0$; $\bar{H}_s$ is the applied static field; $\bar{h}_1 = 2\bar{H}_1 \cos\omega_0 t$ is the applied HF field; $H_0 \equiv \omega_0/\gamma$; $\delta \equiv 1/\gamma T_2$; $\zeta \equiv 1/\gamma T_1$.
In cases 2. and 4. it is said that the spin system responds in a "non-linear" way to the exciting HF field. In case 3. the system is, strictly speaking, linear, but the treatment given by Bloch, neglecting one of the circular components of $\bar{h}_1$ and assuming relaxation of $\bar{M}$ towards $\bar{M}_s$ is no longer valid, even not for $p \to 0$. The "non-linearities" in cases 2. and 4. are of somewhat different nature. In case 2. the transfer function of the system, χ_{HF}, is not constant - it depends upon H_1 - but no higher harmonics of ω_0 show up in the response, i.e. in the magnetization. In case 4, however, such harmonics appear, as we will see. Although both responses may be called "non-linear", we propose - in order to avoid confusion - to call the response in case 2 <u>non-linear</u> and the response in case 4 <u>polychromatic</u>. Both effects can, of course, occur at the same time.
A thermodynamic, statistical treatment provides, in many cases, solutions for the problems raised in these "non-ideal" non-linear circumstances. We want, however, to try to solve these problems in a rather magnetodynamic way. Let us assume, first of all, an exponential relaxation function. The linearly polarized HF field $\bar{h}_1$ is applied in the x-direction, perpendicular to $\bar{H}_s$, which is in the z-direction. The case $(q, q/\eta \not\ll 1)-(p, p/\eta \ll 1)$ has been investigated before (see e.g. Van Gerven, 1963).
Let us consider now the more general case $(q, q/\eta \not\ll 1)-(p, p/\eta \not\ll 1)$; of course, we have also $p^2 \not\ll qr$. No doubt, the response $\bar{M} = M_x\bar{u}_x + M_y\bar{u}_y + M_z\bar{u}_z$ of the spin system to $\bar{h}_1 = h_1\bar{u}_x$ will be polychromatic and has to be described by a Fourier expansion:

$$M_x = \Sigma_{i=0}^{N}[A_{2i+1}\cos(2i+1)\omega_0 t + B_{2i+1}\sin(2i+1)\omega_0 t] \qquad (1,a)$$

$$M_y = \Sigma_{i=0}^{N}[C_{2i+1}\cos(2i+1)\omega_0 t + D_{2i+1}\sin(2i+1)\omega_0 t] \qquad (1,b)$$

$$M_z = \Sigma_{i=0}^{N}[E_{2i+2}\cos(2i+2)\omega_0 t + F_{2i+2}\sin(2i+2)\omega_0 t] + M_s \qquad (1,c)$$

where: M_s = the static magnetization, N = the number of harmonics of ω_0. The absence of even, viz. odd harmonics is explained in (Accou, 1974).
In almost all experimental conditions, one is interested in, and measures only the x-response to the x-excitation, i.e. the xx-component $\tilde{\chi}_{xx} = \chi' - j\chi''$ of the transfer tensor $\tilde{\chi} = \bar{M}/\bar{h}_1$. In particular, as explained below, we prefer to measure and to study the dispersion term χ', which is given by

$$\chi' = LF[\Sigma_{i=0}^{N} A_{2i+1}\cos(2i+1)\omega_0 t/2H_1\cos\omega_0 t] \qquad (2)$$

because of the LF-only response of any spectrometer. Solving eq.(2) yields:

$$\chi' = (A_1 - A_3 + A_5 - A_7 + \dots \pm A_N)/2H_1 \qquad (3)$$

and not: $\chi' = A_1/2H_1$, as claimed by all previous authors in the field, even those studying higher order harmonics response and using eq.(1) (e.g. Runge, 1962; Arimondo, 1967; Accou et al., 1974).
In a first attempt to describe the general magnetodynamic behaviour of a spin system in arbitrary fields, we use the "twice" modified, generalized Bloch equation, proposed by Van Gerven and Talpe (1967):

$$\frac{d\bar{M}}{dt} = \gamma(\bar{M}\times\bar{H}_a) + \chi_s(1-F)d\bar{H}_a/dt + (1/T_1)[\chi_s\bar{H}_a - (\bar{M}.\bar{H}_a/H_a^2)\bar{H}_a] - (1/T_2)[\bar{M} - (\bar{M}.\bar{H}_a/H_a^2)\bar{H}_a] \quad (4)$$

where $\bar{H}_a \equiv \bar{H}_s + \bar{h}_1$ = the actual (effective) field; χ_s = the static susceptibility; $F \equiv (\chi_s - \chi_{ad})/\chi_s$.

It should be pointed out that the modified relaxation terms are constructed in such a way - by replacing $\bar{H}_s$ by $\bar{H}_a$ - that, not only, $\bar{M}$ relaxes towards $\bar{M}_a$, the actual (effective) equilibrium magnetization, but that, moreover, the longitudinal relaxation time T_1 "acts", not on the z-

component of $\bar{M}$ but on the a-component (the "actual" component) of $\bar{M}$. In this way we take into account one of the most fundamental remarks of Redfield's theory (Redfield, 1955), namely that the magnetization component M_1 along $\bar{H}_a(\equiv\bar{H}_{eff})$ has to relax with T_1 and not with T_2, even when $\bar{H}_a = \bar{h}_1$ and $M_1 = M_x$. The second term on the right hand side of eq. (4) is an "adiabatic" or "rotational" term explained in (Van Gerven et al., 1967).
Eq. (4) - with eq. (1) - cannot be solved analytically of course. In favorable cases a solution in the form of continued fractions is possible (Accou, 1974). We performed numerical solutions on an IBM 370 computer (using the double precision subroutine DGELB) for a large number of different values of the parameters p, q and r, while F is taken either 0 (zero and low field-case) or 1 (high field-case) and N=0, 1, 2 or 5. χ', as defined in eq. (3), is calculated as a function of $\eta\equiv H_s/H_o$:

1. F=0: we find $\chi' = \chi_s$ for all values of all parameters, whatever the value of N, and at all H_s. This result is in agreement with the Casimir-Dupré relation $\chi' = F\, f(\omega_o, T_1)\chi_s + (1-F)\chi_s$.
2. F=1: a typical set of results is given in table 1, for p=1.110, q=1.235, r=0.0309, the last two being typical values for ESR in DPPH at 3 MHz and 4 K.

Table 1: χ'/χ_s

η \ N	0	1	2	5	η \ N	0	1	2	5
0.0000	0.00100	0.00100	0.00100	0.00100	3.7350	0.6738	0.7122	0.7103	0.7104
0.9338	0.03574	0.04592	0.05041	0.05224	5.6026	0.9100	0.9355	0.9362	0.9361
1.8675	0.17014	0.17967	0.18192	0.18223	7.4701	0.9716	0.9853	0.9857	0.9857
2.8013	0.42421	0.44224	0.44213	0.44224	9.3377	0.9896	0.9980	0.9982	0.9982

It is clear that the contribution of the first harmonic $3\omega_o$ to $\bar{M}$ is important, but that higher harmonics are not needed at all (except around η=1) to get a reasonably accurate solution, even not for quite high values for p and q.
In order to verify our theoretical results we measured $\chi'(\eta)$ of the electron spin system in DPPH.$(CHCl_3)_{1/3}$ by means of a spectrometer of own design, consisting essentially of an LC-autooscillator, whose frequency shift is measured and plotted accurately as a function of H_s (Accou et al., 1973): $\chi' = -2\Delta\omega_o/\xi\omega_o$. The filling factor ξ of the coil is determined experimentally, using a superconducting filling of exactly the same size as the DPPH sample, so that direct absolute measurements of χ' are possible. One single DPPH sample has been used. Measurements were carried out from 2 to 12 MHz (to vary H_0) and from 300 to 2 K (for sensitivity reasons and to vary δ). We measured the static susceptibility χ_s of the electron spin system in our sample by means of the selective resonance dispersion method described in (Talpe and Van Gerven, 1966).
Let us discuss here only the H_s=0 results: F=0. Our experiments yield $\chi'(H_s=0)\cong 0$, in striking disagreement with theoretical result no 1. of the numerical calculations, mentioned above, and with the Casimir-Dupré relation. This is most probably due to the fact that neither in the original Bloch equation, nor in the modified Bloch equations, nor in the Casimir-Dupré relation, spin-spin interaction is taken into account. That the once modified Bloch equation (which is identical to the twice modified equation for F=1) gives a theoretical result in agreement with our experiments (at least for H_s = 0) is likely to be considered as purely accidental. Hence, the conclusions of most previous studies about (once) modified Bloch equations (Van Gerven, 1963; Garstens et al., 1955; and many others) should be seriously reconsidered.
In order to solve this problem, theoretical work, introducing internal fields not only in the active term of the Bloch equation (as done by Van Gerven and Talpe, 1967), but in the relaxation terms as well, is in progress. Preliminary solutions of this so called "thrice modified" Bloch equation yield $\chi'(F=0, H_s=0)=\chi_s/(1+\omega_o^2T_2^2)\cong 0$, in agreement with our experimental results. The authors gratefully acknowledge financial support of the Belgian "Interuniversitair Instituut voor Kernwetenschappen" to this project.

References

Accou, J., Van Hecke, P. and Van Gerven, L., 1973, Magnetic Resonance and Related Phenomena, Proc. XVIIth Congress AMPERE, North Holland Publ. Co, Amsterdam, 379.
Accou, J. and Van Gerven, L., 1974, to be published in Proc. ICM-73, Nauka, Moscow.
Accou, J., 1974, Dr. Sc. Thesis (unpublished), Leuven.
Arimondo, E., 1967, Nuovo Cimento X 52B, 379.
Bloch, F., 1946, Phys. Rev. 70, 460.
Garstens, M.A. and Kaplan, J.I., 1955, Phys. Rev. 99, 459.
Redfield, A.G., 1955, Phys. Rev. 98, 1787.
Runge, R.J., 1962, J. Math. Phys. 3, 1267.
Talpe, J. and Van Gerven, L., 1966, Phys. Rev. 145, 718.
Van Gerven, L., 1963, Lijnvormen in Paramagnetische Resonantie, Monogr. 11, I.I.K.W., Brussels.
Van Gerven, L. and Talpe, J., 1967, Magnetic Resonance and Relaxation, Proc. XIVth Coll. AMPERE, North Holland Publ. Co, Amsterdam, 845.

B NOT H IN MAGNETIC RESONANCE!

E.E.Schneider
School of Physics, University of Newcastle upon Tyne.

Abstract. Current-like point sources interacting with B are the only models of spin magnetic moments consistent with their electrodynamic origin in relativistic quantum mechanics. The field truly inside a distribution of such point magnets, a concept of wider usefulness, yields the correct 'contact' term of spin-spin interaction.

1. Introduction. Backed by the axiomatic relationship between electrodynamics and relativity* the rational approach to, or interpretation of, current magnetism as an interaction between moving charges is now well established and the field B, alias flux density, alias induction, generally recognized⌿ as the physically relevant 'intensive' quantity. The treatment of magnetic matter, however, is still haunted by poles and inverse square laws, particularly those aspects associated with electron or nuclear spin. The following is an attempt to lay this old ghost - albeit by the new magic of relativistic quantum mechanics.

2. Spin and Fine Structure. As far as the fine structure-split energy levels of hydrogenic systems are concerned the Dirac theory of 1928 (Di 28, Da 28) and Sommerfeld's relativistic theory of precessing orbits, presented in his monumental paper of 1916 (So 16), give identical results: $E(n,k) = E_0 - RZ^2/n^2 + \alpha^2 RZ^4/n^3(3/4n - 1/k)$. The subtle difference lies in the degeneracy of the $k < n$ levels (lifted by the Lamb shift) which appears in Dirac and has no counterpart in the old theory. We know now that this degeneracy, symptomatic for the theory of fermions, can be interpreted in terms of spin S and that k is not L + 1 but equal to J + 1/2 where $J = L \pm S$.

Indeed Dirac's theory of the electron represents the formal canonisation of Uhlenbeck and Goudsmit's "spinning electron" (UG 25,26) and its acceptance into a partnership of spin and non-relativistic quantum mechanics which is still the only practicable scheme for dealing with atomic and molecular phenomena beyond the few simple cases which can be handled with the rigorous Dirac formalism. In one of the earliest efforts on these lines Heisenberg and Jordan (HJ 26) established the salient features of the anomalous Zeeman effect and, using the semi-classical form of the spin-orbit interaction with the relativistic Thomas correction (Th 26, Fr 26) for the $\underline{S}\cdot\underline{L}$ term, obtained the energies of the hydrogenic system directly in the form given above.

3. Spin a Relativistic Effect. To explain fine structure Uhlenbeck and Goudsmit (UG26) appealed to the model of an electron magnet precessing around the field which it sees by virtue of its own motion in the orbit. In the old theory the precession of the perihelion of elliptical orbits is the significant effect of relativity responsible for fine structure. Thus if spin precession is presumed to occur rather than perihelion precession it must be spin which is the significant effect of relativity!

It is just this which Dirac demonstrated, less naively, as a byproduct of his theory. He showed that the four-component linear wave equations, fulfilling all the requirements of relativistic invariance and four-dimensional symmetry and containing apart from the electromagnetic potentials only the charge and rest mass, can be transformed into a second order equation for the two large components which has exactly the form of the spin-modified non-relativistic quantum-mechanical one mentioned in §2. That is to say there are terms additional to those of the non-relativistic Hamiltonian which have the same effect as, or are equivalent to, or can be interpreted as the interactions of a magnetic moment of one Bohr magnetom with an external magnetic field, the $\underline{S}\cdot\underline{B}$ term, and with an internal field arising from the orbital motion, the spin orbit coupling $\underline{S}\cdot\underline{L}$ ready-made with Thomas correction. We can say now with some confidence: the intrinsic magnetic moment of the electron is a relativistic quantum effect associated with the electronic charge. The natural model, the only model consonant with this strictly electrodynamic origin, is therefore a solenoidal point source, topologically equivalent to the limiting case of a current loop. Essentially, the intrinsic moment is an artefact, a concession to our limited conceptual and mathematical powers to deal with four-dimensional spacetime. It has its parallel in magnetism itself when, as an aspect of the axiomatic relationship between relativity and electrodynamics touched on in the beginning, the interaction between moving charges is obtained as a second order relativistic effect (Ro68, Ch.2-3) and the solenoidal field B has to be introduced to allow a sane description in the laboratory system of what the observer riding on the charges experiences in terms of Coulomb's law and irrotational source-sink electric fields.

circular current

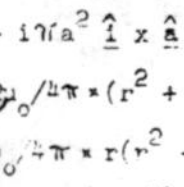

extended dipole

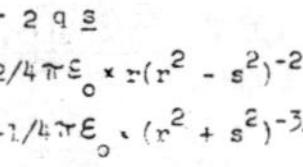

	circular current	extended dipole
	$\underline{m} = -i\pi a^2\, \hat{\underline{i}} \times \hat{\underline{a}}$	$\underline{p} = -2q\underline{s}$
axial	$\underline{B} = 2\mu_0/4\pi \times (r^2 + a^2)^{-3/2}\, \underline{m}$	$\underline{E} = 2/4\pi\varepsilon_0 \times r(r^2 - s^2)^{-2}\, \underline{p}$
equatorial	$\underline{B} = -\mu_0/4\pi \times r(r^2 - s^2)^{-2}\, \underline{m}$	$\underline{E} = -1/4\pi\varepsilon_0 \times (r^2 + s^2)^{-3/2}\, \underline{p}$
centre	$\underline{B} = 2\mu_0/4\pi \times a^{-3}\, \underline{m}$	$\underline{E} = -1/4\pi\varepsilon_0 \times s^{-3}\, \underline{p}$
Point S's	$\underline{m} = -\pi\, \hat{i} \times \hat{a} \lim_{a \to 0} (i\, a^2)$	$\underline{p} = -2\hat{s} \lim_{s \to 0} (q\, s)$
external	$\underline{B} = \mu_0/4\pi \times (-\underline{m} + 3(\underline{m}\cdot\hat{\underline{r}})\hat{\underline{r}})/r^3$	$\underline{E} = 1/4\pi\varepsilon_0 \times (-\underline{p} + 3(\underline{p}\cdot\hat{\underline{r}})\hat{\underline{r}})/r^3$
	$\underline{B} = \mu_0/4\pi \times (\nabla\nabla\cdot - \nabla\cdot\nabla)\underline{m}/r$	$\underline{E} = 1/4\pi\varepsilon_0 \times \nabla\nabla\cdot \underline{p}/r$

* The monograph of Rosser (Ro 68) provides an exhaustive discussion.
⌿ even if it is a case of "thinking B and writing H", see my preface in Ko 58.

4. Spin Without Spinning and the Spin Hamiltonian. As foreshadowed by Heisenberg's notion of an "unmechanischer Zwang" (non-mechanical constraint BL 24) the spin or intrinsic angular momentum of the electron established in the same "as if" sense from the Dirac equations, is an irreducible quantum effect and has no classical model. Any idea of actual rotations is certainly in blatant conflict with relativity. It is therefore not surprising that the ratio of magnetic moment to angular momentum for spin is twice the classical value for orbital motion which is unaffected by non-relativistic quantization*. This leads then to the general magnetic moment operator $\beta(\underline{L} + 2\underline{S})$ in the spin-modified quantum mechanical scheme and with a suitable $\underline{S}\cdot\underline{L}$ term eventually to the spin Hamiltonian in terms of which spin resonance is interpreted. In this a g-tensor represents the effective relationship between spin magnetic moment and spin angular momentum.

5. Spin-Spin Interaction. The theory of hyperfine interaction leads to a useful extension of the model of current-like point magnets. Apart from higher order corrections the Dirac theory and spin modified non-relativistic quantum mechanics (Sp + n.r. Q.M.) lead for non-S states to the same Hamiltonian. The spin-spin part of this can be interpreted as an interaction of the nuclear spin I with the magnetic field B_I arising from the distribution of the intrinsic moment $2\beta\underline{S}$ over the wave function, equivalent to a continuous magnetization $\underline{M}(\underline{r}) = 2\beta\psi^2(\underline{r})\underline{S}$.

If we maintain this interpretation for S-states with their spherically symmetrical wave functions**, we need only consider the field $\underline{B}_I = 2/3\,\mu_0 \underline{M}(0)$ at the centre of a sphere sufficiently small for the wave function to be constant and the magnetization to be uniform (the field inside uniformly magnetized spherical shells vanishes) and obtain the correct 'contact' term first derived by Fermi (FS 33) from the Dirac theory. This lends further support to our model of solenoidal point sources≠. The same holds for the mathematical way of dealing with the singular internal field arising for S-states by using the differential operator for the field of a solenoidal source instead of the multiplicative one (e.g. Si 66). This amounts in effect to using $\underline{p}\cdot\underline{A}/m$ as the spin-spin term in place of $2\beta\,\underline{S}\cdot\underline{B}_I$.

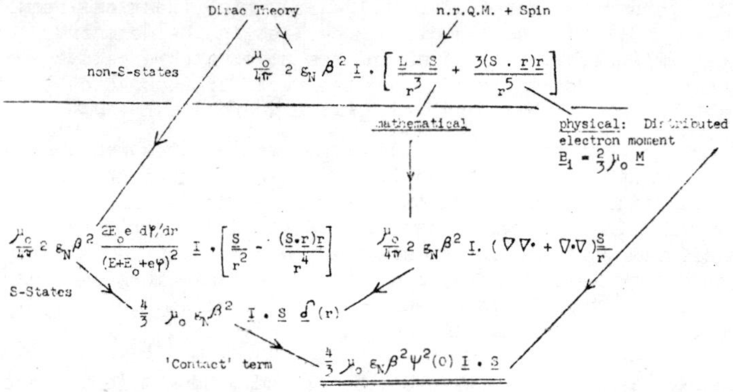

6. Electric and Magnetic Worlds. The concept of a true internal field, the field truly inside magnetized matter, interpenetrated by a probing magnet (in HFS the nuclear one) should have wider applications, at least as a guide, to other problems in magnetism, particularly to phenomena at domain or crystal boundaries. To describe these in terms of free poles is not only in flagrant conflict with the intellectually so satisfying conception of our world as a coherent electrical one but confuses the issue of the existence elsewhere of actual magnetic monopoles.

It was, ironically, Dirac (Di 31) who envisaged the possibility of opposite magnetic charges Φ_0/μ_0 = 20.4 GeV/Tm as the dangling ends of broken flux loops of one flux quantum $\Phi_0 = h/e$. The search for these monopoles or for theoretical grounds inhibiting their existence is still going on unabatedly and, so far, unsuccessfully. Their eventual discovery as whiffs of magnetic worlds or remains of the primordial bang will be as fascinating and far reaching in its consequences for high energy physics and cosmology as a fundamental theory explaining why flux rings are unbreakable or magnetic charges otherwise unobservable.

For us the enjoyment and excitement of watching these momentous developments from the sidelines will be all the more intense if in our mundane occupation of researching and teaching down-to-earth magnetism and magnetic resonance we keep conscientiously within the confines of electrodynamics.

References. BL 24 Back & Lande, "Zeeman-Effect und Multiplettstruktur der Spektrallinien" pp.55 ff, 1924. Da 28 Darwin, Proc.Roy.Soc.A118, 654, 1928. Di 28 Dirac, Proc.Roy.Soc.A117, 610; A118, 351, 1928. Di 31 Dirac, Proc.Roy.Soc. A133, 60, 1931. EdH 15,16 Einstein & de Haas, Verhdgn Deutsch.Physik.Ges. 17, 152, 1915; 18, 173, 423, 1916. Fr 26 Frenkel, Zeits.Physik. 37, 243, 1926. FS 33 Fermi & Segre, Zeits.Physik. 82, 729, 1933. HJ 26 Heisenberg & Jordan, Zeits.Physik. 37, 263, 1926. Ko 58 Kopfermann "Nuclear Moments", 1958. Ro 68 Rosser "Classical Electromagnetism via Relativity", 1968. Si 66 Sillescu "Kernmagnetische Resonanz", p.45, 1966. So 16 Sommerfeld, Ann.Physik, 51, 1, 1916. Th 26 Thomas, Nature, 117, 514, 1926. UG 25,26 Uhlenbeck & Goudsmit, Naturwiss. 13, 953, 1925; Nature, 117, 264, 1926.

* It is interesting that it was Einstein in his experimental (!) work with de Haas in 1915 (EdH 15,16) on the rotation by magnetization, an effect proportional to the reciprocal of the g-factor who provided evidence, at the time unrecognised, for this "anomalous ratio". The work is also remarkable for the primitive form of resonance lock-in used for detection.

** This was first mentioned as one of my additions to Ko 58, page 121.

≠ The source sink dipole would lead to a B_I half as large and opposite in sign.

SECTION V

Heterogeneous Systems

THE ROLE OF CRYSTAL SURFACES AND OTHER IMPERFECTIONS IN THE NMR OF AgBr PRECIPITATES

L. G. Conti and F. Di Piro
Laboratorio di Metodologie Avanzate Inorganiche del C.N.R., Istituto di Chimica Generale e Inorganica, Università di Roma, 00185 Roma, Italy

Abstract. The method of detection of surface effects in NMR proposed by O'Reilly (1960) has been applied to AgBr precipitates previously studied by NMR (Conti and d'Alessandro, 1971a). Alternatively, a new method has been developed which takes into account Ostwald ripening and relaxation of dislocations both in the precipitate and in the recrystallizate.

1. Introduction. An attempt to discover motional narrowing of the ^{81}Br line at a temperature of 18°C in AgBr freshly precipitated from aqueous solutions led to an unsuccessful result (Conti and d'Alessandro, 1971a). Since the first AgBr sample examined in the Pound-Knight-Watkins spectrometer was 10 min old, the simplest explanation is that complete annealing of excess Br^- vacancies occurs in a time less than 10 min. If one extrapolates the data of Tannhauser (1958), who relies also on previous work of Kurnick, then it can be easily shown that, for AgBr microcrystals whose size is about 5×10^{-5} cm, practically complete annealing by diffusion to the surface at room temperature occurs in a time well below 1 min, if a vacancy mechanism is postulated for the diffusion of Br^- ions. In the same experiments, however, Conti and d'Alessandro found an ageing effect, namely, an increase of the ^{81}Br peak-to-peak intensity $\hat{S}$, normalised to unit mass and relative to a reference sample, with the time of ageing in mother solution (Fig. 2). We wish to discuss the causes of this effect.

2. Crystal surfaces. A very detailed study of fine powders of NaCl has been carried out by Hughes et al. (1967). A simpler method is that of O'Reilly (1960) which is based on a linear relationship between integrated intensities $\mathcal{J}$ and specific surface areas. In Fig. 1 the ^{81}Br peak-to-peak intensities $\hat{S}$ determined by Conti and d'Alessandro (1971a) are plotted against specific surface area values as determined by electron microscopy after taking into account Ostwald ripening. The slope of the straight line drawn through the experimental points is 0.13, which means that $\sim$70 layers of Br^- ions of a $\{100\}$ face, corresponding to a depth of $\sim$200 Å, do not contribute to the measured intensity because of 2nd order quadrupole interactions (in AgBr only the $1/2 \leftrightarrow -1/2$ line is observed). Such values seem exceedingly high in the light of the findings of Hughes et al. (1967) on NaCl. Further experiments and independent calculations on AgBr would be worthwhile.

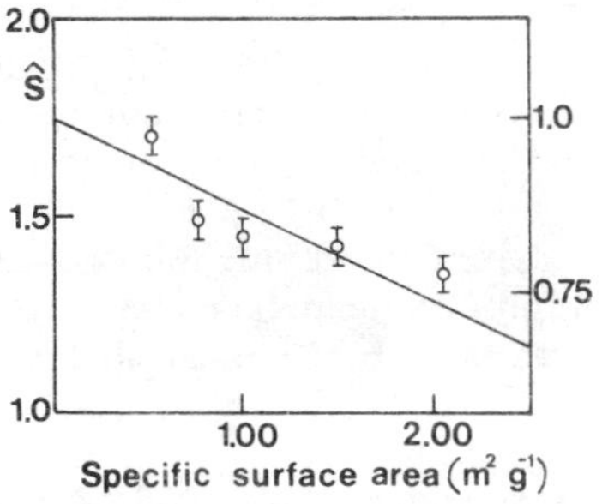

Fig. 1. Ageing of AgBr in mother solution. ♁ expl. data. See text.

3. Ostwald ripening and dislocations. Let's assume equal ^{81}Br integrated intensities for the pure precipitate before recrystallization and for the pure recrystallizate. The last one is formed through Ostwald ripening. Then, assuming gaussian lines, the following equation can easily be derived

$$\hat{S} = \left(1 - \frac{V_o}{V_o+Kt}\right)^2 (\hat{S}_r^{1/2} - \hat{S}_o^{1/2})^2 + 2\left(1 - \frac{V_o}{V_o+Kt}\right)(\hat{S}_r^{1/2}\hat{S}_o^{1/2} - \hat{S}_o) + \hat{S}_o \quad (1)$$

where V_o is the initial AgBr microcrystal volume (6.4×10^{-14} cm^3) and K is the growth constant (3.4×10^{-15} cm^3 min^{-1}). According to Eq. (1), the peak-to-peak intensity $\hat{S}$ approaches asymptotically $\hat{S}_r$ with time, starting from the initial value $\hat{S}_o$. An obvious semplification of Eq. (1) would be to substitute the maximum value of the intensity, $\hat{S}_{dip} = B/\Delta\nu^2_{dip} = 3.8$ (Conti and d'Alessandro, 1971a), for the $\hat{S}_r$ value ($\Delta\nu$ is the peak-to-peak line width) and, for $\hat{S}_o$, the value 1.4, taken from Fig. 2, or a lower one. This would mean that the recrystallizate is almost free of dislocations soon after recrystallization. The results are given in Fig. 2 by the solid curves 1 & 2. As can be seen a different approach must be sought. In particular, Eq. (1) should be modified by considering the relaxation of dislocations both in the original precipitate and in the recrystallizate.

To this end, following Bloembergen (1955) and Otsuka and Kawamura (1957), the following relationship can be written for the recrystallizate

$$(\langle\Delta\nu^2\rangle_r - \langle\Delta\nu^2\rangle_{dip})\langle\Delta\nu^2\rangle_r^{-1/2} = Ac = He^{-t/\tau} \quad (2)$$

where A and H are constants related to second order broadening for a given ratio of edge and screw dislocations, c is the total dislocation density, and τ is the relaxation time. A simple exponential decay has been assumed. Since approximately $\Delta\nu = 2\langle\Delta\nu^2\rangle^{1/2}$, and B = 4.0, the above expression becomes

$$(1/\hat{S}_r - 1/\hat{S}_{dip})\hat{S}_r^{1/2} = He^{-t/\tau} \qquad (3)$$

from which

$$\hat{S}_r^{1/2} = \left[(H^2e^{-2t/\tau} + 4/\hat{S}_{dip})^{1/2} - He^{-t/\tau}\right]\hat{S}_{dip}/2, \qquad (4)$$

$$\hat{S}_r(t=0) = \left[2H^2 + 4/\hat{S}_{dip} - 2H(H^2+4/\hat{S}_{dip})^{1/2}\right]\frac{\hat{S}^2_{dip}}{4} \qquad (5)$$

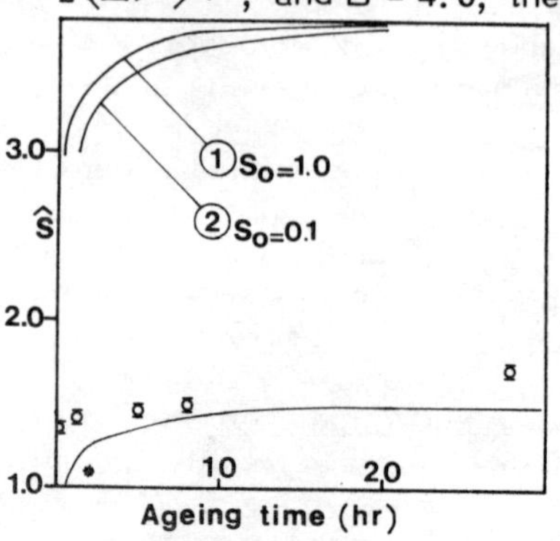

Fig. 2. Ageing of AgBr in mother solution. ▫ expl. data. ——— Eq.(1) See text.

Equations (2 – 5) are valid also for the original precipitate, not considering Ostwald ripening. In such a case we have the constants H_1 and τ_1. If the diffusion data obtained in this la boratory (Conti and Cammarata, 1974) for Br^- ions in fresh AgBr at 25°C are utilized, the ratio H/H_1 can be approxima tely determined. This is found to be 9.5×10^{-2}.
The final three-parameter equation for the ageing of AgBr in mother solution (^{81}Br resonance) is

$$\hat{S}_{nt_0} = \left[2H_1^2 + 4/\hat{S}_{dip} - 2H_1(H_1^2+4/\hat{S}_{dip})^{1/2}\right]\hat{S}^2_{dip}/4 + \sum_{p=n\;q=0}^{p=1\;q=n-1}\left\{\left(\left[(H^2e^{-2pt_0/\tau} + \frac{4}{\hat{S}_{dip}})^{1/2} - He^{-pt_0/\tau}\right]\frac{\hat{S}_{dip}}{2}\right.\right.$$
$$\left. - \left[(H_1^2e^{-2(q+1)t_0/\tau_1} + \frac{4}{\hat{S}_{dip}})^{1/2} - H_1e^{-(q+1)t_0/\tau_1}\right]\frac{\hat{S}_{dip}}{2}\right)^2\left[(1 - \frac{V_0}{V_0+K(q+1)t_0})^2 - (1 - \frac{V_0}{V_0+Kqt_0})^2\right]$$
$$+ 2(\frac{V_0}{V_0+Kqt_0} - \frac{V_0}{V_0+K(q+1)t_0})\left(\left[(H^2e^{-2pt_0/\tau} + \frac{4}{\hat{S}_{dip}})^{1/2} - He^{\frac{-pt_0}{\tau}}\right]\left[(H_1^2e^{-2(q+1)\frac{t_0}{\tau_1}} + \frac{4}{\hat{S}_{dip}})^{1/2} - H_1e^{-(q+1)\frac{t_0}{\tau_1}}\right]\frac{\hat{S}^2_{dip}}{4}\right.$$
$$\left.\left. - \left[(H_1^2e^{-2(q+1)t_0/\tau_1} + 4/\hat{S}_{dip})^{1/2} - H_1e^{-(q+1)t_0/\tau_1}\right]^2\hat{S}^2_{dip}/4\right)\right\} \qquad (6)$$

for t_0 small enough. $\hat{S}_{nt_0}$ is the peak-to-peak intensity of the complex system at time nt_0. Extensive calculations have not yet been carried out, but a preliminary investigation has shown that the ageing curve in its initial part should have a shape not very different from that marked by an asterisk in Fig. 2. If this will be confirmed it means that the diffusion data are not indicative of the dislocation concentration, but also of a higher state of dislocation motions in the original precipitate (the ratio H/H_1 should be much nearer to unity). Such motions had been postulated by Conti et al. (1971b) in their study of $PbCrO_4$ precipitates.

4. References

Bloembergen, N., 1955, Defects Cryst. Solids, Rep. Conf., The Physical Society, London.
Conti, L. G., and Cammarata, S., 1974, J. Phys. Chem. Solids, 35, 125.
Conti, L. G., and d'Alessandro, R., 1971a, J. Phys. Chem. Solids, 32, 1092.
Conti, L. G., d'Alessandro, R., and di Napoli, V., 1971b, J. Phys. Chem., 75, 350.
Hughes, D. G., Benson, G. C., and Freeman, P. I., 1967, J. Phys. Chem. Solids, 28, 2305.
O'Reilly, D. E., 1960, Advan. Catal. Relat. Subj., 12, 31.
Otsuka, E., and Kawamura, H., 1957, J. Phys. Soc. Japan, 12, 1071.
Tannhauser, D. S., 1958, J. Phys. Chem. Solids, 5, 224.

NUCLEAR MAGNETIC RESONANCE LINE SHAPES IN LIPID BILAYER SYSTEMS*

M.I. Valic, E. Enga, E.E. Burnell and M. Bloom
The University of British Columbia, Vancouver, B.C., Canada V6T 1W5.

Abstract. The present work represents a straightforward, qualitative interpretation of the existing experimental proton magnetic resonance (PMR) data in both vesicular and lamellar systems of lipid bilayers. The explanation is a result of some previously overlooked basic nuclear magnetic resonance (NMR) properties of having a lipid bilayer structure.

1. Introduction. The current efforts in understanding the details of structure and function of biological membranes have prompted many studies of model membranes. The lamellar, liquid crystalline phase of lecithin in water is an often used model system. The evidence that the fluid phase is a close analogue to real membranes is derived from results obtained with numerous techniques amongst which NMR is particularly useful as it is sensitive to details of both molecular order and molecular motions.

There is a considerable amount of NMR literature on the lecithin/water model system. The PMR spectra of the lamellar phase reveal weak, 'sharp' peaks corresponding to choline, methylene and methyl groups of the lipid molecules superimposed on broad lines. The spectra of ultrasonically irradiated coarse aqueous liquid crystalline phase dispersions also give rise to 'sharp' PMR lines with line shapes and intensities being strongly dependent upon temperature and the vesicle size.

Two theories have been put forward to explain the PMR observations in vesicles. According to Finer (1974) this difference can be explained solely by different vesicle tumbling rates while Sheetz and Chan (1972) argue that there are considerable structural differences in vesicular bilayers relative to those of lamellar phase. Since the vesicular system is simple, well defined, and particularly suitable for transport studies, the resolution of the above controversy is of extreme importance.

In attempting to understand the PMR spectra, we have been led to consider an important feature of the vesicular system which has previously been overlooked. This new feature enables one to understand quantitatively the existing experimental PMR data in vesicular as well as in lamellar systems in a straightforward, unified manner consistent with the basic NMR properties of these systems.

2. Lamellar Systems. The dipolar Hamiltonian which is time dependent because of the molecular motions, can be divided into its time average and fluctuating components

$$H_d(t) = \langle H_d \rangle + \{H_d(t) - \langle H_d \rangle\}. \qquad (1)$$

The above decomposition is useful in the short correlation time limit when $M_2\tau_c^2 \ll 1$ (Abragam, 1960). Under these conditions the experimental line shape is governed by $\langle H_d \rangle$ for all cases in which $\langle H_d \rangle \neq 0$. Usually, the second moment M_2 can be calculated and a measure of $\langle H_d \rangle$ can be obtained by comparing M_2 with the residual (observed) second moment M_{2r} due to $\langle H_d \rangle$.

In principle, the NMR line shape associated with $\langle H_d \rangle$ contains information on the manner of ordering of the lipid chains in the liquid crystalline phase. When the lateral diffusion constant D is large, the effectiveness of the intermolecular dipolar interaction is severely reduced and its contribution to $\langle H_d \rangle$ may be completely neglected. This is certainly true in the fluid phase of lamellar systems where $D \geqslant 10^{-8}$ cm^2/sec. The residual dipolar interactions, due completely to the interactions among nuclear spins on the individual lipids, are projected along the perpendicular $\vec{n}$ to the lamellae so that the average interaction becomes (Bloom, 1973)

$$\langle H_d \rangle_\theta = \langle H_d \rangle_o \cdot \left(\frac{3\cos^2\theta - 1}{2}\right) \equiv \langle H_d \rangle_o \cdot P_2(\cos\theta), \qquad (2)$$

where θ is the angle between $\vec{n}$ and the external magnetic field.

The relationship (2) is now well established experimentally in a number of ways. For example, it predicts that the free induction decay (FID) for a domain oriented at an angle θ to the magnetic field is given by

$$f_\theta(t) = f_o\{P_2(\cos\theta)t\}. \qquad (3)$$

Indeed, our results on oriented samples of soap/water solutions, using soaps of different chain lengths (Burnell et al.), manifest the predicted $P_2(\cos\theta)$ dependence remarkably well. This idea can be further extended to the more frequently studied unoriented samples for which the resulting FID is an average over all possible orientations. A simple calculation shows that $f(t) \propto t^{-1}$ and $\propto \exp(-\frac{1}{2}M_{2r}t^2)$ for long and short times respectively (Bloom, 1973). As seen from Fig. 1, our experiments (Burnell et al.) on the proton FID in 70% potassium palmitate—30% D_2O lamellar phase are in excellent agreement with these predictions.

Eqn. (2) also predicts an absorption line shape which varies as $\log|\omega-\omega_o|$ near the center, fitting very well the often quoted absorption line of Lawson and Flautt (1968). This result has been derived independently by Wennerström (1973). The main feature to note here is the extremely sharp line at the center which represents the divergence of $\log|\omega-\omega_o|$ at ω_o. In the actual experiment the height is of course limited by several factors, such as the distribution of chemical shifts, magnet inhomogeneity, intensity of the rf field, and relaxation processes. When the applied magnetic field becomes sufficiently large to separate the lines of individual groups on the lipid molecule due to their different chemical shifts by an amount greater than the other line width contributions, sharp lines are revealed on top of the broad logarithmic background. It is for this reason that high resolution PMR peaks have been observed (Chapman et al., 1972) in the liquid crystalline phase of lamellar systems. An example is shown in Fig. 2a[+].

3. Vesicular Systems. It is a straightforward matter to extend the above results and ideas to the case of vesicles. The adopted model is as follows (Bloom et al.): (a) For a lipid having a polar angle θ in the vesicle, the effective dipolar interaction gives rise to a line shape

$F_\theta(\omega) = F_0\{\omega|P_2(\cos\theta)|^{-1}\}$. (b) We assume an additional broadening of the line by an amount $1/T_2$, i.e., the PMR line shape of a lipid at θ is given by the convolution $F_\theta(\omega)$ with a Lorentzian line characterized by $1/T_2$ due to the relaxation associated with fluctuations of H_d about $\langle H_d \rangle_\theta$.
(c) The angle θ is assumed to change with time for each lipid in a stochastic manner governed by a correlation time τ_c associated with the rotational diffusion equation. The origin of $\theta(t)$ can be either reorientation of the entire vesicle in the solution (tumbling) or diffusion of the lipids around the surfaces (inner and outer) of the vesicle. It is $\theta(t)$ which leads to the 'motional narrowing' (Abragam, 1960) of the PMR lines. The resultant line shape for the case $M_{2r}\tau_c^2 << 1$ is a superposition of Lorentzians given by

$$F(\omega) = \frac{T_2}{\pi}\int_0^1 d\mu \frac{1 + Q.|P_2(\mu)|^2}{\{1 + Q.|P_2(\mu)|^2\}^2 + \omega^2 T_2^2}, \qquad (4)$$

where $Q = M_{2r}\tau_c T_2$, $\mu = \cos(\theta_0)$ and $1/T_2$ is estimated to correspond to a width of ≈10 Hz in the liquid crystalline phase of the bilayers. The 'narrow lines' mentioned in the introduction have widths of the order of $1/T_2$ while the 'broad lines' predicted for rotational diffusion of large vesicles in aqueous solution are of order Q/T_2. This is schematically depicted in Fig. 3. The main point to note from Eqn. (4) is that even when $Q >> 1$, $F(\omega)$ has a narrow component due to those lipids which are initially near the magic angle. For example, when $Q = 100$ the integrated intensity of the narrow line within $\pm 5/T_2$ of the peak is still about 30% of the total and it will appear as a pip on a broad line as observed experimentally (Fig. 2b†). Thus, it is possible to explain the appearance of 'sharp' lines in vesicles without invoking additional structural disorder (Sheetz and Chan, 1972).
In conclusion, we note that with the interpretation given here, the experimental PMR data can also yield quantitative information on the orientational order and lateral diffusion constants in bilayer systems.

4. References.

Abragam, A., 1960, Principles of Nuclear Magnetism, Oxford University Press, London.
Bloom, M., 1973, Proc. First Specialized "Colloque Ampère", Ed. J.W. Hennel, Krakow, Poland.
Bloom, M., Burnell, E.E., Valic, M.I., and Weeks, G., (to be published).
Burnell, E.E., Valic, M.I., Enga, E., and Bloom, M., (to be published).
Chapman, D., and Chen, S., 1972, Chem. Phys. Lipids, 8, 318.
Finer, E.G., 1974, J. Magn. Resonance, 13, 76.
Lawson, K.D., and Flautt, T.J., 1968, J. Chem. Phys., 72, 2066.
Sheetz, M.P., and Chan, S.I., 1972, Biochemistry, 11, 4573.
Wennerström, H., 1973, Chem. Phys. Letters, 18, 41.

* Research supported by the National Research Council of Canada.
†The authors are very thankful to Professor D. Chapman for permission to reproduce some of his results prior to their publication, and for several stimulating and valuable discussions.

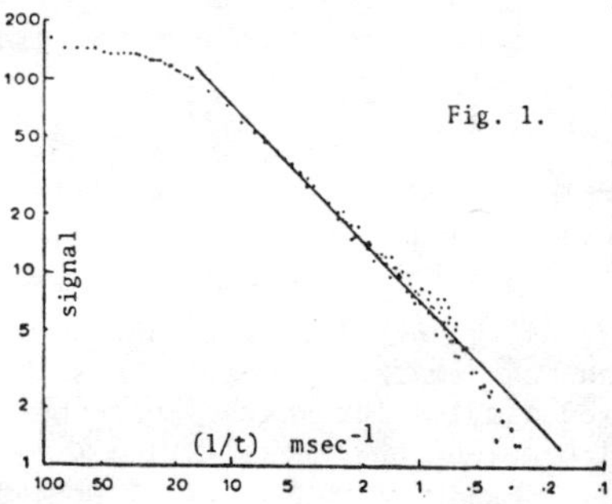

Fig. 1.

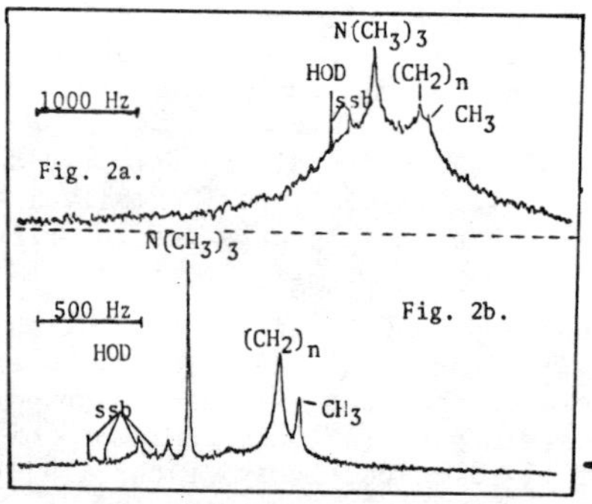

Fig. 2a.
Fig. 2b.

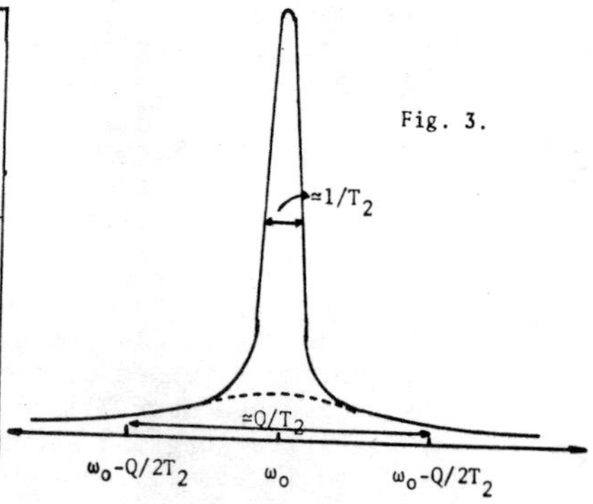

Fig. 3.

Fig. 1. The proton FID at 30 MHz in 70% potassium palmitate—30% D_2O as a function of t^{-1}. The initial part of FID is $\alpha \exp(-\frac{1}{2}M_{2r}t^2)$. The departure from t^{-1} at longest times is due to the differences in chemical shifts of individual groups on the lipid molecule.
Fig. 2. PMR spectra† of dimyristoyl lecithin (DML) in D_2O at 220 MHz and 30°C: (a) 30 wt% DML, unsonicated, (b) 10 wt% DML, sonicated.
Fig. 3. A schematic representation of the line shape predicted by Eqn. (4) for lipid vesicles

SECTION V

Nuclear Polarization

FIRST MOMENT OF THE NMR LINES FOR HIGHLY POLARIZED NUCLEI

Y. Roinel and V. Bouffard
Service de Physique du Solide et de Résonance Magnétique, Centre d'Etudes Nucléaires de Saclay, BP n° 2 - 91190 Gif-sur-Yvette (France)

Abstract. We present a theoretical and experimental study of the first moment of the NMR lines in solids as a function of nuclear polarization. In addition, we have measured the concentration and the relaxation time of paramagnetic impurities, by observing the shift they produce on the NMR lines.

1. Introduction. In solid samples where high nuclear polarizations can be obtained (as a consequence for example of dynamic nuclear polarization) important changes happen to the NMR lines. In particular, since the time average of the dipolar fields created by the spins is proportional to their polarizations, one can observe a shift of the lines with respect to their Larmor frequencies. It is possible to relate these shifts to the magnetizations of the spins by means of the classical magnetostatic formulae, and this has already been used in order to get an intrinsic calibration of the nuclear polarizations (Method of the Lorentz field) [Abragam et al.,1962]. However, this method is not rigorous as it does not take into account two fondamental features :
a) First, it is well known that the effect of "like" spins on their own shift is 3/2 times greater than their contribution to the Lorentz field [Kittel, 1948].
b) Secondly, not only the position but also the shape of the lines varies as a function of the polarizations.

The rigorous treatment must take into account these two effects, and this can be done through the study of the successive moments of the lines [Abragam et al., 1973]. We have studied the variations of the first moments of the NMR lines as a function of nuclear polarizations, and shown that this can provide an intrinsic and accurate calibration of the polarizations.

2. Theory. We consider a sample containing two spin species I and S with polarizations p_I and p_S uniform and much larger than their thermal equilibrium values. The formal expression of the first moment of the absorption line of spins I (in frequency units) is given by [Abragam et al., 1973] :

$$M_{1I} = \frac{\mathrm{Tr}\{\sigma[I_+,[\mathcal{H}'_D,I_-]]\}}{\mathrm{Tr}\{\sigma[I_+,I_-]\}} , \qquad (1)$$

where σ is the density matrix of the system and $\mathcal{H}_D'$ the truncated dipolar Hamiltonian describing the dipolar interactions between all the spins :

$$\mathcal{H}'_D = \frac{1}{2}\sum_{ij} A_{ij}\,[3I_z^iI_z^j - \underset{\sim}{I}^i\underset{\sim}{I}^j] + \frac{1}{2}\sum_{\lambda\mu} B_{\lambda\mu}\,[3S_z^\lambda S_z^\mu - \underset{\sim}{S}^\lambda\underset{\sim}{S}^\mu] + \frac{1}{2}\sum_{i\lambda} C_{i\lambda}\times 2I_z^iS_z^\lambda . \qquad (2)$$

A_{ij} , $B_{\lambda\mu}$ and $C_{i\lambda}$ are the usual dipolar coefficients.

A simple calculation shows that M_{1I} is the sum of two terms :

$$\text{a)}\quad M_{1I}^{II} = -3I\,p_I\sum_{ij}A_{ij}/N_I \qquad (3)$$

$$\text{b)}\quad M_{1I}^{IS} = -2S\,p_S\sum_{i\lambda}C_{i\lambda}/N_I .$$

The dipolar sums $\sum_{ij} A_{ij}/N_I$ and $\sum_{i\lambda} C_{i\lambda}/N_I$ can be calculated analytically in the case of a rectangular sample whose one side ℓ_z is parallel to the magnetic field. For a crystal of cubic symmetry, one finds :

$$\sum_{ij} A_{ij}/N_I^2\gamma_I = \sum_{i\lambda} C_{i\lambda}/N_IN_S\gamma_S = \frac{2\pi}{3}\,\xi\,\hbar\,\gamma_I , \qquad (4)$$

where the dimensionless factor ξ satisfies the conditions :

$$-2 \leq \xi \leq +1 , \qquad (5)$$

and is an analytical function of ℓ_x/ℓ_z and ℓ_y/ℓ_z only which is symmetrical on interchanging these two quantities.

3. Variations of the first moments with nuclear polarizations. Expressed in magnetic field units, the first moments m_{1I} and m_{1S} take the simple form :

$$m_{1I} = \frac{4\pi}{3}\,\xi\,(\frac{3}{2}M_I + M_S)$$

$$m_{1S} = \frac{4\pi}{3}\,\xi\,(M_I + \frac{3}{2}M_S) , \qquad (6)$$

where M_I and M_S are the magnetizations of spins I and S .

In order to check the validity of formulae (6) we have measured the first moments of ^{19}F and ^{7}Li nuclei in a rectangular sample of LiF doped with F centers, where high nuclear polarizations were attained by means of the Solid Effect. The dimensions of the sample were : 0.295×2.32×3.74 mm and its factor ξ was equal to : 0.775 ± 0.005 . During the experiments, the polarizations of the paramagnetic impurities was always equal to unity so that they made no contribution to the variations of the nuclear first moments. The magnetizations of ^{19}F and ^{7}Li nuclei were calibrated against thermal equilibrium values at 4.2 K. Figure 1 shows the experimental values of m_{1F} and m_{1Li}

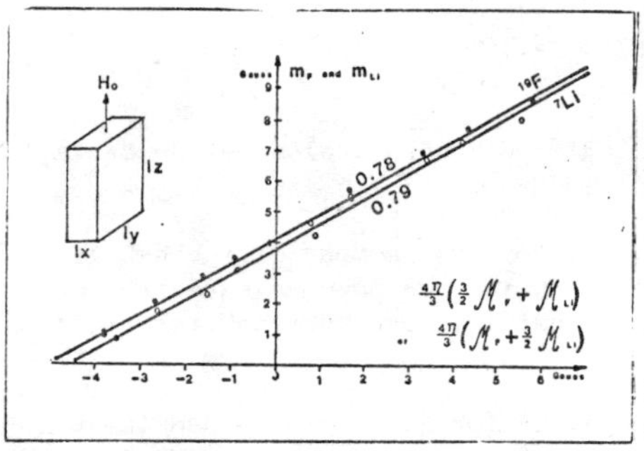

Fig.1 : First moments of ^7Li and ^{19}F absorption lines in LiF, as a function of $\frac{4\pi}{3}(\mathcal{M}_F + \frac{3}{2}\mathcal{M}_{Li})$ and $\frac{4\pi}{3}(\frac{3}{2}\mathcal{M}_F + \mathcal{M}_{Li})$.

as a function respectively of $\frac{4\pi}{3}(\frac{3}{2}\mathcal{M}_F + \mathcal{M}_{Li})$ and $\frac{4\pi}{3}(\mathcal{M}_F + \frac{3}{2}\mathcal{M}_{Li})$. The curves fit straight lines whose slopes are equal to 0.78 ± 0.01, in excellent agreement with theory. Since the first moments are measured with respect to an arbitrary point, the ordinate of the lines have no particular significance, and only the slopes of the curves are meaningfull.

Because of the presence of the "3/2 factor" for like spins, system (6) is non-degenerate and can yield and intrinsic calibration of the magnetizations, thus of the polarizations, once the first moments are determined experimentally. We have found by this method polarizations in agreement with the values determined through the observation of the natural signal at 4.2K, within a few per cent. Similar experiments have been performed on a rectangular sample of CaF_2 doped with paramagnetic Tm^{2+} impurities, for which the ξ factor was of order of -0.94. The results were also in agreement with theory.

In all these experiments, great care must be taken to control any dispersive component of the signal because odd moment measurements are very sensitive to dispersion. It is also necessary that NMR frequencies and magnetic field remain very stable. For that purpose, use has been made of RF quartz generators and a superconducting solenoïd operating in the persistent mode.

4. Indirect observation of the paramagnetic impurities("anti-ENDOR"). The paramagnetic impurities also contribute to the first moment of the nuclear lines. However, if these impurities are dilute enough, their main effect is a small shift of the lines, without appreciable change of their shape. In this case, the Lorentz field approximation can provide satisfactory results. The experimental procedure consists in setting the magnetic field on a wing of the NMR absorption signal $\chi''(H)$, and observing the change of this signal under saturation of the ESR line and its subsequent relaxation towards thermal equilibrium value. As a function of the electronic polarization P, the change $d\chi''$ of the signal is given by :

$$(7) \qquad d\chi'' = \frac{\partial\chi''}{\partial H} dH = \frac{\partial\chi''}{\partial H} \times \frac{4\pi}{3} N_S \hbar \gamma_S S(1-P) \quad ,$$

where S is the electronic spin. $\partial\chi''/\partial H$ is calibrated by applying a known, small shift of the static magnetic field, and formula (7) allows one to determine the concentration of the paramagnetic impurities. By this method, we have found that the concentration of F centers in the sample of LiF was $1.3\times10^{19}cm^{-3} \pm 10\%$.

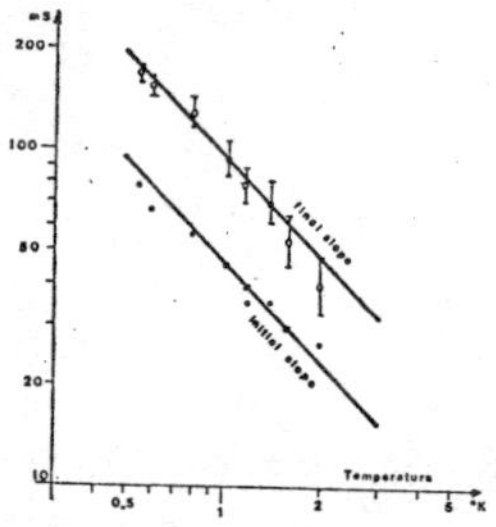

Fig.2 : Relaxation of F centers in LiF as a function of temperature.

The relaxation of F centers was not exponential, and was also temperature dependent, in constrast from what is expected for a direct process at hight field and low temperature. It is well known however that the relaxation of F centers is very strongly concentration dependent and not understood quantitatively. From the relaxation curve in semi-log scales, we determine the initial and final slopes. On Fig.2, we report the variations of these two parameters as a function of temperature.

In the $CaF_2 : Tm^{2+}$ sample, it has been possible to demonstrate clearly the presence of phonon bottleneck in the spin lattice relaxation of Tm^{2+} impurities. Theory predictes that the evolution of the electronic polarization P towards its thermal equilibrium value 1, must obey the equation [Abragam et al., 1972] :

$$(8) \qquad dP/dt = -(1/T_{1e})(1-P)/(1+\sigma P) \quad ,$$

where T_{1e} is the intrinsic electronic relaxation time, and where the phonon bottleneck coefficient σ is proportional to the electronic concentration N_S. According to equation (8), the behaviour of the relaxation for $1-P \ll 1$ is roughly exponential with an "apparent" time constant :

$$(9) \qquad T^*_{1e} = (1+\sigma)T_{1e} = (1+\alpha N_S)T_{1e} \quad .$$

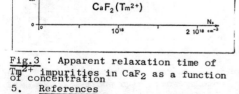

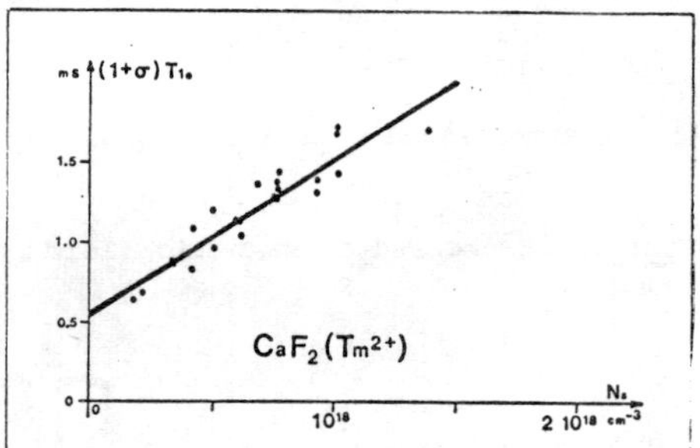

Fig.3 : Apparent relaxation time of Tm^{2+} impurities in CaF_2 as a function of concentration

The ESR spectrum of $CaF_2 : Tm^{2+}$ is composed of two well resolved hyperfine lines, the relative intensity of which can be varied at will by using the phenomenon of diagonal relaxation [Abragam et al.,1972]. We have studied the spin lattice relaxation of one hyperfine line as a function of its population (Fig.3). The experimental points fit a straight line whose y intercept yields, according to formula (9), the intrinsic relaxation time T_{1e}.

The value we found, $T_{1e} = 0.6 \pm 0.1$ m.sec., is in good agreement with previous experimental data [Sabisky et al. 1970] [Abragam et al., 1972].

5. References

Abragam, A., Borghini,M., Chapellier,M., 1962, C.R. Acad. Sciences, France, 255, 1343.
Abragam, A., Jacquinot,J.F., Chapellier,M., and Goldman, M.,1972, J.Phys.C. Solid St.Phys.vol5 2629.
Abragam, A., Chapellier,M., Jacquinot,JF., Goldman, M., 1973, Journ. of Magn. Res. 10, 322.
Kittel, C., 1948, Phys. Rev. 73, 155.

SUPERPOSITION OF SOLID EFFECT AND NEGATIVE OVERHAUSER EFFECT IN PLASTIC CRYSTALS: CASE OF CYCLOHEXANE

J. Avalos and B. Marticorena
Universidad Nacional de Ingenieria, Departamento de Fisica, Casilla 1301, Lima, Peru.
F. Volino
Universidad Nacional de Ingenieria, Lima, Peru*, and Institut Laue-Langevin, B.P. 156, 38042 Grenoble Cedex, France.

Abstract. It is shown theoretically that solid effect and negative Overhauser effect may exist simultaneously in a plastic crystalline phase. This theory is verified both qualitatively and semi-quantitatively in the case of cyclohexane doped in low concentration with a nitroxyde free radical.

1. Introduction. The aim of this work is to show theoretically and for one experimental case that dynamic nuclear polarization (DNP) may occur in a crystalline plastic phase containing paramagnetic impurities in low concentration and may be analyzed in terms of the superposition of a solid effect and a negative Overhauser effect. The basic reason is that the intermolecular dipolar interaction H_d between the electronic spin of the impurities and the nuclear spins of the host molecules may be split into two parts: a static part $\bar{H}_d$ corresponding to the fact that the centers of the molecules are fixed in space and a fluctuating part $H_f(t) = H_d - \bar{H}_d$ corresponding to the fact that the molecules reorient rapidly. For the first term, one can generally assume that the spins of the molecules are concentrated at their center of gravity and apply the spin diffusion theory (Khutsisvili, 1965) to calculate the nuclear relaxation and the solid effect, as was done in a previous work (Verdun et al., 1972). If the correlation time τ_R of the second term is sufficiently short ($\omega_e\tau_R \lesssim 1$) the relaxation due to the relative motion of electronic spins and nuclear spins may appear as in a liquid and consequently an Overhauser effect be induced. The main difference here is that the magnetization is transported through the sample via spin diffusion rather than via molecular diffusion. In what follows, we analyze the relaxation and the Overhauser effect due to $H_f(t)$ in such systems and apply this analysis to new experimental data on plastic cyclohexane.

2. Theory of the Overhauser Effect in a Plastic Crystal. The formalism to be used here is exactly the same as that employed for liquids (Hubbard, 1966), the main difference lying in the manner in which the relaxation probabilities due to the various terms of $H_f(t)$ are calculated. This point is discussed now. Let us consider two reorienting spheres, fixed in space, one containing on its surface spin $\underset{\sim}{S}$ and the other spin $\underset{\sim}{I}$. Let $\underset{\sim}{d}$ and $\underset{\sim}{d}'(t)$ be the vectors joining the centers of the spheres and the spins, respectively, and let us define the (convenient) following quantity:

$$\Lambda_m(t) = (4\pi)^{1/2} d^3 d'(t)^{-3} Y_2^m[\theta'(t),\phi'(t)] \qquad [1]$$

where $\theta'(t)$, $\phi'(t)$ are the spherical coordinates of $\underset{\sim}{d}'(t)$ with respect to a laboratory frame with $\underset{\sim}{H}_0 \equiv Oz$, and Y_2^m the normalized spherical harmonics of order 2. The relaxation rates are proportional to the Fourier transform of quantities of the form $d^{-6}C_m(t)$ where $C_m(t)$ is the self correlation function of $\Lambda_m(t) - \overline{\Lambda_m(t)}$. The bar denotes an average over the motion. Assuming an exponential law characterized by τ_R, we can write

$$C_m(t) = Z_m(d) \exp(-t/\tau_R) \qquad [2]$$

with

$$Z_m(d) = <|\Lambda_m(0) - \overline{\Lambda_m(t)}|^2> = <|\Lambda_m(0)|^2> - <|\overline{\Lambda_m(t)}|^2> \qquad [3]$$

The brackets indicate a powder average and the second equality is valid only if $\Lambda_m(t)$ is real (m=0). Note that $<|\overline{\Lambda_m(t)}|^2> = 1$ if the motion is isotropic (Thibaudier et al., 1972). The relaxation rate of the whole sample will depend on the nature of the spin diffusion regime. If the latter is assumed to be rapid, as is the case in cyclohexane (Verdun et al., 1972) the bulk relaxation rate due to $H_f(t)$ can be calculated by averaging $d^{-6}Z_m(d)$ over a sphere of radius R such that $(4/3)\pi R^3 n_e = 1$ where n_e is the impurity concentration. Clearly, $Z_m(d)$ is maximum for $d \approx d_0$, where d_0 is the distance between the sphere $\underset{\sim}{S}$ and its nearest neighbour spheres $\underset{\sim}{I}$, and zero for $d \gg d_0$. If we assume that its variation is much slower than d^{-4}, we get

$$[d^{-6} Z_m(d)]_{Av} \approx (4/3)\pi\, n_e\, d_0^{-3}\, Z_m(d_0) \qquad [4]$$

Taking into account properly all the coefficients and assuming that $Z_m(d_0) = z(d_0)$ independent of m, we finally can write, in the same manner as Webb et al. (1969)

$$2q = 4n_eJ(0)j(\omega_n) \qquad r = (4/3)n_eJ(0)j(\omega_e + \omega_n) \qquad s = 8n_eJ(0)j(\omega_e - \omega_n) \qquad [5]$$

with

$$J(0) = \pi\gamma_e^2\gamma_n^2\hbar^2 z(d_0)\,(10d_0^3)^{-1} \qquad [6]$$

$$j(\omega) = \tau_R(1 + \omega^2\tau_R^2)^{-1} \qquad [7]$$

and

$$\eta = \gamma_e\gamma_n^{-1}(r-s)(2q+r+s)^{-1}(1-T_1T_B^{-1})\,KP(1+KP)^{-1} = \eta_\infty KP(1+KP)^{-1} \qquad [8]$$

where P is the microwave pumping power and K a constant. In Eq. [8] the quantity $f = 1-T_1T_B^{-1}$ is a a leakage factor. Calling $T_{OV} = (2q+r+s)^{-1}$ [9] the relaxation due to $H_f(t)$, T_s the relaxation time due to $\bar{H}_d$ (which is also directly related to the solid effect) and T_{10} the relaxation time due to all the other mechanisms, we have

$$T_1^{-1} = T_{OV}^{-1} + T_s^{-1} + T_{10}^{-1} \quad [10], \quad T_B^{-1} = T_s^{-1} + T_{10}^{-1} \quad [11] \quad \text{and} \quad f = T_{OV}^{-1} \cdot T_1 \quad [12]$$

We see here that T_s acts to decrease the maximum enhancement by Overhauser effect and that its influence can be very strong as we shall see now.

3. Experimental Results and Analysis. The figure represents the enhancement η, of the proton magnetization of cyclohexane frozen in its plastic phase at -40°C, as a function of the departure ΔH from the center of the EPR line of the pumping microwave frequency. The paramagnetic impurity is the radical TANO and its completely deuterated derivative TANO-D, with a concentration $n_e \approx 6.67 \times 10^{17}$ cm^{-3} (10^{-3}M). Only the DNP associated with the high field EPR line has been considered here. The experimental device is described elsewhere (Verdun et al., 1972). The most striking feature is the great similarity of the spectra for TANO and TANO-D, indicating that the protons of the radical molecule play a minor role in this phenomenon Both spectra present (i) the typical solid effect with maxima at $\Delta H \approx \pm 4.8G \approx \pm\omega_n/\gamma_e$, (ii) a hump at $\Delta H = +9.6G \approx 2\omega_n/\gamma_e$ corresponding to induced transitions involving the electronic spin and two nuclear spins, and (iii) a well marked asymmetry of the solid effect around $\Delta H = 0$. Assuming that this asymmetry is due to a (negative) Overhauser effect, we have broken up the central part of the spectra into an asymmetric part (solid effect) and a symmetric part (Overhauser effect), as shown in the enlarged inserts C and D. It is seen that the Overhauser spectra, (drawn in dashed lines) has the same width as the solid effect. This suggests that the two effects come from the same EPR line and strongly supports, at least qualitatively, our hypothesis. We now show that this support is also semi-quantitative. The insert E represents the enhancement versus P, for TANO-D at $\Delta H = 0$ (in fact η^{-1} versus P^{-1}). The curve is a straight line whose intercept with the ordinate axis gives $\eta_\infty \approx -3.1$. The total nuclear relaxation time was measured and found to be $T_1 \approx 0.9$ sec. Assuming that $\omega_n \ll \omega_e$, τ_R^{-1}, and taking into account relations [5] to [12], we can obtain a relation between $z(d_o)$ and τ_R. With $d_o = 6.6$ Å, $\omega_e = 5.7 \times 10^{10}$ rd/sec and $\gamma_e\gamma_n^{-1} = 660$, we get $J(0)n_e \approx 1.8 \times 10^8$ sec^{-2} and

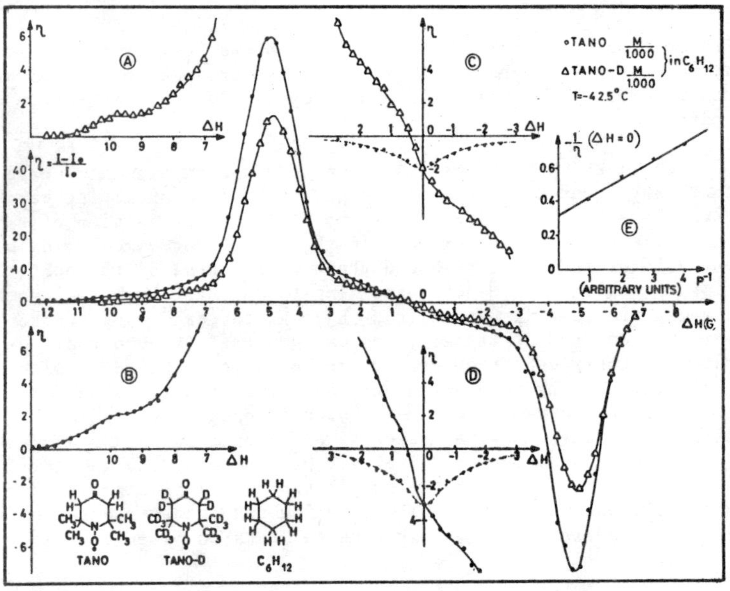

$$z(d_o) \approx 0.25\,(\omega_e\tau_R)^{-1}\,(1 + \omega_e^2\tau_R^2) \qquad [13]$$

$T(d_o)$ is a geometrical quantity which characterizes the configuration between the impurity and its nearest neighbours. It is quite similar to the quantity $(M_e^2 - M_d^2)/M_d^2 = Q$, used in the study of nuclear relaxation in plastic crystals (Resing, 1969). For cyclohexane, we have $M_e^2 \approx 10\ G^2$, $M_d^2 \approx 1.4\ G^2$ giving $Q \approx 6$ (Andrew et al., 1952). τ_R characterizes the relative motion. The rotational correlation time of the radical itself is $\tau_1 \approx 4.5 \times 10^{-11}$ sec (Volino et al., 1972), whereas that of the bulk cyclohexane molecule is $\tau_2 \approx 10^{-12}$ sec. If the motions are independent we should have $\tau_R^{-1} = \tau_1^{-1} + \tau_2^{-1}$. Although it is not clear that τ_2 for the neighbouring molecules is as short as in the bulk, taking $\tau_R \approx \tau_2$, we get $z(d_o) \approx 4.4$ from relation [13], in very good agreement with the above Q value. Using now the above values we get from [9] $T_{OV} \approx 100$ sec, much longer than $T_s \approx T_1 \approx 1$ sec. This last result justifies a posteriori that the influence of $H_f(t)$ was neglected in a previous study of nuclear relaxation in this system (Verdun et al., 1972).

4. Acknowledgements. The authors are indebted to J.P. Kraut and P. Chenavas for their contribution in preparing the nitroxyde radicals.

5. References.

* While Visiting Professor (July - October, 1973), funded by Organization of American States under the Multinational Project of Physics.

Andrew, E.R. and Eades, R.G., Proc. Roy. Soc. A216, 398 (1958).
De Graaf, L.A., Physica (The Hague) 40, 497 (1969).
Hubbard, P.S., Proc. Roy. Soc. London A291, 537 (1966).
Khutsisvili, Sov. Phys. Usp. 8, 743 (1973).
Resing, H.A., Mol. Cryst. Liq. Cryst. 9, 101 (1969).
Thibaudier, C. and Volino, F., Mol. Phys. 25, 1037 (1973).
Verdun, H., Meerovici, B. and Volino, F., J. Chem. Phys. 57, 2414 (1972); ibid 58, 5769 (1973).
Volino, F. and Rousseau, A., Mol. Cryst. Liq. Cryst. 16, 247 (1972).
Webb, R.H., et al., J. Chem. Phys. 50, 4408 (1969).

SECTION X

Microwave Measurements

FREQUENCY DEPENDENCE OF MAGNETO-MICROWAVE FARADAY EFFECT RELATED TO ELECTRON SPIN RESONANCE

Y. Servant
Laboratoire d'Optique Ultra-Hertzienne, Equipe de Recherche Associée au C.N.R.S., Université de Bordeaux-I, 33405 Talence, France.

Abstract. The frequency dependence of ESR Faraday effect has already been studied on line shapes and line widths by comparing experimental data at 3 GHz and 9 GHz. This comparison is now extended to angular amplitudes. In agreement with theoretical formulas, the specific effects of rotation and ellipticity are typically proportional to the square of the microwave frequency.

1. Introduction. The magneto-microwave Faraday effect related to ESR has been foreseen by Kastler (1949), and was then studied by several physicists. The first frequency dependence study of ESR Faraday effect was presented by Soutif-Guicherd (1958), who compared ESR Faraday rotations of a few compounds at 3 and 30 GHz. Unfortunately, this comparison suffered from a very bad homogeneity of the magnetic field in the set-up at 30 GHz, resulting in the distorsion of line shapes, the broadening of line widths, and the reduction of angular amplitudes. It was therefore interesting to present the same type of frequency dependence study, under improved homogeneity conditions. This was done by Servant (1969), who compared ESR Faraday rotations and ellipticities of several compounds at 3 and 9 GHz, with a special attention to line shapes, line widths, and g-factors. In the present paper, the comparison is applied to the angular amplitudes of rotations and ellipticities.

2. Derivation of formulas. After the theory of Soutif-Guicherd (1956, 1958) on ESR Faraday rotation α in a circular waveguide, and the remarks of Servant (1969) on the associated ellipticity β, we may write

$$\alpha = k(\ell/\lambda_g)\xi' \ , \quad \beta = k(\ell/\lambda_g)\xi'' \ , \qquad (1)$$

where k is a constant factor, ℓ the sample length, λ_g the wavelength in the circular guide with sample filling, and $\pm j\xi = \pm j(\xi' - j\xi'')$ are the off-diagonal terms of the magnetic susceptibility tensor. Let us now define specific angular terms and normalized susceptibility terms

$$\alpha_s = \alpha/\ell \ , \quad \beta_s = \beta/\ell \ , \quad \xi'_n = \xi'/(\chi_0\omega T_2) \ , \quad \xi''_n = \xi''/(\chi_0\omega T_2) \ , \qquad (2)$$

where T_2 is the spin-spin relaxation time of the sample, χ_0 its static susceptibility, and ω the angular frequency of microwaves. From Eqs. (1) and (2), we obtain

$$\alpha_s = k(\chi_0\omega T_2/\lambda_g)\xi'_n \ , \quad \beta_s = k(\chi_0\omega T_2/\lambda_g)\xi''_n \ . \qquad (3)$$

When plotted against the steady magnetic field H, the rotation α gives a dispersion-type curve, whereas the ellipticity β gives an absorption-type curve. From Eqs. (3), the specific effects of rotation and ellipticity will be

$$\alpha_{spp} = k(\chi_0\omega T_2/\lambda_g)\xi'_{npp} \ , \quad \beta_{smx} = k(\chi_0\omega T_2/\lambda_g)\xi''_{nmx} \ , \qquad (4)$$

where α_{spp} and ξ'_{npp} are the peak-to-peak (pp) amplitudes of the α_s vs. H and ξ'_n vs. H curves, while β_{smx} and ξ''_{nmx} are the maximum (mx) amplitudes of the β_s vs. H and ξ''_n vs. H curves. In order to compare ESR Faraday effects at two angular frequencies ω_A and ω_B, we define the ratio R between the specific effects of Rotation, the ratio E between the specific effects of Ellipticity,

$$R = \alpha_{sppB}/\alpha_{sppA} \ , \quad E = \beta_{smxB}/\beta_{smxA} \ , \qquad (5)$$

and we obtain from Eqs. (4)

$$R = (XPQ)U \ , \quad E = (XPQ)V \ , \qquad (6)$$

with

$$X = \chi_{0B}/\chi_{0A} \ , \quad P = \lambda_{gA}/\lambda_{gB} \ , \quad Q = (\omega T_2)_B/(\omega T_2)_A \ , \qquad (7)$$

$$U = \xi'_{nppB}/\xi'_{nppA} \ , \quad V = \xi''_{nmxB}/\xi''_{nmxA} \ . \qquad (8)$$

3. Application to experimental data. In all our comparisons between Faraday effects at angular frequencies ω_A and ω_B, both samples A and B contained the same compound of molecular weight M, at the same density d, and at the same (room) temperature, resulting in the same value of the molar susceptibility χ_M, so that the static susceptibility $\chi_0 = (\chi_M d)/M$ was also the same for both samples. The first Eq. (7) gives $X = 1$, and Eqs. (6) are reduced to

$$R = (PQ)U \ , \quad E = (PQ)V \ . \qquad (9)$$

The wavelength λ_g in the circular guide with sample filling is related to the free-space wavelength λ_0 by the formula $(\lambda_g/\lambda_0)^{-2} = \varepsilon - (\lambda_0/\lambda_c)^2$, where ε is the dielectric constant of the (powder) specimen, and λ_c the cut-off wavelength of the TE_{11} mode. First, the frequency dependence of ε is weak enough to be neglected; the density dependence of ε is important, but as d was the same for both samples, ε was also the same. In both waveguides, we had practically the same value of the ratio λ_0/λ_c , and its square was much smaller than the typical values of ε, so that the difference $\varepsilon - (\lambda_0/\lambda_c)^2$ was practically the same for both experiments. From the formula, we conclude that λ_g/λ_0 was also the same ($\lambda_{gA}/\lambda_{oA} = \lambda_{gB}/\lambda_{oB}$), and we find for P defined in Eqs. (7)

$$P = \lambda_{oA}/\lambda_{oB} = \omega_B/\omega_A . \quad (10)$$

By comparing the experimental β vs. α plots to the theoretical ξ''_n vs. ξ'_n plots corresponding to the Lorentzian or Gaussian cases, and to various values of ωT_2, we determine first the exact or approximate line shape (e.g. Lorentzian or pseudo-Gaussian), and then the values of $(\omega T_2)_A$ and $(\omega T_2)_B$, and therefore the value of Q defined in Eqs. (7). Practically, we often find at a good approximation

$$Q = \omega_B/\omega_A , \quad (11)$$

which means that the frequency dependence of T_2 may be neglected. However, in the case of $MnCl_2.4H_2O$, we have found $Q > \omega_B/\omega_A$, which means that $T_{2B} > T_{2A}$, as a consequence of the so-called 10/3 effect. Knowing the line shape and the values of $(\omega T_2)_A$ and $(\omega T_2)_B$, we may also determine from the convenient ξ''_n vs. ξ'_n plots the theoretical values of U and V defined in Eqs. (8). Practically, as a result of having normalized the susceptibility terms, and provided that $(\omega T_2)_A$ and $(\omega T_2)_B$ should exceed about five, we find at a good approximation

$$U = V = 1 . \quad (12)$$

When Eqs. (11) and (12) are valid, they permit with Eq. (10) to reduce Eqs. (9) to

$$R = E = \omega_B^2/\omega_A^2 , \quad (13)$$

showing that the specific effects of rotation and ellipticity are simply proportional to the square of the microwave frequency. When comparing the experimental data of Servant (1969) at 8.78 and 3.32 GHz, Eq. (10) gives $P = 2.65$, and Eq. (13) would give $R = E = 7.02$.
In the following table, we have kept $P = 2.65$, but we have reported the precise values of Q, U and V in each case. According to Eqs. (9), PQU and PQV are considered as theoretical values of R and E, which are compared to the experimental values R_{exp} and E_{exp} , determined on the β vs. α (experimental) plots, according to the definition of Eqs. (5).

Compound	Line Shape	P	Q	U	V	PQU	R_{exp}	PQV	E_{exp}	$\frac{R_{exp}}{PQU}$	$\frac{E_{exp}}{PQV}$
$Fe_2(SO_4)_3$	Lorentzian	2.65	2.56	1.00	1.00	6.78	6.93	6.78	6.93	1.02	1.02
MnF_2	Lorentzian	2.65	2.63	1.02	1.01	7.11	7.63	7.03	7.63	1.07	1.08
$MnCl_2.4H_2O$	Lorentzian	2.65	2.93	1.16	1.13	9.00	9.58	8.77	9.58	1.06	1.09
$Fe_2(SO_4)_3.(NH_4)_2SO_4.24H_2O$	Pseudo-Gaussian	2.65	2.73	1.04	1.00	7.52	9.53	7.23	8.66	1.27	1.20

4. Discussion. We may estimate to ±1% the measurement error on each effect, and so to ±2% the error on the ratio between two effects; to ±2% of the single passage effect, the spurious effects resulting from residual multiple reflections or from accidental birefringences in the set-up; to ±1% the approximations made when writing $X = 1$ and Eq. (10). Finally, when comparing experiment to theory, the best agreement to be expected is of the order of 10%. This is confirmed by the last two columns of the table, for the first three compounds, which are perfect illustrations of the Lorentzian case. For the last compound, the line shape is only Pseudo-Gaussian, so that Gaussian plots are not a perfect fit for the experimental plots. Accordingly, an agreement of the order of 25% may be still considered as satisfactory for this last case.

5. References.

Kastler, A., 1949, C. R. Acad. Sci. (Paris), 228, 1640-2.
Servant, Y., 1969, Ann. Phys. (Paris), 4, 579-615.
Soutif-Guicherd, J., 1956, C. R. Acad. Sci. (Paris), 242, 1868-71.
Soutif-Guicherd, J., 1958, Ann. Télécommunic. (Paris), 13, 169-85 and 222-38.

HYPERSONIC ATTENUATION IN VANADIUM DOPED MAGNESIUM OXIDE

P.J. King and S.G. Oates

Department of Physics, University of Nottingham, University Park,
Nottingham, England.

Abstract. The effects of transition metal ions on the hypersonic attenuation in MgO are studied at low temperatures.

1. Introduction. By studying the attenuation of longitudinal waves in MgO doped with a nominal 450 p.p.m. of vanadium we have obtained information on various relaxation times as a function of temperature, and by assuming Orbach type thermal activation processes[3] we have obtained values for low lying exited states.

2. Method. The ends of a single crystal specimen of MgO, cylindrical and aligned along the (100) direction were polished flat and parallel to optical tolerances. Standard pulse echo and cryogenic techniques were then used to measure the microwave acoustic attenuation. Measurements were made on both the longitudinal and transverse modes at 580 MHz, and on the longitudinal mode at 998 MHz, between 1.5K and 300K. A thin Cadmium Sulphide film overlaid on a thin aluminium ground plane was used to generate the acoustic waves.

3. Results. The attenuation measurements were made by displaying the echo patterns on an X-Y recorder and comparing pairs of echoes at different temperatures. The transverse wave attenuation showed no peaks over the observed temperature range to the limit of the experimental accuracy which was of order 0.03 dB cm^{-1}. Our measurements on longitudinal waves are shown in Fig. 1. Three peaks were found in

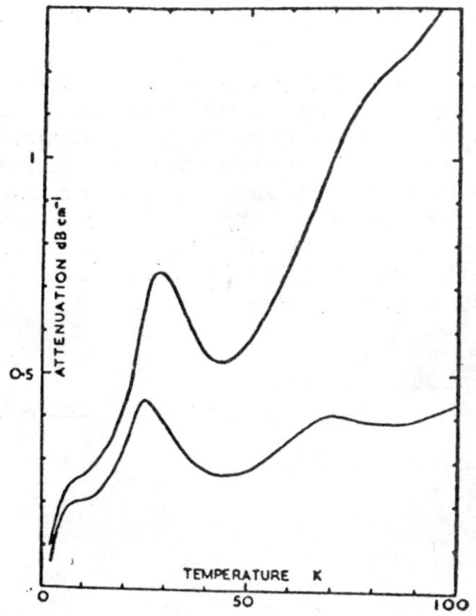

Figure 1. The attenuation of longitudinal waves along the (100) direction as a function of temperature. Upper curve 998 MHz, lower curve 580 MHz.

the attenuation occuring at 6K, 25K and 67K at 580 MHz and these moved upwards in temperature as the frequency was raised. The heights of the peaks are close to being linearly proportional to frequency and the results are summarised in Table 1.

Frequency in MHz	Low Temperature peak (~6K)	Intermediate peak (~25K)	High Temperature peak (~67K)
580	0.13 ± 0.05	0.42 ± 0.06	0.20 ± 0.03
998	0.19 ± 0.05	0.66 ± 0.06	0.35 ± 0.03

Table 1. Heights of the observed relaxation peaks (in dB cm^{-1})

where the peak heights corrected for an estimated phonon contribution are given. Since these peaks exhibit the features of relaxation phenomena a relaxation model was used to analyse them, which in its simplest form supposes that the absorption is due to a collection of systems which can exist in one of two degenerate ground states. The presence of a strain ε causes one of the levels to be raised by an amount $A\varepsilon$ and the other to be depressed by the same amount. If the strain is provided by an acoustic wave there will be a sinusoidal time dependence at frequency ω. The relaxation between the levels leads to an attenuation of the acoustic wave of magnitude α given by

$$\alpha = 4.34 \frac{A^2 N v^{-3}}{\rho K T} \frac{\omega^2 \tau}{1+\omega^2\tau^2} \quad \text{dB/unit length}$$

Where N is the number of absorbing systems per unit volume, τ is the relaxation time, ρ the density of the medium, v is the acoustic propagation velocity. The expression may be used to derive the dependence of τ on temperature from the shape of the peak. It is useful to attempt to fit the relaxation times to a thermal activation expression of the form $\tau(T)^{-1} = \nu_0 \exp(-\Delta/KT)$, where ν_0 and Δ often reveal the physical mechanism involved. In the case under discussion a fit was possible for the low temperature peak above 6K giving $\Delta = 13 \pm 1$ cm^{-1} and $\nu_0 = 4 \times 10^{10}$ Hz. An excellent fit was obtained for the high temperature peak giving $\Delta = 290 \pm 10$ cm^{-1}, $\nu_0 = 4 \times 10^{10}$ Hz. A reasonable fit was made for the intermediate peak below 28K giving $\Delta = 45 \pm 10$ cm^{-1} and $\nu_0 = 33 \times 10^{10}$ Hz, but there were signs here of a smaller narrow peak and work is in progress on a more detailed analysis. These values of ν_0 and Δ are all characteristic of relaxation between the energy levels of magnetic impurities, the thermal activation process being referred to as an "Orbach process". Such processes are in competition with "direct processes" between the relaxing levels but tend to dominate at higher temperatures. A further complication is that the relaxing levels may not be the actual ground state levels as supposed in the simple model, making the expression for the attenuation incorrect by a depopulation factor, the influence of which becomes smaller as the temperature is raised. As a result one would expect the Orbach relaxation theory to provide a good fit for the higher temperature peaks but to provide a poorer fit for the lower temperature parts of low temperature peaks. This is what is observed.

4. Conclusions. The longitudinal wave propagating along the (100) direction showed three relaxation peaks at low temperatures. The lower peak is perhaps due to Cr^{2+}, good agreement for ν_0 and Δ being shown with Lange[1] 1973. It is odd that no peaks were observed in the shear wave attenuation, Lange[2] 1972 having observed peaks due to Fe^{2+}, which was present in somewhat lower concentrations than in our specimen. It is fairly certain that the other two peaks are due to Vanadium with perhaps an Iron contribution to the middle peak. Work is progressing at the moment to confirm this with optical, infra red and E.S.R. studies and by changes of valence state using γ irradiation.

5. References

1. Lange, J.N. Phys Rev. B., 8, No. 12, p5999, (1973)
2. Lange, J.N. International conference on Phonon Scattering, Paris, p238, July 1972
3. Sturge, M.D. Phys. Rev., 155, No. 2, p218, (1967)

LIST OF PARTICIPANTS

(WITH PAGE NUMBERS OF INVITED AND CONTRIBUTED PAPERS)

ABE, Professor H.	Tokyo, Japan	127
ABOU-GHANTOUS, Dr. M.	Grenoble, France	565
ABRAGAM, Professor A.	Saclay, France	
ACHLAMA, Dr. A.	Heidelberg, FGR	541
ADENIS, Mr. J. C.	Sevran, France	
AILION, Professor D. C.	Utah, USA	235
AL-BAGHDADI, Mr. S. M.	Nottingham, UK	285
ALEKSANDROV, Professor I. V.	Moscow, USSR	463
ALEWAETERS, Mr. G.	Brussels, Belgium	
ALLEN, Dr. P. S.	Nottingham, UK	383, 375
AL-MOWALI, Mr. A. H.	Glasgow, UK	
ALQUIE, Mr. G.	Paris, France	99
AMITY, Mr. I.	Jerusalem, Israel	315
ANDERSSON, Dr. L. O.	Varian, Switzerland	129
ANDREW, Professor E. R.	Nottingham, UK	325, 271, 269, 241
ARDELEAN, Professor I.	Cluj, Romania	
ARIF, Dr. S.	Manchester, UK	85
ARIMONDO, Dr. E.	Pisa, Italy	189
ARMSTRONG, Professor R. L.	Toronto, Canada	459
ARONS, Dr. R. R.	Julich, FGR	343, 81, 75
AZZONI, Professor C.	Pavia, Italy	
BABERSCHKE, Dr. K.	Berlin, FGR	347, 321, 311
BACQUET, Professor G.	Toulouse, France	161
BAILEY, Mr. R. F.	Newport Instruments Ltd, UK	
BAINES, Mr. T.	Nottingham, UK	
BAKER, Dr. J. M.	Oxford, UK	
BARAK, Dr. J.	Grenoble, France	83
BARJHOUX, Mr. Y.	Grenoble, France	89
BATES, Dr. A. R.	Cardiff, UK	393
BATES, Dr. C. A.	Nottingham, UK	565, 441, 417
BATES, Professor L. F.	Nottingham, UK	
de BEER, Dr. R.	Delft, Netherlands	95
BENE, Professor G. J.	Geneva, Switzerland	185, 179
BENNETT, Dr. J. E.	Shell Ltd, UK	
BENOIT, Professor H.	Paris, France	409
BERGLUND, Mr. B.	Uppsala, Sweden	
BERTHIER, Dr. C.	Grenoble, France	
BERTHIER, Dr. Y.	Grenoble, France	83
BEYNON, Dr. P.	London, UK	
BIEHL, Dr. R.	Berlin, FGR	499
BIESIOT, Mr. W.	Delft, Netherlands	
BILL, Professor H.	Geneva, Switzerland	155
BITTNER, Professor H.	Vienna, Austria	
BLANC, Dr.	Romgant, France	
BLEANEY, Professor B.	Oxford, UK	
BLICHARSKA, Dr. B.	Krakow, Poland	281
BLICHARSKI, Dr. J. S.	Krakow, Poland	473
BLINC, Professor R.	Ljubljana, Yugoslavia	295
BLOYET, Mr. D.	Orsay, France	363
BOAS, Dr. J. F.	Monash, Australia	277
BODEN, Dr. N.	Leeds, UK	481
den BOEF, Dr. J. H.	Eindhoven, Netherlands	197, 117
BOEYKENS-JANSSENS, Mrs. G.	Leuven, Belgium	
BOHN, Dr. H. G.	Julich, FGR	81, 75, 343
BONERA, Professor G.	Pavia, Italy	301
BONORI, Dr. M.	Venice, Italy	193
BOROSKE, Mr. E.	Berlin, FGR	
BOROVIK-ROMANOV, Professor A. S.	Moscow, USSR	97, 5
BOSCAINO, Dr. R.	Palermo, Italy	139
BOUFFARD, Dr. V.	Saclay, France	575
BOVEE, Dr. W. M. M. J.	Delft, Netherlands	247
BOWDEN, Dr. G. J.	Manchester, UK	85, 77
BOWKER, Mr. K. J.	Nottingham, UK	105
BRAI, Dr. M.	Palermo, Italy	139
BRANSON, Mr. P.	Nottingham, UK	
BRODBECK, Dr. C. M.	St. Louis, USA	171
BROEKAERT, Dr. P.	Brussels, Belgium	

BROM, Dr. H. B.	Leiden, Netherlands	377
BROWN, Dr. M. A.	Loughborough, UK	
BRUN, Professor E.	Zurich, Switzerland	
BRUNNER, Dr. H.	Heidelberg, FGR	
BUDAK, Professor H.	Istambul, Turkey	
BURGET, Dr. J.	Rez, Czechoslovakia	
BURZO, Dr. E.	Bucharest, Romania	71, 151
BUTTERWORTH, Dr. J.	Berkshire, UK	
CAPELIN, Mr. S. C.	Nottingham, UK	
CARTER, Dr. G. C.	Nat. Bureau of Standards, USA	
CARTLEDGE, Mr. G.	Sheffield, UK	
CAVENETT, Dr. B. C.	Hull, UK	213
CHABRE, Professor Y.	Grenoble, France	351
CHALLIS, Professor L. J.	Nottingham, UK	
CHATELAIN, Dr. A.	Lausanne, Switzerland	
CHELKOWSKI, Dr. A.	Poland	
CICCARELLO, Dr. I.	Palermo, Italy	139
CLARK, Mr. I. A.	Nottingham, UK	565, 559
CLERJAUD, Mr. B.	Paris, France	563
CLIFFORD, Mr. J.	Unilever, UK	
CLOUGH, Dr. S.	Nottingham, UK	389, 387, 379
CONARD, Dr. J.	Orleans, France	507
CONTI, Dr. L. G.	Rome, Italy	571
CONWAY, Dr. T. F.	CPC International Inc, USA	183
CORRADI, Mr. G.	Budapest, Hungary	
COSTLEY-WHITE, Mr. A.	Oxford, UK	
CRANSHAW, Dr. T. E.	Harwell, UK	
CRICK, Mr. I. L. A.	Cardiff, UK	393
CUSUMANO, Dr. C.	Palermo, Italy	449
DALGLEISH, Mr. W.	Glasgow, UK	
le DANG, Dr.	Orsay, France	261, 135, 111
DARVILLE, Mr. J.	Liege, Belgium	365
DAVIES, Mr. J. J.	Hull, UK	213
DAVIES, Mr. J. T.	Swansea, UK	
DEEGAN, Mr. M. F.	Coventry, UK	
DEENEY, Dr. F. A.	Cork, Eire	
DEPIREUX, Professor J.	Liege, Belgium	403
DERBYSHIRE, Dr. W.	Nottingham, UK	285
DESCOUTS, Dr. P.	Geneva, Switzerland	335
DIETRICH, Mr. W.	Stuttgart, FGR	215
DINSE, Dr. K. P.	Berlin, FGR	499
DIXON, Dr. J. M.	Warwick, UK	
DOKTER, Mr. H. D.	Leiden, Netherlands	
DRAIN, Dr. L. E.	AERE, Harwell, UK	357
DUGDALE, Dr. D. E.	Keele, UK	
DUPAS, Mr. A.	Orsay, France	111
DURAN, Dr.	Paris, France	217
van DUYNEVELDT, Dr. A. S.	Leiden, Netherlands	421, 119
EDEL, Dr.	Grenoble, France	
EDMONDS, Dr. D. T.	Oxford, UK	13
EGUCHI, Mr. K.	London, UK	
ELEY, Professor D. D.	Nottingham, UK	
ELVEY, Miss S.	Inst. of Physics, London, UK	
EMID, Dr. S.	Delft, Netherlands	373
ENGEL, Professor U.	Berlin, FGR	347, 321, 311
van ENGELEN, Mr. P.	Eindhoven, Netherlands	445
ESCRIBE, Miss C.	Toulouse, France	159
ESTRADE, Dr. H.	Orleans, France	507
FARCAS, Mr. S.	Cluj, Romania	131
FEINSTEIN, Dr. C.	Oxford, UK	
FLETCHER, Dr. J. R.	Nottingham, UK	565
FORIAN-SZABO, Dr.	Budapest, Hungary	
FRADIN, Dr. F. Y.	Argonne, USA	355
FUJIWARA, Professor S.	Tokyo, Japan	165
FUKUSHIMA, Dr. E.	Los Alamos, USA	239
GADSDEN, Mr. C.	Nottingham, UK	
GAITE, Dr. J. M.	Orleans, France	
GALE, Mr. H. J.	Nottingham, UK	271
GARROWAY, Dr. A. N.	Nottingham, UK	539, 435
GASSNER, Mr. M.	Heidelberg, FGR	
GELINEAU, Dr. A.	Paris, France	563
GERSONDE, Professor	Aachen, FGR	263
VAN GERVEN, Professor L.	Leuven, Belgium	567, 501, 401

GIBSON, Dr. A. A. V.	Florida, USA	
GIORJADZE, Dr. N. P.	USSR	429
GOLENISHCHEV-KUTUZOV, Professor V.	Kazan, USSR	299
GOLTZENE, Dr.	Strasbourgh, France	
GORDON, Mr. R. E.	Aberdeen, UK	
GOREN, Professor S.	Beer-Sheva, Israel	291, 209
GOURDJI, Mr. M.	Paris, France	
GOZZINI, Professor A.	Pisa, Italy	189
GRANDJEAN, Dr. F.	Liege, Belgium	
GRANNELL, Dr. P. K.	Nottingham, UK	431
GRAPENGETER, Mr. H. H.	Aachen, FGR	
GRECU, Dr. V. V.	Bucharest, Romania	163
GREENSLADE, Dr. D. J.	Essex, UK	
GRIFFIN, Dr. R. G.	MIT, USA	555
GROBET, Dr. P.	Leuven, Belgium	501
GROSESCU, Dr.	Bucharest, Romania	
GUETTLER, Mr. W.	Stuttgart, FGR	227
GUIBE, Dr.	CNRS, France	
GUNSSER, Professor	Hamburg, FGR	143
HAAS, Dr. H.	Berlin, FGR	259
HACKELOER, Dr. H. J.	Hombruch, FGR	233
HAHN, Professor E. L.	Berkeley, USA	30
HALLSWORTH, Mr. R. S.	Ontario, Canada	381
HALSTEAD, Dr. T. K.	York, UK	
HARDIMAN, Dr. M.	Geneva, Switzerland	447
HARRIS, Dr. E. A.	Sheffield, UK	
HARROWFIELD, Dr. B. V.	Nottingham, UK	147
HASELER, Mr. S.	Nottingham, UK	
HAUPT, Dr. J.	Euratom, Italy	371
HAUSMANN, Professor A.	Aachen, FGR	
HAUSSER, Professor K. H.	Heidelberg, FGR	229
HAVILL, Dr.	Sheffield, UK	
HEATH, Dr. M.	Nottingham, UK	105
van HECKE, Dr. P.	Leuven, Belgium	513, 401
HEIGLE, Mr. T.	Freiburg, FGR	
HEINEKEN, Professor F. W.	Beirut, Lebanon	503
HEISS, Dr.	Coventry, UK	
HELD, Dr. G.	Stuttgart, FGR	461
HENKENS, Mr. L. S. J. M.	Leiden, Netherlands	69
HENNEL, Professor J. W.	Krakow, Poland	
HEY, Dr. M. J.	Nottingham, UK	
HILL, Dr. H. D. W.	Varian, USA	
HILL, Dr. J. R.	Nottingham, UK	389, 379
HILTBRAND, Dr. E.	Jussy, Switzerland	185
HINSHAW, Dr. W. S.	Nottingham, UK	433, 325, 271, 269
HOBSON, Mr. T.	Nottingham, UK	387
HOCH, Dr. M. J. R.	Johannesburg, South Africa	391, 307
HODGES, Dr. J.	Sceaux, France	153
HOHENER, Dr. A.	Zurich, Switzerland	469
HOOGSTRAATE, Dr. H.	Leiden, Netherlands	427
HUGHES, Professor D. G.	Alberta, Canada	
HUNDHAUSEN, Dr. R.	Stuttgart, FGR	
HUNDT, Dr. E.	Siemens AG, FGR	73
HUT, Mr. G.	Groningen, Netherlands	233
HUTCHINSON, Dr. J. M. S.	Aberdeen, UK	283
HYDE, Dr. J.	California, USA	
INDOVINA, Dr. P. L.	Rome, Italy	483
IOFFE, Professor V. A.	Leningrad, USSR	123
JACCARD, Professor C.	Neuchatel, Switzerland	219
JACKOWSKI, Dr. K.	Warsaw, Poland	529
JANOSSY, Dr. A.	Budapest, Hungary	313
JANSSENS, Dr. L.	Leuven, Belgium	401, 317
JAUSSAUD, Dr. P. C.	Grenoble, France	565
JEENER, Professor J.	Brussels, Belgium	
JEZIERSKI, Dr. A.	Wroclaw, Poland	437
JOKISAARI, Dr. J.	Oulu, Finland	531, 527
KACIMI, Dr. M.	Algiers, Algeria	419
KALVIUS, Dr. G. M.	Munich, FGR	87, 67
KAMMER, Dr. H.	Zurich, Switzerland	
KANERT, Professor	Dortmund, FGR	233
KAPLAN, Professor N.	Jerusalem, Israel	
KASTLER, Professor A.	Paris, France	1

KEDEM, Mrs. D.	Yavne, Israel	345
KELLERSHOHN, Professor C.	Paris, France	289
KHUTSISHVILI, Professor G. R.	Tbilisi, USSR	471, 17
KIELMAN, Dr. H. S.	Leiden, Netherlands	515
KIMBALL, Professor C. W.	N. Illinois, USA	355
KIND, Dr. R.	Zurich, Switzerland	297
KING, Dr. P. J.	Nottingham, UK	581
KINNEAR, Mr. R. W. N.	New South Wales, Australia	
KISPERT, Professor L. D.	Alabama, USA	385, 279
KLEIN, Professor M. P.	California, USA	
KNOWLES, Mr. A.	Nottingham, UK	559
KOCHELAYEV, Professor B. I.	Kazan, USSR	23
KOOPMANN, Mr. G.	Berlin, FGR	321, 311
KORB, Mr. J. P.	Paris, France	415
KORN, Dr. C.	Beer-Sheva, Israel	341, 209
KRALIK, Mr. M.	Brno, Czechoslovakia	
KREISLER, Dr. A.	Paris, France	99
KRISHNAN, Dr. R.	CNRS, France	135
KRUGER, Dr. G. J.	Euratom, Italy	339
ḰRYNICKI, Dr. K.	Kent, UK	517, 511
LACROIX, Professor R. P.	Geneva, Switzerland	
LAINE, Dr. D. C.	Keele, UK	199
LAMOTTE, Dr. B.	Grenoble, France	493
LASSMANN, Dr. G.	Berlin, DGR	265
LAU, Mr. P.	Heidelberg, FGR	229
LAUTERBUR, Professor P. C.	Stony Brook, New York, USA	27
LEE, Professor S.	St. Louis, USA	171, 169
LEGGETT, Dr. A. J.	Sussex, UK	31
LEIBLER, Dr. K.	Paris, France	
LENK, Dr. R.	Geneva, Switzerland	191
LEVANON, Dr. H.	Jerusalem, Israel	225
LEYTE, Dr. J. C.	Leiden, Netherlands	523, 515
LOCHER, Dr. P. R.	Eindhoven, Netherlands	
LOPEZ, Professor P.	Albi, France	423, 413
LOSCHE, Professor A.	Leipzig, DGR	201
van der LUGT, Dr. W.	Groningen, Netherlands	329
LUND, Dr. A.	Studsvik, Sweden	173
LUPEI, Dr. V.	Bucharest, Romania	177, 175
LUTGEMEIER, Dr. H.	Julich, FGR	343, 81, 75
MAAN, Dr. J. C.	Delft, Netherlands	
McBRIERTY, Dr.	Dublin, Eire	
McCAUSLAND, Dr. M. A. H.	Manchester, UK	113
McGEEHIN, Dr. P.	Berks, UK	
MACKENZIE, Dr. I. S.	Manchester, UK	113
MANUEL, Dr. A. J.	Leeds, UK	
MANSFIELD, Dr. P.	Nottingham, UK	539, 431
MARCHAND, Dr. A.	Grenoble, France	439
MARCHANT, Mr. J. L.	Nottingham, UK	237
MAROOF, Mr. A. K. H.	Keele, UK	199
MARTIN, Dr. J.	Caen, France	
MARUANI, Dr. J.	Paris, France	425, 415
MAUDSLEY, Mr. A. A.	Nottingham, UK	431
MAXWELL, Dr. K. J.	Nottingham, UK	
MAYAS, Mr. L.	Berlin, FGR	
MEHRING, Professor M.	Dortmund, FGR	35
MEIER, Dr. P. F.	Villigen, Switzerland	489
MERCADER, Dr. R. C.	Harwell, UK	
MERKS, Mr. R. P. J.	Delft, Netherlands	95
MERLE D'AUBIGNE, Professor Y.	Grenoble, France	39
MESSER, Dr. R.	Stuttgart, FGR	327
MICHELL, Dr. R.	Essex, UK	
MILIA, Dr. F.	Athens, Greece	
MILLET, Mr. J. L.	Lausanne, Switzerland	319
MIMS, Dr. W. B.	Bell Labs., USA	275
MOBBS, Dr. D. W.	Surrey, UK	
MOBIUS, Professor K.	Berlin, FGR	499
MOGNASCHI, Professor E. R.	Pavia, Italy	
MOORE, Dr. W. S.	Nottingham, UK	565, 559
MOREHOUSE, Dr.	Ankara, Turkey	407
MORRIS, Mr. P. G.	Nottingham, UK	
MOSSBAUER, Professor R. L.	Grenoble, France	43
MUELLER, Mr. L.	Zurich, Switzerland	557

MULLER, Professor K. A.	IBM, Switzerland	309, 145
MUTABZIJA, Dr. R.	Zagreb, Yugoslavia	195
MUTTA, Mr. K. S.	Nottingham, UK	
MYERS, Professor R. J.	Berkeley, USA	443
NAHMANI, Dr. A.	Grenoble, France	
NAVON, Professor G.	Tel-Aviv, Israel	
NEURINGER, Dr. L. J.	MIT, USA	555
NICOLLIN, Dr. D.	Geneva, Switzerland	
NIJMAN, Mr. A. J.	Amsterdam, Netherlands	253
NISTOR, Dr. S. V.	Bucharest, Romania	491, 149
NOSEL, Dr. W.	Krakow, Poland	
NUGENT, Mr. S. M.	Redhill, UK	
OATES, Mr. S. G.	Nottingham, UK	581
OHYA, Dr. H.	Kyoto, Japan	495
OLKHOV, Professor O. A.	Moscow, USSR	
ORBACH, Professor R.	Tel-Aviv, Israel	47
van ORMONDT, Dr. D.	Delft, Netherlands	
OUDET, Dr. X.	Meudon, France	109
PAJAK, Dr. Z.	Poznan, Poland	405, 403
PANEK, Mr. P.	Brno, Czechoslovakia	
PAPON, Professor	Paris, France	305
PENEAU, Dr. A.	Boulogne, France	
PERNESTAL, Mr. K. M.	Uppsala, Sweden	
PETERSON, Professor G. E.	Bell Labs., USA	475
PETRINOVIC, Professor M.	Zagreb, Yugoslavia	187
PETRAKOVSKY, Professor G. A.	Krasnoyarsk, USSR	115
PFEIFER, Professor H.	Leipzig, DGR	51
PIEJUS, Dr. P.	Orsay, France	363
PILAR, Dr. J.	Prague, Czechoslovakia	411
PINES, Professor A.	Berkeley, USA	547, 203
PINTAR, Professor M. M.	Ontario, Canada	381
PIRNAT, Mr. J.	Ljubljana, Yugoslavia	243
PLOIX, Mr. J. L.	Grenoble, France	487
POLAK, Mr. M.	Tel-Aviv, Israel	479
POLLAK-STACHURA, Dr. M.	Heidelberg, FGR	
PONTNAU, Dr. J.	Orsay, France	133
PORTE, Dr. A. L.	Glasgow, UK	
POULIS, Professor N. J.	Leiden, Netherlands	427, 353, 69
POUW, Mr. C. L. M.	Leiden, Netherlands	421
POWLES, Professor J. G.	Kent, UK	517, 511
PREUSSER, Dr.	Dortmund, FGR	
PRINS, Dr. D. Y. H.	Leiden, Netherlands	
PRINS, Dr. K. O.	Amsterdam, Netherlands	255
PROVOTOROV, Professor B. N.	Moscow, USSR	55
PUMPERNIK, Dr. D.	Ljubljana, Yugoslavia	207
PUNKKINEN, Dr. M.	Turku, Finland	389, 383
QUITMAN, Professor D.	Berlin, FGR	525, 349
RABII, Professor M.	Tehran, Iran	409
RAHKAMAA, Dr. E. J.	Oulu, Finland	
RAMPTON, Dr. V. W.	Nottingham, UK	
RAOULT, Professor G.	Clermont-Ferrand, France	
RAPPAZ, Professor M.	Lausanne, Switzerland	
READ, Dr. M.	Liege, Belgium	
READ, Professor S. F. J.	Rutherford Lab., UK	
REITER, Dr. G.	IBM Zurich, Switzerland	309
RENARD, Dr. J. P.	Orsay, France	261, 111, 93
RIGAMONTI, Professor A.	Pavia, Italy	301
RINNE, Dr. M.	Liege, Belgium	
ROCKENBAUER, Dr. A.	Budapest, Hungary	125
ROGERSON, Mr. A.	St. Andrews, UK	367
ROINEL, Dr.	Saclay, France	575
ROOS, Mr. J.	Zurich, Switzerland	297
ROSS, Dr. J. W.	Manchester, UK	77
RUSHWORTH, Dr. F. A.	St. Andrews, UK	
SALUVERE, Dr. T.	Tallin, USSR	519, 509
SALVADOR, Dr.	Heverlee, Belgium	
SANDARS, Dr. P. G. H.	Oxford, UK	59
van SANTEN, Mr. J. A.	Leiden, Netherlands	119
SARDOS, Professor R.	Bordeaux, France	
SASANE, Dr. A.	London, UK	
SHAAFSMA, Professor T. J.	Wageningen, Netherlands	497
SCHIRMACHER, Mr. W.	Berlin, FGR	331

SCHMID, Dr. D.	Stuttgart, FGR	221, 215
SCHNABEL, Dr. B.	Jena, DGR	553, 551, 545
SCHNEIDER, Dr. E. E.	Newcastle, UK	569
SCHOENBERG, Dr. A.	Staffs, UK	
SCHRIEVER, Dr. J.	Leiden, Netherlands	523
SCHRITTENLACHER, Mr.	Berlin, FGR	347
SCHUCH, Dr. H.	Berlin, FGR	223
SCHWEITZER, Dr. D.	Heidelberg, FGR	543, 521
SCOTT, Professor T. A.	Florida, USA	
SECHEHAYE, Dr. R.	Jussy, Switzerland	185
SERVANT, Dr. Y.	Bordeaux, France	579
SERVOZ-GAVIN, Dr. P.	Grenoble, France	
SERWAY, Dr. R. A.	IBM Ruschlikon, Switzerland	145
SHAHAM, Mr. M.	Yavne, Israel	
SHAPOSHNIKOV, Professor I. G.	Perm, USSR	419
SHAW, Mr. K.	Manchester, UK	
SHIELDS, Dr. L.	Bradford, UK	
SIEBENTHAL, Mr.	Geneva, Switzerland	
SJOBLOM, Mr. R. O. I.	Nottingham, UK	485, 269
SLOTFELDT-ELLINGSEN, Mr. D.	Oslo, Norway	
SMALLCOMBE, Dr. S. H.	California, USA	
SMITH, Professor J. A. S.	London, UK	
SMOLUCHOWSKI, Professor R.	Venasque, France	
SOLTWISCH, Mr. M.	Berlin, FGR	525
SPARTALION, Dr.	Pittsburgh, USA	293, 267
SPIESS, Dr. H. W.	Heidelberg, FGR	543, 521
STALKER, Mr. D. C.	Nottingham, UK	539
STANDLEY, Professor K. J.	Dundee, UK	107
STANKOWSKI, Professor J.	Poznan, Poland	395, 303
STEER, Dr. I.	Cambridge, UK	
STEHLIK, Dr. D.	Heidelberg, FGR	230, 229
STEPISNIK, Dr. J.	Ljubljana, Yugoslavia	181
STESMANS, Mr. A.	Leuven, Belgium	317
STEVENS, Professor K. W. H.	Nottingham, UK	441, 63
STIBBE, Mr. F. S.	Amsterdam, Netherlands	137
ST. JOHN, Mr. M. R.	California, USA	443
STILES, Dr. J.	UBC, Canada	323
STOCKHAUSEN, Professor M.	Mainz, FGR	535
STOHRER, Mr. M.	Stuttgart, FGR	245
STOPPELS, Dr. D.	Groningen, Netherlands	
STRANGE, Dr. J. H.	Kent, UK	
SUSS, Dr. J. T.	Yavne, Israel	121
SVECHKAREV, Dr. I. V.	Kharkov, USSR	359
SVENNINGSEN, Mrs. M.	Parhus, Denmark	
SYMONS, Professor M. C. R.	Leicester, UK	505
TAYLOR, Mr. D. G.	Nottingham, UK	383, 375
TEGENFELDT, Dr. J.	Uppsala, Sweden	549
TERENZI, Dr. M.	Venice, Italy	273
THEVENEAU, Miss	Paris, France	305
TITMAN, Dr. J. M.	Sheffield, UK	
TOMCHUK, Professor E.	Winnipeg, Canada	205
TOMPA, Professor K.	Budapest, Hungary	337, 79
TOTH, Mr. F.	Budapest, Hungary	79
TRONTELJ, Dr. Z.	Ljubljana, Yugoslavia	243
TROUP, Dr. G. J.	Monash, Australia	449
TUNSTALL, Dr. D. P.	St. Andrews, UK	367, 333
TUOHI, Dr. J. E.	Turku, Finland	399
TZALMONA, Dr. A.	Nottingham, UK	397, 241
UNDERHAY, Mr. A.	Nottingham, UK	
URSU, Professor I.	Bucharest, Romania	533, 491, 177
VALIC, Dr. M. I.	UBC, Canada	573, 369
van der VALK, Mr. H. J. L.	Groningen, Netherlands	157
van der VALK, Mr. P. J.	Amsterdam, Netherlands	
VARACCA, Professor V.	Parma, Italy	451
VASSON, Mr. A.	Clermont-Ferrand, France	417
VASSON, Dr. A.-M.	Clermont-Ferrand, France	417
VAUGHAN, Dr. R. A.	Dundee, UK	
VAUGHAN, Dr. R. W.	Caltech., USA	537
VEGA, Professor A.	Rehovot, Israel	477, 457
VENNART, Mr. W.	Nottingham, UK	271
VIETH, Dr. H. M.	Heidelberg, FGR	
VILLA, Dr.		

VOLINO, Dr. F.	Grenoble, France	577
WALDNER, Professor F.	Zurich, Switzerland	
WALSH, Professor D.	McGill, Canada	561
WARDELL, Dr. G. E.	Dublin, Eire	
WASIELA, Mr. A.	Grenoble, France	217
WASSALL, Mr. S. R.	Nottingham, UK	
WATTERICH, Miss A.	Budapest, Hungary	
WAYSBORT, Mr. D.	Tel-Aviv, Israel	
WEBER, Mr. E.	Essen, FGR	453
WEEKS, Dr. R. A.	Oak Ridge, USA	167
WEHRLI, Dr. F.	Varian, Switzerland	
WEIDEN, Dr. N.	Darmstadt, FGR	257
WENCKERBACH, Dr. W. Th.	Leiden, Netherlands	427
WHITING, Dr. J. S. S.	York, UK	103
WILLIAMS, Mr. D. L.	Nottingham, UK	
WILLIS, Dr. M. R.	Nottingham, UK	
WIND, Dr. R. A.	Delft, Netherlands	373
WINDSCOM, Dr. C. J.	Berlin, FGR	
WITTERS, Professor J.	Leuven, Belgium	317
WOOD, Mr. P. H.	Nottingham, UK	441
WOODHOUSE, Mr. D. R.	Nottingham, UK	285
WOLF, Dr. D.	Stuttgart, FGR	467, 327, 251
WOLF, Mr. E.	Dortmund, FGR	
WULFFERS, Dr. L. A. G. M.	Leiden, Netherlands	353
van WYK, Dr. J. A.	Wittwatersrand, South Africa	
YABLOKOV, Dr. G. E.	Kazan, USSR	455
YLINEN, Mr. E. E.	Turku, Finland	399
ZABEL, Mr. M.	Bruker, FGR	
ZAHRA, Mr. Y.	Nottingham, UK	
ZARIPOV, Professor M. M.	Kazan, USSR	455
ZEIGER, Dr. H.	Bruker, FGR	
ZEVIN, Professor V.	Jerusalem, Israel	141
ZINGG, Dr. W.	Geneva, Switzerland	447
ZOGAL, Dr. O. J.	Wroclaw, Poland	91
ZUMER, Dr. S.	Ljubljana, Yugoslavia	465

Author Index to Volume 2

G

H

I

J

K

L

M